International Conference on Civil Engineering and Building Services (CIBv 2020)

AA007449

IOP Conference Series: Materials Science and Engineering
Volume 1138

Brasov, Romania
5 - 6 November 2020

Editors:

Ioan Tuns
Radu Muntean
Dorin Radu
Teofil-Florin Galatanu
Christiana Emilia Cazacu

ISBN: 978-1-7138-3413-7
ISSN: 1757-8981

Printed from e-media with permission by:

Curran Associates, Inc.
57 Morehouse Lane
Red Hook, NY 12571

Some format issues inherent in the e-media version may also appear in this print version.

Institute of Physics
Dirac House, Temple Back
Bristol BS1 6BE UK

Phone: 44 1 17 929 7481
Fax: 44 1 17 920 0979

techtracking@iop.org

Additional copies of this publication are available from:

Curran Associates, Inc.
57 Morehouse Lane
Red Hook, NY 12571 USA
Phone: 845-758-0400
Fax: 845-758-2633
Email: curran@proceedings.com
Web: www.proceedings.com

TABLE OF CONTENTS

IOP Conf. Series: Materials Science and Engineering 1138 (2021) 011001 doi:10.1088/1757-899X/1138/1/011001

International Conference CIBv2020
Civil Engineering and Building Services

5th – 6th November 2020, Brasov, Romania

Selected, peer reviewed papers from the International Conference Civil Engineering and Building Services CIBv2020

Edited by:

Ioan TUNS
Transilvania University of Brasov, Romania

Radu MUNTEAN
Transilvania University of Brasov, Romania

Dorin RADU
Transilvania University of Brasov, Romania

Teofil-Florin GĂLĂȚANU
Transilvania University of Brasov, Romania

Christiana Emilia CAZACU
Transilvania University of Brasov, Romania

IOP Conf. Series: Materials Science and Engineering 1138 (2021) 011001 doi:10.1088/1757-899X/1138/1/011001

FOREWORD

The International Scientific Conference CIBv represents one of the most important and traditionally scientific events of Civil Engineering and Building Services in Romania, enjoying increasingly higher visibility at the international level.

The aim of the CIBv conference is to provide a platform for researchers, engineers, academicians as well as professionals from all over the world to present their research results and development activities. It also provides opportunities for the participants to exchange new ideas and application experiences, to establish business or research relations and to find other partners and even friends for future collaboration.

The Conference has been organized each year at Transilvania University of Brașov, since 2000. In November 2020, Faculty of Civil Engineering was honoured to organize, in partnership with Shenyang Jianzhu University from China and to host the 15th edition of "International Scientific Conference – Civil Engineering and Building Services CIBv2020" in Brasov. More than 100 participants from China and Romania, as well as from Serbia, Croatia, Slovakia, Japan, Germany, Russia, Finland and Norway submitted over 70 papers to the scientific board which were presented during the conference. After a peer-review process (which included for each article at least two reviewers), 47 papers were finally accepted for publication in the CIBv2020 proceedings volume.

We would like to address special thanks to all those who were involved in the success achieved by the CIBv2020 International Conference, as well as to our distinguished publisher.

Prof. Eng. Ioan TUNS, PhD

Dean of Faculty of Civil Engineering,
Transilvania University of Brasov, Romania

International Conference CIBv2020
Civil Engineering and Building Services

Faculty of Civil Engineering, Transilvania University of Brasov, Romania in collaboration with **Shenyang Jianzhu University, China**

Honorary Committee

Ioan Vasile ABRUDAN, Rector of Transilvania University Braşov

Weidong YAN, President of Shenyang Jianzhu University

Yupeng LI, Vice president of Shenyang Jianzhu University

Dan STEMATIU, President of Civil Engineering and Urbanism section of the Romanian Academy of Technical Sciences Professor of Technical University of Civil Engineering of Bucharest

International academic scientific committee

Radu BĂNCILĂ, Politehnica University Timişoara, Romania
Miroslav BEŠEVIĆ, University of Novi Sad, Serbia
Alina BĂRBULESCU, Technical University C.E. Bucureşti, Romania
Lucia Maria BOERIU, Transilvania University of Braşov, Romania
Marius Florin BOTIŞ, Transilvania University of Braşov, Romania
Andrei BURLACU, Technical University „Gh. Asachi" Iaşi, Romania
Natalia BREDIKHINA , Southwest State University, Kursk Russia
Ján BUJŇÁK, Technical University Žilina, Slovakia
Sorin BURCHIU, Technical University C.E. Bucureşti, Romania
Mircea Ion BUZDUGAN, Technical University Cluj-Napoca, Romania
Alin Andrei CĂRSTEANU, Instituto Politécnico Nacional, Mexico City, Mexico
Ioan Lucian CÎRSTOLOVEAN, Transilvania University of Braşov, Romania
Nicolae CHIRA, Technical University Cluj-Napoca, Romania
Vasilica CIOCAN, Technical University „Gh. Asachi" Iaşi, Romania
Mircca Radu DAMIAN, Technical University C.E. Bucureşti, Romania
Sorin DAN, Universitatea Politehnica Timişoara, Romania
Liviu Geo DRUGHEAN, Technical University C.E. Bucureşti, Romania
Adam DOŞA, Transilvania University of Braşov, Romania
Liviu DUMITRESCU, AIIR, Romania
Wolfgang FRANCKE, University of Applied Sciences Konstanz, Germania
Radomir FOLIC, University of Novi Sad, Serbia
Mariana FRATU, Transilvania University of Braşov, Romania
Sorin HERBAN, University Politehnica Timişoara, Romania
Mircea HORNEŢ, Transilvania University of Braşov, Romania
Nicolae Fani IORDAN, Transilvania University of Braşov, Romania
Danijel KUKARAS, University of Novi Sad, Serbia
Ilinca NĂSTASE, Technical University C.E. Bucureşti, Romania
Dorina ISOPESCU, Technical University „Gh. Asachi" Iaşi, Romania
Dong LI, Shenyang Jianzhu University, China
Guochang LI, Shenyang Jianzhu University, China
Ming LIU, Shenyang Jianzhu University, China

IOP Conf. Series: Materials Science and Engineering 1138 (2021) 011001 doi:10.1088/1757-899X/1138/1/011001

Organizing committee

IOP Conf. Series: Materials Science and Engineering 1138 (2021) 011002 doi:10.1088/1757-899X/1138/1/011002

Peer review declaration

All conference organisers/editors are required to declare details about their peer review. Therefore, please provide the following information:

- **Type of peer review: Single-blind / Double-blind / Triple-blind / Open / Other (please describe)**
  The peer review process was a single blind review. Each manuscript was reviewed by at least two reviewers.

- **Conference submission management system:**
  The submission management was an on website uploading type one. Each abstract was uploaded on the conference website and the papers were sent directly via conference e-mail.

- **Number of submissions received: 76**

- **Number of submissions sent for review: 54**

- **Number of submissions accepted: 47**

- **Acceptance Rate (Number of Submissions Accepted / Number of Submissions Received X 100): 61.8**

- **Average number of reviews per paper: 2**

- **Total number of reviewers involved: 54**

- **Any additional info on review process:** during the review process, all papers were checked for compliance with the topic of the conference, acceptable English language and the quality of images/figures, tables, equations and general aspect.
  The review process took into account the following key points:
  - Each reviewer was asked to fill in a paper review form containing: an **overall evaluation** (the manuscript was assessed by grade), an **evaluation** in which the reviewer was asked to grade the paper following "Contribution to existing knowledge", "Quality of the presentation", "Scientific soundness", "If the conclusions are supported by data" and "Relevance to the literature review" and a **detailed evaluation** in which the reviewer was asked to fill the "strengths" and the "weaknesses" of the paper. Additional comments to the author(s) and to the editors (optional) were implied. The grades were from 1 – Poor to 5 – Excellent.
  - The criterion for accepting the paper was to have a minimum grade 2 from at least two reviewers. After the receiving of paper reviews, each author was asked to revised the manuscript according with the reviewer's (and, where was the case, editor's) comments.
  - For each revised paper, editors checked the respecting of the reviewer's remarks and, where was the case, asked a second opinion from the reviewer(s).

IOP Conf. Series: Materials Science and Engineering 1138 (2021) 011002 doi:10.1088/1757-899X/1138/1/011002

- **Contact person for queries:**
  Radu MUNTEAN (radu.m@unitbv.ro), Dorin RADU (dorin.radu@unitbv.ro)

Please submit this form along with the rest of your files on the submission date written in your publishing agreement.

The information you provide will be published as part of your proceedings.

IOP Conf. Series: Materials Science and Engineering 1138 (2021) 012001 doi:10.1088/1757-899X/1138/1/012001

Researches on energy-efficient plasma conversion of fibreglass waste into new ecological materials for building industry

I Antonescu[1*], D N Isopescu[1], I Cucoş[1], V Caunii[1], I C Alecu[2] and M L Lupu[1]

[1] "Gheorghe Asachi" Technical University of Iaşi, România

[2] "Gheorghe Zane" Economic and Social Research Institute, Romanian Iași Academy Branch

* ion.antonescu@academic.tuiasi.ro

Abstract. The paper presents the researches, prototype laboratory equipment and technologies developed by the authors for the plasma conversion of fiberglass waste into useful materials for building industry. The aim was to increase the energy efficiency of conversion processes and to obtain technologies and materials with better proprieties for civil engineering appliances and reduced impact on the environment.

1. Introduction

Until recent years, fiberglass waste was considered very hard or even impossible to recycle and the only method of disposal was storing in a landfill, having thus a 400 years lifecycle. Nowadays there are three recycling methods for fiberglass waste: mechanical, thermal and chemical [1].

The mechanical recycling process [2] consists in reducing the size of the fiberglass waste by cutting or grinding in hammer mills, followed by centrifugal size sorting into a cyclone, which separates the fine particles in the recycling process.

Incineration, or thermal oxidation, burns the material generating heat which can be used for other purposes, such as producing steam for power plants using turbines that generate electricity. A byproduct of fiberglass incineration is ash, which is usually stored in a landfill. The heat content of fiberglass waste comes from organic materials such as resins. Usually, the organic material content is only of 25% to 30%, so the heat production is low and the resulted quantity of ash is high. The ash composition is largely of calcium oxide, coming from calcium carbonate, boron and other glass components.

Pyrolysis is the process of chemical decomposition of fiberglass waste into several recoverable substances by heating, at high temperatures, in oxygen-free environment. The fiberglass waste, subjected to pyrolysis, decomposes into three recoverable substances: pyrolysis pyro-gas, pyrolysis oil and a solid by-product, all of which are recyclable.

There are various devices and installations for the recycling of fiberglass waste [3], that use a melting furnace which has as heating source electricity (electric furnace with resistors) or chemical energy (flame burners) because in order to melt glass, in general, and fiberglass in particular, temperatures between 1200-1600° C are required. Also for the treatment of the exhaust gases from the reactor, the solution of station of the gas in an enclosure at a high temperature is widely used. Another process of thermal treatment of fiberglass waste is the use of microwaves, as described in [4]. Generally the existing solutions are high energy consuming and pollutant.

The authors have researched and developed a plant intended for use in the waste recycling industry as well as in the construction materials industry, as the material obtained from the destruction of fiberglass waste, with and without adhesive, with the help of hydrogen plasma can be used, as an additive, in the production of bricks, mortar or other materials in the construction industry.

2. Materials and methods

The plant developed by the authors has, in addition to the conversion system of waste in a reactor with the help of hydrogen plasma, a treatment system of the gas resulting from the operation of the reactor with the help of plasma, as well as an energy recovery system. The gases resulting from the exit of the gas treatment system are used to drive a power generator.

The main advantage of this plant is the destruction of fiberglass waste, with or without an adhesive, at a much higher speed than existing installations, due to the use of hydrogen plasma torches. Also, the amount of pollutants released into the atmosphere and the amount of residues resulting from the treatment of fiberglass waste is much lower, and the energy recovery system makes it attractive from an energy and financial point of view.

The installation, according to figure 1, consists of the plasma reactor 1, the feeding and rolling device 2, the hydrogen plasma torches 3 and 6, the tank 4, the gas treatment system 5, the cogeneration system 7, the industrial numerical control system 8, the sensor blocks 9 and 11, the hydrogen plasma torch power supply block 10, the cooling system 12, the terminal equipment 13, the mixing block 14 and the additivation device 15.

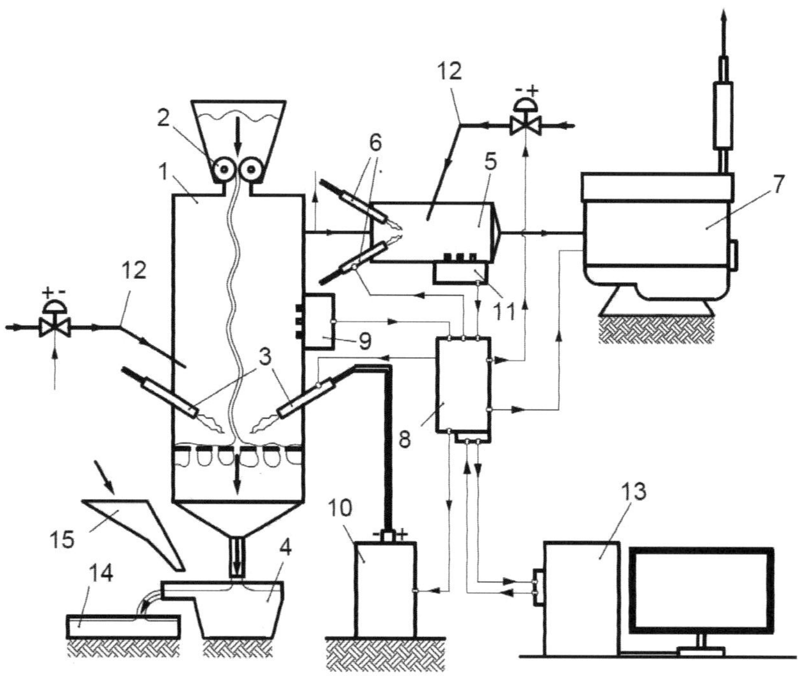

Figure 1. Diagram of the installation for plasma conversion of fiberglass waste.

The plasma reactor, according to figure 2, consists of the reactor shell 1, the steam feeding pipe 2, the plasma torch 3, the refractory grate 4, the refractory concrete layer 5, the reinforcing wire net 6, the refractory ceramic masonry 7, the mineral wool insulation 8, the collecting tank 9, the feeding device 10.

IOP Conf. Series: Materials Science and Engineering 1138 (2021) 012001 doi:10.1088/1757-899X/1138/1/012001

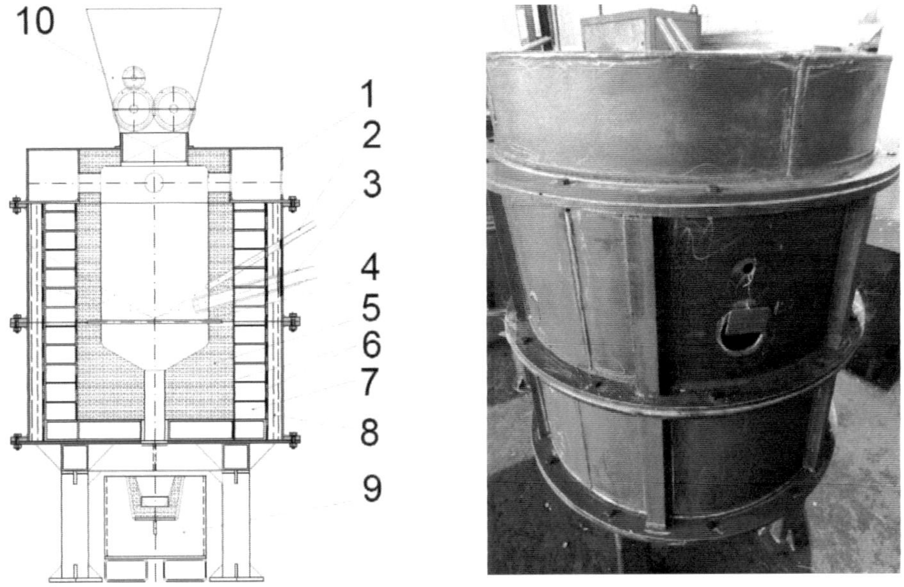

Figure 2. Assembly drawing and picture of the plasma reactor developed by authors.

The authors have used two research directions in the development of hydrogen plasma torches for the fiberglass recycling plant. First was the modifying of an existent plasma cutting torch which is presented in figure 3. The main challenge of this direction is the transforming of the convergent plasma jet of such a cutting torch into a divergent jet needed for plasma treatment of fiberglass waste, which can be accomplished by modifying the shape of the electrodes and adding of an ancillary water circuit.

Figure 3. Plasma torch modified by authors for melting of fiberglass waste.

The second direction was the research and development of a novel plasma torch, dedicated for the treatment of fiberglass waste, using the rapid prototyping techniques. This novel plasma torch is presented in figure 4.

Figure 4. Novel plasma torch researched and developed by authors for fiberglass waste treatment.

3. Experimental results and discussions

The modification of the existing cutting plasma torch consisted in the adjustment of the shape of the two electrodes and in a study of the changing of the shape and size of the plasma plume with the variation of the distance between the electrodes. The study consisted in the variation, step by step, of the distance between the electrodes, in the range from 1 to 6 mm, and the recording of the plasma plume shape and dimensions, as presented in figure 5.

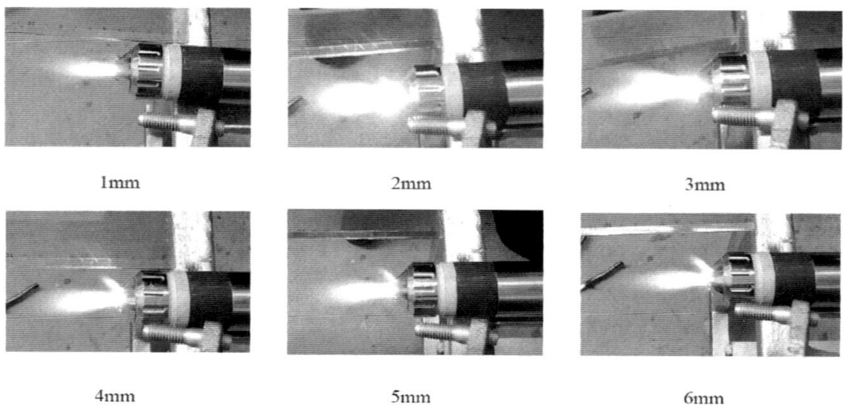

Figure 5. Recordings of the shape and size of the plasma plume as function of electrodes distance.

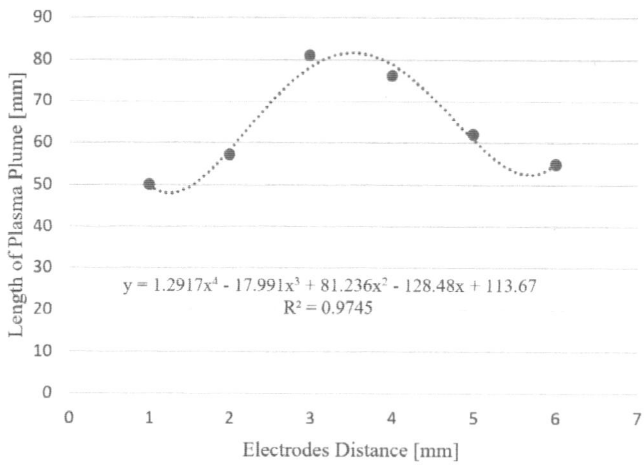

Figure 6. Variation of the plasma plume length with the distance between the electrodes.

The variation of the length of the plasma plume with the distance between the electrodes, resulting from the measurements, is presented in figure 6. It was noticed that the maximum length was obtained for a distance of 3 mm between the electrodes.

In figure 7 is recorded the variation of the maximum width of the plasma plume, as a function of the distance between the electrodes, as it resulted from the measurements. It can be perceived that the maximum width of the plasma plume was obtained for a distance of 2 mm between the electrodes.

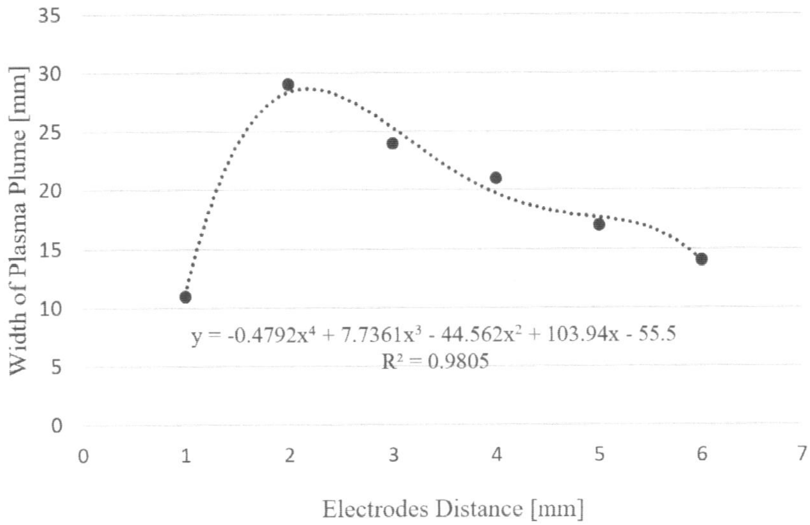

Figure 7. Variation of the plasma plume width as function of the distance between the electrodes.

It was also observed, as results from the figure 5, a separation of a secondary plasma jet from the main plasma plume, starting from a distance of 3 mm between the electrodes.

4. Conclusions

In this study was presented a novel plant for plasma treatment of fiberglass waste, developed by the authors, and two directions for developing plasma torches for this plant. First was the modifying of an existent plasma cutting torch, the second was the research and development of a novel plasma torch, dedicated, using the rapid prototyping techniques.

A research was conducted for the transforming of the convergent plasma jet of a cutting plasma torch into a divergent jet needed for plasma treatment of fiberglass waste, which was accomplished by modifying the shape and the distance between the plasma electrodes.

A study of the changing of the shape and size of the plasma plume with the variation of the distance between the electrodes was also done. This study reveals that the maximum length of the plasma plume was obtained for a distance of 3 mm between the electrodes and the maximum width of the plasma plume was obtained for a distance of 2 mm between the electrodes. It was also observed a separation of a secondary plasma jet from the main plasma plume, starting from a distance of 3 mm between the electrodes.

The advantages of the treatment plant for fiberglass waste, using hydrogen plasma torches, researched and developed by the authors are:

- efficient operation, with a low staff compared to existing facilities;
- fiberglass waste, with and without adhesive, treatment times are much lower than existing installations;
- the treatment of the resulting gases is made with a hydrogen plasma system, resulting in lower emissions;

- using of the syngas, resulting from hydrogen plasma treatment, to drive an electric generator for energy recovery;
- from the treatment process of the fiberglass waste, with and without adhesive, in the plant can be obtained, by future researches, valuable materials such as:
- empty glass microspheres [5] with porous walls or ceno spheres can be used as lightweight fillers in composite materials such as lightweight concrete, the storage and slow release of pharmaceutical drugs or radioactive markers for research, and in the field of controlled storage and release of hydrogen;
- aerogels [6] that can be used as thermal insulants, catalysts or carriers of catalysts, imaging, optical and light guides, filter materials, thickeners in some paints and cosmetics, and as components in solar energy absorbers;
- sealing glass [7] that has gluing and sealing applications between metal and ceramic parts.

Acknowledgment
This paper was realized with the support of project EFECON – Eco-innovative products and technologies for energy efficiency in construction, POC/71/1/4 - Knowledge Transfer Partnership, Code My SMIS: 105524, ID: P_40_295, Project co-financed by the European Regional Development Fund.

References

[1] Yang Y, Boom R, Irion B, Van Heerden DJ, Kuiper P and Wit H 2012 Recycling of composite materials*Chemical Engineering and Processing: Process Intensification* **51** 53-68
[2] Sponberg E W 1999 Recycling dead boats *Professional Boatbuilder* **60** 82-95
[3] Starr T F1997*Glass-Fibre Directory and Databook, 2nd ed.*(London: Chapman & Hall)
[4] De Moraes V T, Luiz Alberto Jermolovicius L A, Tenório J A S, Lebrão S M G and Lebrão G W 2020 Microwave-assisted recycling process to recover fiber from fiberglass polyester composites *Materials Research* **22** (suppl. 1) 1-8
[5] Isobe H, Tokunaga I, Nagai N and Kaneko K 1996Characterization of hydrated silicate glass microballoons*J. of Materials Res.*11 (11)2908–15
[6] Gurav J L, Jung I K, Park H H, Kang E S and Nadargi Y 2010 Silica Aerogel: Synthesis and Applications *J. of Nanomaterials* 1–11
[7] Dumesnil et al. 1980 *U S Pattent 4,186,023*

The use of programs complexes “Ansys” and "Plaxis 3D" in assessing the influence of the density of sand base on the modulus of deformation

V Antonov[1], I Al-Naqdi[1], P Monastyrev[1], E Pakhomova[2] and V Amelin[2*]

[1] Tambov State Technical University, Tambov, Russia
[2] Southwest State University, Kursk, Russia

* amelin-46@yandex.ru

Abstract. The article presents a comparison of the values of the soil deformability characteristics obtained at different densities of the sand base based on the results of laboratory tests[3] and the results of computer modeling, using the software systems "ANSYS" and "Plaxis 3D".It was found that for loose sandy soils, at densities of $1.48 - 1.53\ g/cm^3$, close to experimental values of the deformation modulus were obtained using a three-dimensional finite-element model of polylinear isotropic hardening "PLAS (Miso)," in the program "ANSYS." For average density bases at $\rho = 1.55 - 1.66\ g/cm^3$ minimum differences between experimental and numerical values were obtained using the «Plaxis 3D» complex.

1. Introduction

The widespread method of compression gives underestimated values of the deformation modulus; therefore, stamping tests are widely used in practice to determine the main characteristics of deformability [1, 5]. In this case, a graph of the dependence of settlement on pressure is used, on which a linear section is selected and the deformation modulus is calculated using the Schleicher formula. The secant modulus of deformation is defined as the ratio of pressure to settlement in the selected pressure interval, and the tangential modulus is defined as the ratio of the pressure increment to the slump increment for each pressure interval under the die base. At the same time, the obtained deformation modules reliably characterize the stiffness of only surface layers of soil.

2. Materials and methods

The tests were carried out in a metal tray with rigid side walls measuring 70 cm x 55 cm x 53 cm. Soil - finely dispersed sand, homogeneous. In the experiments, the density of the base was changed ($\rho = 1.663; 1.620; 1.590; 1.530; 1.487\ g/cm^3$). The angle of internal friction for sand in an air-dry state was determined at a density of $\rho = 1.62\ g/cm^3$ and amounted to $39°$. The sand base was formed by layer-by-layer compaction using metal ramming to a given density. The thickness of each layer was 5 cm. During the tests, precipitation (s) was determined in the entire load range by hour-type indicators ICH-10 mounted on the reference frame (figure 1). As a model of the foundation, a rigid metal stamp with a diameter of 120 mm was used. The load on the die was transmitted using a system of levers with a gear ratio of 1:5. Loading stages were taken equal to 0.1 of destructive load. Each stage was held until conditional stabilization of the precipitate (20 min). Loading was carried out either until destruction, at which precipitation grew without increasing the load, or until a conditional maximum settlement of the foundation was obtained $[s]_u$.

Figure 1. Laboratory installation.

Numerical simulation of the experiments was performed using the software package "ANSYS "and "Plaxis 3D". The development of displacements was limited in the x and y directions, i.e. the displacement could only develop in the vertical direction $z(x=0, y=0)$. Two models were used to model the base:

1) Mohr-Coulomb Model.

The simplest known condition is the strength condition formulated in 1773 by Sh. Coulomb, using the Mohr method which can be presented as:

$$\left(\sigma_1-\sigma_3\right)/(\sigma_1+\sigma_3+2cctg\varphi)=\sin\varphi \tag{1}$$

Using this strength condition, the theory of the ultimate stress state of a loose and cohesive medium base and its application to solving a number of engineering problems has been developed. In the stress invariants j_1, j_2, θ the strength condition (1) can be represented as:

$$f(I_1,J_2,\theta)=-I_1\sin\varphi+(\frac{3(1+\sin\varphi\sin\theta+\sqrt{3}(3-\sin\varphi)\cos\theta}{2})\sqrt{J_2}-3c\cos\varphi=0 \tag{2}$$

$$f(\xi,\rho,\theta)=-\sqrt{6\zeta}\sin\varphi+(\frac{3(1+\sin\varphi\sin\theta+\sqrt{3}(3-\sin\varphi)\cos\theta}{2})\rho-3\sqrt{2}c\cos\varphi=0 \tag{3}$$

In the main stress space, the Coulomb-Mohr condition represents an irregular hexagonal pyramid, as shown in figure 2 (a, b).

This condition is widely applied in practice mainly because of its simplicity and acceptable accuracy in solving specific engineering problems [2, 3].

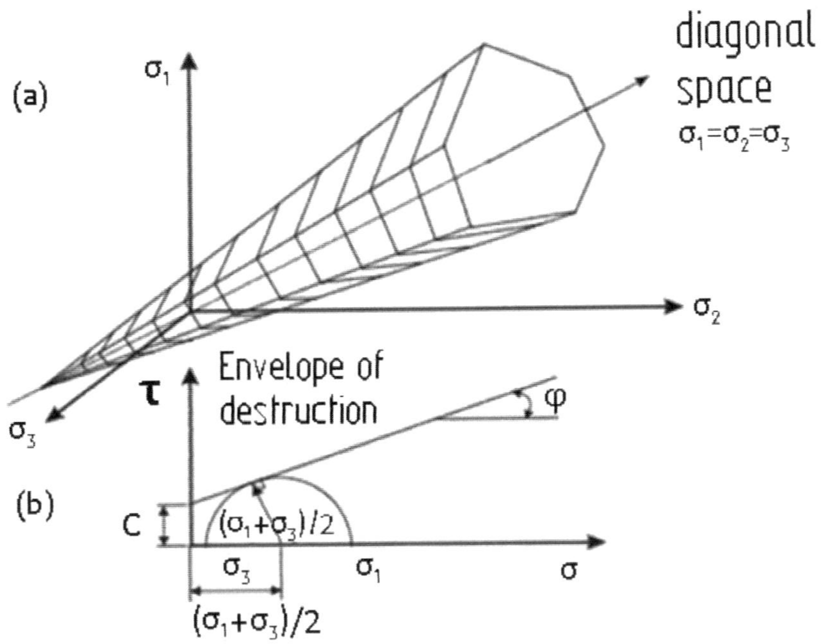

Figure 2. Coulomb-Mohr strength condition: a - in the space of main stresses; b - in the plane of the main stresses [7].

To obtain numerical solutions using a finite element model of the Mohr-Coulomb material, the 10 nodes elements in Plaxis were used in modeling the soil matrix. The modulus of deformation used in the program was calculated using the Schleicher formula based on experimental data, the Poisson's ratio $v = 0.3$ the plate model had dimensions $D = 120\,mm$, thickness 10 mm, elastic modulus $E = 2.06\,x\,10^5\,MPa$. The plate was modeled using a linear isotropic model.

2) A three-dimensional finite element model of PLAS (Miso) multilinear isotropic hardening. The Miso material model is used to model linear and nonlinear soil behavior [4]. The nonlinear behavior of the soil is modeled using a multi-line model of isotropic hardening (MISO) material. This model includes the von Mises yield criterion [4]. When estimating the stress-strain state for sand using the PLAS model (Miso) in the ANSYS program, the first point of the stress-strain state curve must be the yield stress. Subsequent points determine the elastic-plastic response of the material.

The equivalent stress equation has the form:

$$\sigma_e = [3/2\{s\}^T[M]\{s\}]^{(1/2)} - \sigma_k = 0 \tag{4}$$

where $\{s\}$ is the deviator stress; σ_e – is the current yield stress σ_K – is a function that depends on the plasticity of the material. For the case of isotropic hardening, σk can be determined from the uniaxial strain stress curve [7].

When using PLAS (Miso) according to ANSYS APDL, the finite element (discrete) model consisted of the elements Solid186 (a 20-node hexagonal volumetric element for modeling a soil matrix, which has three degrees of freedom per node - this is the recommended type of elements for 3D models) and Solid187 (the element is defined by 10 nodes having three degrees of freedom at each node and was used to model the stamp). All other source data is the same as in the first model.

3. Results

Tables 1 - 3 and figures 3 - 8 show the experimental values and results obtained by ANSYS APDL, using the viscoplastic PLAS model (Miso) and the Mohr-Coulomb elastic-plastic model for Plaxis 3D.

The values of the modulus of deformation were determined using the Schleicher formula [6].

$$E = \omega * D * (1 - v^2) * \Delta p / \Delta s \tag{5}$$

Where ω is the coefficient taken for round plate of 0.8; $D-$ is the diameter of the plate; $v-$ is Poisson's coefficient taken 0.3 for sand; Δp – increment of the average pressure on the plate foundation on a plot of the linear relationship between pressure and settlement; $\Delta -$ - settlement increment of the plate when the pressure changes by Δp.

Table 1. Dependence of sand base modulus of deformation *E* on the pressure at a density of 1.487 g/cm³.

№	*P (MPa)*	Experimental values		Plaxis 3D		PLAS (Miso) ANSYS	
		s, mm	*E, MPa*	*s, mm*	*E, MPa*	*s, mm*	*E, MPa*
1	0.0053	0.67	0.6960	0.225090099	2.056989632	0.74478	0.6261
2	0.0106	1.07	0.866	0.893502196	1.03689171	1.4393	0.6479
3	0.0160	1.96	0.7137	2.0304736	0.688391122	2.13347	0.6557
4	0.0213	3.635	0.5131	3.593894259	0.517758138	2.82729	0.6597
5	0.0266	5.67	0.4112	5.600636916	0.41491281	3.52076	0.6622

Table 2. Dependence of the modulus of deformation on the density of the sand base under ultimate load.

Density g/cm³	1.663	1.620	1.590	1.530	1.487
P, MPa	0.27	0.11	0.08	0.07	0.03
E, Experimental	5.4318	2.0335	0.8665	0.5912	0.4112
E , Plaxis 3D	4.57	2.12	0.925	0.665	0.459
E, PLAS (Miso) ANSYS	9.4017	3.4138	1.5239	0.9891	0.6622

Table 3. Dependence of the modulus of deformation on the density of the sand base at *P=0.03 MPa*.

Density g/cm³	1.663	1.620	1.590	1.530	1.487
P, MPa	0.03	0.03	0.03	0.03	0.03
E, Experimental	5.588	2.440	1.879	0.822	0.4112
E , Plaxis 3D	7.02	2.46	1.03	0.7	0.459
E , PLAS (Miso) ANSYS	7.713	3.187	1.479	1.018	0.6622

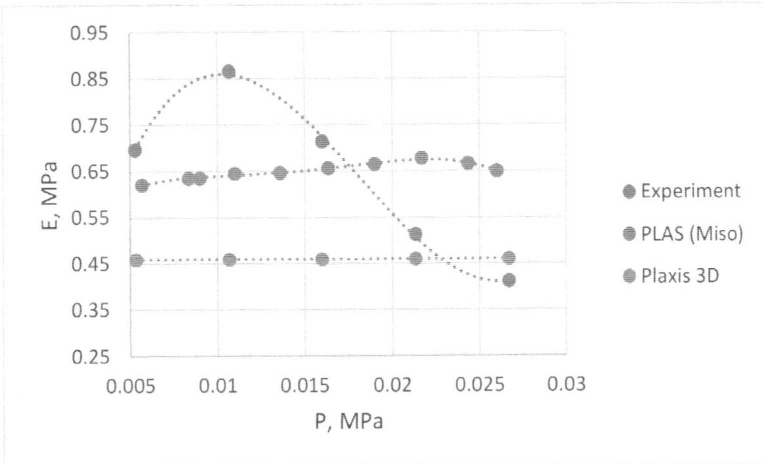

Figure 3. E - P Dependence at a density 1.487 g/cm³.

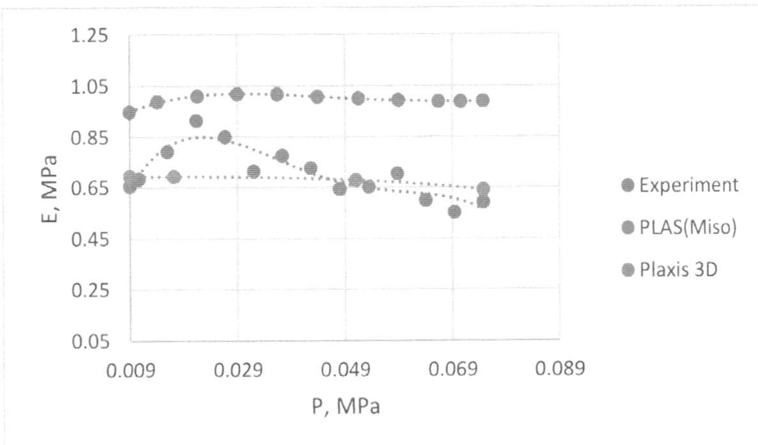

Figure 4. E - P Dependence at a density 1.53 g/cm³.

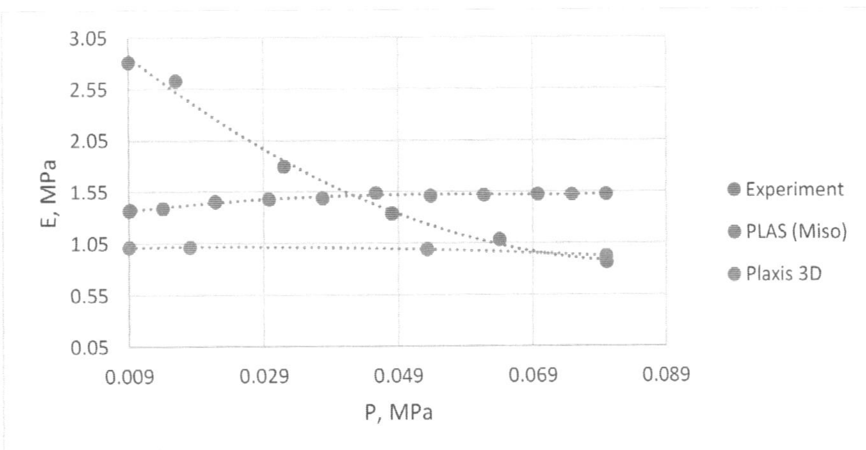

Figure 5. E - P Dependence at a density 1.59 g/cm³.

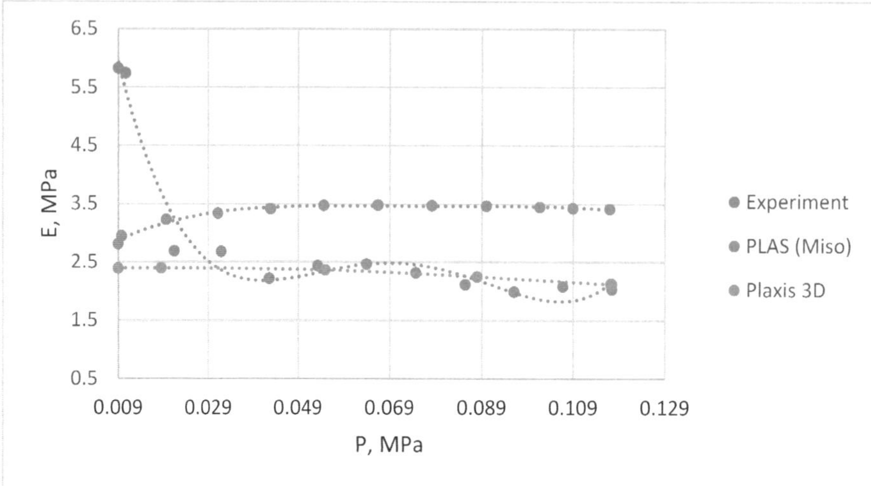

Figure 6. E-P Dependence at a density 1.62 g/cm^3.

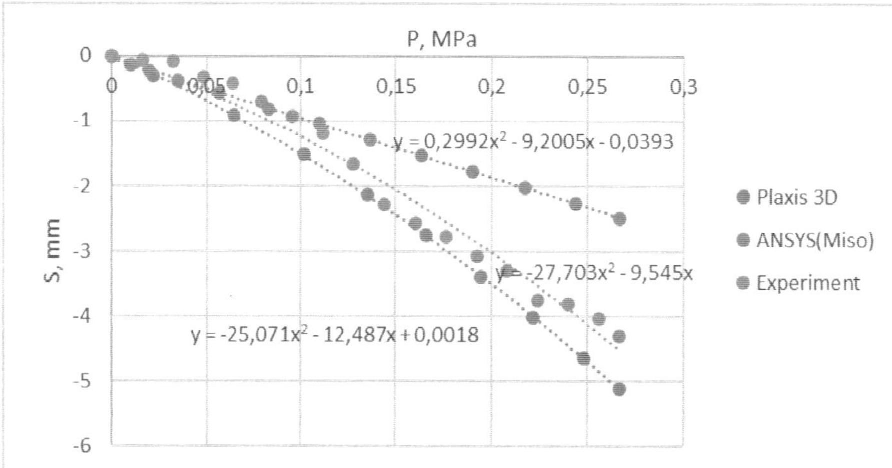

Figure 7. a) Dependence of the plate settlement on the pressure at a density 1.66 g/cm^3.

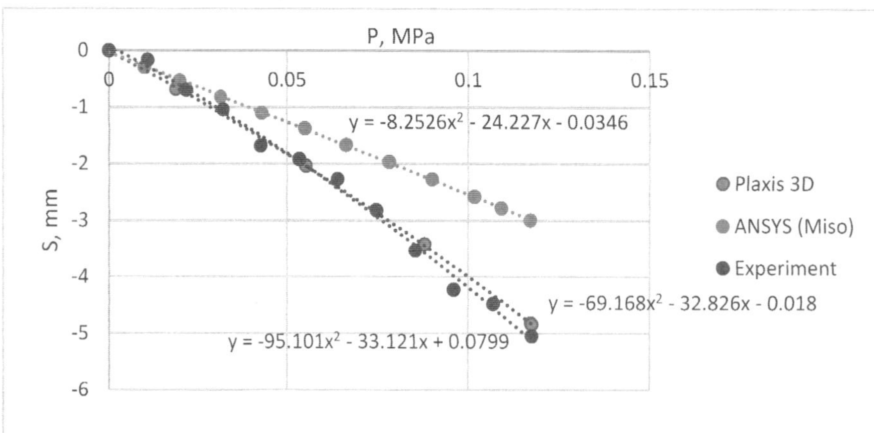

Figure 7. b) Dependence of the plate settlement on the pressure at a density 1.62 g/cm^3.

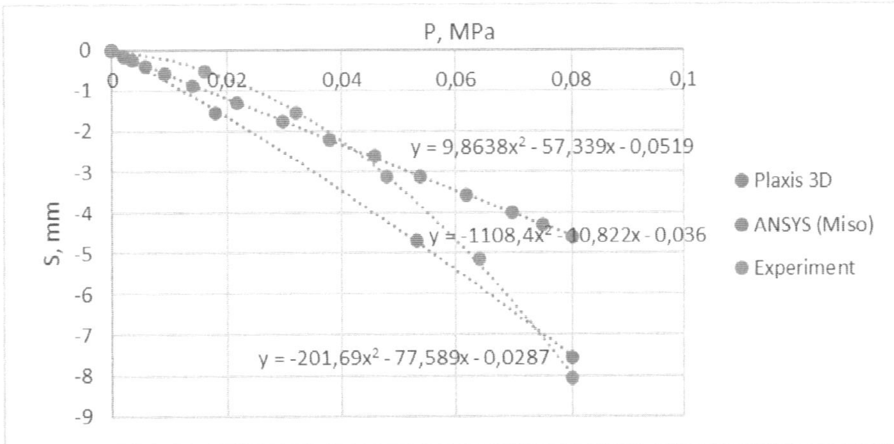

Figure 8. a) Dependence of the plate settlement on the pressure at a density of: density 1.59 g/cm^3.

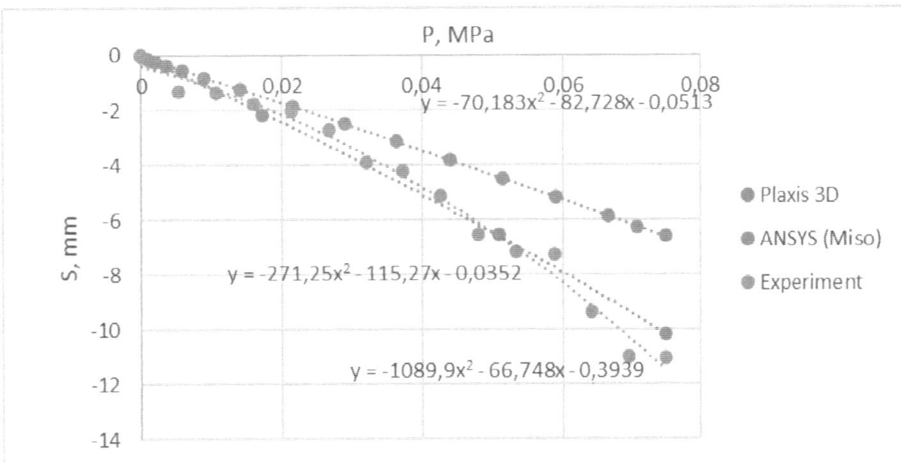

Figure 8. b) Dependence of the plate settlement on the pressure at a density of: density 1.53 g/cm^3.

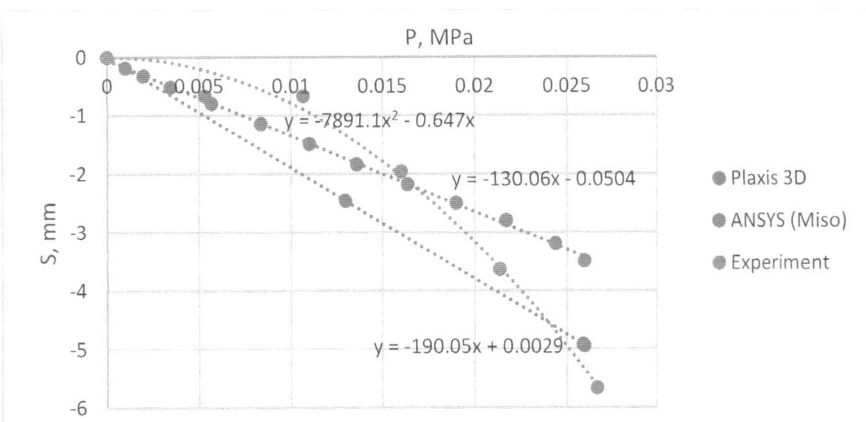

Figure 8. c) Dependence of the plate settlement on the pressure at a density of: density 1.487 g/cm^3.

4. Conclusions

At density values $1.48 - 1.53\ g/cm^3$, the minimum discrepancies between the experimental and numerical values were obtained using the model of multilinear isotropic hardening "PLAS (Miso)", in the program "ANSYS".

For sands of average density at $1.55 - 1.62\ g/cm^3$ the minimum discrepancies between the experimental and numerical values were obtained using the Plaxis 3D complex.

A change in density from $1.53\ g/cm^3$ to $1.62\ g/cm^3$according to the results of experiments led to an increase in the breaking load by 1.57 times, and a change in density from $1.62\ g/cm^3$ to $1.66\ g/cm^3$ by 2.42 times.

The values of the deformation modulus when changing the density from$1.48\ g/cm^3$ to $1.66\ g/cm^3$ according to stamp tests and according to the results of modeling using the "Plaxis 3D" software complex increased almost 10 times.

References

[1] GOST 20276-2012, Soils. Field methods for determining the strength and strain characteristics, 2013
[2] Bezgodov M, Kaloshina S 2012 Vybor modeli grunta pri chislennom modelirovanii vlijanija razrabotki glubokih kotlovanov na sushhestvujushhuju zastrojku // Vestn. Perm. nac. issled. politeh. un-ta. Prikladnaja jekologija. Urbanistika. No. 2. pp. 17–27
[3] Orehov V, Orehov M 2014 Ispol'zovanie modeli uprochnjajushhegosja grunta dlja opisanija povedenija peska razlichnoj plotnosti pri nagruzhenii // Vestn. MGSU. № 2. pp. 91–97
[4] ANSYS Manuals version 18.1
[5] Pantelidis, Lysandros. 2008 Determining of the soil strength characteristics through the plate bearing test. *Foundations of Civil and Environmental Engineering* 11: 55-65
[6] Lavrov S 2011 Metodika issledovanij deformacionnyh svojstv dispersnyh gruntov rasklinivajushhim dilatometrom v polevyh uslovijah. PhD diss., Avtoreferat dissertacii na soiskanii uchenoj stepeni kandidata tehnicheskih nauk. Barnaul.
[7] Das B 1983 Advanced soil mechanics 511 p

IOP Conf. Series: Materials Science and Engineering 1138 (2021) 012003 doi:10.1088/1757-899X/1138/1/012003

Freezing water simulations in isochoric systems – preliminary analysis

G A Beşchea[1]*, Ş I Câmpean[1], L M Scutaru[2], L Costiuc[2] and A Şerban[3]

[1] Department of Building Services, Faculty of Civil Engineering, Transilvania University of Brasov, Romania,
[2] Department of Mechanical Engineering, Faculty of Mechanical Engineering, Transilvania University of Brasov, Romania,
[3] Thermotechnics, engines, thermal and refrigeration equipment Department, Faculty of Mechanical Engineering and Mechatronics, University Politehnica of Bucharest, Romania

* george.beschea@unitbv.ro

Abstract. Freezing liquids into ice is a complex simulation and is called a phase transformation or phase change. Isochoric (constant volume) preservation is a new thermodynamic concept that takes advantage of an until now ignored aspect of thermodynamics, life under conditions of constant volume as opposed to the conventionally studied life conditions of constant pressure. Isochoric system are simple constant volume devices, made of steel, capable of withstanding the pressures that develop in the system, with minimal deformation. Because ice-I is less dense than water, the formation of an ice nucleus in an isochoric (constant volume) chamber will cause an increase in pressure. Traditionally, ice formation is performed in an isobaric process (constant pressure) at 1 atm, because this is our natural environment and it is most convenient experimentally. The goal of this study is to introduce the fundamental thermodynamic principles and boundary conditions of isochoric (constant volume) simulations at low temperatures. In this analysis we intend to shows that homogeneous ice nucleation can be thermodynamically improbable in an isochoric system.

1. Introduction

Constant volume (isochoric) study at negative temperatures for different type of liquids become in the last period very interesting by the point of view of different industries. For medicine, food preservation or tissue preservation it is very important to develop new ways of preservation, because in many cases the traditional, isobaric preservation reached his limits and we can't preserve for long periods biological matter, without a change of his molecular structure. For example, the human organs, collected for transplant could resist only a couple of hours with actual common accepted preservation method, but with isochoric preservation we try to find a way to extend this time to 48 or 72 hours. This means that an organ collected from a donor on any continent from the Word could be implanted to a compatible person all around the world using the current way of transport disponible for medical services.

Recent studies done on potatoes [1], fish muscle [2] or even on living biological matter [3], showed promising results for using constant volume preservation as a viable alternative [4] for isobaric preservation. The only disadvantage of using this method, is that is obvious that the intern energy of the system will increase (according that all liquids began to freeze at different temperatures), and because of this the inevitable formed ice will determine that in the constant volume chamber the pressure will increase. The studies done until now, led to recognition of the isochoric preservation as a pressure related cooling and freezing technique for food industry [5] and studies ongoing at the time of writing this article will develop theoretical methods of using this type of preservation for other interested industries. Every liquid solution has a different behavior and it is early to estimate for each used type a common accepted theory.

Different types of aqueous solutions can be used to preserve in constant volume conditions, but it is important to try with the most common used liquids. One of that is the water (H_2O).

The behavior of the water under isochronous conditions was well established already, and we could use it as thermodynamic and mathematic verification of the experimental data we collect, to explicit our systems [6].

With the data sets and the theoretical calculations, using Comsol software, we created a model and we start to simulate the same conditions.

2. Experimental model

The isochoric freezing system is a simple constant volume chamber capable to resist to the pressure from the inner of the reactor, with minimal deformation. The chamber is instrumented with a pressure transducer.

The isochoric chamber is a 2 mL 316 stainless steel micro-reactor MS-1 with maximum working pressure of 60,000 psi and a total of inner volume of 3 ml. The micro-reactor is capable to resist at liquid nitrogen temperatures. The inside diameter is 3/16’ and the outside diameter is 9/16’. Inside depth is 4’ and outside length is 7’. The microreactor is connected to a HP1100 0±4000 bar pressure transducer, connected to DATAQ Instruments Hardware Manager to a laptop.

The micro-reactor is filled with water and immersed in a water–ethylene glycol bath (50/50) cooled by means of a Lauda RE 1225 S cooling system (LAUDA DR. R. WOBSER GMBH & CO. KG, Lauda-Königshofen, Germany). This device can control temperatures to -25°C.

With this protocol, we are able to read the temperature from the inside of the bath (assuming that for temperature we set -20°C and the pressures from inside the microreactor. Some of the collected data sets, are presented in the tables below. We exemplify only the sets interested for our simulations.

Table 1. Time – Average temperature

Time [min]	Avg. temp [°C]
0	19.99
30	-6.67
60	-14.21
90	-19.56

Table 2. Pressure – Average temperature

Pressure [bar]	Avg. temp [°C]
0	19.99
776	-6.67
1475	-14.21
1905	-19.56

3. Theoretical model

For our simulations we used a theoretical 316 stainless steel 57 mL microreactor, with an internal diameter of 30 mm, outer diameter of 50 mm and an internal height of 80 mm. The nominal conditions, a view are presented in 'figure 1'. Initially in the isochoric reactor we have liquid water at ambient temperature and atmospherically pressure. We assume that the volume difference between the chamber we used for measurements and the chamber we used to simulate, did not affect our results.

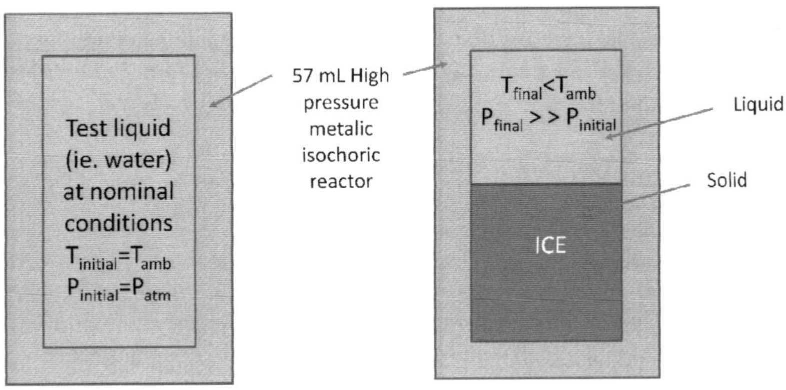

Figure 1. Initial conditions (left) and expected behaviour of water in isochoric conditions (right).

The expected behavior of water is well known under big pressures [7] and the translation to isochoric model (which has the same pressure conditions and newly introduced "constant volume" model) is also know [8] and the theoretical and experimental graph it is shown in figure 2.

The triple point of the water in isochronous conditions is at -21.985°C and 209.9 MPa. The interval for this test is below the triple point, so theoretically we will have water and ice in our system together during all measurements.

For the temperature sets tested our expectation is that not all the water will freeze during the simulations and we will have a pressure increase. However, we could simulate the displacement of the theoretical model, to see if its value affects our measurements. According that the reactor is made from Stainless steel 316 and the manufacturers data sheet provided, the expectation is that the change caused by the heat transfer and the pressure increase it will be minimal. With all these correlations we could have a real expectation for the behavior of our model during the simulation.

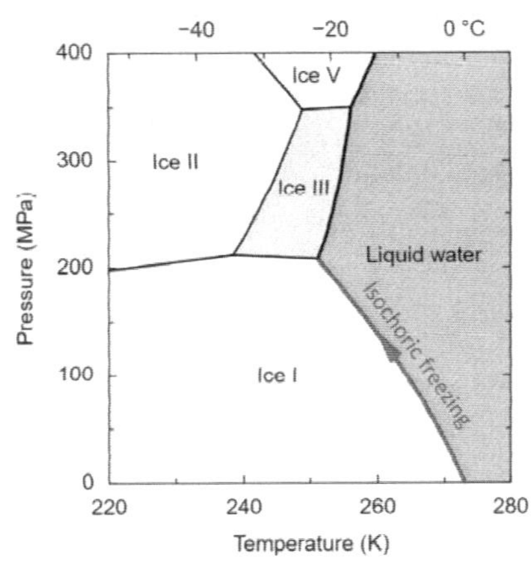

Figure 2. Phase change diagram for water under isochronous conditions.

4. Comsol simulations

We implemented the model presented in figure 1 in Comsol Simulation Software. According this our model is a cylinder, we used 2D Axisymmetric Geometry. Our problem uses coupled physics. First of all, we need Heat Transfer in Solids because we want to see the behavior of our solid and liquid part of the model during the freezing process. Also, we want to see the possible deformation of the cylinder meantime, so we need to use Solid Mechanics. Because we need to study the behavior of our system in a certain period of time, we need to use a Time Dependent Study. The difference between this type of study and a Stationary study is that the software allows us to see the transient steps of the simulation. With these two types of physics, we consider that our problem could be explained.

The exterior conditions are set as boundary conditions of temperature. On the water inside, we set the Phase Change temperature at 0°C. The total phase change interval was set from 0°C to -21.985°C.

For simplicity we used 3-time steps in our captions to exemplify the what the simulation shows us.

The initial time is "0" (figure 3a). The heat transfer started from cooling bath to the reactor. We could see and compare it from the different colorization of the steel part of the reactor.

We expect the heat transfer in the 316 stainless steel reactor to be fast and when the heat transfer starts from the stainless steel to the water, we have a moderate temperature evolution. After 30 minutes (figure 3b), the reactor reached the temperature of the cooling bath -20°C, but the water inside has an average temperature of -6,67°C. After 60 minutes (figure 5c) all of our domain is on the negative field of temperature (below 0 °C)

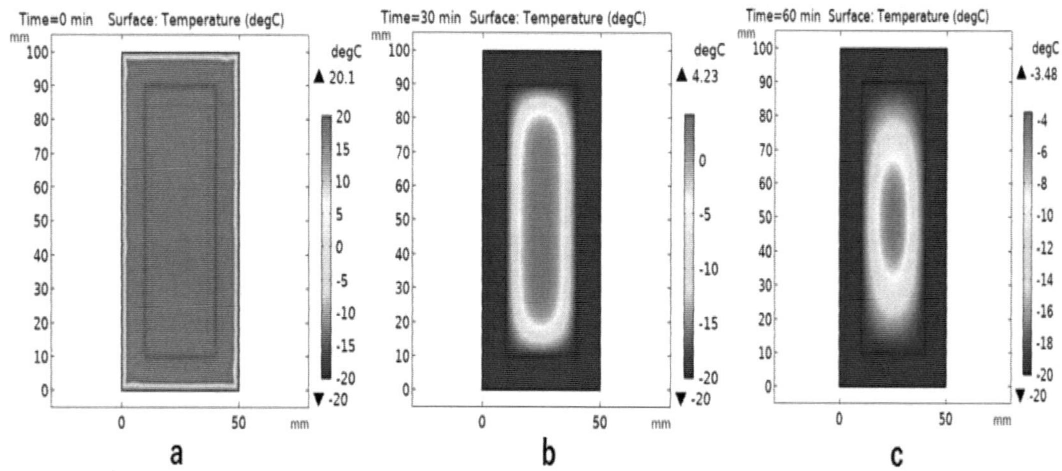

Figure 3. Heat transfer in solid and liquid at the beginning (a), after 30 min (b) and after 60 min (c).

If we are looking to the phase change, in the same time steps. At moment "0" (figure 4a), it is obvious that is impossible that the water starts to freeze, because the cylinders inner side has a temperature of 20°C. The water started to pass from liquid to ice, but after the first 30 minutes in the reactor we still have 26.23% of liquid (figure 4b). After 60 minutes of constant expose to -20°C figure 4c, just a small amount of water remains in liquid form. That is because of the set temperature for phase change is -21.985°C, so the results look correct. Simulation done for a longer period of time, shows that after 60 minutes, the system reached an equilibrium a further data to present is not relevant for this preliminary analysis.

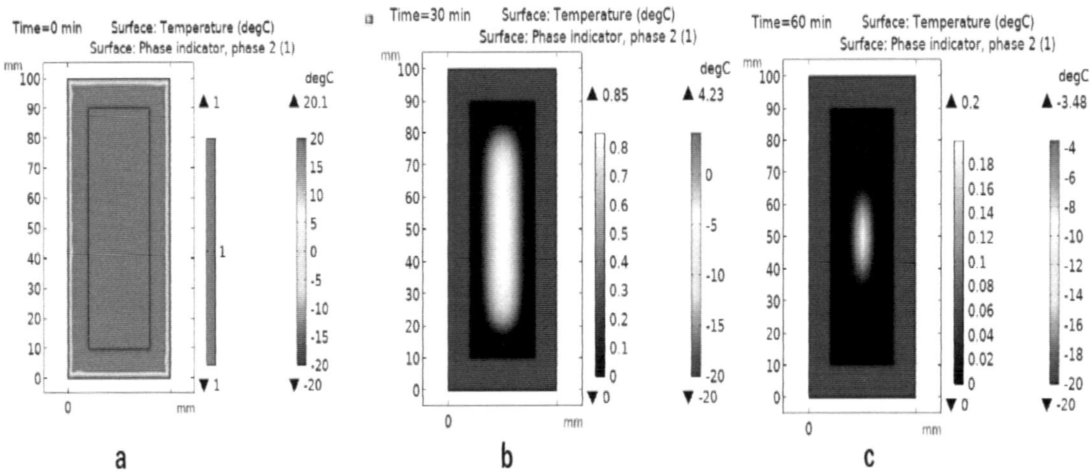

Figure 4. Phase change indicator and temperature of the reactor at the beginning (a), after 30 min (b) and after 60 min(c).

With this two sets of data, we generated a graph (figure 5). The percentage of ice I formed during the freezing is in inverse proportion with the temperature, so as the temperature become lower, the quantity of ice increases. After 60 minutes, we could see that our system reached an equilibrium, and the quantity of ice from the inside of the reactor remains constant.

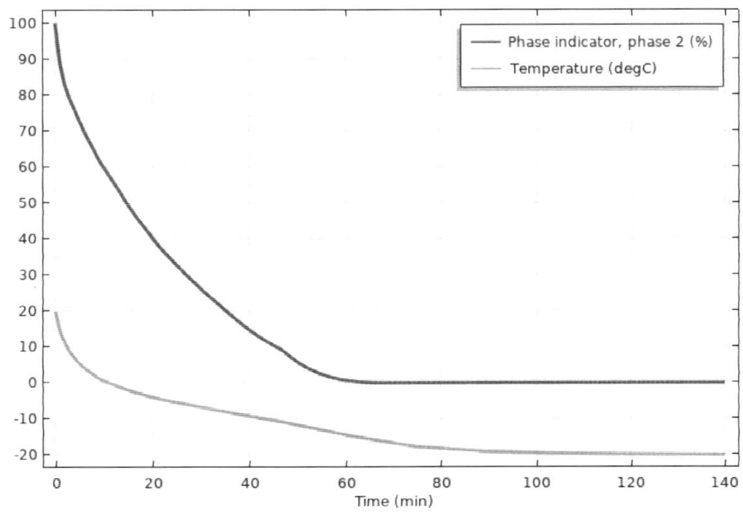

Figure 5. Phase indicator [%] and temperature values [°C].

The total displacement of the reactor gap (on Y direction), after 60 minutes, is around 0.075 mm. This value is not affecting results of the simulation from the heat transfer and phase transition point of view, but is important to observe the thermal stress effects on the reactor due the negative temperatures (figure 6).

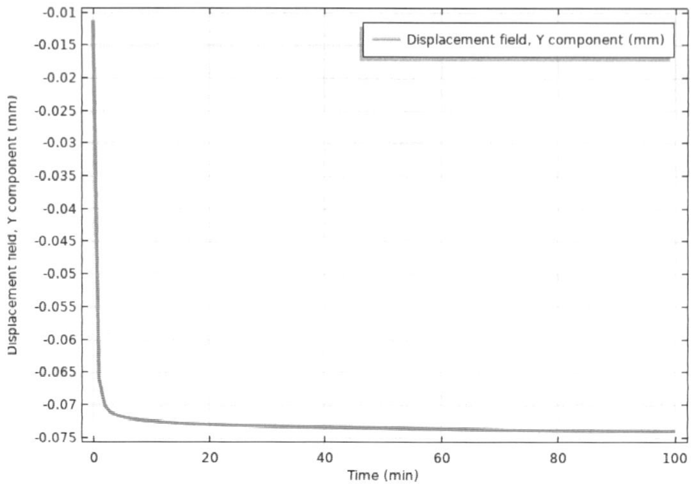

Figure 6. Total displacement of the superior part of the reactor on Y direction [mm].

5. Conclusions

We consider that further studies about the behavior of the liquids in isochoric conditions are important because of the possibility of use them in different industries. A new way of conservation could be the conservation in isochoric conditions and we try to contribute on this field with our studies. This type of conservation was recently recognized as a possible way to preserve in food industry [8], so the first steps all already done for using it on bigger scales.

References

[1] Lyu C, Nastase G, Ukpai G, Serban A and Rubinsky B 2017 A comparison of freezing-damage during isochoric and isobaric freezing of the potato. *Peer J.* **2** (5):1–15

[2] Năstase G, Lyu C, Ukpai G, Şerban A, Rubinsky B. 2017 Isochoric and isobaric freezing of fish muscle. *Biochem Biophys Res Commun* **485** (2):279–83

[3] Mikus H, Miller A, Nastase G, Serban A, Shapira M and Rubinsky B 2016 The nematode Caenorhabditis elegans survives subfreezing temperatures in an isochoric system. *Biochem Biophys Res Commun* **477** (3):401–5 http://dx.doi.org/10.1016/j.bbrc.2016.06.089

[4] Năstase G, Perez P A, Şerban A, Dobrovicescu A, Ştefănescu M F and Rubinsky B 2016 Advantages of isochoric freezing for food preservation: A preliminary analysis. *Int Commun Heat Mass Transf.* 78:95–100

[5] Zhu Z, Li T and Sun D W 2020 Pressure-related cooling and freezing techniques for the food industry: fundamentals and applications. *Crit Rev Food Sci Nutr.* https://doi.org/10.1080/10408398.2020.1841729

[6] Rubinsky B, Perez P A and Carlson M E 2005 The thermodynamic principles of isochoric cryopreservation. *Cryobiology*. **50** (2):121–38

[7] Chizhov V E and Nagornov O V 1991 Thermodynamic properties of ice, water and their mixture under high pressure. *Glaciers-ocean-atmosph Interact.***463** (208):463–70

[8] Kalichevsky MT, Knorr D amd Lillford P J. 1995 Potential food applications of high-pressure effects on ice-water transitions. *Trends Food Sci Technol.* **6** (8): 253–9. http://www.sciencedirect.com/science/article/pii/S0924224400891098

IOP Conf. Series: Materials Science and Engineering 1138 (2021) 012004 doi:10.1088/1757-899X/1138/1/012004

Petrographical characteristics as an indicator of physico-mechanical properties of concrete aggregate

L Bianco[1,2*] **and G Micallef**[3]

[1] Faculty for the Built Environment, University of Malta
[2] Faculty of Architecture, University of Architecture, Civil Engineering and Geodesy, Sofia, Bulgaria
[3] CVC Architecture Studio, Valletta, Malta

* lino.bianco@um.edu.mt

Abstract. The physical and mechanical properties of concrete designed to a C40 mix using 6.5 mm and 10.0 mm aggregate from Montenegro and Malta respectively were investigated. Ultrasonic pulse velocity (UPV) and compressive strength (CS) tests corroborated findings in the literature. The coarseness of aggregate was found to be of primary technological importance: CS increased with increasing aggregate size. Although UPV in all the tests was between 3.5 km/s and 4.5 km/s – that is the concrete quality was of good grade – the CS of the control cubes for the 6.5mm and 10.0mm coarse aggregate mixtures was 27 N/mm^2 and 39 N/mm^2 respectively. The homogeneity of the concrete was affected by the physical and mechanical properties of the aggregate, including its size, but its petrological texture was also a significant parameter. The petrological characteristics of the aggregate severely affected the load-bearing capacity of the concrete. These characteristics had a significant bearing on the physico-mechanical properties of the hardened concrete.

1. Introduction

Concrete is a widely used construction material [1], a statement upheld by [2] in their recent study on the properties of multi-recycled concrete. Aggregate is the main ingredient which accounts for nearly 75% of the volume, with the coarser fraction, depending on the mix proportions, accounting for 50 to 60% of the mix.

Recent research addressed the process of designing a prototype coffee table, from drawing to the production of an industrial item [3]. The final product made extensive use of concrete, walnut-finished timber components and black-finished steel elements (figure 1). Concrete was utilised both in plane and edge orientations, the latter being useful to enhance the minimum thickness possible in order to reflect the concept of lightness. Whilst weight was a major drawback, concrete was nevertheless selected as the dominant product material. The surfaces were produced from freshly poured concrete in 25mm deep moulds designed for the panels. There were 5 moulds, each varying in form to correspond to the panels. They were constructed from melamine laminated particle board for (i) the production of a smooth finish and (ii) ease of demoulding. The sides of the moulds were manufactured in sanded pine wood as mitre edges could not be produced in melamine.

(a) (b)

Figure 1. Coffee table: conceptual impression (a) and finished product (b).

The design strength for the required concrete was set between 35 MPa and 40 MPa and thus a C40 concrete mix design was utilised. Two mixtures were used, differentiated by the type and size of the coarse aggregate: 6.5 mm aggregate from Montenegro and 10 mm aggregate from Malta. Whilst the panels manufactured from the 10 mm aggregate were not damaged on demolding after the 7-day curing period (figure 2a), the 6.5 mm aggregate panels were damaged at the edges (figure 2b) and one failed at the corner (figure 2c). The aim of this paper is to investigate why these panels failed when, at face value, they had the most appropriate aggregate available on the market for the design mix. Given that all the basic components were uniform except for the type and size of aggregate, the study investigated their texture.

(a) (b) (c)

Figure 2. Concrete panels after a 7-day curing period using 10.0 mm (a) and 6.5 mm (b and c) aggregate.

2. Methodology

2.1. Materials

Portland limestone cement CEM II/A-LL 42.5 R and local potable tap water were used in all the mixtures. Silica fume was introduced to improve the strength of the concrete. The coarse aggregate was of two sizes: 6.5 mm and 10.0 mm. According to the aggregate supplier, the former was from Volujica Quarry in Port of Bar, Montenegro. The latter was from Wied Filep Quarry HM22 located at the limits of Naxxar, Malta. The aggregate from Montenegro fell within the 4 to 8 mm size range supplied by the company processing the limestone. The characteristics of the aggregate met the specifications of the European Directive for building materials and CPR 305/2011 standard EN 12620: 2002 + A1: 2008 (http://www.nimont.me/quarry.php). The fine aggregate used in all the mixes was from the quarry in Malta. The Volujica Formation outcrops at Bar and is composed of Late Cretaceous sediments of banked and stratified limestone with dolomite interbeds. Light brown banks, thick bedded rudist bioclastic biomicrites, lumakeles, and konkins are intercalated with dark brown medium beds of wackestones [4, 5]. The formation which outcrops at Naxxar is the Oligocene Lower Coralline Limestone [6-8], specifically the Magħlaq Member which is the oldest limestone quarried and distinguished for its high crushing value [9].

2.2. *Type and size of aggregate*
Particle density and water absorption were determined as per [10] whilst the Aggregate Impact Value (AIV) was determined as per [11], now replaced by [12]. To study the texture at micro level, two samples of each type of aggregate were studied in thin section under a petrological microscope. The samples were selected through quartering.

2.3. *Mix design and casting of concrete*
The mix design for the C40 concrete is given in Table 1. Silica fume was added at 5% by weight of cement. To ensure consistency in the quality, quartering was used for both the fine and coarse aggregate. The mixture was produced by an electricity-driven concrete mixer and the water-cement ratio was kept at 0.55.

Two mixtures were cast: one for each size of the coarse aggregate. For a given mixture, the 5 moulds for the panels were cast at the Laboratory of Civil and Structural Engineering of the University of Malta. Following mixing for a duration of 15 minutes, the concrete was poured into each mould. The moulds were on a vibrating table to aid compaction, to limit entrapped air and to the fill the moulds to the edges. Trowelling was used to give an even finish to the panels. Four 100 x 100 x 100 mm concrete cubes were cast as a control unit.

The melamine-laminated moulds were thoroughly cleaned between each cast; no lubricant was utilised on their surface. Both the panels and cubes were cured for 7 days under plastic sheet covering until they were demoulded and left to air-cure for the remaining 28-day period.

Table 1. C40 mix design.

	Cement	Water	Aggregate [kg]		Silica fume
	[kg]	[l]	fine	coarse	[kg]
Per m^3 (to nearest 5 kg)	345	190	761	929	17.25
Per trial mix of 0.034 m^3	11.73	6.46	25.87	31.59	00.59

2.4. *Testing regimes*
A Zeiss Axioskop 40 transmitted light-microscope was used to analyse the thin sections of the coarse aggregate samples to establish the texture, porosity and permeability of the material:

- 6.5 mm aggregate: Sample I-1 and Sample I-2; and
- 10.0 mm aggregate: Sample C-1 and Sample C-2.

A ProgRes GT3 digital camera was used to capture images in both plane-polarized and cross-polarized light. The petrographical analysis and microphotographs were undertaken at the Department of Palaeontology, Stratigraphy and Sedimentology of the Geological Institute Strashimir Dimitrov, Bulgarian Academy of Sciences, Sofia.

To evaluate the physical-mechanical properties of the concrete cubes, two testing regimes were used on the 28th day:

- ultrasonic pulse velocity (UPV): in terms of [13], using the PUNDIT model and
- compressive strength (CS): in terms of [14], now superseded by [15], using Universal Testing Machine CONTROLS S.p.A. compression machine.

The cast panels were only subjected to the first of these regimes as it is a non-destructive test. Both tests were undertaken at the laboratory of the Department of Civil and Structural Engineering, Faculty for the Built Environment, University of Malta.

3. Results and discussion

3.1. Physical properties of aggregate

The particle density, water absorption and AIV for both aggregate sizes are given in Table 2. Although the 6.5 mm aggregate is marginally denser (3.8%) than the 10.00 mm aggregate, the latter is 21 times more water absorbent. The AIV of the larger-sized aggregate is 12.2% higher.

Table 2. Particle density, water absorption and Aggregate Impact Value of aggregates

	Coarse aggregate	
	6.5 mm	10.0 mm
Particle density – oven-dried (mg/m^3)	2.67	2.22
Particle density – saturated and surface dried (mg/m^3)	2.68	2.36
Apparent particle density (mg/m^3)	2.70	2.60
Water absorption (%)	0.30	6.30
AIV (%)	23.70	26.60

3.2. Petrographical characteristics of aggregate

Both Samples I-1 and I-2 are completely recrystallized limestone consisting of sparry calcite mosaics (figure 3a-f). The sizes of the calcite crystals in the former vary from 0.10×0.10 to 3.0×1.5 mm whilst in the later they vary from 0.10×0.10 to 1.5×0.5 mm. With respect to Sample I-1, the limestone was most probably originally composed of fossils and/or bioclasts, which are of aragonite and high-Mg calcite composition. The aragonite and high-Mg calcite are unstable polymorphs and easily recrystallize into stable low-Mg calcite during diagenesis. The recrystallization completely obliterates the original limestone texture. Single and poorly expressed stylolites can be observed (figure 3b). These were formed as a result of a pressure-dissolution process arising from overburden sedimentation or tectonic pressure. The stylolites are outlined by the presence of a dark brown to black insoluble residue that contains single, very fine-sized quartz grains (figure 3b). In Sample I-2, dispersed black organic matter is observed in the mosaics. Several terrigenous quartz grains are recognized (with a maximum size of 0.3×0.2 mm) (figure 3c and figure 3d), as well as rare well-formed crystals of diagenetic quartz. Single dissolution seams (figure 3e and figure 3f) and stylolites can be observed (figure 3f). These are also impregnated (lined) with a dark brown to black insoluble residue containing sporadic very fine-sized quartz grains (figure 3f).

Samples C-1 and C-2 are micrite limestones with fine bioclasts. According to Dunham's textural classification [16] they represent fine bioclastic wackestones because the matrix to allochems ratio is 80:20 and 85:15 for Sample C-1 and Sample C-2, respectively. In Sample C-1 micrite crystals (sizes < 4 μm) that construct micrite matrix are irregularly recrystallized to form microspar (min. ≥ 4 μm; max. ≤ 10 μm) and rarely to spar (> 10 μm) (Figure 4a). Fine bioclasts are the main allochems. Their sizes are ≤ 0.5 mm. Most bioclasts are completely recrystallized (figure 4a, RB) and some are highly altered and difficult to recognize (figure 4b, RB). Occasionally, preserved fragments of echinoids (with maximum size 0.4x0.2 mm), bryozoa (Figure 4c, BR) and bivalve shells (figure 4c, BSH) are present. Single benthic foraminifera occur (figure 4a, F). Some fragments resemble algae (figure 4a, A). In places in the matrix, peloids < 0.15 mm in size (figure 4b, MP), which are composed of dark micrite, are present. There are single quartz grains (figure 4a, Q) which are silt sized (< 0.063 mm).

In Sample C-2, the micrite in the matrix is often recrystallized into microspar and spar (figure 4d). Fine bioclasts (with sizes predominantly < 0.5 mm) are the main allochems (figure 4e, RB). Most often they are recrystallized and/or altered and are difficult to determine. One more preserved fragment of bivalve shell (1.5x0.01 mm in size) (figure 4e, BSH) can be seen. Some bryozoa bioclasts (up to a maximum length of 1.5 mm) are found. Benthic foraminifera are sporadically represented (figure 4d, F). Interparticle porosity, mainly observed in the matrix, is about 5% (figure 4d).

IOP Conf. Series: Materials Science and Engineering 1138 (2021) 012004 doi:10.1088/1757-899X/1138/1/012004

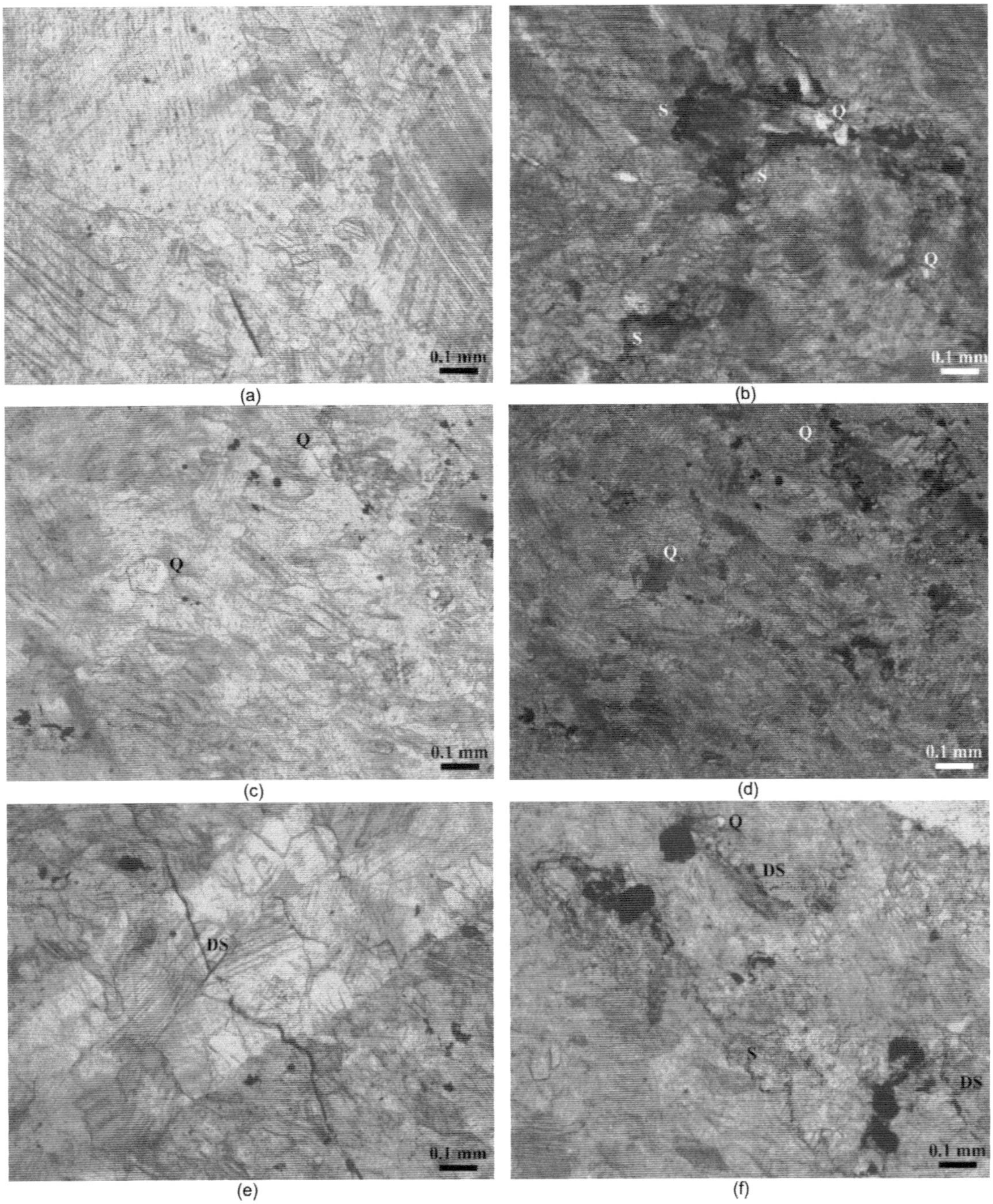

Figure 3. *Sample I-1:* plane-polarized light: (a) sparry calcite mosaics, calcite crystals are of various sizes and (b) plane-polarized light: sparry calcite mosaics with poorly expressed stylolites (S) and single quartz grains (Q); *Sample I-2:* (c) plane-polarized light: sparry calcite mosaics, calcite crystals are of various sizes and single quartz grains (Q), (d) cross-polarized light: same view as (c), (e) plane-polarized light: sparry calcite mosaics crossed by dissolution seam (DS) and (f) plane-polarized light: sparry calcite mosaics with poorly expressed stylolites (S), dissolution seams (DS) and single quartz grains (Q).

IOP Conf. Series: Materials Science and Engineering 1138 (2021) 012004 doi:10.1088/1757-899X/1138/1/012004

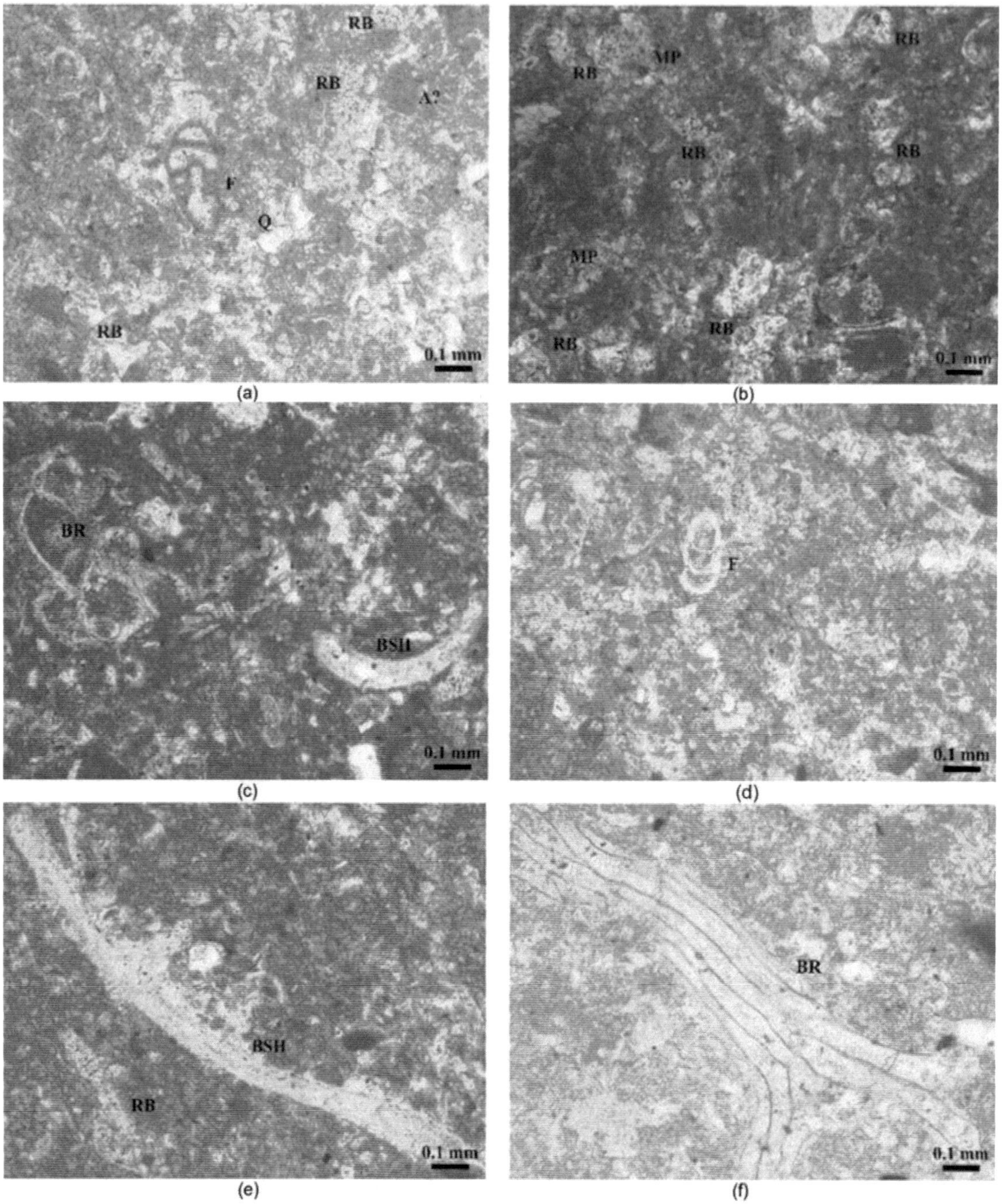

Figure 4. *Sample C-1:* plane-polarized light: (a) micrite matrix is irregularly recrystallized to microspar and spar, recrystallized bioclasts (RB), alga fragment (A?), foraminifera (F) and quartz grain (Q), (b) completely recrystallized and highly altered bioclasts (RB) and micrite peloids (MP) and (c) fragments of bryozoa (BR) and bivalve shell (BSH); *Sample C-2:* plane-polarized light: (c) micrite matrix is irregularly recrystallized to microspar and spar, foraminifera (F), (e) recrystallized bioclasts (RB) with part of more preserved fragment of bivalve shell (BSH) and (f) a part of more preserved fragment of bryozoa (BR).

3.3. *UPV and CS*

Figure 5 illustrates the effect of varying coarse aggregate size on the UPV and CS after a 28-day curing process. The UPV for the cast panels is given in Table 3. The UPV and the CS of the control cubes are given in table 4. The UPV for concrete manufactured with 6.5 mm and 10.0 mm aggregate is plotted for the panels in figure 5a and for the cubes in figure 5b; the CS for the cubes is plotted in figure 5c. Strong correlations exist between UPV and CS for rocks and concrete; UPV increases with the increase in CS [17]. Regression analysis shows a reliable correlation of UPV with compressive strength for high porous limestone in dry and wet conditions, both lowered in the later conditions [18]. Given that the UPV for both the panels and the cubes is between 3.5km/s and 4.5km/s, the concrete quality is of good grade in terms of uniformity and homogeneity. There were no significant variations between the two aggregates [19].

Table 3. UPV of the concrete panels at 28th day

		Length l_1	Breath l_2	Height	Weight	Time* [ms]		UPV [km/s]	
		[mm]	[mm]	[mm]	[g]	t_1 across l_1	t_2 across l_2	V_1	V_2
6.5mm aggregate	Panel 1	392.00	502.00	25.93	10351.50	103.3		3.795	0.000
	Panel 2	392.00	504.00	25.27	09540.00	103.9		3.773	0.000
	Panel 3	391.00	502.00	26.55	10565.50		136.7	0.000	3.672
	Panel 4	334.00	501.00	26.14	09041.00		138.8	0.000	3.610
	Panel 5	104.00	503.00	25.09	02613.00	26.3		3.954	0.000
	Mean							3.841	3.641
10.0mm aggregate	Panel 1	390.00	504.00	24.19	8900.50	105.0		3.714	0.000
	Panel 2	390.00	502.00	24.88	9509.00	104.1		3.746	0.000
	Panel 3	391.00	500.00	24.77	9102.00		130.5	0.000	3.831
	Panel 4	335.00	504.00	24.00	7486.00		133.0	0.000	3.789
	Panel 5	103.00	503.00	24.60	2500.50	25.1		4.104	0.000
	Mean							3.855	3.810

* Measured along the parallel non-bevelled sides.

Table 4. UPV and CS of the concrete cubes at 28th day.

		l_1	l_2	Height	Weight	Time [ms]		UPV [km/s]		Loading*	Area	CS*
		[mm]	[mm]	[mm]	[g]	t_1 across l_1	t_2 across l_2	V_1	V_2	[kN]	[mm^2]	[N/mm^2]
6.5mm aggregate	Cube 1	100.48	100.32	95.49	2055.00	24.6	25.0	4.085	4.013	272.5	10080	27.03
	Cube 2	100.42	100.47	99.17	2098.50	24.8	25.1	4.049	4.003	260.5	10089	25.82
	Cube 3	100.47	100.51	99.30	2099.50	25.0	25.1	4.019	4.004	264.2	10098	26.16
	Cube 4	100.45	100.48	95.91	2043.50	24.9	25.0	4.034	4.019	276.7	10093	27.41
	Mean							4.047	4.010			26.61
10.0mm aggregate	Cube 1	100.38	100.40	100.98	2088.50	24.3	24.8	4.131	4.048	390.5	10078	38.75
	Cube 2	100.35	100.77	103.69	2092.50	25.0	24.2	4.014	4.164	400.3	10112	39.59
	Cube 3	100.63	100.68	101.19	2086.00	24.8	24.4	4.058	4.126	402.4	10131	39.72
	Cube 4	100.84	100.32	100.19	2054.50	24.9	24.6	4.050	4.078	396.6	10116	39.20
	Mean							4.063	4.104			39.31

* At crushing.

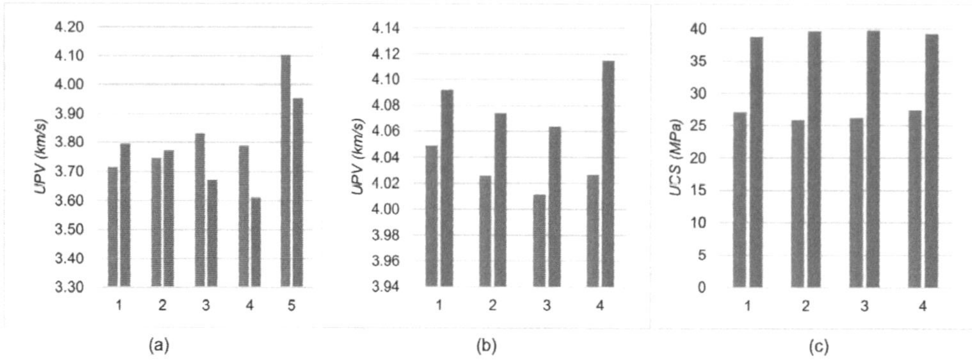

Figure 5. UPV after 28 days (air-cured) for panels (a) and for control cubes (b); CS after 28 days (air-cured) for control cubes (c). Blue and red refer to 6.5mm and 10.0mm aggregate respectively.

In the case of the 6.5 mm aggregate, the resultant mean CS of the concrete was 27 MPa, just above two-thirds the grade of the mix design; the resultant mean CS of the concrete using 10.0 mm aggregate was 39 MPa, close to the grade of the mix design. All cubes typically had a bell-shaped profile on crushing except for cubes 1 and 2 constructed in 6.5 mm aggregate, which exhibited shear failure [figure 6). Grade C40 is structural concrete. The strength of the concrete is dependent on its coarse aggregate. The homogeneity of the concrete decreases as the aggregate size increases [20]. Coarseness of aggregate is a parameter of primary technological importance. Although the UPV results met the standard of concrete quality, the concrete manufactured using the less coarse aggregate had significantly lower strength than that produced with the coarser aggregate.

Both aggregate types are supplied on the building construction market as having similar physico-mechanical properties. However, they are petrologically different and this directly affects the resulting compressive strength of the hardened concrete. For the same mix design, the 6.5 mm aggregate is derived from recrystallized limestone whose original texture has been obliterated, resulting in an almost impervious rock. This produces a significantly weaker concrete grade than the 10 mm aggregate from micrite limestone with fine bioclasts, whose micrite matrix has been recrystallised into microspar and spar.

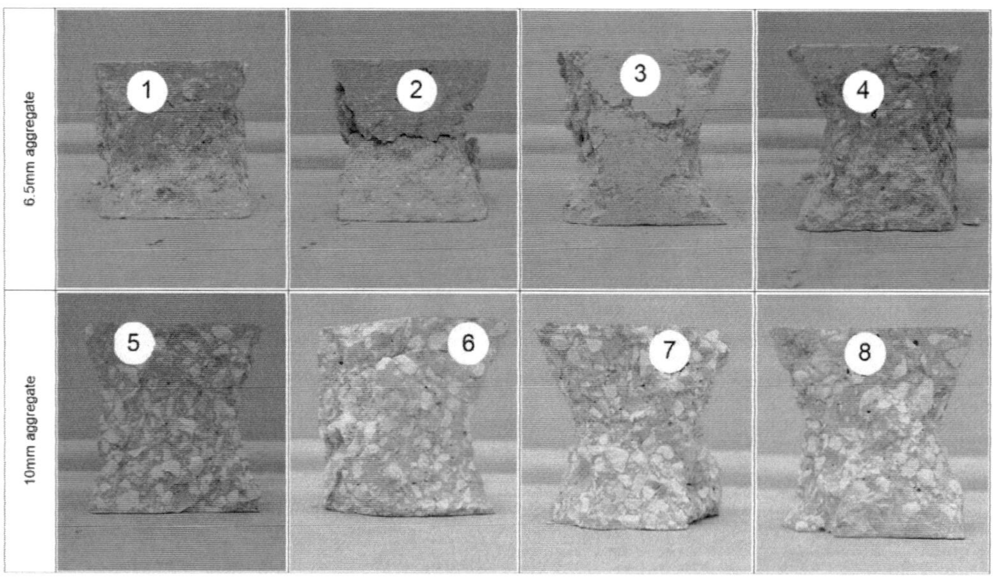

Figure 6. Uniaxial compressive test at 28 days: All cubes had a typically bell-shaped profile on crushing except for cubes 1 and 2, constructed in 6.5 mm aggregate, which exhibited shear failure.

4. Conclusions

Petrological characteristics are essential elements in aggregates because they have a significant bearing on the physico-mechanical properties of the hardened concrete. This study, which investigated the failure of concrete panels cast during the course of the manufacture of a prototype coffee table, confirmed that compressive strength increased with coarse aggregate size:

- For a C40 mix design, the grade of concrete attained by the 6.5 mm and 10.0 mm coarse aggregate, from limestones quarried from different lithostratigraphic formations, was 27 MPa and 39 MPa respectively. The 10.0 mm aggregate met the desired ultimate strength requirement but the 6.5mm aggregate reached just two-thirds of the mark.
- The apparent particle density of the 6.5 mm aggregate is higher than that of the 10.0 mm aggregate. In addition, it has lower AIV and significantly lower water absorption.
- The UPV test did not provide important data on the nature of the rock and the CS. Variation in the petrological and textural content of the rocks impacted the CS.

These findings have a broader impact. Although the size of the aggregate has a bearing on the CS of concrete, its petrological characteristics are indicative of its behaviour in the hardened concrete. The homogeneity of the concrete is affected by the physico-mechanical properties of the aggregate, including particle size, but its petrological texture significantly affects the load-bearing capacity of the hardened concrete. Thus, specifications of aggregate for concrete manufacture, which are normally stated in terms of physico-mechanical properties, should specifically include the geological provenance of the raw material used, as petrological properties vary between limestones from differing geological formations and origins.

5. Limitations

The study by [3] was undertaken to explore the shift from design to production phases in lieu of finding a better mode of communication between product design and manufacture, in this case using a prototype of a coffee table manufactured partly in concrete; it was not a study on concrete technology. The objective was to produce a mixture using 6.5 mm aggregate from Montenegro. After obtaining CS values well below the mix design the mixture was repeated using a 10.0mm aggregate from Malta, this being a limited sample from the last batch available at the Civil and Structural Engineering Laboratory of the University of Malta. The findings would have been expedient if the mixtures in both aggregate sizes were undertaken in material resulting from both sources.

Acknowledgements

The authors would also like to thank Professor Elena Koleva-Rekalova from the Geological Institute of the Bulgarian Academy of Sciences, Sofia, for the microphotographs and her advice on petrographical interpretations, and Nicholas Azzopardi, concrete technologist and senior laboratory officer at the Civil and Structural Engineering Laboratory, University of Malta, for his support in undertaking the physical testing of the materials focused on in this study.

The authors would like to acknowledge their gratitude to Martin Djakovic and Mileva Milić from the Geological Survey of Montenegro for professional support in describing the geology and petrography of the Volujica formation respectively.

References

[1] Nagaratnam S, Carthigesu T, Rabin T and Bobby K 2017 *Civil Engineering Materials* (Queensland, Australia: Cengage Leraning US).

[2] Azzaz R A, Chemrouk M and Ammar-Boudjelal A 2020 Rheological, physico-mechanical and durability properties of multi-recycled concrete *Advances in Concrete Construction* **9** 1 9-22.

[3] Micallef G 2018 *An Architecture of Communication: Exploring the gap between design and production* [University of Malta: M Arch dissertation).

[4] Cadjenovic D, Radulovic N, Milutin J. Ostojic Z, Cepic M and Dakovic M (2002-2013) *Geological Map of Montenegro: Podgorica-Sheet 3* (Podgorica Montenegro Geological

IOP Conf. Series: Materials Science and Engineering 1138 (2021) 012004 doi:10.1088/1757-899X/1138/1/012004

Survey of Montenegro).
[5] Cadjenovic D T 2020 *Terain's geological composition* (https://www.academia.edu/31543907/TERAINS _GEOLOGICAL_COMPOSITIONOURCES, retrieved on 08 March 2020)
[6] Oil Exploration Directorate 1993 *Geological map of the Maltese Islands: Shhet 1:Malta* (Malta: Oil Exploration Directorate, Office of the Prime Minister).
[7] Pedley H M, House M R and Waugh B 1976 The geology of Malta and Gozo. *Proceedings of the Geologists' Association* **87** (3) 325–341.
[8] Pedley H M 1978 *A new lithostratigraphical and palaeoenvironmental interpretation for the coralline limestone formations (Miocene) of the Maltese Islands* (London: H M S O).
[9] Bianco L 1995 The industrial minerals of the Maltese Islands *Hyphen* **7** *3) 111–118.
[10] BS EN 1097-6:2013 *Tests for mechanical and physical properties of aggregates. Determination of particle density and water absorption.*
[11] BS 812-111:1990 *Testing aggregates. Methods for determination of ten per cent fines value.*
[12] BS EN 1097-2:2010 *Tests for mechanical and physical properties of aggregates. Methods for the determination of resistance to fragmentation.*
[13] BS EN 12504-4:2004 *Testing concrete, Part 4: Determination of ultrasonic pulse velocity.*
[14] BS EN 12390-3:2009 *Testing hardened concrete. Compressive strength of test specimens.*
[15] BS EN 12390-3:2019 *Testing hardened concrete. Compressive strength of test specimens.*
[16] Dunham R J 1962 Classification of carbonate rocks according to depositional texture *Classification of carbonate rocks* ed W E Ham (Tulsa, Oklahoma: American Association of Petroleum Geologists) Memoir 1, pp 108-121.
[17] Sertçelik I, Kurtuluş C, Sertçelik F, Pekşen E and Aşçı M 2018 Investigation into relations between physical and electrical properties of rocks and concretes *Journal of Geophysics and Engineering* **15** 142.
[18] Vasanelli E, Colangiuli D, Calia A, Sileo M and Aiello M A 2015 Ultrasonic pulse velocity for the evaluation of physical and mechanical properties of a highly porous building limestone *Ultrasonic*s **60** 33-40.
[19] IS 13311 1992 Part I *Standard Code of Practice for Non-Destructive Testing of Concrete: Part 1 - Ultrasonic Pulse Velocity* (New Delhi: Bureau of Indian Standards).
[20] Tsiskreli G D and Dzhavakhidze A N 1970 The effect of aggregate size on strength and deformation of concrete *Hydrotechnical Construction* **4** 448-453.

IOP Conf. Series: Materials Science and Engineering 1138 (2021) 012005 doi:10.1088/1757-899X/1138/1/012005

High performance buildings with optimized energy systems based on exergy principles

I Boian[1] and I Tuns[2*]

[1] Department of Building Services, Faculty of Civil Engineering, Transilvania University of Brașov, Romania
[2] Department of Civil Engineering, Faculty of Civil Engineering, Transilvania University of Brașov, Romania

* ioan.tuns@unitbv.ro

Abstract. Optimising buildings and energy systems with a focus on mitigation of greenhouse gas emission is not an easy task: first, principles are necessary to be established and then the steps to be followed. Exergy efficiency can show the best issues: building envelope insulation and air tightening, low-valued energy sources, low-exergy heating and cooling systems, avoiding destructive exergy-processes like boilers burning fossil fuels, systems and grids for low-temperature heating and for high temperature cooling, renewable energy sources, daylight, and passive solar energy use supported by control strategies of building systems can contribute to a better future when considering global warming and comfort.

1. Introduction

The energy demand for heating and cooling in the building sector is responsible for more than one third of the final energy consumption in Europe and worldwide, energy commonly resulting from combustion processes using different fossil fuels. The resulting greenhouse gas (GHG) emissions are the core challenges in fighting climate change considering its large contribution from the built environment. Two issues exist: a more efficient use of energy and burning fewer fossil fuels. Increased efficiency is related to optimization and the concept of exergy can help a better understanding and evaluation of the processes and systems quality. An increased efficiency in energy use is possible when investigating the exergy flows in buildings. While energy is conserved – First Law of Thermodynamics – the exergy, as the entirely convertible energy into other types of energy, can be destroyed, i.e., converted in anergy, as shown in figure 1. The heat transferred through a wall separating the warm space from the cold exterior is the same on the inside face and the outside one, 1035 W. But the exergy entering the inside-warm face, 93,15 W is different from the exergy leaving the outside-cold face, 0 W, meaning that the entire exergy was destroyed during the heat transfer through the wall.

The destroyed exergy is:

$$\dot{I} = \dot{Q}_{wall}\left(1-\frac{T_0}{T_1}\right) - \dot{Q}_{wall}\left(1-\frac{T_0}{T_2}\right) == 1035\left(1-\frac{273}{300}\right) - 1035\left(1-\frac{273}{273}\right) = 93.15\ \text{W} \quad (1)$$

Buildings destroy more input exergy than they lose and produce, (84....93% of exergy consumed by buildings are destroyed by irreversibility) [1].

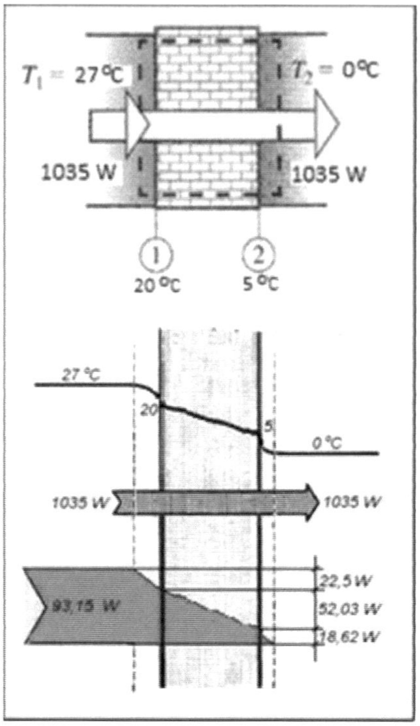

Figure 1. The flow diagram for the energy (a) and for the exergy (b).

Minimizing the irreversible dissipation of low-value energy into the environment is possible if the heat transferred through the wall Q_{wall} is reduced: insulating the wall not only the heat transferred is diminished but the temperature of the inside face of the wall will become higher. In figure 2 is presented an example where the design heat loss for the space is initially 8600 W for an outdoor design temperature of -21°C resulting in a heat loss per degree of 210 W/K in case of an indoor design temperature of 20°C. Considering an average internal heat gain Q_g of 540 W the balance temperature will be diminished, about 2.5°C. Insulating the envelope with 10 cm of expanded polystyrene the balance temperature will be diminished with 5.3°C. In case of a better insulation, say 20 cm and having a better glazing almost a difference of 9°C between the indoor design temperature and the balance one will result. Lowering the balance temperature means shortening the time of the heating period and of course a reduced amount of energy supplied for the house heating.

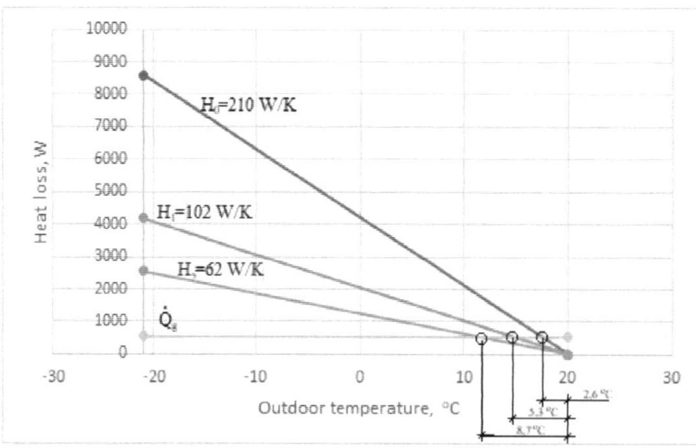

Figure 2. Insulating the envelope, the balance temperature is becoming lower in case of the same internal heat gains.

But insulating the envelope results in a higher temperature on the inside surface of walls, i.e., from 12 to 18 or even 19°C, which has as an effect the decrease of the heat radiated by humans to these surfaces and finally the inside air temperature can be dropped 1…3°C.

The exergy factor $(1 - T_0/T_1)$ resulting if the inside air temperature, T_1 is diminished, caused by the insulation, will be affected 5 to 10% depending on the outside temperature, as shown in figure 3.

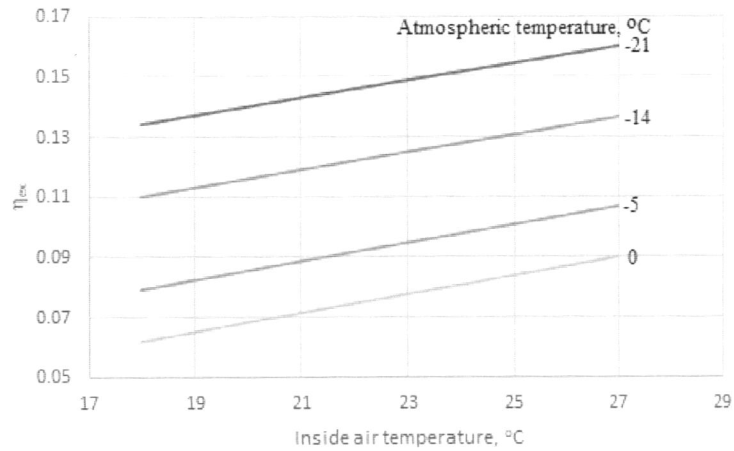

Figure 3. Exergy efficiency as a function of the inside air temperature for some usual atmospheric temperatures.

The exergy factor has the meaning of an exergetic efficiency defined as the ratio of the exergy output, or resulted and the exergy input

$$\left(1 - \frac{T_0}{T_1}\right) = \eta_{ex} = \frac{Ex_{res}}{Ex_{in}} \tag{2}$$

Exergy efficiency illustrates how far the efficiency of a conversion process is from its theoretical maximum.

An exergy-optimized building design must follow the steps:

1. Reduction of the energy use;
2. Temperature levels harmonization;
3. Intelligent storage of exergy;
4. Efficient use of high-quality sources;
5. Avoid destructive exergy-processes;
6. Avoid long components chains.

2. Energy versus exergy efficiency of systems and equipment

A substantial decrease of CO_2 emissions for the building stock can be realized through a better and exergy optimized building design. Additionally, enhanced heating and cooling systems lead to a lower exergy demand of buildings. Figure 4 illustrates the energy efficiency together with the exergy efficiency in case of four usual equipment: the energy efficiency for liquid fuel-based boilers and cogeneration systems are quite similar, less than 100% as it is in case of electrical boilers, in which nearly all electrical energy is converted into heat, while heat pumps can reach values over 300%. On the other hand the corresponding exergy efficiency do not exceed 4 to 5 per cent in case of boilers, fossil fuel based or electrical; these low values are due to the large temperature difference between the flame (more than 1000°C) and the heated water (about 60°C) in case of fossil fuel boilers; the electrical boiler is transforming the extremely high-quality electrical energy into a low-quality thermal energy aiming to cover the gap between the inside temperature (20°C) and that of the outside air (5°C or less).

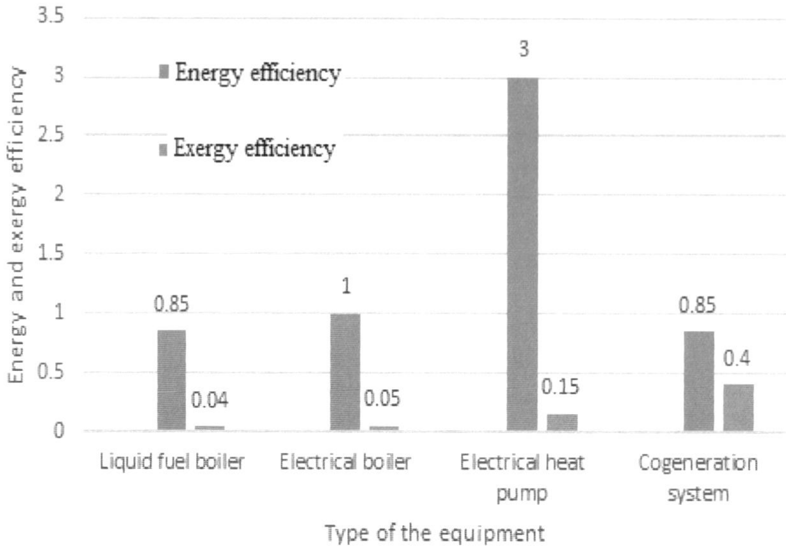

Figure 4. A comparison of the energy efficiency with exergy efficiency for some usual heating systems [2].

The exergy efficiency of a heat pump is significantly greater than that of an electric heater because of the fact it transfers the energy from the environment to the interior space by means of a smaller amount of electrical energy, resulting in a lower cost. Heat pumps can use the low-valued energy delivered by sustainable energy sources (from the environment) to supply the heating energy necessary for the comfort inside buildings. Saving energy is in fact an exergy saving. The focus in case of heat pump systems design for individual buildings is to adjust its capacity to the heating load of the building including the DHW provision.

Additionally, heat pumps can operate in a reversible mode, cooling for comfort in the interior space, working as low-temperature heating and high temperature cooling systems.

Heat pumps and thermal collectors as decentralized supply technologies together with centralized options like thermal grids at different temperature levels can be future issues for communities. Increasing the exergy efficiency because of the temperature lowering of supply and / or return district heating network is a low-exergy approach leading to energy use reduction especially from fossil fuels.

Low-valued local energy sources and renewable ones are a flexible alternative able to meet different demands. Low-temperature heat from geothermal sources supplying communities can be a possible low-exergy approach to be solved at a local level.

3. Envelope insulation or renewable energy sources use

Minimizing GHG emissions being directly associated with the energy performance of the building and with the use of renewable energy sources a question arises: what the contribution of each of the two components is.

An exergy analysis can show that a house with 20 to 30 cm thick insulated walls heated with a heat pump having a coefficient of performance COP = 3…4, supplying a low temperature radiant system will have about the same energy use as that required by a passive house, as shown in figure 5. The development of low temperature heating and high temperature cooling systems is a prerequisite for the usage of alternative energy sources. Low thermal losses of buildings with high standards having a low energy demand can benefit the renewable energy with a better exergy efficiency as that specific for traditional thermal systems. Old district heating systems can be improved in terms of exergy efficiency if they are transformed in ultra-low temperature ones being assisted by local electric boosters.

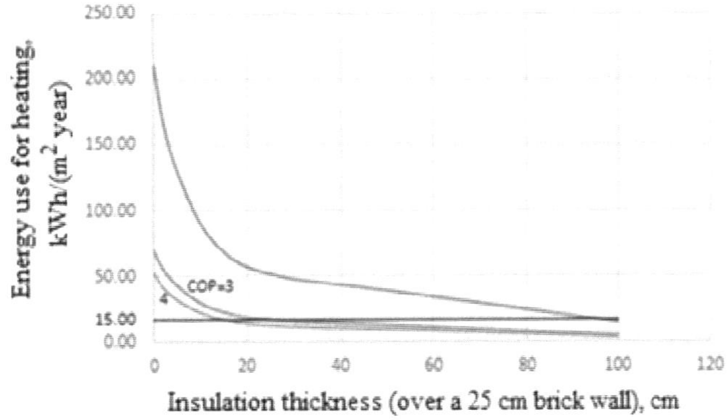

Figure 5. A heat pump supplying a 20…30 cm insulated house needs almost the same energy for heating as a passive house.

4. Low temperature heating and high temperature cooling systems

A wide variety of possibilities exists for low exergy heating and cooling systems, but heat sources of lower temperature for heating require a thermally well-insulated envelope making use of building materials with an appropriate heat capacity. For the cooling season a combination of night ventilation with shading devices for windows and a mitigation of internal heat gains added to previous mentioned insulation materials for walls having a corresponding heat capacity represent the prerequisite condition for high temperature cooling systems. The exergy factor (efficiency) for an isothermal heating process lies between 1.4 and 22 per cent, but is only about half if the heat carrier does not remain at a constant temperature, as can be seen in figure 6 a) and b). The same for a cooling process, figure 7 a) and b).

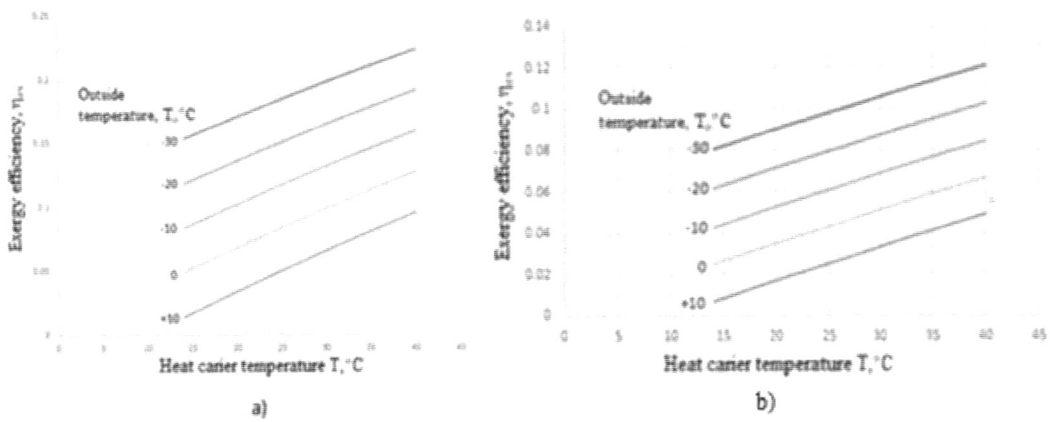

Figure 6. Exergy efficiency as a function of fhe heat carrier temperature for isoyhermal processes (a) and for non-isoyhermal ones, (b) in case of heating.

Buildings having a low energy demand for space heating / cooling require increased demand for domestic hot water, DHW. It must be noticed that the exergy factor for DHW is about twice as high as that of space heating systems, i.e., 13%.

Buildings with systems working with low-level temperatures have advantages:

- Integration of renewable heat sources is much easier;
- Pumping costs can be reduced as a decreased mass flow in the network, for a constant power input. Additionally, transport lines having smaller dimensions will result in a lower investment cost.
- Lower temperature results in reduced heat losses to the surroundings.

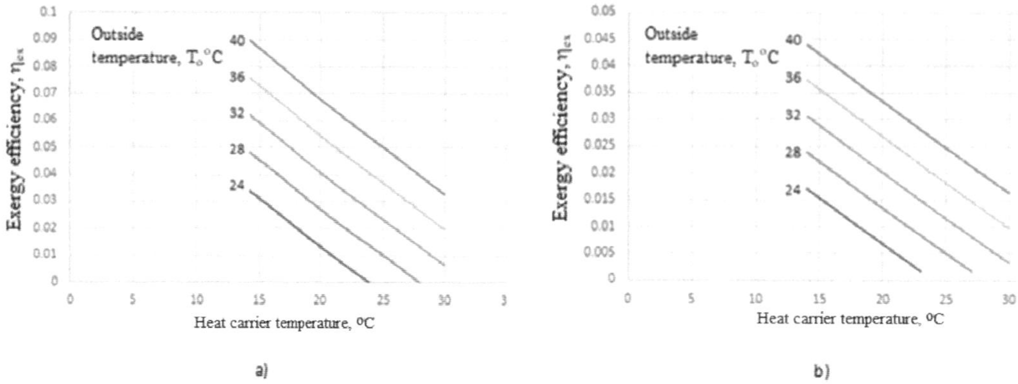

Figure 7. Exergy efficiency as a function of fhe heat carrier temperature for isoyhermal processes (a) and for non-isoyhermal ones, (b) in case of cooling.

5. Conclusions

The optimization process as a combination of energy efficiency measures and renewable energy use is aimed to lead to the nearly-zero energy target, i.e., a reduction of approx. 90% of CO_2 emissions, considering the cost-effectiveness rather than cost-optimality. The overall exergy efficiency based on the analysis of every component losses is the way to improve de performance of the system leading to a progressive mitigation of GHG emission. The exergy concept used when developing a sustainable community offers a holistic understanding of the optimization process resulting from successive improvements. An exergy efficient building supply depends primarily on low exergy sources to be exploited. First-of-all, combustion processes should be avoided in an exergy efficient energy system based on low temperature heat. Low exergy sources, like solar thermal heat, geothermal-heat, or process waste heat (wastewater or exhaust ventilation air) should be used for space heating and cooling applications in buildings via heat recovery systems. Making use of more effective low exergy sources surface heating and cooling systems operates at lower temperature levels than conventional units like radiators or fan coils. Low exergy emission systems being more flexible, as they can be used for higher supply and return sources or can be coupled with low exergy systems like ground source heat pumps must be integrated in building for a better efficiency. Electricity required by pumps and fans being a high-exergy resource can be minimized by heat recovery systems connected to highly efficient energy systems, such as heat pumps. Chances offered by the renewable electricity production, PV or wind, [3] must be considered too as a challenge resulted from its fluctuating supply. The high potential of buildings to lessen the input exergy destroyed by irreversibility (84 to 93% of the exergy consumed) can be exploited through low-exergy heat-sources like the geothermal ones. Lowering the temperature of the heating agent will improv the exergy efficiency of the system.

A reduced energy demand of the building resulting from a good envelope insulation and an improved air tightness, daylight, and passive solar energy use supported by control strategies of building systems are important factors to minimize exergy losses.

An improved thermal comfort and indoor air quality together with a reduced energy consumption are feasible when applying low-exergy systems.

References

[1] LOWeX communities - optimised performance of energy supply systems with exergy principles. *In EBS News Issue 69*, Pag 83

[2] Dincer I and Yunus A. 2001 Energy, Entropy and Exergy Concepts and Their Roles in Thermal Engineering *in Entropy*, *3*, 116-149

[3] Dragomir G, Şerban A, Năstase G and Brezeanu A I 2016 Wind energy in Romania: A review from 2009 to 2016, *In Renew. Sustain. Energy Rev.*, vol. 64, pp. 129–143, ISSN: 1364-0321

IOP Conf. Series: Materials Science and Engineering 1138 (2021) 012006 doi:10.1088/1757-899X/1138/1/012006

Intelligent buildings for energy consumption management

I Bucur[1]*, P Axinte[1], C Pleșcan[1] and A Șerban[2]

[1] Transilvania University of Brasov, Faculty of Civil Engineering, Romania
[2] University Politehnica of Bucharest, Faculty of Mechanical Engineering and Mechatronics, Romania

* irina.bucur@student.unitbv.ro

Abstract. The main concern of the paper is the energy consumption management, in relation to the advantages and disadvantages of an intelligent building. The building sector consumes about one third of primary energy. An intelligent use of energy could have an impact on the energy consumption and cost. Energy management is the process of controlling, conserving and monitoring the energy in a building or recently in a home. With the use of sensors, all systems are connected and useful data is gathered to improve the building's efficiency. Intelligent building design is the future of building and building services sector. Enhancing energy preservation strategies and using sustainable design approaches are necessary in developing intelligent buildings. This implies a more effective approach to reaching the green goals, and it also implies financial savings. Building an intelligent building involves a larger investment, but in the long run the usual expenses will decrease. National grid energy consumption will decrease with the use of renewable energy sources, such as solar or wind energy, if the energy is produced on site. In order to make a difference in the climate change, intelligent buildings depend on the non-fossil sources, which will affect the energy usage. Furthermore, an intelligent building will increase the comfort and productivity of its occupants, for either a workspace or a home.

1. Introduction

Defining smart buildings has been done in countless ways, in connection with many topics, such as energy consumption, technology, equipment, user interaction [1], but a clear definition is given by the UK-based European Intelligent Building Group: an intelligent building 'creates an environment which maximizes the effectiveness of the building's occupants, while at the same time enabling efficient management of resources with minimum life-time costs of hardware and facilities' [2]. In this paper we analyze the advantages and disadvantages of the intelligent buildings regarding energy consumption and energy management.

As it is shown in figure 1, the households energy consumption represents almost one third of the final energy consumption in European Union, 2018. Nowadays, the goal is to have a building with minimal energy consumption and maximum comfort for the occupants. An intelligent building can decrease its energy consumption by integrating a building energy management system, which is a system that controls the occupants' energy needs. In order to make a difference in the building sector (and as a consequence, in the climate), intelligent buildings depend on the non-fossil energy sources.

The usual renewable energy sources are solar, wind, water (hydro), biomass and geothermal. The most used renewable energy sources in the intelligent building sector are solar, wind, and geothermal. Photovoltaic panels are a great way for the intelligent building to produce electricity, due to the fact that they are easily installed on the roof, on the facades or in the nearby land. If mounted on the roof or on the facades, the PV panels take little to no usable space. The great advantage they have is that they usually do not require useful space. The geothermal energy is used for heating or cooling the space by using a heat pump. Heat pumps are a great option to produce thermal energy on the scene, and on the long run they are a way to save money regarding heating or cooling the spaces. The heat pumps are very popular, not only in the intelligent building sector, and their demand is constantly growing due to the advance of the technology. With the use of non-fossil energy sources (regarding on-site energy production), the energy consumption costs will decrease.

Furthermore, monitoring the energy consumption will have a positive impact over the energy consumption, while following the user's needs and by taking measures to control and to limit the energy consumption. Energy consumption monitoring can mean logging into an application to see if there are appliances that can be turned off or even programming them to work according to a preset schedule. For example, the heating or the cooling installation can be set up depending on the user's schedule. If the user has a classic 9 to 5 job, the heating/cooling installation can be set up to start one hour before the user will be back home. Therefore, there will be a minimal energy consumption while he is at work and also the comfort levels will be met by the time the user will be back home. This could be a use even in larger or public buildings, by setting up the heating/cooling installation to start right before the building program will start.

Beside the energy sources, the building's insulation is another important step to take into consideration, since the energy loss dictates the energy consumption. Insulating the building efficiently increases the energy efficiency. In order to keep the building green, more and more engineers choose to use bio-based insulation materials. The bio insulation is made out of animal or plant-based materials, they require less energy in the production process and also, they have a positive impact on the healthy indoor climate.

Furthermore, a thermal, auditory, natural light and natural ventilation comfort is given by double-skin facades. [3] Double-skin facades started to be a trend among office buildings and public buildings. Besides the modern look they give to the building, they keep up with the insulation standards for the heating and for the cooling seasons.

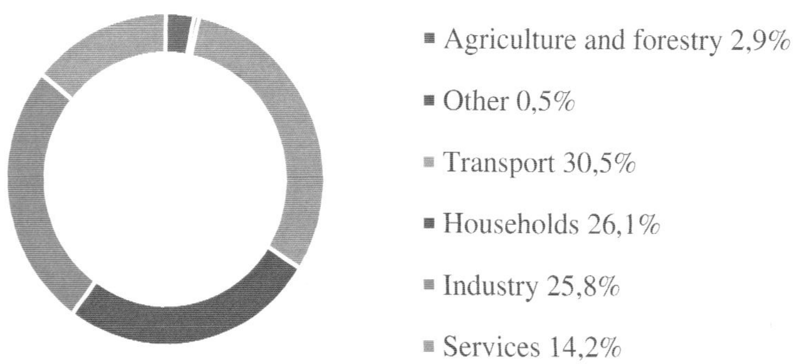

Figure 1. Final energy consumption, EU, 2018 [4].

2. Methods and procedures

This paper analyses the new concept of intelligent buildings and traces the link between comfort, smart systems and energy efficiency. An intelligent building allows its occupants to control the appliances and the energy consumption. The energy management system that every intelligent building has (even if it is an office building, a house or a commercial building) can control many aspects of the building, such as thermal comfort, lighting smart system or water supply. Designing a good management system for the building is crucial to keep track of the energy consumption. The user's comfort and the increasing need of energy in the household/work place will increase the energy consumption, but with a smart configuration, the energy levels will be met and the energy consumption will decrease. This paper displays the advantages and disadvantages that an intelligent building has, as well as it traces the link between the advantages/disadvantages and the energy consumption management. The paper was realised by gathering information about household energy uses, renewable energy sources and technology. The research shows how the energy use can be monitored and controlled using the technology of an intelligent building and gives perspective on how renewable energy sources cand be used in an intelligent building household.

3. Advantages and disadvantages of intelligent buildings

Intelligent buildings have many advantages (figure 2). The most important one is the comfort (which translates to a satisfaction about the surrounding environment) that they bring to the occupants. The goal to an intelligent building is to create a safe, healthy and comfortable place for its occupants and this goal is achieved when the users can control the building's systems and appliances, even remotely. Smart buildings have an intelligent set up that allows its owners to control the appliances remotely, using an internet connection. In addition to this, the comfort in a smart building can be achieved through a good design.

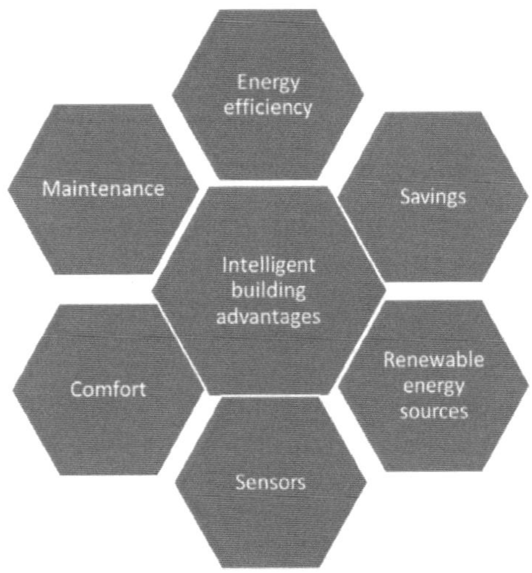

Figure 2. Advantages of the intelligent buildings.

The thermal comfort (the user satisfaction with the thermal environment) can be achieved by maintaining a good thermal environment, regarding the temperature, air quality, heat loss, humidity etc. Besides the thermal comfort, the acoustic comfort is also important, and it is achieved with a good acoustic insulation. The lighting affects the comfort directly and it is taken into consideration when designing an intelligent building. Lighting is also important in the comfort of an intelligent building, since different spaces (designed for different activities) require different lighting levels.

Tab 1. Energy consumption in households [5].

Space heating	63.6%
Water Heating	14.8%
Lighting an appliance	14.1%
Cooking	6.1%
Space cooling	0.4%
Other	1%

IOP Conf. Series: Materials Science and Engineering 1138 (2021) 012006 doi:10.1088/1757-899X/1138/1/012006

In the European Union, the energy consumption in 2018 in households are as shown in table one. The energy consumption in an intelligent building is an important step that is taken into consideration in the designing and construction stages. It is important to meet the occupants comfort level, but also have a building design with an energy consumption rate as small as possible. First of all, there is the thermal insulation that reduces the heat transfer, both during winter and summer, by keeping the temperatures between a suitable range. The building envelope should be designed in such a way that the heat transfer is minimal. The building envelope is the direct link between energy loss and energy consumption. The maximum energy efficiency is achieved as long as the heat transfer is minimal. Second of all, energy efficiency is achieved also when providing a significant reduction of electrical energy consumption. This means controlling the appliances energy consumption and using low energy light bulbs.

In the European Union, in 2015, only 15.7% was covered by renewable energy sources (figure 3). An important advantage of an intelligent building is the use of the renewable energy sources. An intelligent approach to an intelligent building is integrating the energy production by using renewable sources, such as solar, wind and geothermal. The most used technologies that can be integrated in the building's energy system are photovoltaic systems, solar thermal, geothermal heat pump and wind turbines. Usually for the electrical energy production, the photovoltaic systems are the most used [6-8]. The small wind turbines that can be installed near the building are less used, since they take up valuable space and they can make too much noise. For heating the space, the geothermal heat pump is the go-to-choice.

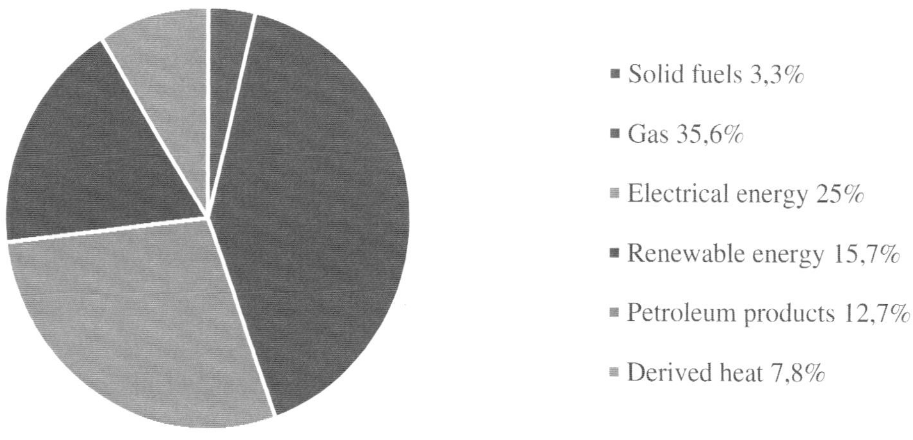

Figure 3. Final energy consumption of households by energy product, EU, 2015 [9].

A great advantage of an intelligent building is the fact that they have a hands-free convenience and that they save time with the automated tasks. For example, the sensor system that the building has can detect many of the malfunctions of the systems, alert the owner and therefore the failure can be fixed in time, with a minimal investment. This system offers a simpler maintenance and less expensive interventions taking into account the fact that the defects are signaled in time.

On the other hand, one of the main disadvantages (figure 4) is the higher investment that an intelligent building requires. Some of the technology mentioned is usually more expensive. Also, a smart building integrates the energy production systems that have a higher price-tag, but over time, the investment pays off. For example, the investment using a heat-pump for space heating instead of a classic methane gas heating system would be higher. However, over the years savings will be made because the user will have the needed energy produced on site, instead of paying for non-fossil fuels consumptions (mains gas). Finally, by designing an intelligent building with an on-site renewable energy sources system, the the necessary energy will be produced and there will be little to no need to pay for electricity or heat(gas).

Another disadvantage is the dependency on the professionals. Usually if there is a malfunction in the smart-system, there are only a handful of people that can take care of it. Compared to conventional buildings, which also need professionals, here it takes skilled people who have the necessary training to solve problems in terms of new technologies that are implemented in the intelligent buildings. Usually, the companies that installed the system provide help and handle the problems. The last disadvantage refers to the fact that an intelligent building needs a reliable internet connection for the remote use and access. Therefore, if technology fails the user could be helpless.

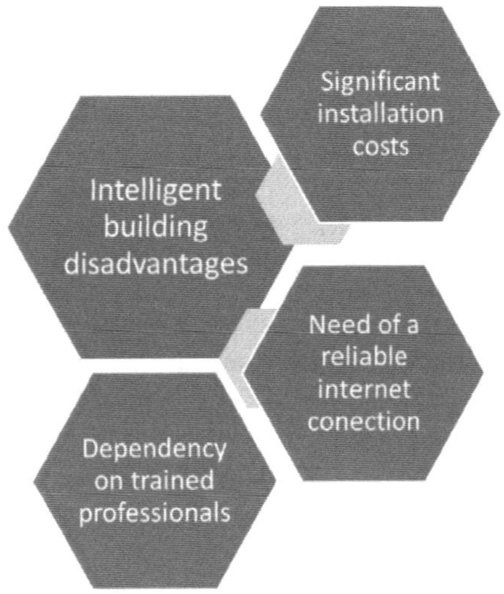

Figure 4. Disadvantages of the intelligent buildings.

4. Conclusions

This paper summarizes general ideas about intelligent buildings, regarding the link between advantages and disadvantages and the energy consumption and management. As a future research direction, the study will continue with the idea of reducing the energy consumption and reducing the carbon footprint, with references to other studies.

The intelligent building systems are the future of the building sector. Investing in an intelligent building can affect the energy consumption of the household, therefore on the long run there will be significant savings. Most of the energy in households is consumed by heating the spaces, so significant savings can be seen by insulating the envelope and by designing the space heating properly. The use of renewable energy sources provides lower running costs, reduce carbon dioxide emissions, which meets with the green and sustainable directives. Switching to renewable energy sources when designing an intelligent building decreases the fossil fuels usage for power generation in the building sector. Also, switching to renewable energy sources will help decrease the energy consumption costs over time for the building. Usually the investment is rather higher, but usually it pays off over time, since the needed energy will be produced on site. Smart homes are set up with wireless or hardwiring sensor systems and they are built in such a way that they have a low energy consumption and rely mostly on renewable energy sources. Also, air pollution in human activities is increased by the use of fossil fuels in households and in the building sector for power generation [10].

An intelligent building has an automation system which is able to provide information about the linked systems by using sensors. An intelligent building offers a higher level of comfort, both for a home or a work space, provides enough information in order to prevent energy losses, it can detect early malfunctions and allows users to access the system remotely.

Optimal energy management and energy efficiency in the building sector is a valuable tool for conserving natural resources. In addition, the financial benefits of using renewable energy are a major motivation for building owners. As a future research direction, energy efficiency and environmental protection measures must be treated differently in the design phase and in the operation phase.

Given the inherent limitations of traditional design methods, are we still surprised that our energy systems often fail to achieve their performance? Many of the buildings in the world, but especially in Romania, have an annual energy consumption of around 300 kWh/m^2.year, the energy conversion and transport systems work with a much lower efficiency than their optimum, and the needs of human comfort and health are seldom fully met.

Architects and investors are indirectly involved in the development of energy efficient systems, but can play a very important role in the widespread adoption of these systems or technologies.

To overcome the simple role of thermal/sound insulation system, a building must be controlled by a BMS (Building Management System) which allows, in addition to monitoring comfort and energy parameters, a complex control and efficiency.

References

[1] Wong J K W , Li H and Wang S W 2005 *Intelligent building research: A review*, Automation in Construction 14 143 – 159

[2] Wigginton M and Harris J 2002 *Intelligent Skin*, Architectural Press, Oxford, UK

[3] Năstase G, Şerban A, Dragomir G, Bolocan S and Brezeanu A I 2016 *Box window double skin façade. Steady state heat transfer model proposal for energetic audits*, Energy Build., vol. 112, pp. 12–20, ISSN: 0378-7788

[4] https://ec.europa.eu/eurostat/statistics-explained/index.php?title=File:Final_energy_consumption_in_the_residential_sector_by_use_EU-27,_2018.png (accessed on the 1st of October)

[5] https://ec.europa.eu/eurostat/statistics-explained/index.php/Energy_consumption_in_households#Energy_consumption_in_households_by_type_of_end-use (accessed on the 1st of October)

[6] Năstase G, Şerban A, Dragomir G, Brezeanu A I and Bucur I 2018 *Photovoltaic development in Romania. Reviewing what has been don*, Renew. Sustain. Energy Rev., vol. 94, pp. 523–535, ISSN: 1364-0321

[7] Dragomir G, Şerban A, Năstase G and Brezeanu A I 2009 *Wind energy in Romania: A review from 2009 to 2016* Renew. Sustain. Energy Rev., vol. 64, pp. 129–143, 2016, ISSN: 1364-0321

[8] Năstase G, Şerban A, Năstase A F, Dragomir G, Brezeanu and Ioardan N 2017 *Hydropower development in Romania. A review from its beginning to the present* , Renew. Sustain. Energy Rev., vol. 80, pp. 297–312, ISSN: 1364-0321.

[9] https://ec.europa.eu/eurostat/web/products-eurostat-news/-/DDN-20170328-1 (accessed on the 1st of October)

[10] Năstase G, Şerban A, Năstase A F, Dragomir G and Brezeanu A I 2018 *Air quality, primary air pollutants and ambient concentration inventory for Romania*, Atmospheric Environment, vol.184C, pp.292-303, ISSN: 1352-2310

IOP Conf. Series: Materials Science and Engineering 1138 (2021) 012007 doi:10.1088/1757-899X/1138/1/012007

Renewable energy sources potential evaluation in Romania

I Bucur[1]*, P Axinte[1], C Plescan[1] and A Şerban[2]

[1] Department of Building Services, Faculty of Civil Engineering, Transilvania University of Brasov, Romania

[2] Thermotechnics, engines, thermal and refrigeration equipment Department, Faculty of Mechanical Engineering and Mechatronics, University Politehnica of Bucharest, Bucharest, Romania

* irina.bucur@student.unitbv.ro

Abstract. Renewable energy sources (wind energy, solar energy, energy hydropower, ocean energy, geothermal energy, biomass and biofuels) are alternatives to conventional fuels in the greenhouse gases emissions reduction process. Because of the EU legislation support, the renewable energy sources have evolved significantly in recent years and in 2009 EU set a target of 20% energy consumption comes from renewable energy sources by 2020. In 2018, the target has been set for 32% of energy consumption from renewable energy sources by 2030. There is currently a debate on the policy framework future for the period after 2030. Greenhouse gas emissions is considered a serious threat regarding climate change, with potentially disastrous effects on humanity. The use renewable energy sources (RES), together with improved energy efficiency, can contribute to the reduction of energy consumption, the reduction of greenhouse gas emissions and, consequently, to prevent dangerous climate change. The potential of biomass, solar, hydro, wind and geothermal energy is very important, especially in recent years, due to financial support mechanisms such as green certificates, which in many European countries helped this sector to developed progressively.

1. Introduction

Greenhouse gas emissions is considered a serious threat regarding climate change, with potentially disastrous effects on humanity. Renewable energy sources (wind energy, solar energy, energy hydropower, ocean energy, geothermal energy, biomass and biofuels) are alternatives to conventional fuels in the greenhouse gases emissions reduction process [1, 2].

Despite Romania generally had the third least energy reliance rate in the EU, because of natural gas and oil reserve and to an oversized energy production network. The use renewable energy sources (RES), together with improved energy efficiency, can contribute to the reduction of energy consumption, the reduction of greenhouse gas emissions and, consequently, to prevent dangerous climate change. The potential of biomass, solar, hydro, wind and geothermal energy is very important, especially in recent years, due to financial support mechanisms such as green certificates, which in many European countries helped this sector to developed progressive. In the last ten years Romania has invested much in renewable energy especially in wind and solar power. In this sense the country boasts as Europe's largest onshore wind park, namely the 600 MW Fântanele-Cogealac, in the region of Dobrogea.

In Romania, grid operators are obliged to connect renewable energy plants to the national grid and to supply electricity from renewable sources as a priority.

IOP Conf. Series: Materials Science and Engineering 1138 (2021) 012007 doi:10.1088/1757-899X/1138/1/012007

2. Methods and procedures

This article reviews the potential of the renewable energy sources exploitation in regard to the European Union legislation and objectives. Until 2020, the European Union had three targets: 20% cut in greenhouse gas emissions, 20% of EU energy to come from renewables, and a 20% improvement in energy efficiency. The EU took action in different areas and now the targets are met. In the future, EU provides a new action plan called the European Green Deal. The new EU target is to be climate neutral in 2050 [3].

During 2016, the European Council (EC) presented two packages of proposals for reforming European energy policies, anticipated in 2015 through the Energy Union Framework Strategy. These packages are defining for the European energy sector, and implicitly for the Romanian one, in the period 2020-2030, being meant to accelerate the energy transition in the EU.

In July 2016, a first package of proposals was published on: reducing non-ETS emissions in each state for the period 2021-2030. In this proposal Romania is allocated a 2% reduction rate, accounting for the resulting GHG emissions from land use, land use change and forestry, as well as a communication on a European strategy for the decarbonization of the transport sector [2, 4].

The EC proposal for updating the RES Promotion Directive (EC 2016b) provides for six directions for action. The first proposes general principles to be followed when Member States define support policies for RES, using the principles of transparency, economic efficiency based as much as possible on competitive market mechanisms. These elements are brought together in the Strategy, under the principle of technological neutrality.

3. Renewable energy sources in Romania so far

In order to decarbonize the power sector while keeping up with the continuously growing energy demand, renewable energy technologies is essential. Renewable energy in Romania had a delayed boom compared to the European Union, yet Romania managed to meet the 2020 targets. So far, the most used renewable energy source in Romania is the hydropower and it is followed by wind, solar, biofuel and biogas (figure 1).

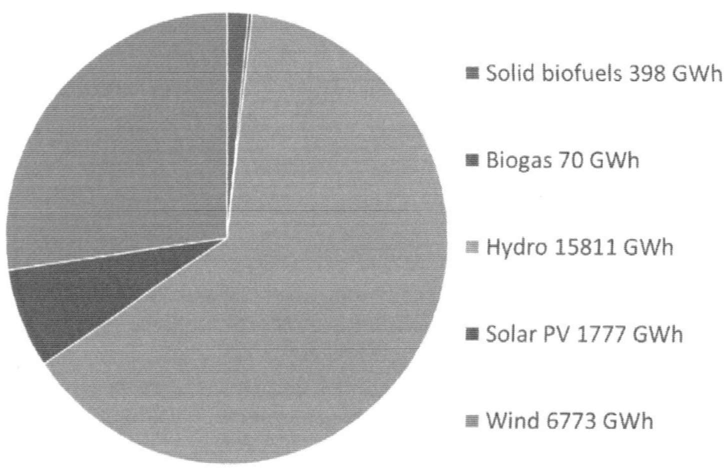

Figure 1. Renewable and waste energy data in Romania in 2019 [5]

4. National strategies

The National Integrated Plan in the field of Energy and Climate Change 2021-2030, provides that the share of renewable energy will reach 27.9% of Romania's gross final energy consumption in 2030, just an idea higher than the level achieved 15 years earlier. According to Eurostat, the statistical office of the European Union, 25% of energy consumption in 2016 was from renewable sources [5].

The Ministry of Energy mentions, in PNIESC, that more production capacities will be installed, but also the final consumption will increase, by approximately 18% in the period 2021-2030. Half (50.9%) of the energy consumed in Romania will be for heating or cooling, according to estimates, a slightly lower percentage than at present (54.6%), but still significant [5].

5. Development projections of the Romanian energy sector

In 2013, Romania was included in Top 15 most attractive countries in the world for investments in renewable energy. Our country is considered to have the highest wind potential in Southeast Europe. The second greatest wind-potential area on the continent, is considered the Romanian region of Dobrogea. Worldwide wind power is estimated at 57 PWh/year whereas the off-shore wind energy, estimated at 25-30 PWh/year. Eolian energy potential in one year, in Romania is estimated at a value of approximately 23 TWh.

The largest contributor Romania's energy potential derived from renewable-energy sources is represented by hydro energy, followed by wind energy. The main advantage of hydroelectric power is that can be stored, and electricity can be produced at a constant rate. The hydropower potential of our country is between 50 and 100 TW h/year.

Romania has also a high potential for harvesting solar energy in significant regions of the country, especially in south and on the Black Sea coast.

The development perspective of the energy sector by 2050 is useful for two main reasons: (1) the energy sector has a high capital intensity and many projects have a long investment cycle, so a large part of the investment decisions that will have place in the near future will continue to take effect in 2050; and (2) EU energy and environmental policies, including targets for 2030, are built around the long-term goal of reducing GHG emissions by at least 80% by 2050.

The global goal of climate change mitigation can only be achieved through transformative actions and measures at the global level. A main direction of action will be to accelerate the energy transition. Many of the long-term transformations of the energy sector can be anticipated, given the slow pace of energy infrastructure replacement.

Development trends refer to: increasing the sustainable role of biomass in the energy mix; the future of electromobility; increasing the share of RES in the electricity mix and the use of CCS technologies; forms of energy storage; energy efficiency, especially of buildings; electric heating based on heat pumps. [6, 7].

Romania ranked 7th in the EU in 2019 in terms of the share of electricity obtained from renewable sources, according to data published by Eurostat, ahead of Germany and above the European average.

With a percentage of 41.71%, is behind Austria (75.14%), Sweden (71.19%), Denmark (65.35%), Portugal (53.77%), Latvia (53, 42%) and Croatia (49.78%), but unfortunately the percentage has been decreasing by about a point and a half over the last four years, from 43.16% in 2015.

The idea is that Romania is the only state that has been satisfied with the progress made initially, and in the last few years has not made the necessary investments and has not taken support measures to sustain the long-term progress. This fact is relevant because it is not valid only in the case of electricity from renewable sources [8].

Regarding the share of energy from renewable sources in total national energy mix, Romania is slightly weaker. It ranks 10th, also above the EU average, but it was content to reach its 2020 target and has taken the step from 2016 onwards (in 2016, we were already at 25.03%, but by 2019 we had a marginal decrease in 24.29%).

Considering the European level, both hydropower and wind farms contributed about 35% to the supply of electricity, followed by solar panels (13%), solid biofuels (8%) and other sources (9%). For Romania that the largest development was in the solar energy sector, which crossed the threshold of 1% of the total energy mix in 2008 [8].

All these developments, although expected to reduce GHG emissions, could have a strong impact on the environment, and the opportunity for the development of new technologies on a large scale needs to be carefully considered. Most likely, new generations of these technologies, more efficient and greener, will be widely adopted.

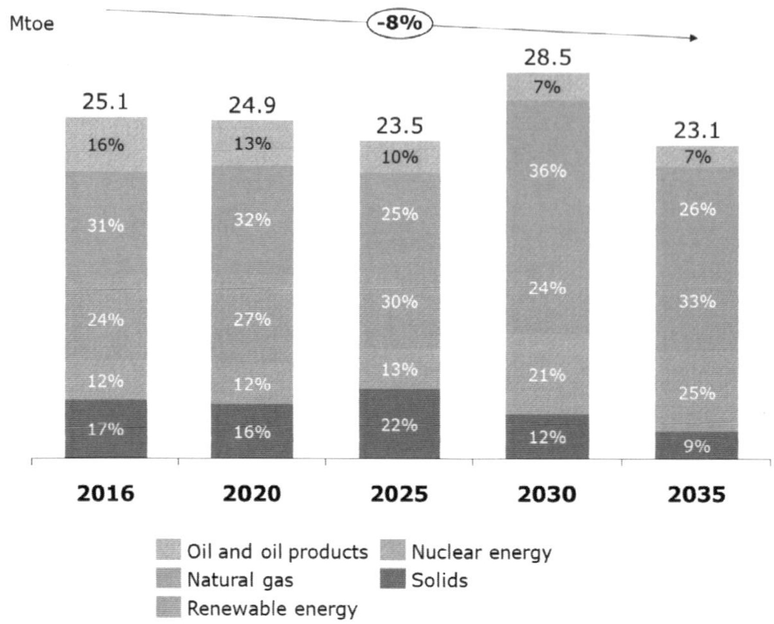

Figure 2. Expected development of the primary energy generation, by energy sources, with existing policies and measures [5]

According to Romanian Wind Energy Association (RWEA), Romania can reach a share of 32.4% conservatively and inertially in 2030, with the more ambitious potential to reach 35%, with additional wind capacities between 3000 and 4000 MW; and this is achievable with significantly lower investments than those considered by PNIESC [3].

With existing policies and measures, the projection shows a decrease of energy generation, which is mainly caused by a reduction of solid fuels and natural gas production. [4]

The use of renewable resources in energy generation has experienced an upward trend in recent decades, the determining factor for this trend being the research and accelerated development of new technologies that were possible only through the financial support of several countries. And, perhaps the most important aspect relates to that energies renewable energy sources lead to a reduction in energy dependence.

6. Conclusions

Sustainable development, involves the use of renewable energy to produce electricity is a clean and environmentally friendly way, and by all means the producers should be encouraged to invest in this sector. Over 1200 wind turbines are spread on the entire surface of Romania, with an installed power of 600 MW. This development of wind farms was achieved primarily in two regions: the core in Dobrogea Plateau with near 78% of the total power installed, and a secondary one in Bârlad Plateau. Romania's largest solar park, with a surface of 200 ha, was inaugurated in 2013 and it is placed in Ucea de Sus, Brasov County. It has an installed capacity of 82 MW and uses 320,000 PV units.

Investors in Romania prefer to develop more photovoltaic and wind facilities rather than other alternatives (i.e. biomass) which pose risks, have climate restrictions and require experience.

IOP Conf. Series: Materials Science and Engineering 1138 (2021) 012007 doi:10.1088/1757-899X/1138/1/012007

References

[1] Dragomir G, Şerban A, Năstase G, and Brezeanu A I 2016 Wind energy in Romania: A review from 2009 to 2016 *Renew Sustain Energy* **64**, pp. 129–143 ISSN: 1364-0321

[2] Năstase G, Şerban A, Năstase A F, Dragomir G, Brezeanu A I 2018 Air quality, primary air pollutants and ambient concentration inventory for Romania, *Atmospheric Environment* **184**, pp. 292-303, ISSN: 1352-2310

[3] https://eceuropa.eu/info/strategy/priorities-2019-2024/european-green-deal_en

[4] Năstase G, Şerban A, Dragomir G, Bolocan S, and Brezeanu A I 2016 Box window double skin façade Steady state heat transfer model proposal for energetic audits *Energy Build*, **112**, pp. 12–20, ISSN: 0378-7788

[5] https://www.euractiv.ro/economic/tinta-deloc-ambitioasa-pentru-energiile-regenerabile-romania-nu-vrea-sa-fie-e2-80-9eprea-verde-13899

[6] https://eceuropa.eu/energy/sites/ener/files/documents/romania_draftnecp_en.pdf

[7] Năstase G, Şerban A, Dragomir G, Brezeanu A I and Bucur I 2018 Photovoltaic development in Romania Reviewing what has been done, *Renew Sustain Energy*, **94** pp. 523–535 ISSN: 1364-0321

[8] PRIMES 2016 scenario, prepared for the Ministry of Energy, EU Commission, Energy datasheets

IOP Conf. Series: Materials Science and Engineering 1138 (2021) 012008 doi:10.1088/1757-899X/1138/1/012008

Processing of exposure results for composite material with the choice of their optimal composition

A Bulgakov[1*], V Afonin[2], I Erofeeva[2] and V Fedortsov[2]

[1] Southwest State University, Russia
[2] Ogarev Mordovia State University, Russia

* natalybredikhin@yandex.ru

Abstract. The article presents the results of processing a number of compositions of composite materials sustained under cyclic temperature changes, as well as changes in humidity and positive temperatures. A method is proposed for determining the optimal composition of materials, taking into account changes in such characteristics as hardness, modulus of deformation of the material and the elastic modulus of the material during exposure.

1. Introduction

At the present stage of studying concretes, composite concretes and composite materials in general, higher attention is paid to their optimization, the study of their characteristics, the dependence of various properties on operating conditions.

Questions of studying the properties and characteristics of composite materials and concretes are widely covered. The problems of modeling the problems of exposing materials in adverse operating conditions are considered. Experiments are conducted to study the properties of building materials, cement and polymer composites [1-30].

Various methods and algorithms for processing experimental data are used, for example, methods of experiment planning [9, 11-14], regression analysis [25-27], decision-making theory [28-33], optimization [11, 12, 15, 19, 20, 22, 23, 28, 29, 30]. To varying degrees, they are used to analyze the results of exposure of composite materials in adverse conditions [7, 9-11, 13, 14, 27]. Various goals are set. In particular, an important task is to determine the composition of composites, the characteristics of which are most resistant to operational impacts. First of all, it is the hardness of the material, then the modulus of elasticity of the corresponding character. In this paper, they are chosen – the hardness (T), the material's modulus of deformation (Ed), and the equilibrium elastic modulus (Eee). As a result of the tests, their absolute changes were recorded during the following time periods: 0 (indicators of control samples), 15, 30 and 45 days. The task is to select the tested composites of the composition that is least susceptible to adverse operational effects by any criteria. A multi-criteria and multidimensional optimization problem arise. This problem is considered and justified by numerical simulation experiments.

IOP Conf. Series: Materials Science and Engineering 1138 (2021) 012008 doi:10.1088/1757-899X/1138/1/012008

2. Materials and methods

The problem statement consists in the optimal choice of those compositions of composite materials, in which their properties – hardness, modulus of deformation and equilibrium modulus of elasticity change at least during the exposure time. The materials were aged under the influence of positive and negative temperatures, and under the influence of high humidity and variable temperatures.

Thermocyclic tests of samples (composites) were carried out in laboratory conditions in the climate chamber of VIAM (all-Russian research Institute of aviation materials) in two stages. At the first stage, a batch of samples was tested with 15 thermocycles, and at the next stage, the second batch of samples was tested with 13 thermocycles. One 24-hour thermal cycle includes four thermal states. Tests of samples under cyclic influence of positive and negative temperatures included the following operations: cooling of samples from room temperature (+23°C) to –50°C – about an hour (50 – 55 min); exposure of samples at a temperature of –50°C – 9 h; natural heating of samples to room temperature (+23°C) with the camera off – at least 5 h; exposure of samples at room temperature (+23°C) – 9 hours [27].

Tests of samples at high humidity (98%) and varying positive temperatures included other parameters: heating of samples from room temperature (+23°C) to +60°C – about 5 minutes; exposure of samples at a temperature of +60° C – 9 hours; natural cooling of samples to room temperature (+23° C) with the camera off – at least 5 hours; exposure of samples at room temperature (+23° C) – 9 hours [27]. Experimental data of the conducted tests are given in table 1, table 2, where A1, A2, ..., B1, B2 – designations of the studied composites.

Table 1. The results of thermo-cyclic testing.

Composition (Alternative)	Duration tests, day	Hardness (T), [MPa]	Module Deformations (Ed), [MPa]	Equilibrium modulus of elasticity (Eee), [MPa]
A1	0	4010.17	92914.38	144975.95
	15	2037.07	33544.78	38384.05
	30	1701.21	26063.26	37544.48
	45	1349.36	18225.47	36664.94
A2	0	7016.08	214592.64	285239.47
	15	3629.11	79711.48	97264.28
	30	2982.76	60725.11	86090.54
	45	2166.32	36742.32	71976.34
A4	0	5476.90	147693.86	183037.86
	15	3966.10	91025.36	103947.96
	30	3643.01	80613.86	89938.26
	45	3194.28	66153.45	70480.34
A5	0	9746.86	354170.95	417368.27
	15	3786.73	85008.56	91850.16
	30	3273.56	69346.72	81875.07
	45	2488.71	45393.31	66619.05
A9	0	7488.59	250219.85	286890.30
	15	1228.19	15722.21	17726.83
	30	1743.69	28292.49	39728.65
	45	2811.52	54330.94	85303.85
A10	0	3503.34	75640.22	111134.81
	15	2728.51	51977.25	63513.81
	30	2641.86	49571.33	65824.93
	45	2441.90	44019.21	71158.28
A11	0	2472.17	44833.37	64208.59
	15	2451.66	44358.11	61429.77
	30	2199.48	38246.00	52036.89
	45	1465.86	20465.32	24712.15

Table 2. The results of tests under cyclic impact high humidity and positive temperatures.

Composition (Alternative)	Duration tests, day	Hardness (T), [MPa]	Module deformations (Ed), [MPa]	Equilibrium modulus of elasticity (Eee), [MPa]
A1	0	4010.17	92914.38	144975.95
	15	4308.81	103122.15	128411.95
	30	4188.12	98968.91	131254.64
	45	4058.15	94496.20	134316.00
A2	0	7016.08	214592.64	285239.47
	15	4926.15	129501.89	165838.56
	30	5604.37	156282.45	192936.22
	45	6590.88	195236.00	232351.00
B3	0	2065.24	34205.20	61323.24
	15	3492.44	75571.86	100653.67
	30	3432.16	73535.70	101391.15
	45	3356.80	70990.50	102313.00
B4	0	9746.86	354170.95	417368.27
	15	5577.39	151786.84	202344.36
	30	8217.58	289799.12	355544.23
	45	12705.90	524420.00	615984.00
B5	0	4089.57	95300.69	111239.79
	15	8337.79	279496.41	327340.94
	30	10339.26	394961.27	451458.63
	45	14342.20	625891.00	699694.00
B6	0	1187.85	14949.93	32376.75
	15	7496.49	236655.80	269806.53
	30	5633.02	171126.11	199181.37
	45	1207.29	15493.10	31446.60
B7	0	7488.59	250219.85	286890.30
	15	3119.91	63500.34	97528.19
	30	3092.42	63674.97	96429.32
	45	3013.89	64173.90	93289.70

As can be seen from the tables 1 and 2, the properties (T, Ed, Eee) of the compositions change relative to their control values during the exposure time. Their relative changes are of interest. In this regard, the values of the tables should be converted to relative values, relative to the control values, relative to the values in the row with the indication zero (0 – the duration of the test). This will allow you to compare the properties of materials with different ranges of changes in a particular property. These two selves may differ by several orders of magnitude.

Table 3. The results of tests under cyclic impact high humidity and positive temperatures.

Composition (Alternative)	Duration tests, day	Hardness (T)	Module deformations (Ed)	Equilibrium modulus of elasticity (Eee)
A1	0	1.000000	1.000000	1.000000
	15	1.074471	1.109862	0.885747
	30	1.044375	1.065162	0.905355
	45	1.011965	1.017024	0.926471
A2	0	1.000000	1.000000	1.000000
	15	0.702123	0.603478	0.581401
	30	0.798789	0.728275	0.676401
	45	6590.88	195236.00	0.814582

For an explanation, see table 3, which summarizes the relative values for the first two compounds (A1, A2) from table 2.

Regardless of the purpose of a particular composite, the natural requirement is primarily the immutability of its properties. In this case, the relative values of the properties will be equal to one. At the same time, the actual state of things may not meet the requirement that the property indicators remain unchanged during the exposure time under the specified operating conditions. For comparison of the desired requirement of immutability of properties and their actual changes in figure 1, figure 2 charts are presented in relative units.

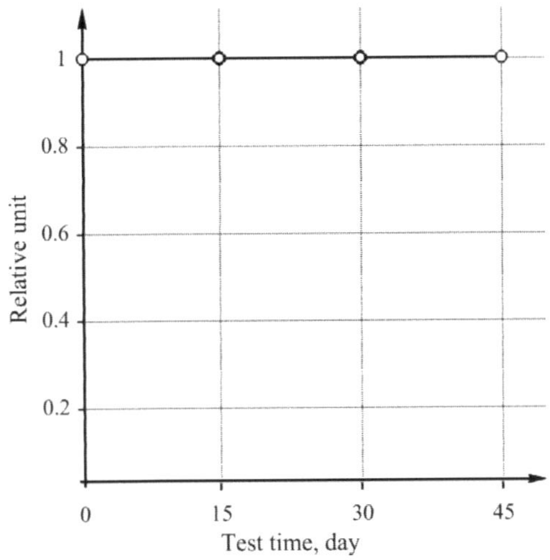

Figure 1. A property is invariant for the tests.

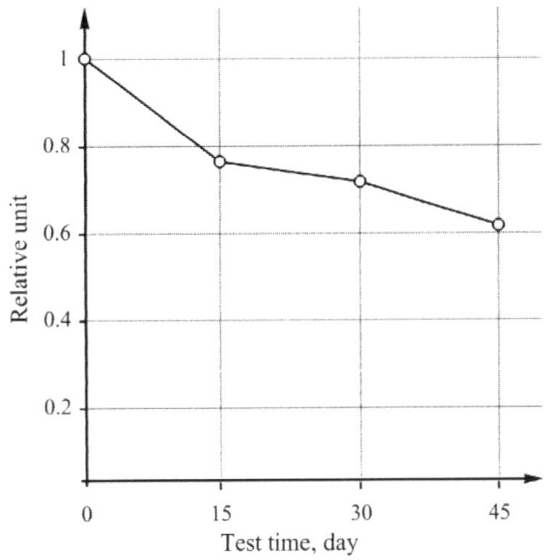

Figure 2. The property depends on the test.

As shown in tables 1 and 2, relative changes in properties may exceed the level of the control sample property during the test. In this regard, it is proposed to calculate the area under the polyline curve (figure 3, figure 4) and compare it with the area of the rectangle S_{rect} with a height equal to one. In the case of figure 1, S_{rect} is equal to 45 square units.

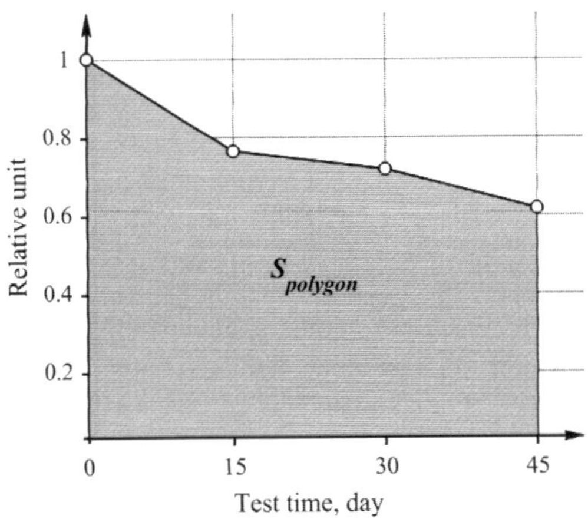

Figure 3. Monotonous decrease of the property value.

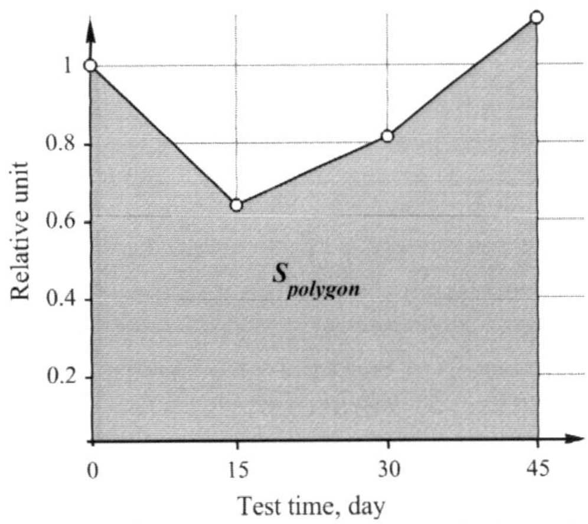

Figure 4. Non-monotonic change in the values of the properties.

When the values of the controlled property (for example, hardness) of the composite are monotonously reduced, one of the criteria for determining the best composition can be the ratio of $S_{polygon}$ / S_{rect} areas, which will tend to zero in the ideal case. If there is a non-monotonic change in the value of the controlled property (figure 4), the area of the polygon $S_{polygon}$ may differ little from the area of the rectangle for the composite property with the unchanged nature of the change from the test time. However, the given area ratio can be taken as a criterion in the problem of determining the optimal composition. In [15], a two-criteria optimization problem was considered, and the criteria used to determine the preferred composition of composites were described. Here we propose an addition to the criteria for searching for the optimal composition of a number of composites that have been tested under test operating conditions. For the first criterion, we use the $S_{polygon}$ / S_{rect} area ratio.

Figure 5 and figure 6 show the construction to determine the criteria K_i, H_i, i = 1, 2, 3.

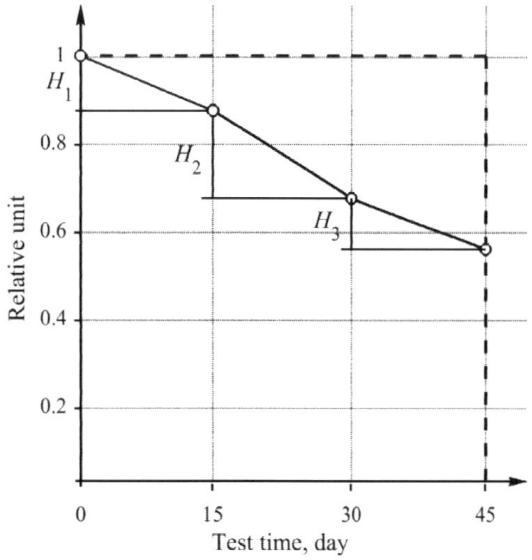

Figure 5. To determine the H_i criterion.

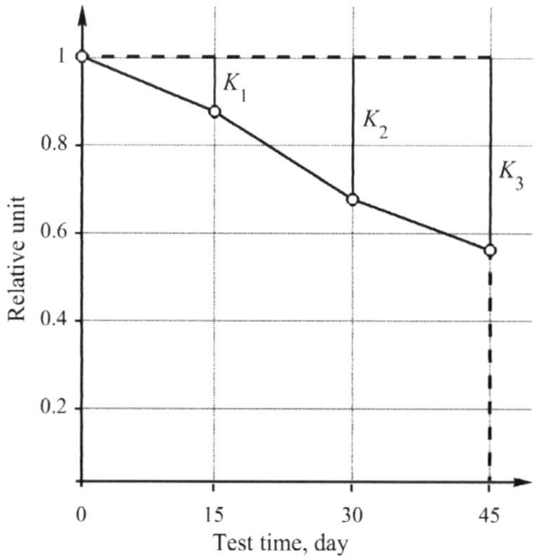

Figure 6. To determine the K_i criterion.

The criterion S, which takes into account the area ratio, is formed in the form $S_i = S_{polygon} / S_{rect}$. Another criterion, the J_i criterion, is defined as the sum of the specified criteria, i.e.

$$J_i = S_i + K_i + H_i,\ i = 1, 2, 3, 4. \tag{1}$$

It is obvious that for a composite whose properties change slightly during the exposure period, the entered criteria will tend to zero. However, the number of exposure points is not limited (within reasonable limits). These ratios will be determined for each composite composition, in particular in accordance with table 1 and table 2. There will be seven for each exposure point, i.e. 28 for each controlled property.

Thus, as in the decision-making theory [24, 25], we accept composite compositions (A1, A2,...) as alternatives, and S_i, K_i, H_i, J_i (i = 1, 2, 3) as criteria. However, the criteria values should be defined for each property. The following tables are compiled in accordance with ELECTRA methods:

- table of values for criteria and alternatives,
- table of ratings for the values of each criterion,
- table matrix of the estimates.

In this case, the task of selecting the composition is to determine the optimal sum of ratings from the rating matrix. The size of the rating matrix is set based on possible changes in each controlled property of the composites. The estimation matrix is formed taking into account monotony and arbitrary changes in a particular property during the exposure period. Summation of scores based on the rating matrix is based on comparison of the values of the rating matrix and criteria for a predetermined value. The value of this value is defined as a half digit of significant digits in the data array. For the data shown in tables 1 and 2, it is equal to 0.005. The comparison is performed based on the absolute value of the nearest values of the rating matrix with the values of the calculated criteria. A cyclic algorithm is used with a nesting value equal to three-the number of criteria (controlled properties of composites), the exposure points, and the number of alternative samples.

There are various ways to choose the optimal compositions [4, 7, 12, 15, 19, 21–24, 28, 30]. Here we propose a developed heuristic approach for processing experimental data in order to select the optimal composition of test-tested composites.

3. Results

Taking into account the fact that the selected criteria should be minimized, then the formation of estimates for the criteria will be arranged in descending order. The size of the matrix of possible criteria values should be assumed to be greater than the number of alternatives – the number of compositions. This will allow you to divide the estimates in a wider range, which, in turn, will almost eliminate the coincidence of the sum of estimates for alternatives. In the conducted experiments, the size of the matrix with estimates is assumed to be equal to twice the size of the number of alternatives.

Table 4 shows a matrix of estimates of composites exposed during cyclic temperature tests.

Table 4. The final matrix of the evaluations for cyclic temperature testing.

Total score	Alternatives	The studied properties of the composites: T, Ed, Eee											
		T				Ed				Eee			
		Criteria: S, K, H, J											
		S	K	H	J	S	K	H	J	S	K	H	J
128	A10	13	11	11	14	14	14	3	3	3	14	14	14
124	A11	14	13	14	11	12	12	3	3	4	12	13	13
105	A4	11	10	9	10	10	10	4	4	5	10	11	11
74	A1	8	6	6	4	4	5	8	8	10	5	5	5
73	A2	8	6	7	4	4	5	8	8	8	4	5	6
70	A5	7	5	5	3	3	3	10	11	12	3	4	4
59	A9	5	4	4	2	2	3	12	13	13	3	2	4

In table 4, the alternatives are the teams ranked by total points. Experimental data are taken from table 1. As can be seen, the optimal composition of the composite corresponds to A10, which more than the rest meets the conditions for optimal selection among a given number of samples that have been tested at a temperature change for 45 days. For visual control of the relative change in the hardness of the first group of composites, a diagram is shown in figure 7, which shows that the performed ranking

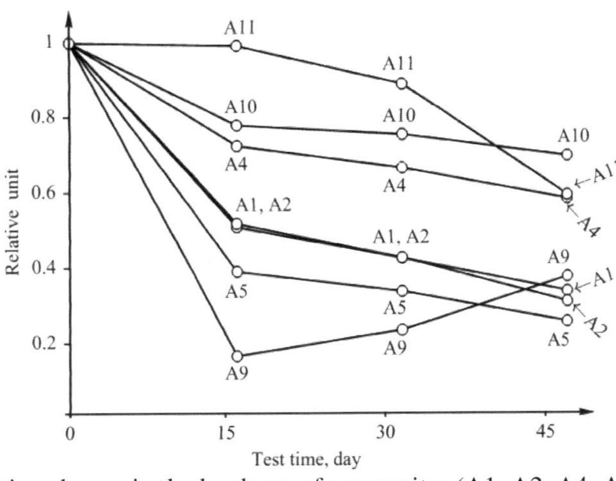

Figure 7. Relative change in the hardness of composites (A1, A2, A4, A5, A9, A10, A11).

of samples (table 4) is consistent with the nature of the change in the hardness of each composition (alternative) during the test time.

A similar table (table 5) is provided for optimal selection of composites that have been maintained in conditions of high humidity and positive temperatures.

Table 5. Final rating matrix for composite exposure in conditions of high humidity and positive temperatures.

Total score	Alternatives	The studied properties of the composites: T, Ed, Eee											
		T				Ed				Eee			
		Criteria: S, K, H, J											
		S	K	H	J	S	K	H	J	S	K	H	J
98	A1	5	2	3	14	14	14	1	1	2	14	14	14
97	B6	14	14	14	1	1	2	14	14	14	2	4	3
66	B5	9	6	10	2	2	2	8	9	10	2	3	3
51	B7	3	1	2	2	1	2	9	12	11	2	3	3
49	B3	6	3	5	3	3	4	4	5	4	3	4	5
45	A2	4	2	3	3	4	4	3	4	4	4	5	5
44	B4	4	2	3	2	2	2	6	7	7	3	3	3

According to the results of table 5, we conclude that the most preferred composite is A1 when tested in conditions of high humidity and positive temperatures.

4. Conclusions

The paper considers the results of experimental studies of various ready-to-use composite materials. In contrast to the known similar studies [7, 9, 11, 13], the paper highlights the results of studies with a periodicity of 15 days: 0-15-30-45. The tests were carried out according to the requirements of climatic tests-from sub-zero temperatures to high positive values. To a certain extent, a heuristic method for solving a multi-criteria multidimensional optimization problem is proposed, which allows it to be used in experimental studies to determine the optimal composition of a number of composites under study. The authors have defined and proposed criteria for finding the optimal composition of composites that have passed test tests with the measurement of three properties of composites. At the same time, the number of properties can be quite arbitrary, according to the authors. The number of exposure points can also be arbitrary. At the same time, the authors are aware that if the results are extrapolated, they may differ to some extent from the experimental results obtained over a longer period of time, more than 45 days.

References

[1] Yatsenko E, Goltzman B, Chumakov A, Smoliy V, Holshemacher K and Bulgakov A 2019 Prospects for the use of thermal disposal products for solid municipal waste in the technology of silicate heat insulation materials *Izvestiya vuzov. severo-kavkazskiy region technical science*. **3**, pp. 77–81. ISSN 0321-2653. DOI: 10.17213/0321-2653-2019-3-77-81

[2] Chumakov A, Yatsenko E, Golovanova S, Bulgakov A and Holshemacher K 2019 Strength and frost resistance of special road concretes materials *Izvestiya vuzov. severo-kavkazskiy region technical science*. **2**, pp. 55–59. ISSN 0321-2653 DOI: 10.17213/0321-2653-2019-2-55-59

[3] Yatsenko E A, Goltzman B M, Bulgakov A G and Holschemacher K 2019 Ways of utilization of mining wastes containing silicates *Journal of Ore-dressing* **2**, pp. 49–54. DOI:

10.17580/or.2019.02.09

[4] Krause M, Otto J, Bulgakov A and Sayfeddine D 2018 Strategic optimization of 3D-concrete-printing using the method of CONPrint3D *Proceedings of the 35th International Symposium on Automation and Robotics in Construction (ISARC 2018).* Berlin, pp. 1–7

[5] Erofeev V, Korotaev S, Bulgakov A, Tretiakov I and Rodin A 2016 Getting fired material with vitreous binder using frame technology *Procedia Engineering* **164** pp. 166–171. DOI: 10.1016/j.proeng.2016.11.606

[6] Bulgakov A, Erofeev V, Bogatov A, Smirnov V and Schach R 2016 Innovative production technology of binding and building composite materials on the basis of glass wastes. In Insights and Innovations in Structural Engineering, Mechanics and Computation (ed. A. Zingoni), Taylor & Francis Group, London, pp. 1583–1586. ISBN 978-1-138-02927-9. *Proceedings of the Sixth International Conference on Structural Engineering, Mechanics and Computation* (SEMC 2016), 5–7 September 2016, Cape Town, South Africa

[7] Erofeeva I V, Afonin V V, Fedortsov V A, Emelyanov D V, Podzhivotov N Yu and Zotkina M M 2016 Investigation of the behavior of cement composites in conditions of high humidity and variable positive temperatures *International Journal for Computational Civil and Structural Engineering* **13** No.4. Pp. 66–81

[8] Erofeev V T, Cherkasov V D, Emelyanov D V and Erofeeva I V 2017 Impact strength cement composites *Academia. Architecture and construction* No. **4**. Pp. 89–94

[9] Afonin V V, Erofeeva I V , Zotkina M M, Emelyanov D V and Podzhivotov N Yu 2019 Reference image quality assessment of composite materials exposed to positive and negative temperatures. *Bulletin of the MSU - Proceedings of Moscow State University of Civil Engineering* **14** 1 pp. 83–93. DOI: 22227/1997-0935. 2019.83-93

[10] Parfenov A A., Sivakova O A., Gusar O A, Balakireva V V. 2019 Work and destruction of concrete in high and low temperature conditions *Construction materials* **3** Pp. 64–66. DOI: https://doi.org/10.31659/0585-430X-2019-768-3-64-66

[11] Dimitrienko Yu I, Drogolyub A N, Sokolov A P, Shpakova Yu V 2013 Method for solving the problem of optimizing the structure of dispersed-reinforced composites with restrictions on thermal and strength properties. Nauka I obrazovanie. *Bauman Moscow state technical University. Electronic journal* **11** Pp. 415–430. DOI: 10.7463/1113.0621065

[12] Kharitonov A M 2015 Development of methods of optimization of construction compositions of multicomponent composites Fundamental research **11** Pp. 520–523. URL: http://fundamental-research.ru/ru/article/view?id=39452

[13] Erofeeva I V, Fedortsov V A, Afonin V V, Emelyanov D V, Podzhivotov N Yu, Moiseev V V, Kremcheyev A N 2018 Studies of the influence of cyclically acting negative and positive temperatures on the damping properties of cement composites *CHEMICAL BULLETIN* **1** 3 Pp. 42–51

[14] Erofeeva I V, Fedortsov V A., Afonin V V, Emelyanov D V, Podzhivotov N Yu, Moiseev V V, Kremcheyev A N 2018 Studies of the influence of increased humidity and variable elevated temperatures on the damping properties of cement composites *CHEMICAL BULLETIN* **1**, No. 3. Pp. 62–71

[15] Afonin V V, Erofeeva I V, Fedortsov V A, Emelyanov D V, Podzhivotov N Yu 2018 Heuristic approach to solving two-criterion problem of optimization of composite materials. Bulletin of the MSU *Proceedings of Moscow State University of Civil Engineering* **13** (11):1357–1366. DOI: 10.22227/1997-0935.2018.11.1357-1366

[16] Lesnov V V, Erofeev V T, Afonin V V. 2018 Investigation of strength properties of filled plasticized cement matrices intended for frame composites *Regional architecture and construction* **4** (37). Pp. 71–79

[17] Bulgakov B I, Alexandrova O V 2017 Mathematical modeling influence of raw materials components on the strength of high-quality fine-grained concrete during compression Bulletin of the MSU. **9** (108) Pp. 999–1008

[18] Soldatenko L V 2009 Introduction to mathematical modeling of construction and technological tasks *Orenburg: GOU OSU* 161 p
[19] Garkina I A, Danilov A M and Korolev E V 2010 Mathematical and computer science modeling in the synthesis of building composites: state and prospects Regional architecture and construction **2** Pp. 9–13
[20] Merkulov D A, Korotaev S A and Erofeev V T 2017 Optimization of the granulometric composition of quartz-filled polyester composites *BST: Bulletin of construction equipment* **5** (993). Pp. 31–33
[21] Yezersky V A, Kuznetsova N In, Bessonov I V and Seleznev A D 2018 Multi-Criteria comparative analysis in assessing the quality of cement composites using waste CPBP Construction and reconstruction **3** (77). Pp. 89–97
[22] Sorokin D S and Danilov A M 2014 Optimization of structure and properties of composite materials Modern scientific research and innovation **5** Part 1 URL: http://web.snauka.ru/issues/ 2014/05/34828
[23] Safronov V N and Kugaevskaya S A 1914 Optimization of the properties of cement composites in various technological methods of preparation of cyclic activation of the closing water *Bulletin of the Tomsk state University of architecture and construction* **1** (42). Pp. 85–99
[24] Gusev N I, Kochetkova M V. and Shcheglova A S 2014 Prerequisites for choosing optimal compositions of polymer-cement compositions for protective and finishing coatings of walls made of foam concrete *Modern equipment and technologies* **12** URL: http://technology.snauka.ru/ 2014/12/5043
[25] Shakhova L D, Chernositov E S, Shchelokov L S and Denisova J V. A 2019 Study of factors affecting the fluidity of cement *Bulletin of BSTU* **11**. Pp. 8–16. DOI:10.34031/2071-7318-2019-4-11-8-16
[26] Stashkov S I. 2019 Dissertation for the degree of candidate of technical Sciences / Perm 212 p.
[27] Erofeeva I V 2018 Dissertation for the degree of candidate of technical Sciences / Penza 318 p.
[28] Yang S Y, Girivasan V, Singh N R, Tansel I N and Kropas-Hughes C V 2003 Selection of optimal material and operating conditions in composite manufacturing. Part II: complexity, representation of characteristics and decision making *International Journal of Machine Tools & Manufacture* **43** 175–184
[29] Xianhong Li, Haibin Yu and Mingzhe Y 2012 Modeling and Optimization of Cement Raw Materials Blending Process Mathematical Problems in Engineering Pp. 1–31. DOI:10.1155/2012/392197
[30] Nitza M G, Hildelix L, Mauricio C R and Suarez O M 2012 Optimal Cement Mixtures Containing Mineral Admixtures under Multiple and Conflicting Criteria *Advances in Civil Engineering* DOI: https://doi.org/10.1155/2018/3780810
[31] Kini R L 1981 Decision-Making under many criteria: preferences and substitutions *Moscow: Radio and communications* 560 p.
[32] Larichev O I. 2002 Theory and methods of decision-making 392 p.
[33] Berdoudi S, Hebhoub H and Djebien R 2017 Valorization and recycling of quarries waste as an addition in cement *Journal of Applied Engineering Science* **15** 2, 420, 122–127. DOI: 10.5937/jaes15-12743

IOP Conf. Series: Materials Science and Engineering 1138 (2021) 012009 doi:10.1088/1757-899X/1138/1/012009

Influence of urban development on change in microclimatic parameters of urbanized area

A Bulgakov[1*] and E Skobeleva[2]

[1] Southwest State University, Russia
[2] Orel State University, Russia

* natalybredikhin@yandex.ru

Abstract. The expediency of quantitative analysis of safety and comfort of the spatial environment when creating master plans of cities was shown by the authors. Predictable thermal comfort should be considered here as one of the significant indicators of the environmental safety of the routes from places of residence to socially significant facilities and recreational zones. It is suggested to use thermal indexes adapted for the conditions of open urban spaces to consider the influence of climatic variables totality on the person's thermal sensation. The physiologically effective temperature (PET) is one such index. This model takes visceral temperature, perspiration rate, the moisture of the skin, and on the other hand the meteorological parameters, including the influence of urban area conditions on them. Criteria values of PET for the middle zone of Russia are determined on the basis of bioclimatic estimation of the territory according to air temperature, cloudiness and wind speed. The authors determined experimentally the parameters of the urban microclimate.

1. Introduction

The main task is the creation of a safe and comfortable urban environment that is being solved by urban planning. At the same time, the security of an urbanized territory is determined not only by the safety of its elements for human life and health, but also by the capacity for self-sustaining development [1]. The requirements for environmental comfort in their turn are based on a complex combination of physiological, psychological and sociocultural responses of individuals and their groups to the state and changes in the spatial environment [2].

The signs of urban microclimate are the indicators of the state of environmental safety and comfort systems built environment as well as the hygienic condition of atmospheric air, insolation, acoustic and vibration pollution.

Global warming and the formation of Urban Heat Islands over densely populated areas and industrial centers make it necessary to revise the policy of long-term urban planning, for example, in accordance with the concept of biosphere-compatible cities that are developing human beings [1, 3].

In the last decades with the purpose of mitigating the climate, ensuring the thermal comfort in open spaces and increasing the energy efficiency of operated buildings, more attention is being paid to the study of the influence of building parameters, planting of greenery and improvement elements on the temperature-humidity and aeration regimes [4-16].

Urban morphology information is an important indicator in urban planning, information management, and urban climatic applications. Some most widely used urban morphological parameters

- building coverage ratio, building height, building volume density and roughness length are employed to generate an Urban Climatic Maps of several cities. The urban microclimate is also influenced by the parameters of planting of greenery, the presence of water objects and sun protection facilities. Thus, for regions with a hot climate, shading of streets and provision of aeration mode is of primary importance, and for regions with continental and mesothermal types of climate, along with ensuring temperature comfort in the summer and winter months, a sufficient level of insolation [17].

The need to mitigate abnormal climatic regimes to ensure environmental safety of habitats in urban areas should become one of the priority tasks of prospective urban planning. However, at the present time, as a rule, from the entire list of climatic and sanitary-hygienic factors in Russian normative documents, priority is given to insolation and wind protection. Normative documents practically do not contain instructions on providing temperature comfort in open urban spaces. Nevertheless, the influence on the urban microclimate of building morphology is significant not only in regions with a hot climate, but also in conditions of a temperate climatic zone [18, 19].

To assess the effect on the combination of climatic parameters of human heat, it is proposed to use universal thermal indexes adapted for the conditions of open urban spaces.

One of the most representative thermal indexes used to assess temperature comfort in open urban spaces is the physiologically effective temperature (PET) [20, 21]. This index is based on the Munich model of the energy balance of the human body, which takes into account the temperature of internal organs, the intensity of sweating, and the moisture of the skin. The thermal index PET was chosen to quantify the human-thermal biometeorological conditions. PET is calculated by RayMan [22, 23] on the basis of the human energy balance and takes the integral thermo-physiological effect of the meteorological parameters air temperature, air humidity, wind velocity and radiation fluxes into account.

The range of comfort PET values can be determined using the Predicted Mean Vote PMV and for the countries of Western and Central Europe it is 18...23°C [23], Taiwan - 26...30°C, Beijing - 22..28°C [12], for conditions of hot dry climate (Negev, Israel) - 21..33°C [24]. In the central (European) part of Russia, on the basis of the results of the bioclimatic assessment of the territory [25], climatic parameters are proposed to be considered comfortable at PET = 19.0...36.1°C [26], which corresponds to the levels "warm", "slightly warm" and "comfortable" in the scale of human perception in Western and Central Europe [13].

The indicator of the openness of the sky - Sky View Factor (SVF) is one of the important factors affecting the thermal sensation of a person who is in open spaces. SVF is a dimensionless parameterization of the quantity of visible sky at a certain location. In open terrain, the SVF value will be close to 1, while in densely built-up or near-tree conditions, SVF will proportionally decrease [27].

2. Methods of research

To assess the urban microclimate from the standpoint of the influence on it of the parameters of building and improvement, the actual values of temperature, air humidity and wind speed in different points of the city were measured. Based on the experimental data obtained, the indicators of the temperature comfort of pedestrians were determined. Studies were carried out in the city of Orel, one of the major cities of the European part of Russia.

The objectives of this study were as:

- experimental determination of urban microclimate parameters (actual air temperature, air humidity and wind speed at various points) in the spring and summer;
- determination of physiologically equivalent temperature at the points of measurement;
- analysis of research results.

Measurements of temperature, humidity and air velocity were carried out using an anemometer-hygrometer Testo 410 with a range of measurements of 0.4-20 m/s, -10 - 50 ° C, 0-100% RH. Measurement mode - Hold Avg - calculation of time-averaged values (average time of measurement - 2 minutes). The height of the device above the earth's surface is 1 m (it is chosen to take into account the influence of various factors on climatic parameters: albedo of surfaces, heat losses through building envelopes, presence of exhaust gases of motor vehicles, etc.).

3. Case of studies

Orel - a city in Russia, the administrative center of the Oryol region, with a population of about 320 thousand people. It is located 368 km southwest of Moscow, on the Central Russian Upland in the European part of Russia, along both banks of the Oka River and its tributary Orlik (figure 1).

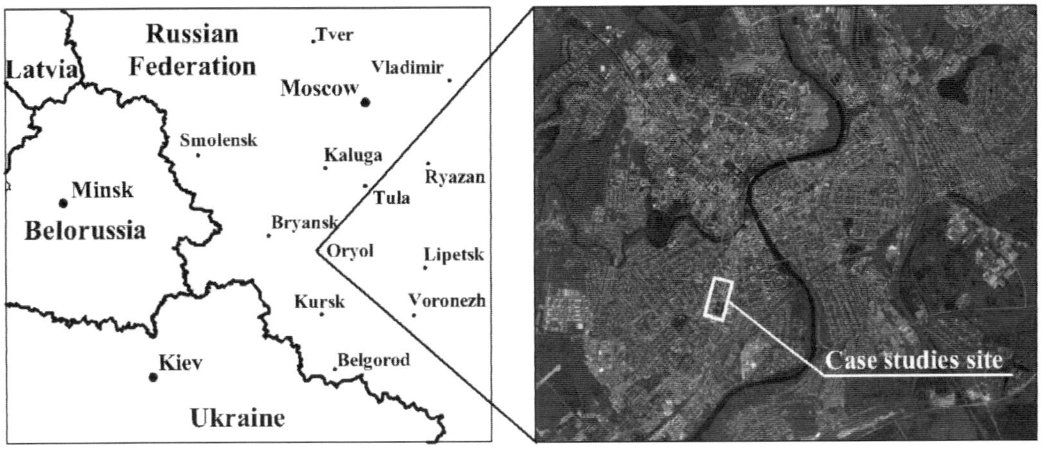

Figure 1. Geographical location of the city of Orel.

3.1. Climate of Orel

The climate of Orelis characterized as moderately continental. The yearly mean temperature is 6.2°C and the yearly average amount of precipitation is 601 mm. The winter is mild, the summer with alternating periods of intense heat and cooler weather (figure 2).

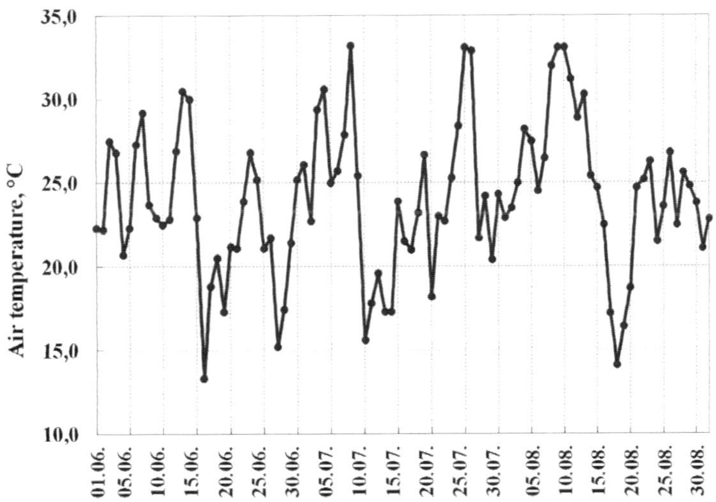

Figure 2. Air temperature during the summer 2015 at 15:00 (data from Meteorological Station, airport).

To study the parameters of the urban microclimate in the spring (08/04/2015) and summer (03/07/2015) period was chosen route from the 1-st Posadskaya Street to the MOPR's Street in the city of Orel. The measurements were carried out on two parallel streets: Str. Komsomolskaya and Str.

Karachevskaya, at five points on each street (figure 3). At each point, three measurements were taken at a distance of 2 m from the buildings facades.

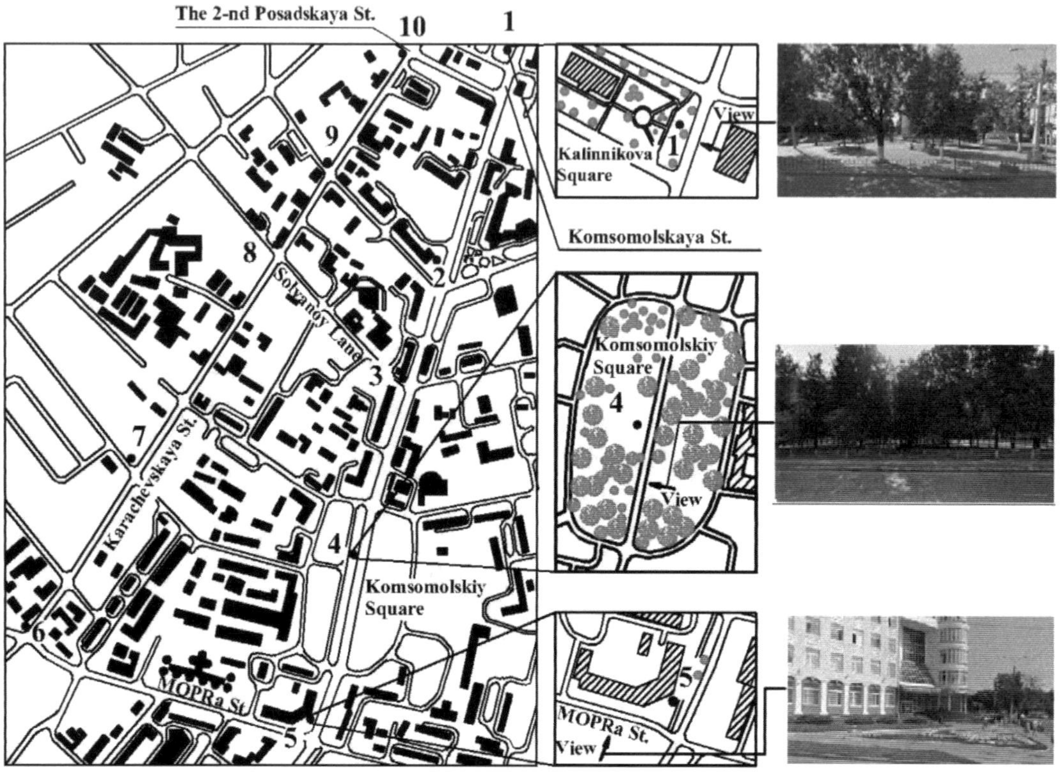

Figure 3. Site maps and existing conditions in some points.

The SVF values are different at the selected waypoints (Figure 4).

Point 4 SVF=0,083 | Point 8 SVF=0,351 | Point 3 SVF=0,668 | Point 10 SVF=0,838

Figure 4. Sky View Factor.

4. Discussion

The averaged measurement results are given in the form of diagrams (figure 5). In this diagram, you can see significant differences between the values of the air temperature measured at a meteorological station located outside the city and measured in the city. This confirms the presence of Urban Heat Island near the measuring points (with a difference of up to 5°C).

The calculation of the physiologically equivalent temperature based on the measurement results was made using the program RayMan [21, 22] for the average pedestrian (a healthy man aged 35 years, weight 75 kg, height 175 cm). The values of the calculated PET index are given for each route point (figure 6).

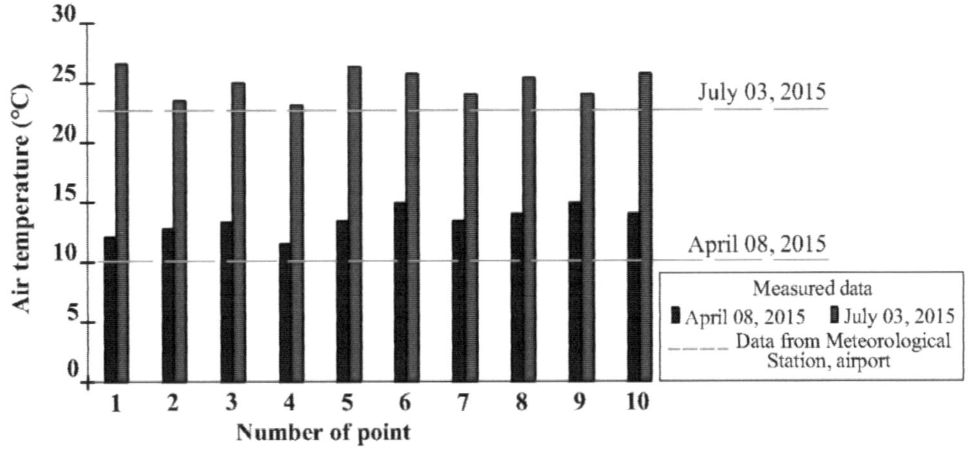

Figure 5. Air temperature at measuring points.

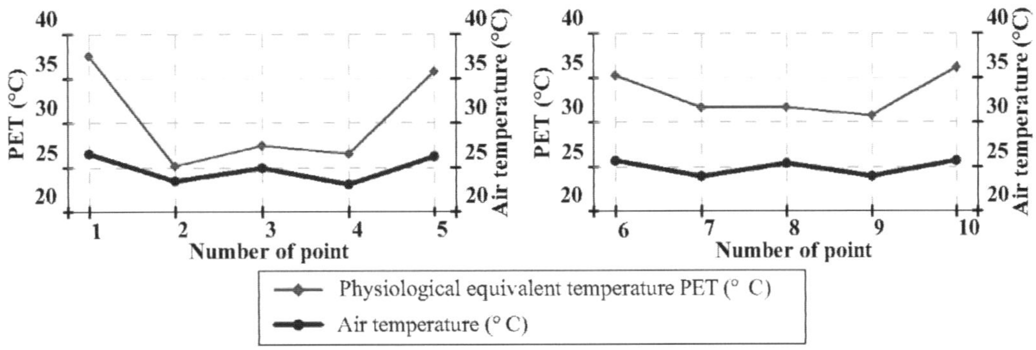

Figure 6. The values of Air Temperature and PET along the Komsomolskaya Street and Karachevskaya Street, corresponding to the route of measurements in the spring and summer periods.

5. Results

The greatest differences between the temperature in the city and at the meteorological station for the spring period are observed for points 6 and 9. This is due to the location of point 6 near the busy intersection, and points 9 - in close proximity to the dilapidated house and significant heat loss through the facade.

In summer, the greatest excess of the actual temperature values over the values obtained at the weather station is primarily due to the location of the measurement points near the intersections with busy traffic.

According to the results of PET calculations, at points 1, 5, 6, 10 the pedestrian will experience thermal discomfort with a strong level of physical stress [23, 26].

It should also be noted the difference in the nature of the distribution of PET for the Komsomolskaya Street and Karachevskaya Street. It is of interest to compare PET values between neighboring measurement points (table 1).

Figure 7. Thermal facade pictures of dilapidated wooden house (near the point 9) (a) and – for comparison - the facade of the new brick building (b).

Table 1. PET difference between adjacent measurement points.

Point spacing	Difference of values PET, °C	
	Spring (08/04/2015)	Summer (07/03/2015)
1-2	0.7	13.0
2-3	2.9	6.4
3-4	4.2	6.2
4-5	3.0	10.2
10-9	2.8	5.5
9-8	2.0	0.6
8-7	3.1	1.0
7-6	1.5	1.6

The most influential factors in PET are air temperature and humidity, as well as wind speed (figure 8, b-d). There was no clear dependence of PET on SVF (figure 8, a).

6. Conclusions

Experimentally determined indicators values of pedestrians' thermal comfort show significant differences in microclimate parameters along one street. Increase PET in the points of selected streets contribute to the surrounding surface with a high albedo, the presence of a lively car traffic. The low number of stories along these streets and the considerable width between them reduce the impact on the PET of the build-up morphology.

At the same time, the presence of dilapidated buildings, large asphalt areas, busy highways contribute to increasing the temperature difference in the center and on the outskirts of the city.

The observed climate change at the scale of a single city, the formation of a thermal island and a particular urban microclimate is one of the problems that must be addressed at the stage of urban planning. At the same time, the creation of a safe and comfortable climate regime will not only promote the preservation of human health, but also strengthen the ability of the anthropogenic and natural environment to self-sustaining development.

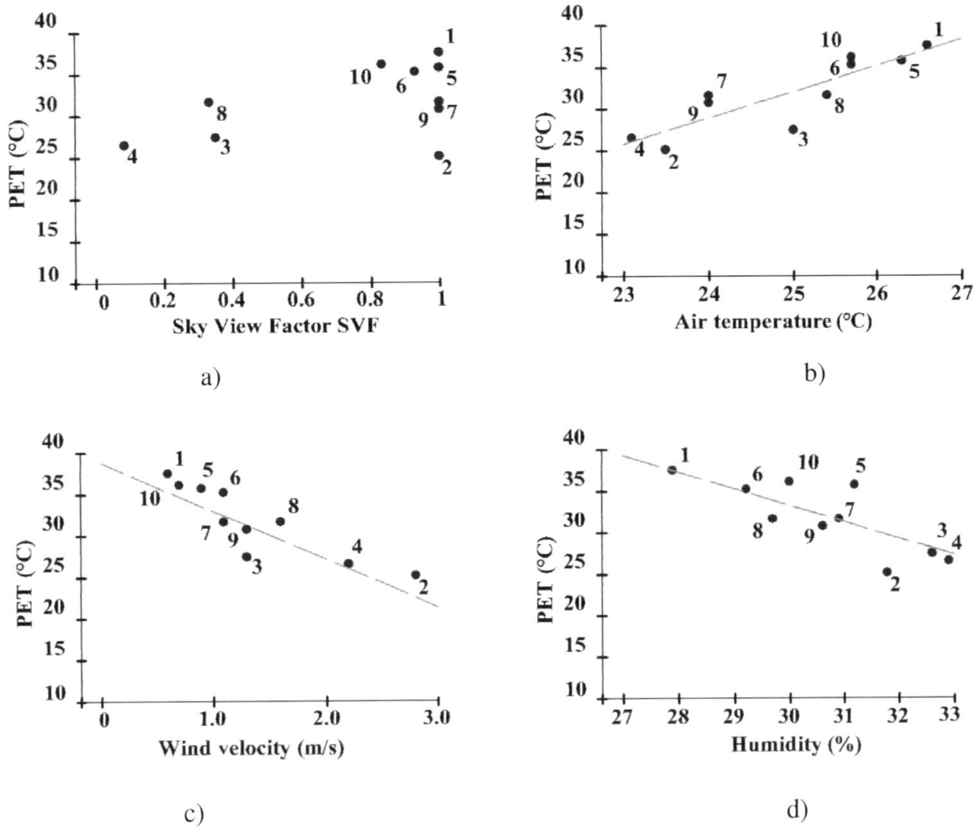

Figure 8. Comparison of PET and SVF (a), Air Temperature (b), Wind Velocity (c), Humidity (d).

References

[1] Ilyichev V, Kolchunov V, Emelyanov S and Skobeleva E 2013 Algorithm for the development of programs for integrated safety and vitality of urban areas *Biosphere compatibility: people, region, technology* **1** pp. 44-52

[2] Kuzmin A 2011 Quality of life and quality of the spatial environment - social standards and standards in urban planning, architecture, construction, *Urban Development* **4** pp. 16-19

[3] Ilyichev V, Emelyanov S, Kolchunov V, Bakaeva N 2014 Innovative practices in cities and the doctrine of town planning, *Biosphere compatibility: people, region, technology* **3** pp. 3-18

[4] Abreu-Harbich L V, Labaki L C and Matzarakis A 2014 Thermal bioclimate in idealized urban street canyons in Campinas, Brazil. *Theoretical and Applied Climatology* **115** pp.333–340, DOI 10.1007/s00704-013-0886-0

[5] Sidorova V, Zhivitsa V and Mosyakin D 2019 The substantiation of architectural-planning transformations of urban public spaces with the account of the principles of biopositivity *Journal of Applied Engineering Science* **17** 2

[6] Erell E, Pearlmutter D, Boneh D and Kutiel P B 2013 Effect of high-albedo materials on pedestrian heat stress in urban street canyons. *Urban climate* **10** pp. 367-386

[7] Gómez F et. al. 2008 Vegetation influences on the human thermal comfort in outdoor spaces: criteria for urban planning,The Sustainable City **V** pp. 151-163

[8] Pilipenko O, Skobeleva E and Bulgakov A 2018 Methodology for assessing the comfort of an urban environment in terms of availability analyzing, *Proceedings of the Creative Construction Conference, 30 June - 3 July 2018, Ljubljana, Slovenia*, pp. 914-921. ISBN 978-615-5270-45-1 DOI: 10.3311/CCC2018-119

[9] Isaa N A, Wan Mohda W M N and Salleha S A 2017 The effects of built-up and green areas on the land surface temperature of the Kuala Lumpur city, The International Archives of the Photogrammetry, *Remote Sensing and Spatial Information Sciences XLII-4/W5, October 2017, Kuala Lumpur, Malaysia*, pp. 107-112

[10] Konarska J, Holmer B, Lindberg F and Thorsson S 2015 Influence of vegetation and building geometry on the spatial variations of air temperature and cooling rates in a high-latitude city, *International Journal of Climatology* DOI: 10.1002/joc.4502

[11] Zvyagintseva M, Skripkina J and Fedorchenko N 2019 Principles and examples of design of rehabilitation centers in Russia and abroad, *Journal of Applied Engineering Science* **17** 4, 646, 567 - 570 doi:10.5937/jaes17-23722

[12] Lin T P and Matzarakis A 2008 Tourism climate and thermal comfort in Sun Moon Lake, Taiwan *Int. J. Biometeorology* **52** pp. 281–290

[13] Martins T, Adolphe L and Krause C 2012 Microclimate Effects of Urban Geometry on Outdoor Thermal Comfort in the Brazilian Tropical Semi-arid Climate, PLEA2012 - 28th Conference, Opportunities, *Limits & Needs Towards an environmentally responsible architecture Lima*, Perú 7-9 November

[14] Moser A, Uhl E, Rötzer T, Biber P, Dahlhausen J, Lefer B and Pretzsch H 2017 Effects of Climate and the Urban Heat Island Effect on Urban Tree Growth, *Houston Open Journal of Forestry* **7**, pp. 428-445

[15] Shashua-Bar L, Pearlmutter D and Erell E 2011 The influence of trees and grass on outdoor thermal comfort in a hot-arid environment, *International Journal of Climatology* **31**, pp. 1498–1506

[16] Yang C, He X, Lingxue Y, Yang J, Yan F, Bu K, Chang L and Zhang S 2017 The Cooling Effect of Urban Parks and Its Monthly Variations in a Snow Climate City, *Remote Sens.* **9**, pp. 1066

[17] Skobeleva E 2015 Analysis of proposals for assessing the microclimate environmentally safe and comfortable urban environment, *Construction and reconstruction*, **4**, pp. 123-132

[18] Ketterer C and Matzarakis A 2014 Comparison of different methods for the assessment of the urban heat island in Stuttgart, Germany, *Int. J. Biometeorology,* **59**, pp. 1299-1309.

[19] Höppe P 1999 The physiological equivalent temperature – a universal index for the biometeorological assessment of the thermal environment *Int. J. Biometeorology* **43** pp. 71-75

[20] Matzarakis A and Mayer H 1996 Another kind of environmental stress: thermal stress, *WHO Collaborating Centre for Air Quality Management and Air Pollution Control* **18**, pp. 7-10

[21] Matzarakis A, Rutz F and Mayer H 2010 Modelling radiation fluxes in simple and complex environments: basics of the RayMan model, *International Journal of Biometeorology*, No. **54** (2), pp. 131–139

[22] Matzarakis A and Rutz F 2007 Rayman: a tool for tourism and applied climatology, *Developments in Tourism Climatology*, pp. 129-138

[23] Matzarakis A, Mayer H and Iziomon M G 1999 Heat stress in Greece. Applications of a universal thermal index: physiological equivalent temperature *International Journal of Biometeorology*, **43** pp. 76–84

[24] Pearlmutter D J and Garb Y 2014 The relationship between bioclimatic thermal stress and subjective thermal sensation in pedestrian spaces *International Journal of Biometeorology*, No. **8** 10, pp. 2111-27

[25] Danilova N 1980 Climate and rest in our country, Moscow, 155 p

[26] Pilipenko O and Skobeleva E 2016 Determination of Thermal Comfort Criteria Person as Indicators of Environmental Safety of Open Urban Space. Biosphere Compatibility: Human, Region *Technologies* **1**, pp.12-21

[27] Svenson M 2004 Sky view factor analysis - implications for urban air temperature differences, *Meteorological Applications* **11**, pp. 201-211

IOP Conf. Series: Materials Science and Engineering 1138 (2021) 012010 doi:10.1088/1757-899X/1138/1/012010

Multi-criteria optimization for a residential horizontal ground heat exchanger

A M Bulmez[1]*, V Ciofoaia[1], C Pleșcan[1], S I Bolocan[1] and A Șerban[2]

[1] Department of Building Services, Faculty of Civil Engineering, Transilvania University of Brasov, Romania

[2] Thermotechnics, engines, thermal and refrigeration equipment Department, Faculty of Mechanical Engineering and Mechatronics, University Politehnica of Bucharest, Bucharest, Romania

* alexandru.bulmez@unitbv.ro

Abstract. Ground source heat pumps (GSHP) are the most efficient heat pump systems currently in use for residential sector today. Although these systems outperform all other heat pump systems when considering the energy source, the installation costs represent an ongoing issue. For the market to increase in this area, multi-criteria optimization for the ground heat exchanger (GHE) is the main research topic today. The main objective of this study is to search ways to optimize the horizontal ground heat exchanger focusing on two distinct viewpoints: physical and energetic optimization. To achieve this goal, the entropy generation number and the coefficient of performance based on several configuration are studied. Using the Pareto optimal set obtained from two objectives provides useful information about the design optimization. In comparison with traditional single objective optimization, multi-criteria optimization avoids cost increase when regarding only the performance or performance decrease when regarding only the cost. The results indicate that the use of a cheaper heat source such as solar panels to mitigate the energy needed for peak loads during the heating and cooling systems can reduce the horizontal ground heat exchanger size and therefore the installation cost. A hybrid solar assisted ground source heat pump system, using more than one renewable energy source, shows a much higher flexibility for residential purposes than a heat pump system that uses only one renewable energy source.

1. Introduction

Energy demand continues to increase as the population grows, with an undesirable effect on the Earth's climate. Around 81% of the electrical energy generated on Earth is obtained from fossil fuels [1]. Renewable energy sources are commonly used today, especially from sources that store energy more efficient [2]. The focus today is on researching efficient ways for reducing the greenhouse gas emissions, such as in [3, 4], by developing more energy efficient systems.

Ground-coupled heat pumps are considered efficient systems that use geothermal energy [5]. Their usage has increased especially in the residential and commercial building sectors [6]. Recently several optimization researches have been performed for horizontal geothermal heat exchangers [7], which focused on the optimization using a single criterion. Single criteria optimization usually results in favorizing a parameter in the detriment of another. The purpose of this study is a multi-criteria optimization for HGHEs.

2. Materials and procedures

To solve these problems for heat pump systems, a multicriteria optimization is required. Establishing the parameters that have the most impact on the performances of the HGHE is required. Thus, three types of parameters were considered: physical, operational, and energetic. Limit intervals were set for this to narrow the results as can be seen in table 1.

Table 1. Parameters considered for multi-criteria optimization.

Type	Parameter	Limit
Physical	HGHE Surface	20÷100 [m^2]
	Pipe depth	-1 ÷ -3 [m]
	Pipe step	0.3 ÷ 1 [m]
	Pipe length	50 ÷ 200 [m]
	Pipe diameter	0.02 ÷ 0.032 [m]
	Soil thermal conductivity	0.5 ÷ 2.5 [W/mK]
Operational	Fluid flow	0,1 ÷ 1 [kg/s]
	Soil temperature with depth	-10 ÷ 30 [°C]
	Exterior temperature	-30 ÷ 30 [°C]
	Fluid temperature	-5 ÷ 40 [°C]
Energetic	Thermal energy extracted/supplied/restored	variable [kWh]

Optimization problems that consider multiple criteria simultaneously provide reliable results. However, not even this optimization strategy can offer the best optimization solution, because some criteria usually cannot be compared. Vilfredo Pareto's research [8] shows us that it is impossible to optimize all parameters simultaneously, because the improvement of one parameter will lead to the worsening of another.

By using several optimization criteria, a set of solutions is searched, that provides a reasonable compromise for several parameters, including the best solution of each parameter. The solutions can be found on the curve called the Pareto Front, among which the best solution resides. To obtain reliable results, in this study two criteria were used. Firstly, the Entropy Generation Number criteria (EGN) that determines the entropy of the system. Secondly, the Coefficient of Performance (COP) was used as a criterion to consider as many parameters as possible.

3. Entropy generation number criteria

Geothermal heat exchangers use the earth soil as a source of renewable energy. Thermodynamic irreversibility is an important criterion when regarding the overall operation of the heat pump system. Entropy, as the unit that measures the thermodynamic irreversibility, can provide a useful view on how the heat pump system impacts the earth.

Assuming this, the calculation of the entropy according to the average temperature of the fluid inside it and the cost of electrical energy in operation offers a perspective on the energy optimization that can be achieved. This optimization criteria is characterized by the function with the same name and is calculated with the formula:

$$N_s = \frac{S_{gen} \times T_{f,a}}{Q} \tag{1}$$

In which, S_{gen} represents the generated entropy, $T_{f,a}$ represents the average fluid temperature and Q represents the energy provided by the HGHE. The average temperature of the thermal fluid is directly influenced by the soil temperature, which is approximately equal to the outlet temperature. To increase this temperature external energy must be used. The cost for the electricity needed to increase the fluid temperature based on the monthly cost in Romania was calculated using the formula:

$$W_e = Q_c \times w_1 \times n_h \tag{2}$$

In which, W represents the electricity cost, w_l the cost for 1 kWh of electricity, Q_c represents the heat flux needed to increase the fluid temperature and n_h represents the average monthly operating hours. In figure 1, the results of all EGN solutions are illustrated based on inlet fluid temperature, average fluid temperature and electrical energy cost.

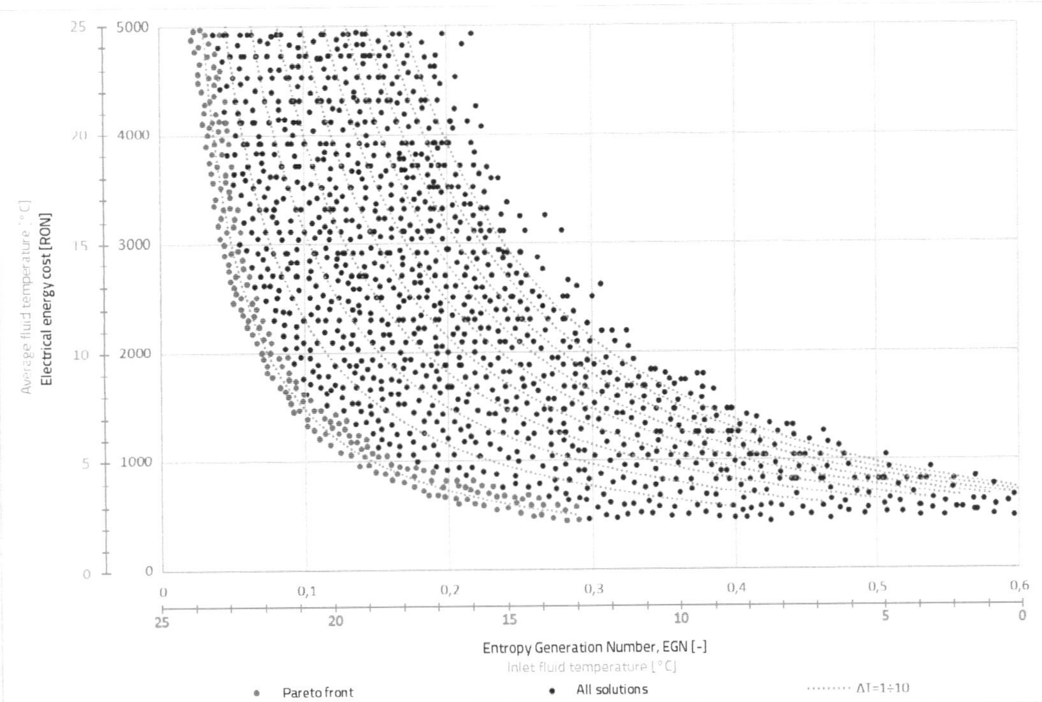

Figure 1. Solutions for the Entropy Generation Number for HGHE.

To highlight only the Pareto front optimum solutions, the rest of the solutions were eliminated as can be seen in figure 2. The optimum desired point is the closest one to 0.

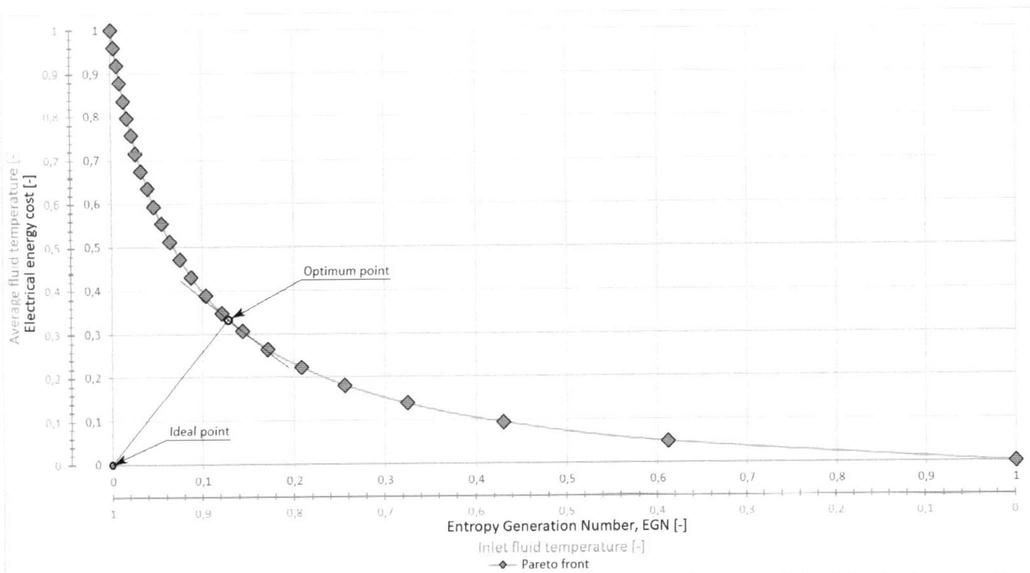

Figure 2. Pareto front of the Entropy Generation Number for HGHE.

4. Coefficient of performance criteria

In use, heat pump systems provide thermal energy at the expense of electrical energy. The ratio between the supplied energy and the energy that is used is called the coefficient of performance. For heat pump systems the COP represents an indicator of its overall performance, meaning that when the ratio has a higher value the system is more efficient and when the ratio has a lower value it is less efficient.

With the COP as a base point, this criteria purpose is to compare the thermal performances generated by a large set of HGHE configurations, for which the COP will be established based on the thermal output of the ground heat exchanger. Afterwards, the installation cost for all these configurations was established, based on average excavation prices, pipe prices and workmanship prices that are common today in Romania.

In order to establish the COP of the large set of configurations the GLHEPro Software was used to simulate the thermal performances of all configurations. After analyzing all the important parameters shown in Table 1, the configuration number was reduced based on common HGHE configurations. This led to configurations for three different diameters at pipe steps between 30cm and 100cm and HGHE surfaces between 20m^2 and 100m^2. The burial depth of the pipes was also considered at first, between 1 and 3m. The results were very similar and so this parameter was considered neglectable.

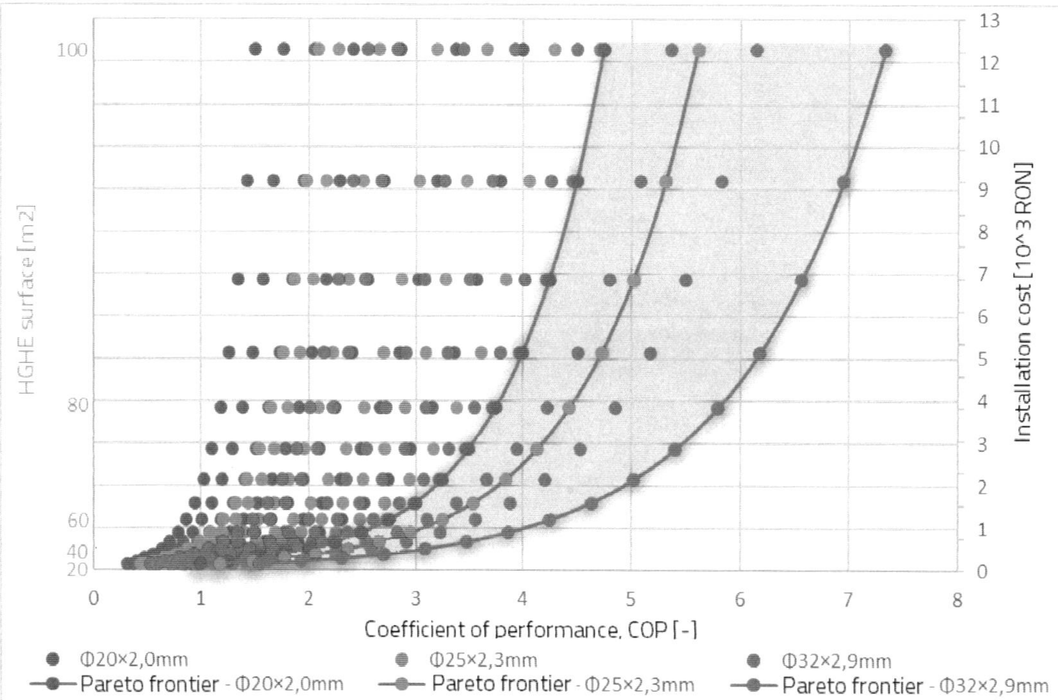

Figure 3. Pareto fronts of the Coefficient of Performance criteria for HGHE with 3 pipe sizes.

In figure 3 all the results simulated and calculated are presented. As this figure shows, the higher the HGHE surface and pipe step the better the COP is, but the installation cost increases. With all the results available and represented in a chart, the Pareto frontier was chosen and corresponds as the highlighted area between results for the 20 mm and 32 mm pipes for the pipe step of 30 cm. It is clear that in order to achieve better COP a higher surface is needed.

In figure 4 the solution outside the Pareto frontier were eliminated. Because all three different pipe diameters offered good results, it was decided that three Pareto frontiers will be represented, one for each diameter. Even if their results are close, the difference is high enough such as pipe diameter must be taken into consideration when designing the HGHE.

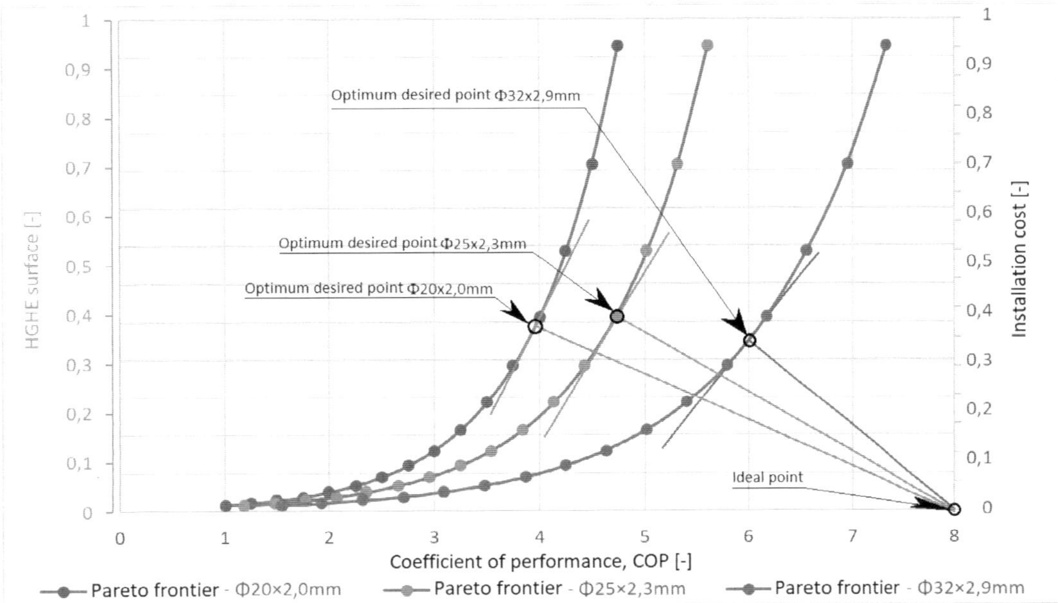

Figure 4. Pareto fronts of the Coefficient of Performance criteria for HGHE with 3 pipe sizes.

5. Discussion

After both criteria were calculated the decision-making process was made. The main topics considered were the reducing of the total entropy, increasing the heat pump performance and reducing the overall installation and exploitation costs. When accounting for all of those, it is difficult to find the optimum desired point and difficult to comprehend how the parameters can be optimized when they affect each other. In figure 5 all Pareto frontiers were overlayed according to the HGHE thermal output and both criteria costs. The optimum EGN point corresponds for a HGHE with a thermal output of approximate 700 W while the optimum COP points correspond to approximate 1900 W, 2400 W, and 3000 W respectively.

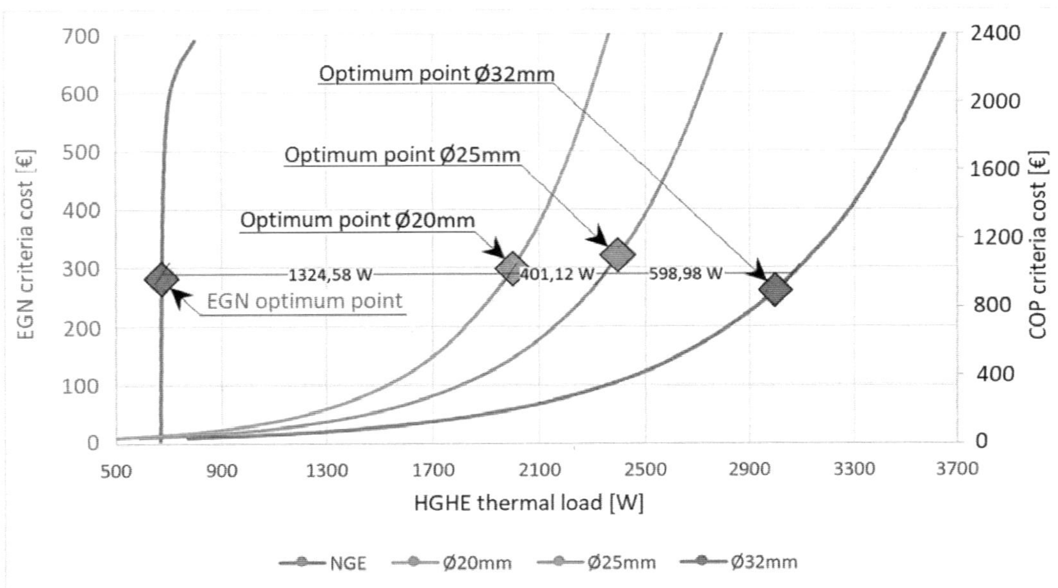

Figure 5. Optimal points from the EGN and COP criteria.

The first criteria, EGN, provides excellent results for reducing thermal irreversibility and maintaining thermal equilibrium in the HGHE but it does not consider the whole system and the thermal loads variation during the year. This means that in order to reduce the total entropy of the system, the HGHE configuration must be as small as possible, but this impacts the thermal output and reduces the heat pump system COP.

The second criteria, COP, provides very good results for choosing the optimal physical configuration, but does not consider both the thermal irreversibility and available HGHE install space. This means that in order for the heat pump system to use less electrical energy, which also translates to less amount of greenhouse gas emissions, the entropy must be neglected.

Therefore, analyzing the results from the two criteria used, the optimal points of the physical configuration of the heat exchanger can be achieved, while maintaining the optimal entropy generation only by providing external energy. This can be achieved by combining the geothermal energy source with another energy source such as solar energy or taking into consideration residual thermal energy from building basements or foundations.

6. Conclusions

Analyzing the results obtained with the two optimization criteria, it was observed that the optimal points, although they are the best solutions, have several disadvantages. The lack of space is one of the biggest disadvantages for physical parameters. Also, limiting the thermal load of the system in terms of thermal irreversibility is a disadvantage, especially when the install space is sparce. The pipe diameter and the pipe step have an important impact, representing the largest share of the thermal load of the HGHE, followed by the material-based parameters. It was also found that the installment depth is not an important parameter, the difference in thermal load of HGHE being insignificant. This aspect being due to the large temperature variation at the soil surface, up to a depth of 7-10 m. The main conclusion of the study is that to optimize the thermal performances of HGHEs, external energy is needed to be supplied to the system from auxiliary sources. The impact of an auxiliary source consists in providing help during peak loads and maintaining thermal equilibrium inside the HGHE.

References

[1] Dragomir G, Şerban A, Năstase G, Brezeanu A I 2016 Wind energy in Romania: A review from 2009 to 2016, *Renewable and Sustainable Energy Reviews*, Romania

[2] Năstase G, Şerban A, Năstase A F, Dragomir G, Brezeanu A I, Iordan N F 2017 Hydropower development in Romania A review from its beginnings to the present, *Renewable and Sustainable Energy Reviews*, Romania

[3] Năstase G, Şerban A, Năstase A F, Dragomir G, Brezeanu A I 2018 Air quality, primary air pollutants and ambient concentrations inventory for Romania, *Atmospheric Environment*, ISSN 1352-2310, pp 16;

[4] Năstase G, Şerban A, Dragomir G, Bolocan S, Brezeanu A I 2016 Box window double skin façade Steady state heat transfer model proposal for energetic audits, *Energy and Buildings*, TMREES15, Romania, pp 4;

[5] Bolocan S, Chiriac F, Şerban A, Dragomir G, Năstase G 2015 Development of a small capacity solar cooling absorption plant, International Conference on Technologies and Materials for Renewable Energy, *Environment and Sustainability*, TMREES15, Romania

[6] Sarbu I and Sebarchievici C 2015 "General review of ground-source heat pump systems for heating and cooling of buildings," *Energy Build*

[7] Congedo P M, Colangelo G, and Starace G 2002 CFD simulations of horizontal ground heat exchangers: A comparison among different configurations *Appl Therm*

[8] M Ehrgott 2005, Multicriteria Optimization Berlin, *Heidelberg: Springer-Verlag*

Investigation of HGHE near building basement – a case study

A M Bulmez[1]*, V Ciofoaia[1] and I Boian[1]

[1] Department of Building Services, Faculty of Civil Engineering, Transilvania University of Brasov, Romania

* alexandru.bulmez@unitbv.ro

Abstract. Ground heat exchangers represent the barrier between installing a ground source heat pump (GSHP) or a heat pump that use another source as water or air. While the vertical ground heat exchangers (VGHE) yield better results than horizontal ground heat exchangers (HGHE), the installation cost is higher than HGHE. Although, HGHE installation cost is a lot more affordable for residential use, it has the disadvantage of requiring a large installation area and lower efficiency than VGHE due to the temperature variation at the surface of the ground throughout the year. This study focuses on investigating possible use of the energy that is lost through the ground from basements in general, to help the heat pump when the HGHE is installed in the proximity of the basement. In order to achieve concluding results, an experiment was conducted at the University of Transilvania from Brașov, at the Faculty of Civil Engineering. In the experiment the HGHE was installed next to the basement exterior wall. In addition, a CFD numerical study was conducted. Firstly, to verify the experimental measurements and secondly, to obtain results with different working parameters. The results indicate that the thermal insulation degree of the basement wall and floor is an important parameter when considering heat loss into the ground. Likewise, the thermal state of the basement, heated or unheated, contributes considerably to the final results. Consequently, positioning the HGHE in the proximity of basements can represent an energetic optimization, especially when the basement is heated and it isn't thermal insulated, as can be the case for an existing building.

1. Introduction

Ground source heat pump (GSHP) systems are becoming more and more attractive as time goes by. As of today, approximate 60% energy consumption in buildings is used for heating, cooling and domestic hot water production [1]. In Romania, almost half of the electrical energy is produced from renewable sources [2]. From that half, around 70% is hydropower and the rest consists mainly in wind, solar and bio renewable energy [3]. Although energy efficient indoor systems [4, 5] contribute a lot for the total energy consumption, the heat source represents the major factor in renewable energy systems. For tapping into geothermal energy stored into the earth, GSHP systems use geothermal heat exchangers (GHE). These are split into two main groups, open loop and closed loop GHEs. The closed loop GHEs also split into two main branches of vertical and horizontal GHEs. While vertical boreholes are the most efficient solution, the drilling cost represents a barrier even today. Horizontal GHEs are cheaper, but have the disadvantage of lower efficiency due to temperature variation at the surface of the earth.

In order to research possible optimizations for HGHEs, a small experiment was conducted at the Faculty of Civil Engineering at University of Transilvania from Brașov. The HGHE was positioned next to the basement wall in order to take full advantage of the heat loss from the heated basement. Few studies have been made to this point considering building foundations and basements as heat sources such as [6-9], but this field still needs more research.

This study focuses on the heat loss from building basements and how HGHEs can take advantage of that. Many existing buildings today have basements that are not thermally insulated. In the future most of these buildings will be required to have renewable energy systems for heating, cooling and domestic hot water production. This in fact means that if the basement residual heat has an impact in the overall heat pump system efficiency it must be taken into account when designing such systems.

2. Materials and procedures

The experiment conducted is represented by a low capacity heat pump that uses as main renewable energy source the ground. For the source a horizontal GHE with the surface of 25 m^2 was proposed. The HGHE is positioned next to the basement wall. The pipe used is a polyethylene Φ20mm pipe buried at the depth of 1,2 m. At that depth 18 temperature sensors were placed, as seen in figure 1.

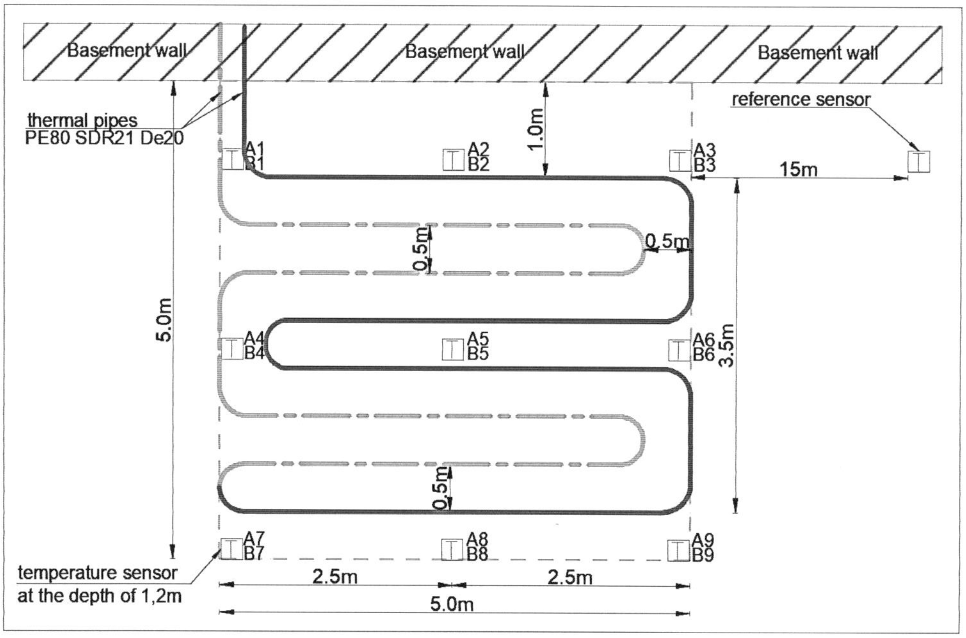

Figure 1. HGHE experiment schematic

Ground temperature at 1,2 m depth, exterior air temperature, interior air temperature, and the heat pump inlet and outlet temperatures were measured for the duration of the experiment. The position of the HGHE was chosen to be next to the heated basement wall in order to use as much energy as possible which is lost from the basement to the ground. The experiment was conducted for the heating season between 15 October 2019 and 15 April 2020, for a duration of a total of 181 days. This experiment was made at a small scale, the heat extracted from the ground was used to heat and maintain the interior comfort temperature for one room. Although solar auxiliary source was also embedded into the experiment, this study focuses on how much energy the basement is losing during the heating season. Similarly, the interest focused on how the heated basement affects the temperature inside the HGHE. In Figure 2 the temperature measurements are represented, along with the reference temperature sensor. It shows that the temperature near the basement wall is higher with around 2-3°C for the duration of the heating season and it lowers as you move away from the basement wall.

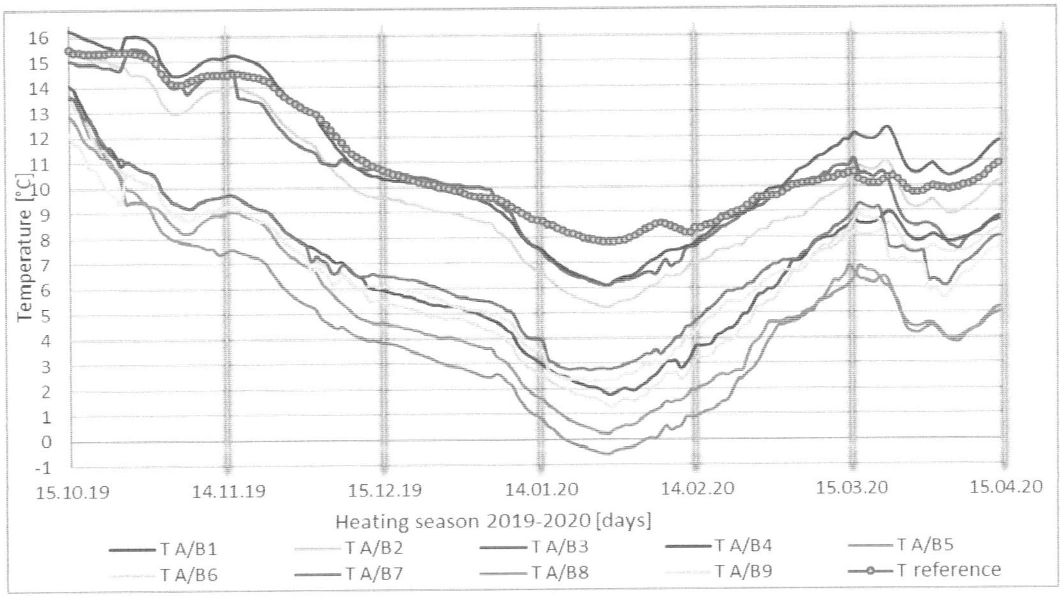

Figure 2. HGHE temperature sensors measurements for the heating season of 2019-2020.

3. Heat loss from the basement

The heated basement can be considered a heat source for HGHEs. The basement room next to the HGHE is the room of the thermal power plant, which operates intermittently during the heating season. To calculate the energy from the heated basement, the standard SR EN ISO 13370/2017 "Thermal performance of buildings. Thermal transfer through the ground. Calculation methods" was used.

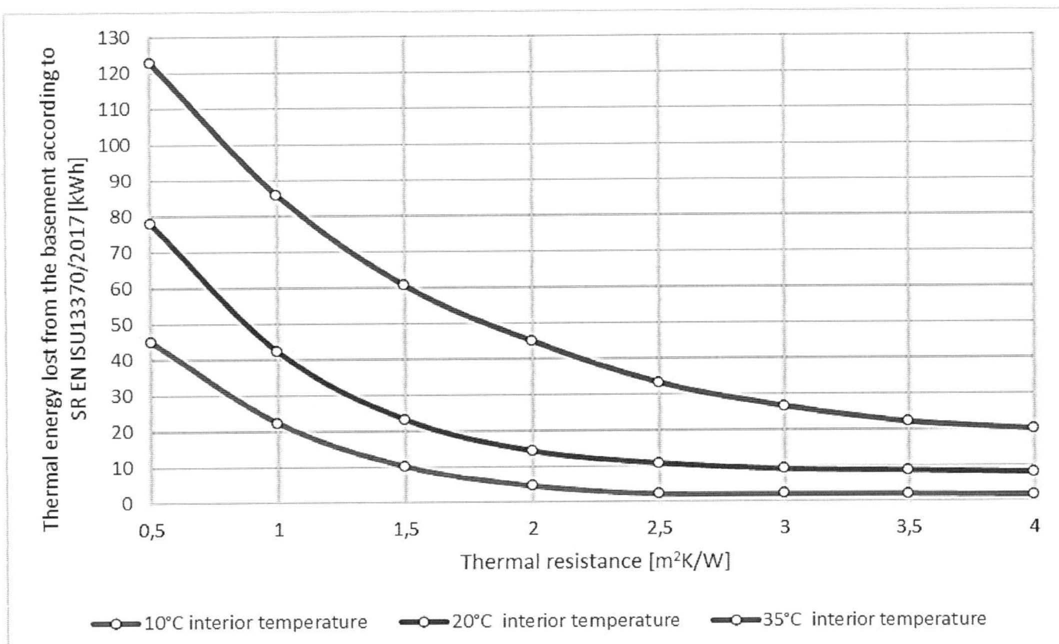

Figure 3. Thermal energy lost from the basement to the surroundings

IOP Conf. Series: Materials Science and Engineering 1138 (2021) 012011 doi:10.1088/1757-899X/1138/1/012011

After that, multiple wall and floor thermal resistances were used, as well as different interior temperatures to obtain the results. The calculation was made for three interior temperatures during the heating season: 10°C, 20°C and 35°C as was the case of the experiment.

Firstly, as seen in Figure 3, the insulation degree of the basement walls and floor represents the most important parameter when considering heat loss. Even if the interior temperature is high, the basements will not lose much heat when properly insulated.

Secondly, the interior temperature has a less important role, as a heated basement has more thermal energy than an unheated one. In the case of the experiment, the average temperature inside the room next to the HGHE is 30-35°C. The calculation results show that 122,75 kWh/heating season are transferred into the ground near that room. The basement wall and floor are not thermal insulated, this being the main reason when the HGHE position was chosen.

4. Numerical simulation results

In the experiment the temperature inside the ground where the HGHE is placed was registered. All the measurements made from all 18 sensors were averaged intro a mean daily temperature for the whole HGHE. To verify the measurements, a virtual model was created in COMSOL Multiphysics in which HGHE temperature was simulated based on experimental system. The conditions used were the same as in the experiment, respectively same outside air temperature, basement room temperature, inlet and outlet temperature as well as pipe geometry, pipe diameter, pipe step and burry depth. Similarly, the numerical simulation was made for the same 181 days period.

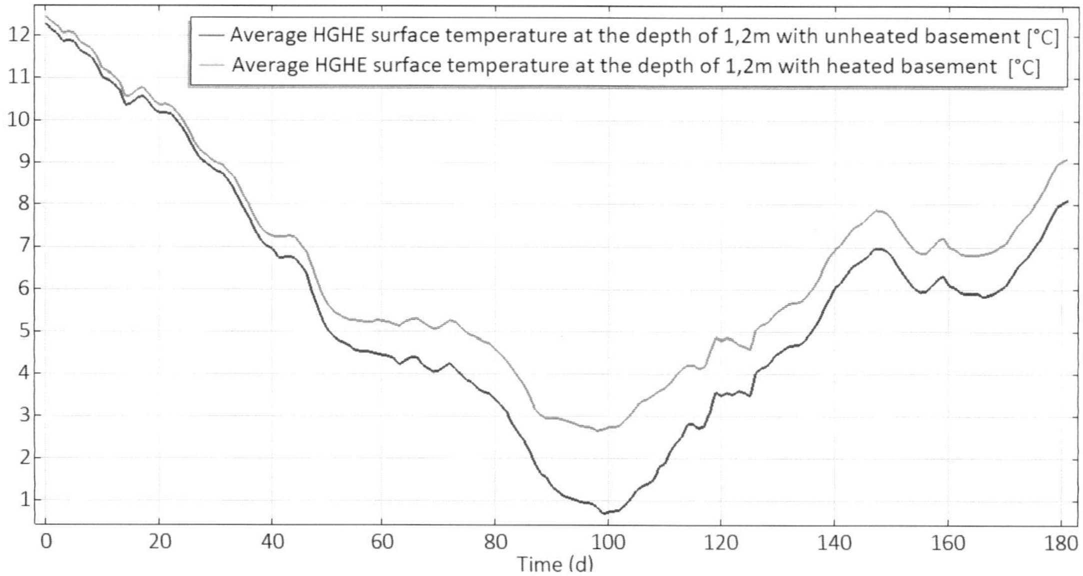

Figure 4. Average HGHE temperature for the experimental case and unheated basement.

In Figure 4, seen with green, is the HGHE average temperature for the heated basement, and the results from the numerical simulation confirm the experimental measurements. Afterwards, a case with the basement being unheated was simulated, showing that the ground cools more during the peak of the heating season. It is interesting to notice that the temperature difference between the two cases increases as time goes by. This reflects the thermal processes that occurs inside and at the surface of the earth. When the exterior temperature drops, more heat is needed for the building but also the ground surface loses more heat in the environment. The heat pump extracts more heat from the HGHE. While the basement provides a good heat buffer for high temperature variations, the heated or unheated state of it has an important role.

5. Discussion

Based on the results that validated the experimental measurements, more cases were simulated, in order to better understand the temperature variation with the use of auxiliary and residual heat sources.

Firstly, a case was simulated that treated the basement as unheated with an average temperature of 10°C inside. This case showed a gradual increase in the temperature difference as time went by and can be seen in both Figure 4 and 5. The temperature difference at the season peak between the experimental case and this simulated one was of approximate 2°C.

Secondly, a case was simulated that eliminated the thermal energy supplied to HGHE from the solar source, but had the temperature input from the heated basement, as the experimental case. This simulation showed further decrease in temperature in the season peak with the difference between it and the experimental case being of approximate 3,5°C.

Thirdly, a case was simulated that eliminated all heat sources, solar and heated basement. The trend of gradual temperature difference increase was confirmed for this case as well. This case showed how the temperature inside the HGHE would be if the system was a regular one, with no additional heat source. The peak temperature difference reached almost 5-6°C.

From analyzing the results, it can be seen that the heated basement has an important role in reducing the ground freeze period from approximate 4 weeks to 3 weeks during the season peak. Similarly, the solar source reduced this period to 2 weeks and both additional heat sources eliminated the freeze period completely. This aspect is important because the ground freeze represents a well-known issue for horizontal ground heat exchangers that can impact other systems nearby such as water or sewerage.

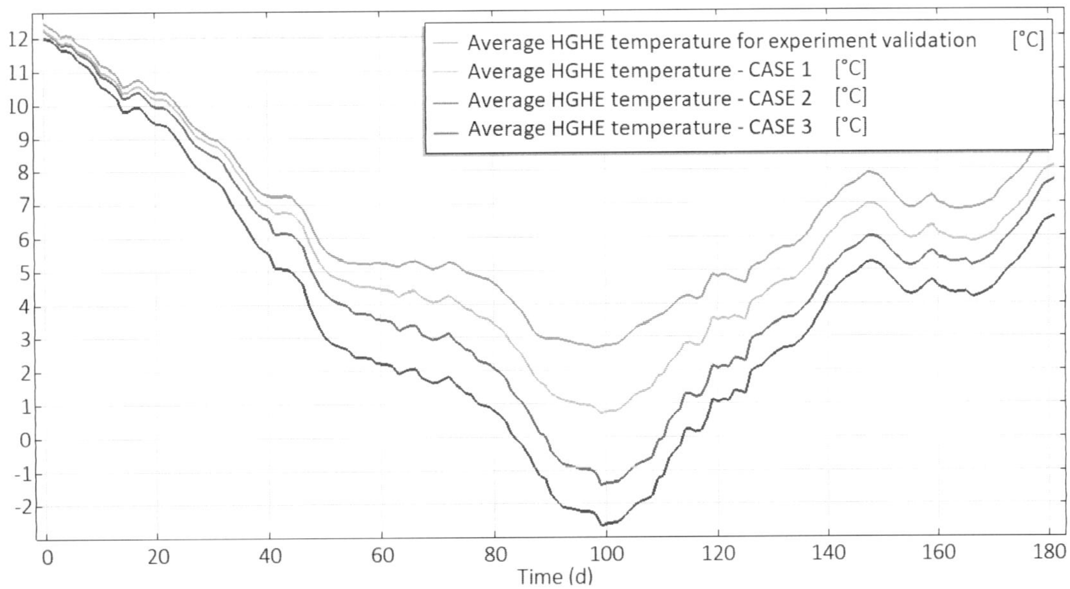

Figure 5. Average HGHE temperature for experimental case and three simulated cases

Another interesting fact that the numerical simulation showed that the temperature inside the HGHE at the end of the season is not the same for the different simulated cases. According to this results, the additional heat sources can maintain a lower temperature variation inside the HGHE. Also, it means that if the thermal energy supplied to the HGHE during the cooling season is not enough, in time the ground will be colder and colder at the start of the heating season, thus reducing the heat pump system performance.

6. Conclusions

The optimization of HGHEs is necessary today for the spread of GSHP systems in order to reduce the greenhouse gas emissions. After analyzing the results several conclusions were established:

- overall conclusion is that the heated basements can be a reliable heat source for the HGHEs, even if it is a residual source;
- the most important parameters for the basement to be a reliable source are a high inside temperature and a low thermal resistance;
- the experimental measurements show that the heated basement rivals the solar panels as auxiliary heat source for the HGHE and the GSHP system overall;
- the solar panels contributed with an approximate 3°C average temperature difference during season peak and reduced the freezing period with approximative 2 weeks;
- the heated basement accounted for approximate 2°C average temperature difference during the season peak and reduced the freezing period from approximate 4 weeks to approximate 3 weeks;
- using both heat sources completely eliminated the ground freezing period during the season peak.

This implies that for existing buildings where space is an issue and the existing basements are not thermal insulated, positioning HGHEs next to them need to be considered. Therefore, as temperature variation is a well-known issue when designing HGHEs, basements provide a reliable heat source towards increased efficiency and thermal equilibrium.

Since HGHE design still base today on empirical results from previously installed heat pump systems, there is untapped potential from using additional heat sources. This can be successfully applied to either prevent the heat pump performance drop or reduce the temperature variation inside the HGHE during the heating season.

References

[1] Bolocan S, Chiriac F, Şerban A, Dragomir G and Năstase G 2015 Development of a small capacity solar cooling absorption plant, International Conference on Technologies and Materials for *Renewable Energy, Environment and Sustainability*, TMREES15, Romania

[2] Năstase G, Şerban A, Năstase A F, Dragomir G, Brezeanu A I and Iordan N F 2017, Hydropower development in Romania A review from its beginnings to the present, *Renewable and Sustainable Energy Reviews*, Romania

[3] Dragomir G, Şerban A, Năstase G and Brezeanu A I, Wind energy in Romania: A review from 2009 to 2016 2016, *Renewable and Sustainable Energy Reviews*, Romania

[4] Năstase G, Şerban A, Năstase A F, Dragomir G and Brezeanu A I 2018 Air quality, primary air pollutants and ambient concentrations inventory for Romania, *Atmospheric Environment*, Available online 23 April 2018, ISSN 1352-2310, pp 16

[5] Năstase G, Şerban A, Dragomir G, Bolocan S, Brezeanu A I 2016 Box window double skin façade Steady state heat transfer model proposal for energetic audits, *Energy and Buildings*, TMREES15, Romania pp 4;

[6] Di A, Loveridge F, M Piemontese, and Barla M, Geomechanics for Energy and the Environment The role of ground conditions on the heat exchange potential of energy walls, *Geomech Energy Environ*;

[7] Kayaci N and Demir H 2020 Geothermics Comparative performance analysis of building foundation Ground heat exchanger *Geothermics*

[8] Nam Y and Chae H, Numerical simulation for the optimum design of ground source heat pump system using building foundation as horizontal heat exchanger *Energy* **73**;

[9] Kayaci N, Demir H, Kanbur B, Ozgur Ş, and Agra O 2018 Experimental and numerical investigation of ground heat exchangers in the building foundation **188**

IOP Conf. Series: Materials Science and Engineering 1138 (2021) 012012 doi:10.1088/1757-899X/1138/1/012012

A review of flash floods in the Dobrogea Region

N Buzgaru[1,2*] **and C Maftei**[3]

[1] Ovidius University of Constanta, Doctoral School of Applied Sciences, Romania
[2] Dobrogea-Litoral Water Basin Administration, Constanta, Romania
[3] Transilvania University of Brasov, Faculty of Civil Engineering, Romania

* nicusor.buzgaru@abadl-rowater.ro

Abstract. The last decade is characterized by an increase in the occurrence of flash floods, fact which is confirmed by the worldwide endeavors undertaken by the WMO (World Meteorological Association) which initiated the signing of a series of memorandums for collaboration with other worldwide institutions meant to lead to the elaboration and implementing of a complex flash flood warning system. The aim of this paper is to analyze the level of implementation of said system in Europe and Romania and the analysis of the flash flood occurrence, specifically in the Dobrogea region of Romania.

1. Introduction

It has become clear that climate change is influencing weather patterns all over the world. Although the impact is not equally distributed or similar everywhere, the effects are noticeable. One of the ways climate change manifests is a change of the rainfall distribution, with vast areas being affected by droughts while others get much more than the average annual rainfall. This is not something we cannot understand or estimate, but the biggest issue is that all previously recorded data is becoming less relevant or reliable as known patterns are becoming no-longer applicable and new issues appear.

This is not an issue for modern weather forecasting since they do not rely that heavily on patterns and have the ability to analyze weather systems in real time based on radar and satellite data, but the same cannot be said about flood protection. Classic flood management design is based on annual average flows in a section of the river bed, and annual average flows are based on previously recorded data. So, if one would be working on the design of a specific river bed calibration, the norms would tell him that he needs to work, starting from the average multi-annual flow, determining the flood return probability (for which the standard across the EU is 1% in localities). The issue here is that the determination of said flood return probability, although mathematically correct, in the context of changing weather patterns, the results will be erroneous for present-day use, and what would have been 1% flood return probability in the past, is no-longer valid. (One could use historical maximums as a guide, but that data is not necessarily available everywhere, depending on whether or not there has been monitoring and recording of the watercourse levels and related rainfall.)

At the same time, human development within the basins is proving to be an important influence in the behavior of rain water that runs off the terrain adding another level of modification of the previously recorded data and thus making it less accurate. Deforestation is a major factor in runoff, but other modifications of parts of the basin are equally important for the changing of runoff behavior, such as landscape management (changing perennial vegetation to grass lawns) or transport infrastructure (concrete/asphalt roads) [1] which may have a strong influence on runoff (concentration times of the flood wave)

All of the above leads to an increase in the occurrence of flash floods or the situations where rainfall is characterized by relatively large flows over short periods of time, causing all sorts of problems in the basin. Flash floods are becoming more and more common in relatively small basins throughout the world, and their parameters vary to such an extent that in some cases even relatively low rainfalls (10 - 20 mm) may cause issues if it rains in a short enough interval. As a reference, most of the flood damage and material losses in the Dobrogea region (Romania) was caused by flash floods which, considering the local geography and thus the typical geometry of the basins combined with the specific vegetation (mostly steppe and shrubs, very little forest coverage), results in a particular flooding pattern. In the past years, most flood damage in Dobrogea was due to runoff concentrated on the slopes, or activation of dry beds (torrents) rather than overflowing of a main (permanent) watercourse.

An analysis of the official flood data for the Constanta County shows that in the past 7 years (2014 - 2020) only 7.22% of total registered flood damage was due to overflowing of rivers, and that is mostly due to the river Danube causing minor floods in 2015. In the last 5 years (2016 - 2020) only 0.89% of total flood damage is due to overflowing of rivers, the rest is all due to flash floods and runoff. At the same time, 100% of flooded homes are due to runoff and not overflowing of rivers. This shows that the riverbed management is relatively efficient, given the weather patterns, and only runoff on the slopes and dry riverbeds (torrents) is problematic and needs to be studied further and managed better. It is thus important to properly understand methods for evaluation of flash floods and the way they adapt to different scenarios in order to get the most accurate results.

2. Description of flash floods

Flash floods are described as flooding that takes place in a short amount of time so short that it may present no warning. It may be the case of heavy or moderate rains depending on the time interval but also the case of broken dams or melting ice or snow. Generally, the accepted notion is that of any flooding that takes less than 6 hours to manifest from the initial cause (rainfall, etc.) to the flooding effects. Flash floods can manifest locally, or where the generating factors take place, or it can manifest downstream as runoff reaches the riverbed naturally or by means of drains and ditches, cumulating to a flow that surpasses the riverbanks and floods the downstream sector. Worst case scenario are natural Flash floods encountered in areas with canyons, where water level may rise suddenly and without the giveaway signs of local rain because the actual rainfall took place several kilometers away (upstream). Similarly, dangerous flash floods are those generated by human objectives such as the breach of a dam which will suddenly release a vast quantity of water in the downstream sector. Of course, there can be mixt scenarios, such as heavy rainfall causing human objectives to fail thus contributing to the creation of a flood wave. One such example in the Dobrogea Region is the flooding of Costinesti/Schitu in 2005. In the context of unusually heavy rains (over 300 mm) a railway embankment acted as a dam collecting water behind it until the embankment gave way to the water pressure and everything downstream was washed by a 2m high flood wave that destroyed homes and other buildings and breached its way to the Black Sea by means of a huge ravine in the beach.

Because heavy rains in a short time period is a characteristic associated with convective rain, worldwide they are implicitly attributed to hurricanes or other types of tropical storms. Still, although tropical storms imply convective rain, that does not mean it is the only possible scenario and convective rain does occur elsewhere.

3. Impact

The impact of flooding is a well-known fact with two major components, respectively the environmental impact and impact on human habitation. Floods can cause severe disturbance in ecosystems or in some cases it can entirely destroy local ecosystems.

Flash floods are often associated with high velocity of water flow which tends to destroy vegetation, carrying logs, mud and sometimes boulders downstream, as well as other debris, and depositing them all around the path of the flooding. This can lead to the destruction of certain specialized species that develop in the floodplain area causing imbalances in the ecosystem. It can take years or decades for the ecosystem to recover and some species might be lost altogether either completely destroyed by the flood, or by the modification of the riverbed (modification of spawning grounds for fish, etc).

The impact on human habitation is also severe leading to loss of property, loss of revenue or famine due to destruction of crops, and worst-case impact results in loss of life. From the loss of life perspective, flash floods are more likely to be associated with casualties due to the speed of manifestation of the flood effects which does not allow enough time for flood defense or evacuation measures to be put in place and sometimes victims don`t even realize the immediate danger.

This study is focused on the evolution of the flash flood in Dobrogea region. Dobrogea is located in south-east of Romania between the lower Danube and Black Sea (figure 1).

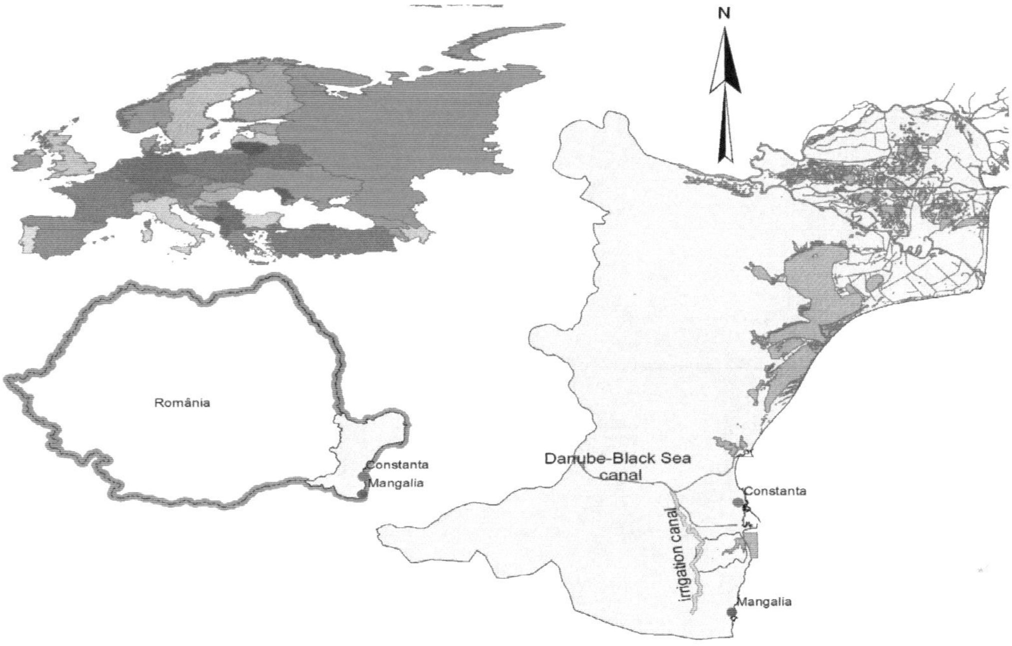

Figure 1. Dobrogea region [2].

From the climatic point of view, Dobrogea region could be divided in two regions [3]: (i) the Eastern region (20 – 50 km on the littoral) which comprises the territory of Danube Delta, its south and the Razim Sinoe lagoons and; (ii) the western region. The climate of first region is continental influenced by the Black Sea and the second one has a temperate continental climate. The average temperature decreases from the Black Sea and Danube towards the interior: the temperature on the coast varies between 11.7°C (Constanta) - 11.6°C (Mangalia) and 9.9°C (Corugea – in interior of Dobrogea). The annual precipitation varies in wide limits (260 – 500 mm approximately), the highest values being registered in the North and center of the region [4]. This area is subject to frequent droughts since it`s average precipitation is much lower than the country average of 600 – 800 mm. The Black Sea presents a strong influence on the air humidity throughout the Dobrogea region, but strongest within the first 15 - 25 km from the shore.

The hydrographic network consists of rivers that are generally under 50 km long with very low multi-annual mean flows of 0.43 m^3/s (Telita) to 0.64 m^3/s (Casimcea). The area, especially in the central and southern sector is littered with hills and hundreds of dry valleys, and generally little forest coverage.

An analysis of the County Operational Reports Regarding Dangerous Meteorological Phenomena [5] for the Constanta County shows that in the past 7 years (2014 - 2020) only 7.22% of total registered flood damage was due to overflowing of rivers, and that is mostly due to the river Danube causing minor floods in 2015. In the last 5 years (2016 - 2020) only 0.89% of total flood damage is due to overflowing of rivers, the rest is all due to flash floods and runoff. At the same time, 100% of flooded homes are due to runoff and not overflowing of rivers. This shows that the riverbed management is relatively efficient, given the weather patterns, and only runoff on the slopes and dry riverbeds (torrents) is problematic and needs to be studied further and managed better. It is thus important to properly understand methods for evaluation of flash floods and the way they adapt to different scenarios in order to get the most accurate results.

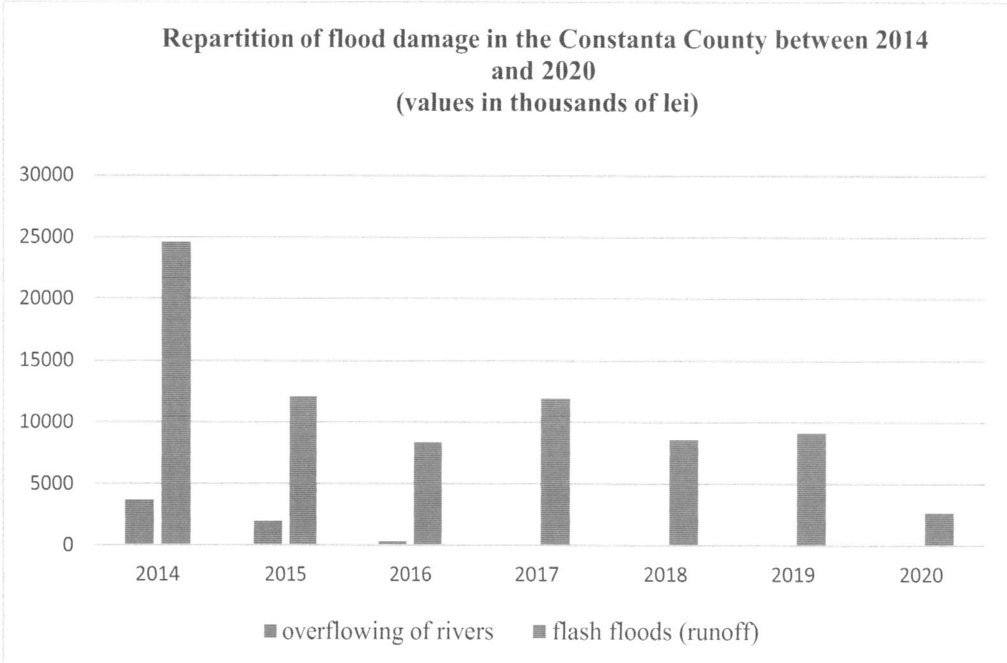

Figure 2. Repartition of flood damage in the Constanta county (2014 – 2020).

Total recorded damages due to flash floods in the Constanta county, for the specified interval was 77.22 million lei. The precipitation context was a normal one, without many record rains being registered in this interval, with one notable exception in October 2015 when the locality of Corbu was affected by flash floods caused by a rainfall of 122 mm in about 3 hours (175.8 mm in the 24h interval). The interesting characteristic of this particular flash flood is that the intense rain only manifested over a small area of about 10 km^2 [6].

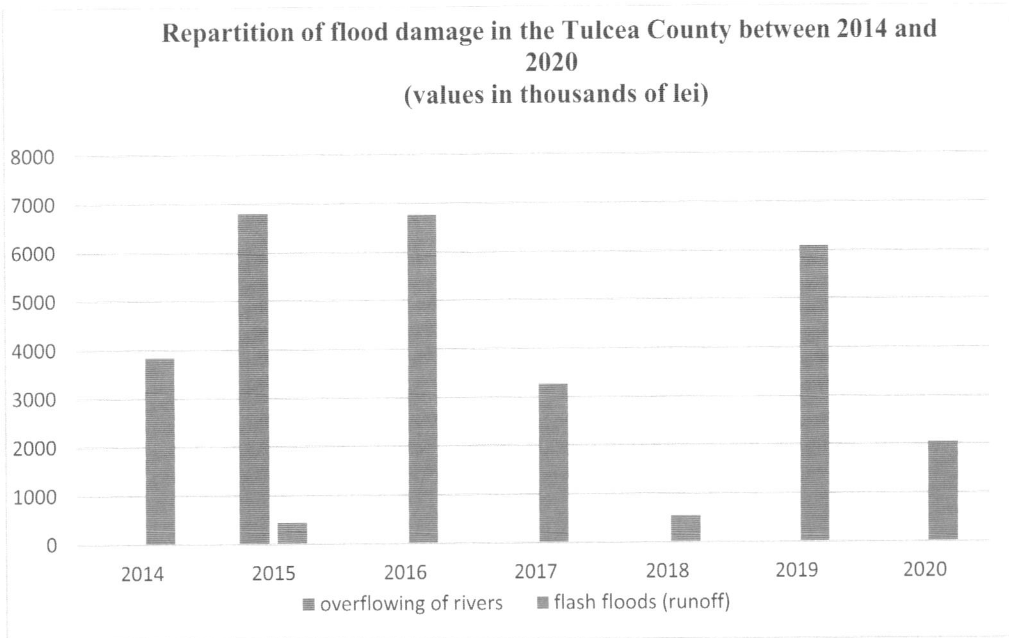

Figure 3. Repartition of flood damage in the Tulcea county (2014 – 2020).

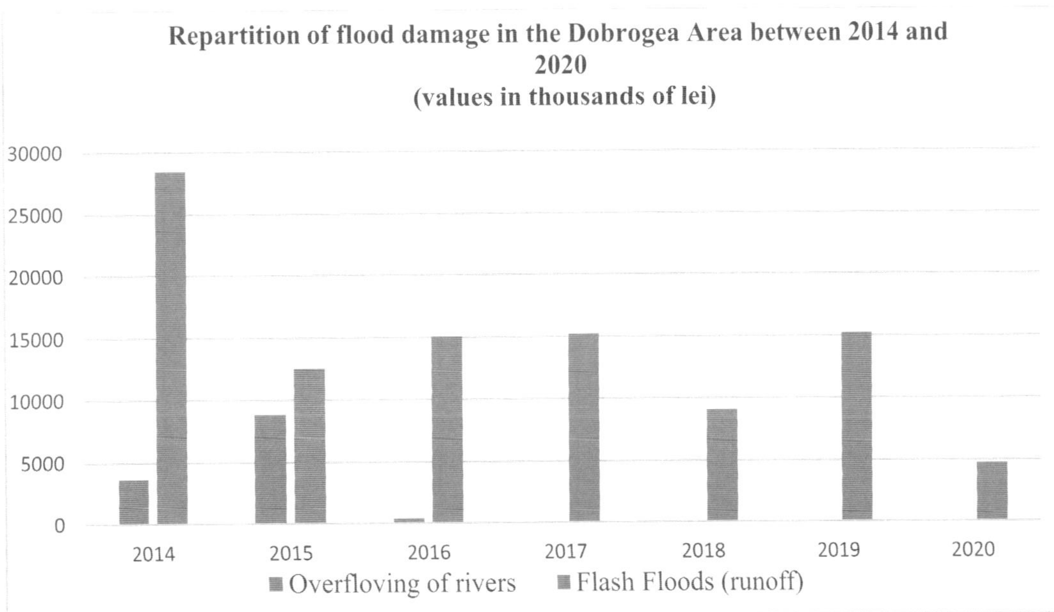

Figure 4. Repartition of flood damage in the Dobrogea region (2014 – 2020).

Total recorded damages due to flash floods in the Tulcea county, for the specified interval was 22.98 million lei. The precipitation context was a normal one, without many record rains being registered in this interval. The year 2015 was a little different because in the winter there were precipitations (snow and rain) that accumulated (an estimated 200 mm in some places) and when the spring thaw came, together with a relatively high level of the river Danube, vast areas of farmland were inundated because of high ground water levels and the soil being saturated. As this is not a phenomenon that occurred fast, it cannot be associated with flash floods and so it was considered similar to an overflowing of the river Danube and categorized as such.

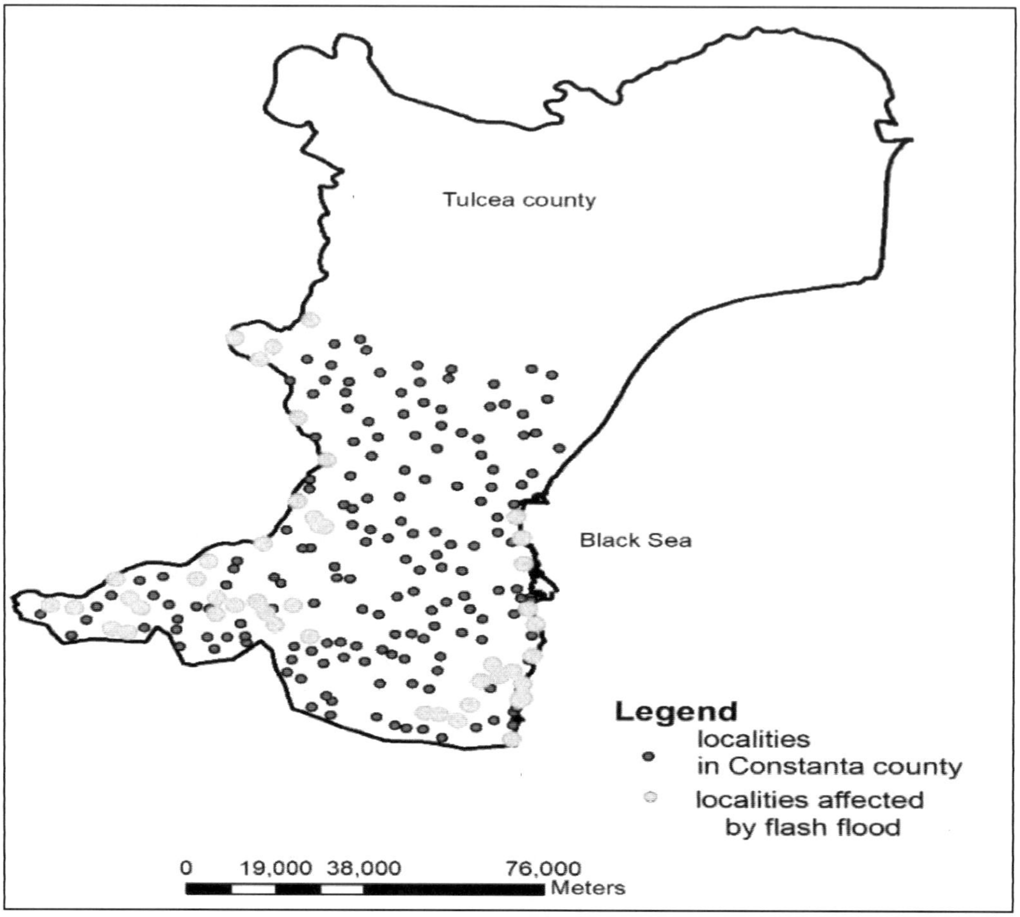

Figure 5. Repartition of localities affected by flash floods in the Constanta county – 2014.

Generally, the coast, the east and south-eastern part of Constanta county are more affected by the flash floods (figure 5).

From a spatial point of view, the available data shows that the flash flood situations tend to manifest over relatively small areas at a time, although about 50% of the time the reports show that multiple localities were affected. The local geometric parameters of the basin are reflected in the reports as some localities situated in sloped areas or surrounded by hills have multiple reports of flood damage. This is consistent with known runoff patterns.

The maximum registered rain for the Dobrogea area in a 24-hour interval, as reported in the flood reports was 175.81 mm, in the Constanta county in the year 2015. Tulcea county has a maximum rain of 82.6 mm in 24 hours, reported in the year 2016. For the specified time interval, Tulcea county reported higher average rains (37.86 mm) than the Constanta county (31.59 mm). The latest research [7] demonstrates a trend in the extreme rainfall frequency occurring which could lead to an increase in the number of flash flood events. In this respect, a strategy for flash flood assessment and management must be created and implemented.

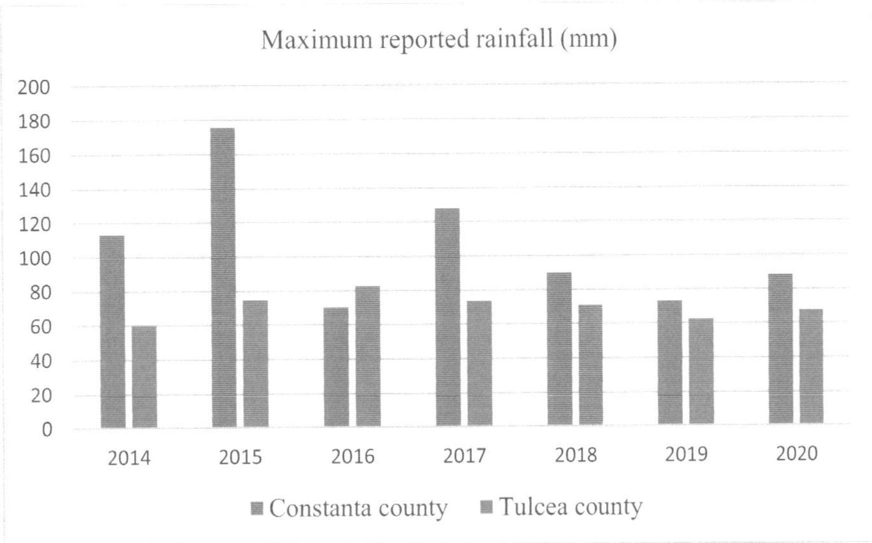

Figure 6. Maximum reported rainfall in flood damage reports.

While the data shows a somewhat constant pattern of flash flood (runoff) damage in the Constanta county, whereas Tulcea county has great variations from one year to the next, we believe this might be due to issues in the reporting process.

Some local authorities have obligations regarding flood damage intervention, and when they can intervene and don`t have to ask for county help, they might not report the flood damage at all. Although reporting is mandatory, we know from experience that some local authorities fail to do so in a proper and timely manner. The two counties have a different local and county level approach to reporting proceedings and we assume that the data for the Constanta county is more accurate, but not by a very large margin. This impacts the whole evaluation giving it an error margin that we estimate to be of up to 30% of reported flood damage, although it is impossible to accurately evaluate the error margin given the lack of available information. In the time interval, a gradual improvement in the detailing and accuracy of the reports is noted which shows a better understanding of the importance of the evaluation and reporting activities. We estimate this to be the result of the awareness campaigns and personnel training led by the D-LWBA in the last 5 years.

4. Conclusions

The financial impact of flash floods is significant, especially due to the fact that when compared to other flood types it represents 88.66% of all registered flood damage. Value-wise, the total registered damage was evaluated at over **113 million lei (a rough equivalent of 23.2 million Euros)**, of which over 100 million lei (20.5 million Euros) are attributed to flash flood damage. The information regarding flash flood damages, unfortunately, is incomplete, as many localities don`t report every event or only report it partially due to lack of trained personnel for the evaluation of flood damage and the proper reporting of flood-related events. The financial evaluation is not very accurate as well since many of the properties situated in the rural area don`t have a clear financial value associated and thus it must be determined indirectly, or by using general values (lump sums). Because of this difficult process, some damage is not even recorded or disregarded as too little to justify an evaluation. While this might be the norm and the populace has learned to cope with minor to medium effects of floods (not flood resilience but rather a traditional acceptance that the effects are impossible to avoid), it generates errors in the appropriate evaluation of damage due to flash floods diminishing the actual value and giving a false sense of a smaller scale. The evaluation itself is not the objective of this paper so we will not study it but rather mention the fact that actual damage and associated values are higher than what is officially recorded.

A secondary issue is the uneven coverage of rain gauge stations and thus potentially incomplete data sets for a relevant analysis. At times, one has to rely on rain data from a rain gauge located over 30km away when reporting, and that data isn`t reliable considering that rainfall cannot be considered uniform at such distances. This limits one`s ability to properly correlate rain intensity to flood damage. The Constanta county, for example, has some 40 permanently monitored rain gauges administered by the Dobrogea-Littoral Water Basin Administration. Considering the county`s total area of 7071 km^2, an average coverage of 176.77 km^2 per rain gauge results. While considered a normal coverage, [8] concludes that *even under ideal situations the representativeness or auto-correlation length of precipitation is surprisingly small when measuring a correlation factor of 0.5 between rain gauges located just 4km apart*. A recent study [9] points that *the optimal rain gauge coverage is of about 25 km^2 per station*. The spatial distribution is less than optimal though and in the Constanta county they provide some cover for the northern and central part of the county, while the south and the south-east of the county has very poor coverage, with vast areas not covered. Although modern methods make use of radar and satellite rainfall data, they can be prohibitively expensive for some areas, so rain gauges are still a necessity, albeit they may be automatized and designed to require as little logistics as possible. As a consequence, in some of the reports we were unable to determine the exact time of the rainfall or the average rainfall. One way this issue is treated is the borrowing of data from private sources, such as local agricultural enterprises which sometimes have their own rain gauges, or local administration (mayor`s office). This data may prove unreliable since the administrators of the rain gauges may have no training in using them and have no permanent monitoring or monitoring protocols, and the whole process of identifying "private" rain gauges is based on luck and the willingness of private entities (3rd parties) to share their data.

A complex monitoring and early warning system under the tutorage of the World Meteorological Organization is the Flash Flood Guidance System, or FFG. This system is based on multiple input sources (radar and satellite forecasting, soil humidity, etc.) in order to constantly assess the flood risk and produce early warnings. This system has proven useful and effective in Turkey according to Ertan Turgu [10]. In Romania, this system is being tested by the National Institute for Hydrology and Water Management but it still needs to be calibrated in the many different basins and is not yet in general use in Romania. While the relevance and necessity of such a system are clear in the context of climate change and the resulting increase in flash flood occurrences, the implementers of the FFG system in Romania will have to find solutions to known issues, shown before, in order for it to represent a reliable information and early warning platform.

References:

[1] Dunne T and Leopold L B 1978 *Water in environmental planning* (W. H. Freeman co San Francisco, CA. USA)

[2] Maftei C et al. 2020 *The Impact of Human Interventions and Changes in Climate on the Hydro-Chemical Composition of Techirghiol Lake (Romania* Water **12**, 2261

[3] Paltineanu C R et al. 2000 *Dobogea. Conditii climatice, consumul si necesarul apei de irigatie pentru principalele culturi agricole (Dobogea. Climatic conditions, consumption and the need for irrigation water for the main agricultural crops)* (Ed. ExPonto Constanta), 258 pp

[4] Maftei C and Bărbulescu A 2008 *Statistical analysis of climate evolution in Dobrudja region* (Proceedings of the World Congress on Engineering vol. 2, London, U.K.) 2 – 4 July, 2008, IAENG p 1082-1087

[5] Official data 2014-2020 *County Operational Reports Regarding Dangerous Meteorological Phenomena* - (provided by the Dobrogea-Litoral Water Basin Administration D-LWBA for the specified time interval)

[6] D-L WBA 2015 *Operational Report Regarding Dangerous Meteorological Phenomena in the Constanta County* (Constanta Water Management System) *10.13.2015*

IOP Conf. Series: Materials Science and Engineering 1138 (2021) 012012 doi:10.1088/1757-899X/1138/1/012012

[7] Cheroy O 2013 *Analysis of hydrometeorological data sets for 1945-2013 and projections of future climate parameters in the Danube Delta region* Expert report

[8] Habib et al. 2001 *Estimation of rainfall interstation correlation* (Journal of Hydrometeorology vol. 2) issue 6 p 621

[9] Gyasi-Agyei Y 2020 *Identification of the Optimum Rain Gauge Network Density for Hydrological Modelling Based on Radar Rainfall Analysis* (School of Engineering and Technology, Central Queensland University)

[10] Turgu E et al. 2017 *A Case Study: Analysis of Flash Flood Using FFGS Products on 17 January 2016 in Çeşme, Dikili, Izmir and Manisa* (Black Sea and Middle East Flash Flood Guidance BSMEFFGS Second Steering Committee Meeting SCM2 in Amman, Jordan from 11-13 April 2017)

IOP Conf. Series: Materials Science and Engineering 1138 (2021) 012013 doi:10.1088/1757-899X/1138/1/012013

Analysis of the most determining factors from Romanian urban regulation, at the state level, in relation to sanitation, hygiene and health throughout recent history

Ch Cazacu[1*], C Chitonu[1] and R Muntean[1]

[1] Faculty of Civil Engineering, Transilvania University of Brasov, Romania

* cazacu.christiana@unitbv.ro

Abstract. Construction represents an important branch of the Romanian economy. The construction branch must be characterized by flexibility and diversity, attributes that favor change, through generation, continuous development and rapid transposition into practice of new ideas. The demands of society are becoming more and more severe throughout the all countries of the world. Expectations and needs are becoming increasingly explicit for issues such as: health and safety of the place where we live or work environmental protection, operational safety.

1. Introduction

This article looks into the past construction sector and at challenges in the modern building mass related to indoor climate, hygiene and occupant wellbeing. As the statistics show, the indoor quality for buildings is directly influenced by the different building materials used in the process of construction and in further maintenance. The indoor environment may vary due to the level or types of human activities, which are carried out inside the buildings. The health of occupants is most related to the air quality, internal atmosphere [1, 2].

The ventilation systems have advantages and also disadvantages. The differences between the architectural programs are mainly determined if ventilation needs to be considered, due to the size of buildings, activities or functionality. Spreading the contaminated air, or impurities and microparticles, brings the high risk of short or long term infections and air borne diseases.

Studies show that this environmental comfort that occupants need is determined by physiological, esthetical and physical factors (microclimate) but also by spatial ones (the shape and dimensions of the interiors, the arrangement of objects) [3]. A building's interior microclimate is influenced by some factors, as it can be seen in figure 1.

- temperature: should be 20 - 21°C, in rooms with low energy consumption activities, and 15 - 18°C, where energy consumption is high;
- the air humidity must be 40 - 50%, because when it grows above these values, microbes and mold appear, and the hot and dry air causes respiratory difficulties;
- a bad relative humidity, temperature or humidification will affect the skin and can give upper airway symptoms, and also the perceptions in the office environment [4]; the optimum composition of the air must be ensured by a good ventilation of the rooms;
- natural light penetrates as much as possible inside the room and has a good distribution if the windows are high and evenly placed;

- in artificial lighting, the position and direction of the light source must have the effect of optimum light intensity and avoid blinding its users;
- the noise produced inside, but also outside the buildings are important for those in the interior spaces; in order not to become annoying, the sound insulation of the building is necessary.

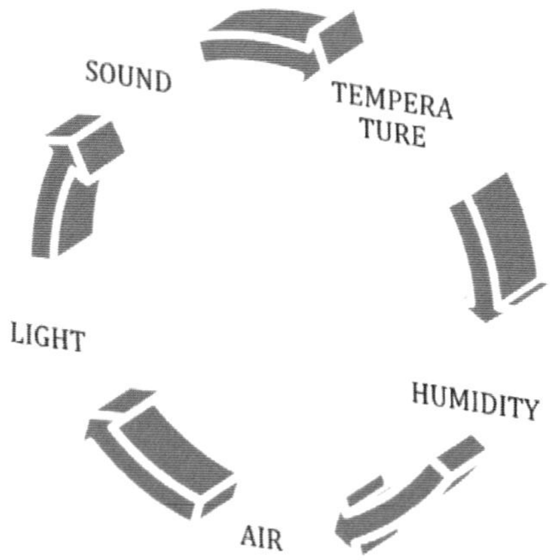

Figure 1. Factors that can influence the microclimate of homes.

1.1. The purpose of this study

The purpose of this paper is a research based on studies and analyses of the most determining factors in relation to sanitation, hygiene and health throughout recent history within Romanian State regulations.

The research is part of BIMHealthy project - Housing as a strategy for health promotion from an intersectoral and multi-disciplinary approach–co-funded by the Erasmus+ Programme of the European Union, [5]. The project started in October 2019 and it will end in September 2021. Six organizations, universities, research and development centers and private companies from three countries, Spain, Poland and Romania, are involved into this project.

The project aims to develop an interoperability between buildings and the health of their users to promote the realization of healthy housing as a global model, both in terms of the built environment and public health, by integrating new BIM technologies (Building Information Modelling) as tools for design and control of buildings [6, 7].

This section analyses the urban regulations, parameter values for Romanian cities, which, sometimes, are freely established in each City Council, but also the most representative parameters that regulate the interior of the homes built in Romania over time.

Through this study answers to the following questions, were sought:

- when did the constructions in Romania start to be build taking into account, apart from the structural strength and appearance, the health and safety of the building's users?
- are there technical and legal rules that regulate the construction of residential buildings from this point of view?
- if so, which are they and when did they appear throughout history?
- which are the main regulations, contained in these norms?
- and how have they changed over time, depending on the users' needs?
- which are their recommendations regarding hygiene, health and safety for a construction?

2. Historical data

In time, in Romanian construction sector, there where important changes, especially in the construction technology, determined by: the introduction in execution of new materials, use of high-performance machinery, machines and equipment on construction sites, introduction of automation in the secondary industrial production of construction, wide range of prefabrication used on a larger scale, the introduction of new structural solutions [8].

The research started in 1935, because it is the first period in which some data on aspects related to the health and safety of constructions were found. In the interwar period and the first years after the Second World War, there is a continuous development of construction activity, in the form of built areas and the number of permits issued nationwide, over 10,000 permits being issued only in 1937, as it can be seen in table 1:

Table 1. Number of building permits between 1935 -1938 [9].

Year	No. authorization	sqm per unit
1935	10728	165
1936	10018	147
1937	10370	131
1938	7776	127

Between 1950 - 1968, the development of constructions was characterized by a rapid dynamic, the production of this branch has been developed in 1968 nine times compared to 1950. The contribution of the construction sector to the creation of total social product and national income increased continuously, occupying in 1968 the third place after industry and agriculture in terms of material production. If in 1938, the construction sector achieved 5.4% of the social product, and 4.4% of the national income, being on the last place within the material production branches, in 1968 its share increased to 10.1% and respectively 9.2% [9].

In 1958 appeared the Decree no 144 regarding the regulation of the issuance of authorizations for construction, repair and demolition of constructions, as well as those related to the alienation and division of lands with or without constructions. Within this norm are presented details regarding:

- number of rooms in a house, provisions between them and minimum dimensions;
- minimum dimensions of rooms;
- insulation and waterproofing conditions;
- quality of the materials used for a building.

Decision no. 1650/1968 on the construction of housing in the urban environment from centralized state funds presents the realization of housing from centralized state funds, differentiated by number of rooms, comfort, endowment and price, apartment areas, facilities and price limits for different categories of housing and apartment sizes like:

- the apartments could have 1 to 4 rooms;
- a normal apartment could contain the following rooms: kitchen with sleeping area (10 - 11 sqm), room with cooking space (10 - 11 sqm), sleeping room (12sqm), bathroom or shared bathroom on the landing;
- for a 1 to 4 room apartment, the kitchen must be at least 3 to 4 sqm;
- common areas like: technical basement, laundry, drier;
- the sanitary objects: W.C. with water, washstand, shower, bath shower;
- the apartments could be centrally heated, or with one or two terracotta stoves.

Decision no. 585/1971 on the construction of urban housing from centralized state funds present data about urban dwellings that are from centralized state funds and put into use since 1972 were differentiated by the number of rooms, comfort, endowment and price, as follows in table 2.

In the past, those who lived in block apartments, had their homes connected to a national gas network and very few used personal apartment heating systems. The rooms were equipped with standardized and pre-dimensioned ceramic radiators, depending on the size of the rooms. For those who lived in private homes, the options were more varied because they could opt for heating with gas or wood stoves, and their size was not limited or special required.

IOP Conf. Series: Materials Science and Engineering 1138 (2021) 012013 doi:10.1088/1757-899X/1138/1/012013

Table 2. Number of rooms, comfort and price of different category of apartment.

Category of apartment	Number of rooms	Room area (sqm)	Kitchen area (sqm)
I	1	14	3
	2	24	3 - 4
II	3	33	4 - 5
	4	42	4 - 5
	1	13	3
III	2	23	3
	3	32	4 - 5
	4	41	4 - 5
studio	1	8.5	-

The change in political regime in 1989 had a significant impact on the field of housing: rules on control of residence and restricted access in major cities were discontinued, ownership of more than one dwelling was allowed, the law on systematization was abolished, access to real estate was liberalized, the number of actors (construction, rental) in housing multiplied and housing in collective blocks was privatized. These measures have determined a fundamental change in the housing sector and the mechanisms of housing construction and administration [10].

In 1996 appeared the Regulation regarding the design of dwelling buildings NP 016 [11]. This normative establishes the location and compliance conditions corresponding to the housing constructions in Romania, as well as the technical conditions, criteria and performance levels corresponding to the quality requirements, to which the housing constructions must meet, in accordance with the provisions of Law 10/1995 regarding the quality in construction. It aims to ensure the necessary protection and comfort, for the users (regardless of age or health) in accordance with specific requirements, by improving the performance of residential buildings. The provisions of this norm apply to the design of new residential buildings, located in urban or rural areas, as well as to the modernization, redevelopment, transformation, repair and consolidation of existing housing constructions.

2.1. Height regime of buildings

The establishment of the height regime of buildings, must be done in accordance with the provisions of PUG (general urban plan), PUZ (zonal urban plan) and PUD (detailed urban plan), as well as indicators of POT (percentage of land occupation) and CUT (land use coefficient) provided in the Urban Certificate, taking into account, at the same time, the influence of the wind and the air currents in the area.

2.2. Choosing the location of residential buildings

The objective of this standard (NP016) is also to establish the minimum acoustic conditions required for buildings, suitable for the use and activity of its occupants. The norm recommends to avoid locations in the vicinity of sources producing noxious substances, loud noises and vibrations (airport, industrial areas, heavy traffic arteries):

- maximum allowed noxious concentrations according to the provisions of STAS 12574/1987;
- permissible noise value: max. 50 dB (A) 2.00 m from the building façade (in case it is not possible to comply with this condition, appropriate insulation measures will be taken) [12].

For an increased thermal comfort inside the houses, it is recommended to orient the houses towards the cardinal points, so as to avoid the north location of the bedrooms.

IOP Conf. Series: Materials Science and Engineering 1138 (2021) 012013 doi:10.1088/1757-899X/1138/1/012013

2.3. Sunshine

The technical condition for sunshine, presented in NP 016, refers to the criteria of direct penetration of sunlight into the building. A home must be located and oriented in such a way that at least one of the living rooms (bedroom or living room) receives direct solar radiation, a certain number of hours per day, at certain minimum angles, throughout the year.

- the duration of sunshine for at least one of the living rooms, on a reference day (February 21, or October 21) must be at least 2 hours;
- the vertical angle of incidence of direct sunlight, on the reference day (February 21 or October 21), must be min. 6°;
- the angle of incidence in the horizontal plane of the direct sunlight, on the reference day (February 21 or October 21), must be min. 20°;
- the avoidance of the blinding phenomenon will be ensured by the correct orientation of the rooms, by the shielding of the glazed surfaces (in the case of buildings located in geographical areas with a high degree of sunshine, or of buildings with improperly oriented rooms).

2.4. Lighting of residential buildings

The technical condition for the lighting of residential buildings involves ensuring the quantity and quality of light (natural and artificial) so that users can carry out their household activities properly, both during the day and at night, in conditions of hygiene and health.

The value of natural lighting (for the winter solstice, overcast sky), should be for:

- living rooms (min. 30 lx),
- kitchens (min. 60 lx),
- bathrooms, hallways, corridors (min. 16 lx).

The ratio between the area of the windows and the area of the floors will be for:

- living rooms 1/6 ... 1/8;
- the other rooms 1/8 ... 1/10;
- stairs 1/10 ... 1/14.

If the natural light does not comply with the conditions specified above, it is necessary that the natural lighting be partially or totally compensated, depending on the destination of the room, with artificial lighting according to the provisions of ongoing standards.

The average lighting level, for normal room lighting at h = 0.85 - 1.0 m from the floor, should be as presented in table 3:

Table 3. The average lighting level for rooms.

	General lighting	Local lighting
Bedroom	50 lx	50 lx
Living rooms	50 + 100 lx	read 300 lx (on the table surface) sewing 500 lx (on the work surface)
Bathroom	75 lx	100 + 200 lx (on the surface of the mirror)
Kitchens	100 lx	300 lx (on the work surface)
Hallway, corridors	50 + 100 lx	50-75 lx (on the surface of the steps)
Garage	50 lx	50 lx
Basements, cellar, garbage	50 lx	50-75 lx

2.5. Hygrothermal hygiene of the indoor environment in homes

Creating a minimum allowable hygrothermal environment involves ensuring a proper thermal environment, both in winter and in summer. It is allowed that these conditions are not met 1 day/year in winter and 5 days/year in summer. Ensuring the hygrothermal environment must be correlated with ensuring air quality and optimizing energy consumption.

The normal global thermal environment comfort index (PMV) (the predictable average option calculated according to SR ISO-7730/2006), should be: PMV = -0.5 ... + 0.5, and the temperature of the indoor environment (Ti,) (according to SR 1907/2-2014):

A. in winter (minimum values):

- in living rooms, halls (20°C), vestibule (18°C), bathroom, showers (22°C), toilets in the apartment (18°C), toilets outside the apartment (15°C), garages under dwellings (10°C), kitchen (18°C), stairs, corridors outside the apartment (10°C);

B. in summer max. 25° C in all rooms.

The relative humidity of the indoor air (Φi) (according to SR ISO-7730/2006), will be between 35% and 60%.

The speed of the air currents (vi) (according to SR ISO 7730/2006), will be:

- in winter: vi = max. 0.15 m/s;
- in summer: vi = max.0.275 m/s.

Ensuring the normal local thermal environment:

- the asymmetry of the radiation temperature of windows or other cold surfaces (calculated according to SR ISO-7730/00), will be: max. 10°C;
- asymmetry of the radiation temperature of a heated ceiling (calculated according to SR ISO-7730/2006), will be: min. 5°C;
- the vertical air temperature difference, between the head and ankle level (according to SR ISO 7730-2006), will be: max. 3°C;
- the amount of heat given off by the foot of the floor, related to the cold-hot sensation (according to STAS 6472/10-85), will be:
  - in living rooms (warm floors):
    $Q1 = 50 \times 103\ J/m^2$ $Q10 = 300 \times 103\ J/m^2$
  - in the other rooms:
    $Q1 = 60 \times 103\ J/m^2$ $Q10 = 400 \times 103\ J/m^2$

2.6. Acoustic hygiene of the indoor environment

The technical condition regarding the acoustic hygiene of the indoor environment, supposes the design and realization of the interior spaces of the residential buildings so that the disturbing noise perceived by the users, to be maintained at a level that cannot affect their health.

Acoustic hygiene refers to the indoor conditions regarding noise, respectively to the indoor acoustic environment. The acoustic ambience in the living rooms coming from sources outside the room it is recommended at the level of max. 35dB.

2.7. Water hygiene

The technical condition, regarding the hygiene of the water, presupposes that the distribution of the water should be done in a sufficient flow, in the conditions of satisfying the purity criteria corresponding to the drinking water.

The water necessary to supply the installations in the residential buildings must have a certain quality, expressed by all its physical, chemical, bacteriological, etc. according to the STAS 1342/1991. The values for the water flow at the point of consumption should be: Q = min. 0.15 l/s (at the kitchen washer), Q = min. 0.05 l/s (at the bathroom sink and bathtub) with an optimal water speed = 1 m/s and the required quantities of drinking water min. 110 l/pers./day.

2.8. Air hygiene

The technical condition related to air hygiene involves ensuring the quality of the air inside residential buildings, respectively ensuring a proper atmospheric environment, so that there are no releases of harmful substances, toxic gases or dangerous radiation emissions, which could endanger the health of occupants. In Romania the required oxygen concentration will be min. 16.3% (of the volume of the room) and the maximum allowed concentrations of carbon monoxide will be: max. 6 mg m^3 [11].

The maximum permitted concentrations of water vapor will be: max. 15.400 mg/m^3 (T_{med} = 25 ± 3°) in summer regime and max. 9.450 mg/m^3 (T_{med} = 20 ± 2°) in winter regime [11].

Natural ventilation of rooms: air exchange (on the whole house) 0.5 ... 1 vol/h.

Organized natural ventilation or mechanical suction ventilation is mandatory for outbuildings without windows to the outside (bathrooms, rooms with shower, closet and washbasin, rooms with closet and pantries), for outbuildings with windows to the outside, but with equipment for cooking food or hot water with open flame (bathrooms, kitchens) and for garbage collection rooms as well as for basements.

All living rooms must be provided with the possibility of natural ventilation.

Mechanical ventilation (individual exhaust ventilation): for kitchens (without windows): min. 120 m^3/h, for bathrooms: min. 60 m^3/h, for shower rooms with toilet and washbasin: min. 60 m^3/h, for rooms with toilet and washbasin: min. 30 m^3/h [11].

2.9. Ensuring the quality of finishes in rooms, in terms of protection against poisoning with toxic substances

- materials that do not contain toxic substances and that do not emit harmful gases, dangerous to health, will be used;
- materials resistant to cleaning and sanitation actions will be used, depending on the destination of the rooms;
- measures will be taken to avoid the formation of fungi, by correcting the external closures correctly and by ensuring adequate ventilation.

2.10. Hygiene of sewage and manure

The technical condition regarding the hygiene of the waste liquids, supposes the assurance of an adequate system of elimination of the used (impure) domestic or meteoric waters, as well as of the manures:

- avoidance of pollution of the natural environment, respectively of groundwater, or of the soil, with wastewater coming from the sewerage system of construction objects;
- avoiding the risk of emitting unpleasant odors - odor level = 0;
- avoiding the interconnection between wastewater and drinking water;
- it is ensured by the correct solution of the sewerage and water supply systems, according to the provisions of regulations I 9/2013, STAS 1795/1987, STAS 3051/1991, NTPA 001/2002, NTPA 002/2002, NP 118/2006.

The regulations also present data about indoor traffic safety, doors, stairs, hallways, floors. These provisions purposes are the realization of buildings that are as safe as possible for the occupants, in order to avoid home accidents.

NP 057-02 [12], the normative regarding the design of residential buildings replaces the old normative NP 016-96, and is valid until today and aims to ensure the necessary protection and comfort for users (regardless of age or health) in accordance with specific requirements, by improving the performance of residential buildings.

The norm mostly includes the provisions of the norm, which it replaces with small modifications brought to some values related to the ventilation of the rooms, humidity or brightness.

3. Conclusion

In the past and also now, financial resources, physical condition and socio-economic circumstances can affect the technical design choices for buildings. However, buildings can present a great number of possible risks both in construction and operation. There are many duties placed on those designing, constructing and operating buildings to control health risks. For a new construction, the hygiene and the health of the occupants must become a priority strategy, and needs to be considered early on in the project because it is the key to securing a safe construction [13 - 15]. The client's role is fundamental to this, to establish and maintain a health and safety approach. The health and safety strategy should set clear health and safety objectives. In the current situation of COVID-19 pandemic, the further urban planning should include some key factors in building design, which can either increase or reduce the spread of diseases in public or private buildings. The size of buildings, height versus smaller buildings, or less densely situated buildings will impact the wellbeing of the occupants and also will benefit their health if the indoor air quality and hygiene factors are considered part of the planning strategy [16, 17].

Recommendations for a hygiene and health construction:

- provide designs that eliminate or reduce hazards in the work or living place;
- prevent occupational injuries and illnesses;
- prevent slips, trips and falls;
- eliminate exposure to hazardous materials;
- provide good indoor air quality and adequate ventilation;
- provide ergonomic workplaces and furniture to prevent work-related musculoskeletal disorders;
- perform proper buildings operations and maintenance.

In a sustainable urban development, it is imperative that the building mass is built for increased hygiene and considering infection control purposes in the future. Making improvements in the use of building materials, shapes, functionality, ability to clean, purposeful ventilation and ensuring smarter and more targeted spatial connections are key factors for improving the wellbeing of occupants. The balance between architectural, urban planning and building services in regards of the energy required for heating and cooling, domestic hot water, controlled ventilation, lighting and other technical facilities in buildings should be designed to optimize the levels of health, indoor air quality and well-being defined by the Member States at national level or regional.

Acknowledgement

The work presented in this paper was funded by the BIMhealty project -2019-1-ES01-KA203-065060, with support from the European Commission. This publication reflects the views only of the authors and the Commission cannot be held responsible for any use which may be made of the information contained therein.

References

[1] Directive 2008/50/EC of the European Parliament and of the Council of 21 May 2008 on ambient air quality and cleaner air for Europe – European Directive or regulation No. 1. (https://eur-lex.europa.eu/legal-content/en/ALL/?uri=CELEX%3A32008L0050)

[2] Report: on social housing in the European Union (2012/2293(INI)) Committee on Employment and Social Affairs. Rapporteur: Karima Delli (https://www.europarl.europa.eu/sides/getDoc.do?pubRef=-//EP//TEXT+REPORT+A7-2013-0155+0+DOC+XML+V0//EN)

[3] Reference Framework for European Sustainable Cities (RFSC) (www.rfsc.eu) Toledo Declaration (http://urban-intergroup.eu/wp-content/files_mf/es2010itoledodeclaration.pdf)

[4] Reinikainen LM and Jaakkola JK 2004 Significance of humidity and temperature on skin and upper airway symptoms, Indoor Air Journal **13** (4):344-52

[5] https://bimhealthy.eu/ (last accessed on 20th of October 2020)

[6] Khoshnava M, Ahankoob A, Nigel C and Rostami R 2012 Application of BIM in Construction

IOP Conf. Series: Materials Science and Engineering 1138 (2021) 012013 doi:10.1088/1757-899X/1138/1/012013

Safety, Conference of Management in Construction Research Association (MiCRA)
[7] Oberste K 2020 How BIM can Improve Health and Safety on Construction Sites
[8] Blaga P 2016 Research on production management and human resources in organization, University of „PetruMaior" from Tîrgu-Mureș
[9] Andrei T 2018 Romania, a century of history, statistical data, Bucharest, National Institute of Statistics;
[10] Eurostat "Eurostat Statistics Explained: Housing Statistics." accessed on August 17, 2015. http://ec.europa.eu/eurostat/statistics-explained/index.php/Housing_statistics
[11] The norm regarding the design of residential buildings, indicative NP 016-97 of 27.03.1997
[12] The norm regarding the design of residential buildings (revision NP 016-96), indicative NP 057-02 of 24.09.2002
[13] https://www.wbdg.org/design-objectives/secure-safe/occupant-safety-health
[14] Directive (EU) 2018/844 of the European Parliament and of the Council of May 30, 2018
[15] National housing strategy, 2017, authors Ministry of Regional Development, Public Administration and European Funds
[16] Statistical Yearbook of Romania, 1990, National Institute of Statistics, page 540
[17] Official Journal of the European Union, 2019, Commission recommendation (EU) 2019/786 on the renovation of buildings (notified under document number C (2019) 3352)

IOP Conf. Series: Materials Science and Engineering 1138 (2021) 012014 doi:10.1088/1757-899X/1138/1/012014

Flood frequency analysis of Casimcea river

C Cerneagă[1]* and C Maftei[2]

[1] Ovidius University of Constanta, Doctoral School of Applied Sciences, Constanta, Romania
[2] Transilvania University of Brasov, Faculty of Civil Engineering, Brasov, Romania

* constantin.cerneaga@gmail.com

Abstract. Flood frequency analysis is the most important statistical technique in understanding the nature and magnitude of high discharge in a river. This technique is used for establishing a relationship between flood magnitude and frequency of occurrence (return period). This study performs the flood frequency analysis of Casimcea river. The Casimcea catchment is located in the Dobrogea-Littoral basin of south-eastern Romania, and covers parts of two counties, Tulcea and Constanta. The main river of Casimcea catchment flows into the Tasaul lake and has two main tributaries (Ramnic and Cartal). In this study, Log-Normal, Log-Pearson Type III and Gumbel probability distributions have been applied on maximum discharge time series data of four hydrological stations, Casimcea and Cheia on Casimcea River and Pantelimonu on its main tributaries Ramnic and Cartal. For this study have been used the annual maximum series data spreading from 29 to 52 years. The frequency analysis will be performed with Hydrognomon. open-source software for the analysis of hydrological data. Hydrognomon is free software licensed under the General Public License (GPLv3).

1. Introduction

Floods occur due to various natural and anthropogenic factors and are most common environmental hazards, affecting people globally. Flooding has been a larger concern today due to rapid development in river catchment area, which increases river runoff and decreases river capacity.

A return period is known as a recurrence interval, which is an estimate of the interval of time between events like an earthquake, flood or river discharge flow of a certain intensity or size [1]. The terms "10-year", "50-year", "100-year" and "500-year" flood are used to describe the estimated probability of a flood event happening in any given years [2].

Flood frequency analysis is the procedure for estimating the frequency of occurrence (return period) of a hydrological event such as flood. The technique involves using observed annual peak flow discharge data to calculate statistical information such as mean values, standard deviations, skewness and recurrence intervals [3].

Various probability distributions are currently used to predict expected rainfall in different return periods, as rainfall varies with time and location [4]. Frequency analyses of rainfall data have been performed for different return periods [5].

The Kolmogorov-Smirnov test and Normal, Log-Normal, Log-Pearson Type III and Gumbel distributions are used to fit the annual extremes rainfall data with 10, 25, 50, 100 and 200 years return periods.

2. Material and methods

2.1. Study area

The Casimcea catchment is located in the Dobrogea-Littoral basin of south-eastern Romania (figure 1), and covers parts of two counties, Tulcea and Constanta.

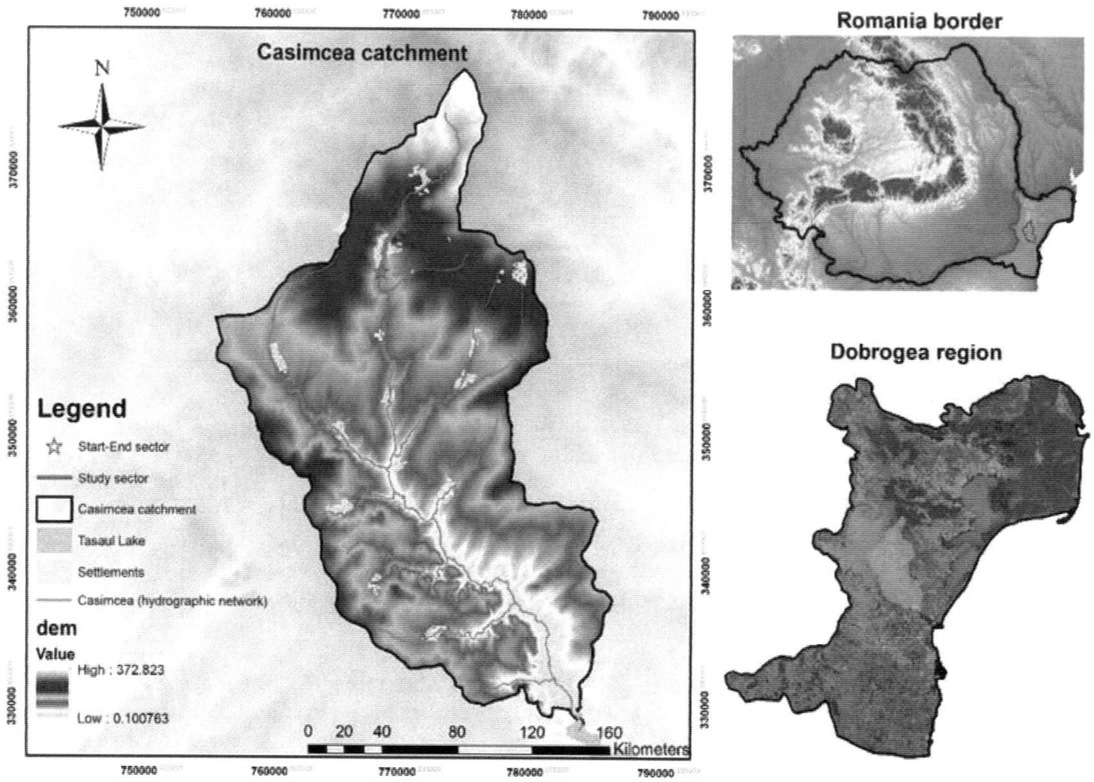

Figure 1. Study area.

In the present study 29 to 52 years of annual maximum peak discharge data was used for analysis (figure 3). The size of the sample data can be considered as reliable for projecting discharge scenarios. „Romanian Waters" National Administration has collected the data from three gauging stations, Casimcea (upstream), Pantelimonu (tributaries Cartal and Ramnic), and Cheia (downstream).

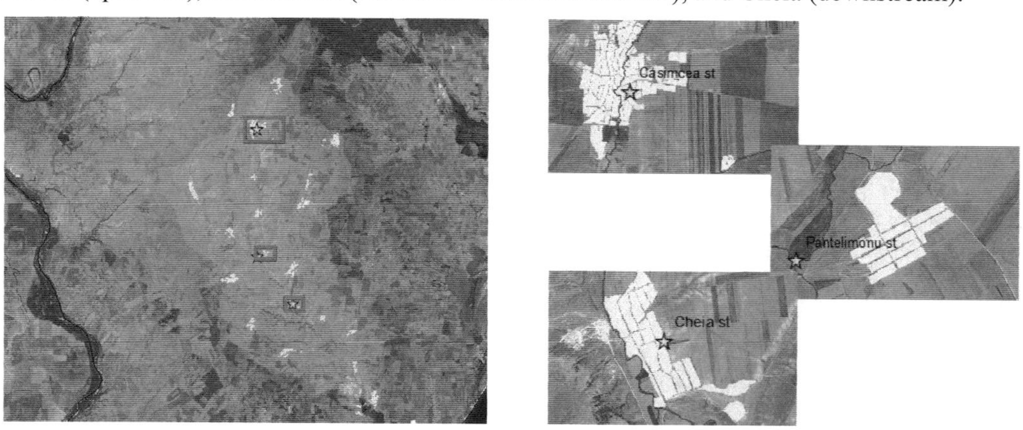

Figure 2. Hydrological stations.

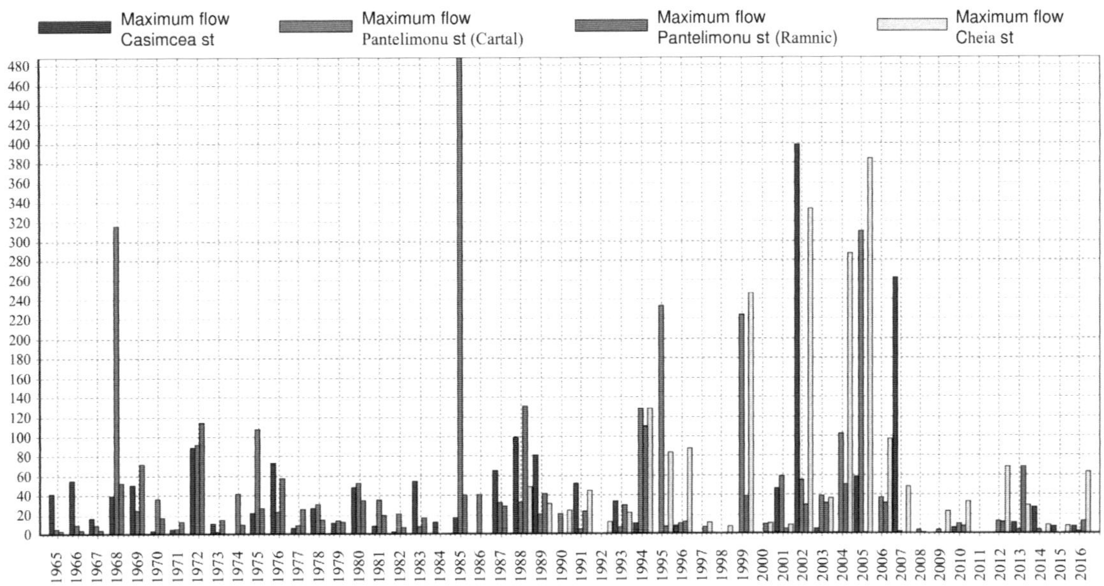

Figure 3. Flow hydrograph (1965-2016).

2.2. Methods

To predict extreme hydrological discharge using frequency analysis, Hydrognomon software [6] and EasyFit software [7] are used. The most common probability distribution functions (PDF) used for extreme discharge data are Log-Normal, Pearson Type III, Log-Pearson Type III and Gumbel distributions [8]. U.S. Water Resources Council [9] recommends Log-Pearson Type III PDF method found to be yielding good results in many applications, particularly for flood peak data. Hosking and Wallis [10] consider that Gumbel PDF procedure is a rational approach to predict the flood recurrence interval.

Kolmogorov-Smirnov (K-S) test from Hydrognomon and Anderson Darling (AD) from EasyFit are used to test if the discharge sample comes from a specific PDF. If the K-S test statistic $\boldsymbol{D}$ is greater than the critical value, the null hypothesis (H_0: the data follow a specified distribution) is rejected. Critical values of $\boldsymbol{D}$ statistic for different significance level, α, depend on number of observations (n) [11]. For example, at α of 5%, and 50 observations, the critical value is 0.190.

The Anderson-Darling (AD) test is similar to K-S but is more powerful especially when all the data values are considered, not just the one that produce the maximum difference. The statistic of AD test, A^2, depends on number of observations (n) and the distribution function (F). If A^2 at a certain significance level, α, is greater than the critical values of the selected distribution function, than the null hypothesis is rejected.

3. Results and discussion

According with Kolmogorov-Smirnov (table 1), the Log-Normal PDF is not accepted for all stations investigated. Pearson Type III is rejected for Pantelimonu station located on Cartal tributary. Log-Pearson Type III is not accepted for Pantelimonu station located on Ramnic tributary. We observe that the Pearson Type III distribution for Casimcea station and Log-Pearson Type III distribution for Cheia station have the best fit with "attained α" of 51.5%, and 66.6% respectively. For Pantelimonu station on Ramnic tributary the better distribution is Gumbel with "attained α" of 15.74%.

IOP Conf. Series: Materials Science and Engineering 1138 (2021) 012014 doi:10.1088/1757-899X/1138/1/012014

Table 1. The results of the Kolmogorov-Smirnov for all stations investigated.

Kolmogorov- Smirnov test	α = 1, 5 and 10%	Attained α (attained threshold)
Casimcea st. on Casimcea		
LogNormal	REJECTED	0.23%
Pearson III	ACCEPT	51.50%
Log Pearson III	ACCEPT	23.28%
EV1-Max (Gumbel)	ACCEPT	0.05%
Cheia st. on Casimcea		
LogNormal	ACCEPT	22.28%
Pearson III	ACCEPT	14.34%
Log Pearson III	ACCEPT	66.61%
EV1-Max (Gumbel)	ACCEPT	13.79%
Pantelimonu st. on Cartali tributary		
LogNormal	REJECTED for 1% and ACCEPT for 5 and 10%	1.37%
Pearson III	REJECTED	0.00%
Log Pearson III	ACCEPT	42.53%
EV1-Max (Gumbel)	REJECTED	0.01%
Pantelimonu st. on Ramnic tributary		
LogNormal	REJECTED for 1% and ACCEPT for 5 and 10	1.44%
Pearson III	ACCEPT	15.74%
Log Pearson III	REJECTED	0.55%
EV1-Max (Gumbel)	ACCEPT	5.11%

Face to the situation presented below, has been performed the Anderson-Darling test under EasyFit in order to choose the best PDF. The A^2 statistic results (table 2) show that the better distribution function for all stations is Log-Pearson Type III because the A^2 value is less than Log-Normal distribution. The Gumbel distribution was rejected for all stations and for all α level of significance.

Table 2. The results perform with EasyFit.

AD test	α = 1, 5 and 10 %	A^2 statistic
Casimcea on Casimcea		
LogNormal	ACCEPT	1.047
Log Pearson III	ACCEPT	0.61
Cheia on Casimcea		
LogNormal	ACCEPT	0.64
Log Pearson III	ACCEPT	0.3
Pantelimonu st. on Cartal		
LogNormal	ACCEPT	1.047
Log Pearson III	ACCEPT	0.89
Pantelimonu st. on Ramnic		
LogNormal	ACCEPT	0.44
Log Pearson III	ACCEPT	0.36

As we can see Log-Pearson Type III was accepted but the values for different return period presented in table 3 are very high compared to those obtained using Hazen empirical formula [12] and with the annual maximum peak discharge data that we use. Hazen formula is calculated with the following equation:

$$P = \frac{2m-1}{2n} \quad (1)$$

where m is rank assigned to the data after arranging them in descending order of magnitude and n represent number of records. Thus, the maximum value is assigned m = 1, the second largest value m = 2, and the lowest value m = n.

Table 3. The maximum discharge for different return period.

T(Max) =	**200 y**	**100 y**	**50 y**	**25 y**	**10 y**
Casimcea st - Casimcea river					
Log Pearson Type III	6624.2	2964.4	1269.6	513.3	136.3
Hazen	-	395.3	324.6	167.6	75.6
Cheia st - Casimcea river					
Log Pearson Type III	4975.2	2635.8	1359.4	676.8	248.8
Hazen	-	-	379.9	350.3	270.6
Pantelimonu st - Cartal tributary					
Log Pearson Type III	7432.8	3448.98	1540.67	654.671	188.877
Hazen	-	484.6	395.1	312.5	156.8
Pantelimonu st - Ramnic tributary					
Log Pearson Type III	10132.5	3913.11	1464.76	526.211	124.026
Hazen	-	130.6	121.8	111.7	60.7

4. Conclusions

In this paper, the main purpose is to understand the frequency analysis of annual maximum discharge on Casimcea River. After frequency analysis, Log-Normal and Log-Pearson Type III are the best distribution functions which are provided by Anderson-Darling test and Log-Pearson Type III by Kolmogorov-Smirnov test. The discharge values obtained for different return period based on those theoretical distribution functions are different from the real ones.

Based on this study, any distribution function should not be considered suitable for analysis of maximum annual data of Casimcea River.

An upcoming paper will assess other distribution functions to fit the best function. The study results lead to the preparation of a flood hazard, which are essential for flood management plan.

References

[1] Rakhecha P R and Singh V P 2009 *Applied Hydrometeorology*. Heidelberg: Springer
[2] Watson D and Adams M 2010 *Design for Flooding*. Canada: John Wiley and Sons
[3] Derman W H A W, Mutalib N J A and Mukhatar N Z 2017 *Determination of return period for flood frequency analysis using normal and related distributions*. IOP Conf: Journal of Physics
[4] Upadhaya A and Singh S R 1998 *Estimation of consecutive day's maximum rainfall by various methods and their comparison*. Indian J. Soil Conserv. 26, 193-201
[5] Vivekanandan N 2012 *Intercomparison of extreme value distributions for estimation of ADMR*. Int. J. Appl. Engg. Technol., 2 (1), 30-37
[6] http://www.hydrognomon.org/, last accessed 2020/11/02

[7] https://easyfit.soft32.com/, last accessed 2020/11/02
[8] Amin M T, Rizwan M and Alazba A A 2016. *A best-fit probability distribution for the estimation of rainfall in northern regions of Pakistan.* Open Life Sciences, 11:432-440
[9] Hammett K M and DelCharo M J 2001 *Estimating the magnitude and frequency of floods for streams in West-Central Florida.* Scientific Investigations Report 2005-5080
[10] Hosking J R M and Wallis J R 1997 *Regional Frequency Analysis – An Approach Based on L-Moment.* Cambridge University Pres, Cambridge, UK
[11] http://www.itl.nist.gov/div898/handbook/, last accessed 2021/02/04
[12] Hazen A 1930 *Flood flows – A study of Frequencies and Magnitudes.* Jhon Wiley & Sons, Inc.

IOP Conf. Series: Materials Science and Engineering 1138 (2021) 012015 doi:10.1088/1757-899X/1138/1/012015

Morocco's housing challenge: Effect of current marketed building materials on the energy performance of residential buildings

M Charai[1, 2]*, H Sghiouri[1], A Mezrhab[1], M Karkri[2] and L Moga[3]

[1] Mohammed First University, Mechanics and Energy Laboratory, Morocco
[2] Paris-Est University, Certes, France
[3] Civil Engineering and Management Department, Technical University of Cluj-Napoca, Romania

* mouatassim.charai@u-pec.fr

Abstract. The paper provides a research on the energy performance of current marketed building materials used for household construction in eastern Morocco. Firstly, the characterization of the thermophysical properties of commonly used building materials (cement, plaster, hollow bricks and insulation boards) was carried out using the Hot Disk method. A local building material database of eastern Morocco region was developed. It was found that the thermal conductivity of the most affordable insulation boards was of 0.03 W/mK, while the thermal conductivity of cement-based products ranged from 0.65 to 0.7 W/mK. The experimental results show that the Moroccan market of building materials cannot participate in promoting eco-construction and designing efficient lightweight building envelopes at local level. Therefore, a strong focus on the development of local low-cost insulation materials for building applications are needed. The numerical building performance analysis shows that available building practices are only efficient for passive heating. Further studied on passive cooling techniques are greatly required for designing low energy buildings in Morocco.

1. Introduction

Over the past decade, Morocco has deployed straightforward strategies to reduce the building energy consumption by implementing radical reforms in favor of sustainable constructions. A thermal regulation standard for buildings, an energy efficiency law (i.e. law 47-09), new energy institutions (i.e. IRESEN - Research Institute for Solar Energy and New Energies, AMEE- Moroccan Agency for Energy Efficiency, MASEN- Moroccan Agency for Sustainable Energy) have been put in place to reach a new target aimed at saving 20% of the energy consumption by 2030, including building sector [1].

In fact, the Kingdom of Morocco put a strong focus on building energy efficiency by updating status quo/go reports on this field and promoting R&D co-operation be-tween Moroccan universities and industrial units for developing green low-cost building materials. For example, the Research Institute for Solar Energy and Renewable Energies (IRESEN) has funded several projects related to construction issues and has enabled the creation of several research laboratories throughout Morocco (figure 1), notably the Green and Smart Building Park (GSBP), which will be interested in green building research. Also recently, the National Centre for Scientific and Technical Research (CNRST) signed a partnership agreement with the Holding AL OMRANE (HAO), which represents the leader-housing operator in Morocco, for the development and promotion of housing and urban planning, through launching the first "*Omraninnov*" research & development call for projects [2].

Such actions are implemented owing to the fact that the national construction market cannot keep pace with current challenges. Several research studies have been conducted to improve the thermal comfort of Moroccan buildings. Sghiouri et al. [3] investigated different passive cooling strategies for reducing overheating for building located in semi-arid climate. A Ketouane et al. [4] studied the potential use of phase change material in hollow brick walls for improving the thermal performance of building envelopes, since almost Moroccan houses are constructed of hollow bricks.

However, this work will characterize the thermophysical properties of current building materials marketed in eastern Morocco to investigate the thermal performance of local residential buildings. Recommendations for the design of thermally compliant houses will be developed.

2. Materials and methods

2.1. Building material specimens

Given the purpose of the paper, we focused on the products most used for the construction of typical Moroccan houses. Figure 1 illustrates the main ingredients of Morocco building envelopes. According to the only supplier of insulating materials located in Oujda, the XPS insulation boards was chosen due to their quality-price ratio. All products are collected from local manufacturers and suppliers to evaluate the energy performance of houses at local level.

Since cement and plaster are sold as powder, the samples were prepared by hand mixing the matrices with water and molded in aluminum mold. A plaster/cement-to-water ratio of 0.7% was used. The final samples were left to dry at room temperature. The thermal characterization was performed when the samples reach a constant weight.

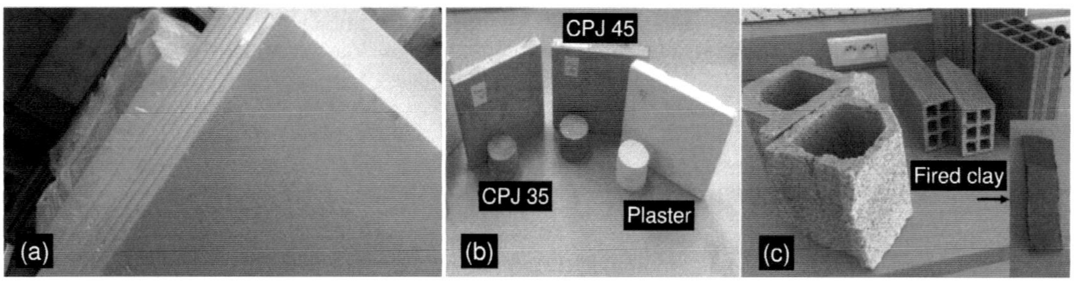

Figure 1. Studied products: (a) insulation boards (b) cement and plaster (c) hourdis and bricks.

2.2. Thermophysical characterization

The Hot Disk (TPS 2200, Sweden) apparatus was used to determine simultaneously the thermal conductivity, the thermal diffusivity and the heat capacity of the studied products. The experimental setup is shown in figure 2. The method is based on a Transient Plan Source (TPS) technique [5]. The measurement consists to sandwich the TPS sensor between to identical samples, as indicated in figure 2b. Direct electrical power is injected through the TPS sensor to heat samples. After the heating phase, the TPS sensor monitors the time-dependent response. Then, the thermal properties of samples are deduced using parameter estimation algorithms based on the recorded experimental temperature data.

The thermal characterization was carried out in accordance with ISO 22007-2 standard.

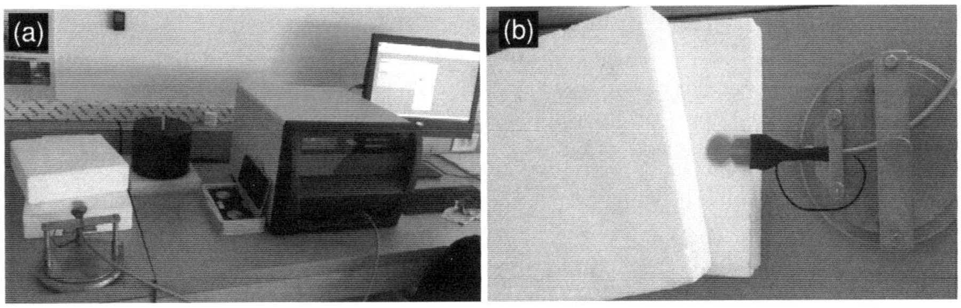

Figure 2. Setup of thermal tests (a) Hot disk apparatus (b) TPS sensor.

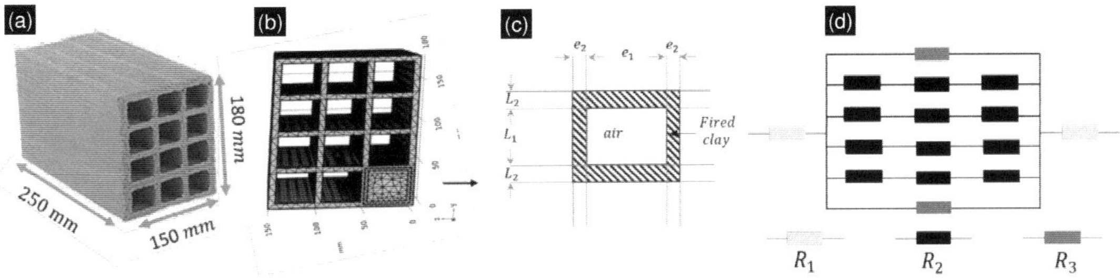

Figure 3. Hollow brick (a) real view (b) sketch view (c) brick-element (d) electrical analogy.

For hollow bricks, we firstly characterized the fired clay constituting the brick because the used method did not take into account the air pores. We then estimated the thermal conductivity of a 12-hole brick using the Electrical Analogy of Heat Conduction, as presented in figure 3. Basing on the thermal circuit shown in figure 3d, the thermal resistance of the hollow brick can be written as follows:

$$R_{hollow} = 2R_1 + (2\frac{1}{R_3} + 4\frac{1}{3R_2})^{-1} \quad (1)$$

Where $R_1 = \frac{e_2}{\lambda_c L l}$, and $R_3 = \frac{(E-2e_2)}{\lambda_c L_2 l}$ are respectively the thermal resistanceof the vertical and the horizontal solid parts of the brick. For the hollow part (figure 3c), and if we assume that that the dimensions are small enough to prevent convective movements, the thermal resistance R_2 can be approximated as follows:

$$\frac{1}{R_2} = \frac{1}{R(L_1)} + \frac{1}{R(2L_2)} \Rightarrow R_2 = \frac{1}{\frac{1}{R(L_1)} + \frac{1}{R(2L_2)}} \quad (2)$$

Where $R(L_1) = R_{air} + \frac{2e_2}{\lambda_c L_1 l} = \frac{1}{L_1 l}\left(\frac{e_1}{h_r e_1 + \lambda_{air}} + \frac{2e_2}{\lambda_c}\right)$ and $R(2L_2) = \frac{2e_2+e_1}{\lambda_c 2L_2 l}$ are respectively the thermal resistanceof the air blade and the solid part. It should be noted that the thermal resistance of the air blade is equal to:

$$\frac{1}{R_{air}} = \left(h_r + \frac{\lambda_{air}}{e_1}\right) L_1 \text{ with } h_r = f(\epsilon_1, \epsilon_2) 4\sigma T_m^4 \quad (3)$$

In which L = 180mm, E = 150 mm, l = 250 mm, L_1 = 37.5 mm, L_2 = 3 mm, e_1 = 40.6 mm, e_2 = 3.5 mm, λ_c = 0.972 W/mK, λ_{air}= 0.025 W/mK, h_r = 6 W/m^{2}K. Hence, we find that the thermal resistance of our hollow brick of 150 mm-thick is 7.8539 m^2 K.W-1.

Hence, the equivalent thermal conductivity could be deduced using the following equation:

$$R_{hollow} = \frac{E}{\lambda_{hollow} \cdot L \cdot l} \quad (4)$$

where E, L and l are the brick dimension.

3. Test results: building material database

Table 1 sums up the thermophysical properties of studied samples. The thermal heat capacity of samples is calculated using the following equation:

$$Cp = \frac{\lambda}{a.\rho} \tag{5}$$

λ is the thermal conductivity in W/mK, a is the thermal diffusivity in m^2/s and ρ is the density of the sample kg/m^3.

According with the analogy of heat conduction, the thermal conductivity of our hollow brick was found to be 0.424 W/mK, representing a gain of 56% in thermal insulation compared to solid bricks. Note that these results were in a good agreement with thermal conductivity value measured on 7-cm-tick hollow bricks using the boxes method. However, the studied hollow bricks in this work present high thermal conductivity compared to those measured by the Moroccan Agency of Energy Efficiency (AMEE) [6]. This is likely due to the fact that the AMEE measurements were carried out on oven dried specimens.

Table 1. Thermophysical properties of building materials in eastern Morocco Market

Product		λ [W/mK]	Thermal diffusivity [mm^2/s]	Heat capacity [KJ/mK]	Density [Kg/m^3]
Insulation boards		0.030 ± 0.001	1.439 ± 0.027	0.992	21
Hourdis		1.744 ± 0.002	0.981 ± 0.025	0.704	2524
Fired clay (solid bricks)		0.972 ± 0.006	0.685 ± 0.020	0.965	2040
Hollow bricks	Elec. analogy	0.424 ± 0.020	0.631 ± 0.050	0.655	950
	AMEE [6]	0.219	0.420	0.729	716
Cement	Cement 35	0.653 ± 0.007	0.447 ± 0.019	0.859	1700
	Cement 45	0.699 ± 0.001	0.322 ± 0.004	1.369	1585
Plaster		0.531 ± 0.001	0.400 ± 0.011	0.954	1392

4. Thermal performance: case study

4.1. Case study building

The thermal simulations of a residential building located in Oujda City for different wall compositions were performed. The building (figure 4) is a two-story villa with a total floor area of 160 m^2. The 2D model of the building is presented in the figure 5. The simulations were carried out using the EnergyPlus™ software, version 9.0.1 [7]. Climate information and the used meteorological data are discussed in [3, 8].

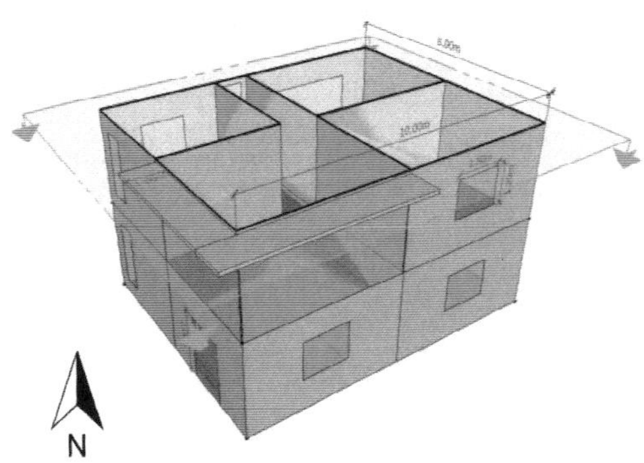

Figure 4. 3D model.

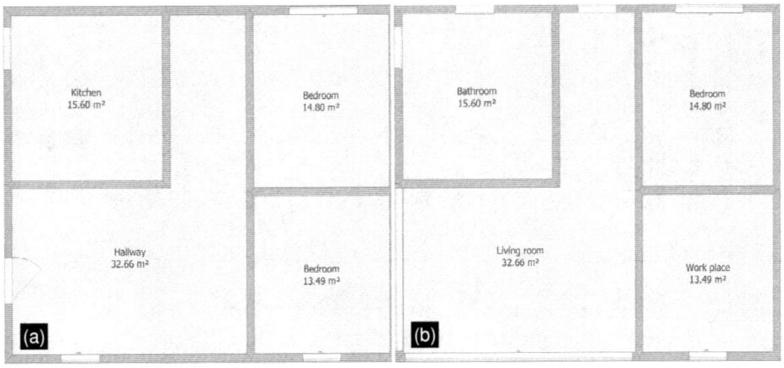

Figure 5. 2D model (a) ground floor (b) 1st floor.

According to building practice in Morocco, four main wall compositions are proposed in table 2.

Table 2. Wall compositions.

Wall code	Description
W1	Simple hollow brick wall
W2	Double hollow brick wall with 100 mm-thick air gap
W3	Double hollow brick wall with 30 mm-thick insulation board
W4	RTCM building envelope = W1 + 6cm-thick insulation board in the roof

The parameter inputs of the dynamic thermal simulation are listed in table 3.

Table 3. Parameter inputs.

Parameter	Value	Schedule
Heating set point	20 °C	On (all days)
Cooling set point	26 °C	On (all days)
People	4 persons	Occupancy schedule (figure 6)
Activity level	120 W/person	Depends on occupancy
Lights	10 W/m^2	Depends on occupancy
Electrical equipment	Fridge 130W	Daily running time 24h
	TV 80 W	Daily running time 4h
	Pc 200 W	Daily running time 3h
	Laptop 60 W	Daily running time 3h
	Phone charger 6W	Daily running time 12h
Infiltration	0.6 h^{-1}	On (all days)

The building was assumed to be occupied by four persons according to the schedule fraction indicated in figure 6. Internal gains were introduced to correspond to an artificial lighting of 10 W/m^2, a desktop (200 W), a fridge (130 W), TV (80 W), Laptop (60 W) and phone charger (60 W).

The walls are coated by a 15mm-thick mortar layer. The properties of mortar are of cement 35 in Table 1. Cement 45 is used for the conception of the floor/roof screed. The roof is a plaster on the inside (4 cm), with hollow-core (hourdis) slab (16 cm), cement coating (4 cm), screed (5 cm) and tile on the outside (1 cm). The ground floor is a reinforced concrete slab (20 cm) with screed (5 cm) and tile inside (1 cm). The glazing is compliant double with a thermal transmittance of 3.33 W/m^2K.

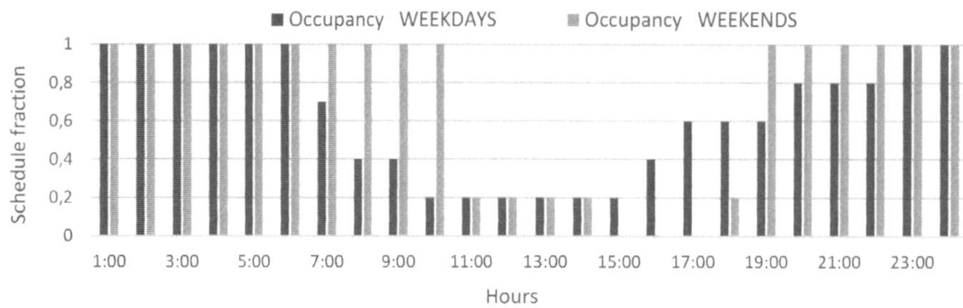

Figure 6. Occupancy schedule.

4.2. Annual loads and discussion

The annual energy consumption of studied envelopes, are presented in figure 7a. The results show that building with simple hollow brick wall consume 69.2 kWh/m^2, exceeding the maximum admissible by 30%. The use of air gap is a good practice for designing efficient wall. Using 10 cm of air blade can reduce the heating/cooling loads by 15%, compared to the reference case (i.e. simple hollow brick wall).

Unlike air gap, the use of insulation boards is better for design efficient walls with small thickness, representing an energy saving of 17.9% compared to the reference. The simulation show that the use of 6cm-thick insulation boards in walls allows the construction of compliant buildings located in cold Moroccan semi-arid climates.

Figure 7b represents the average monthly indoor temperatures of the studied building. The results show that available practices for building construction in Morocco are only efficient for passive heating. Therefore, further studies on passive cooling techniques are greatly required for achieving low-energy buildings in Morocco.

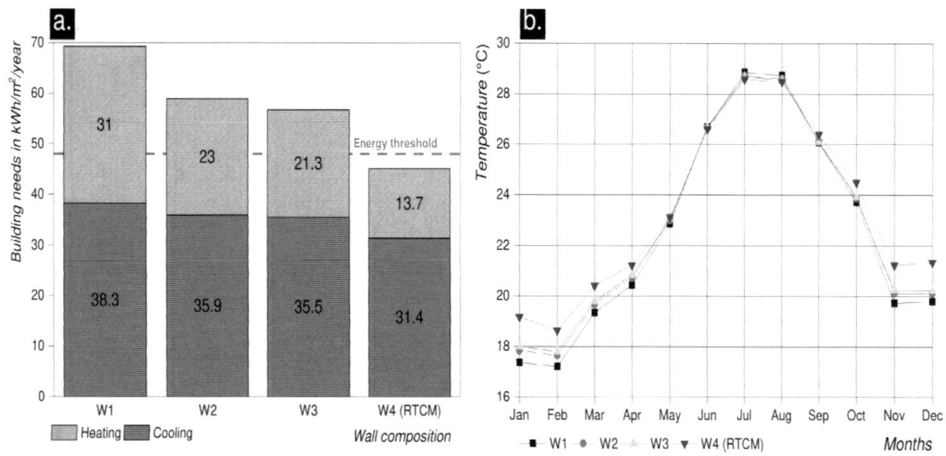

Figure 7. Performance of studied envelopes: (a) annual loads (b) average monthly temperatures.

Besides the high embodied energy of today's industrial building materials, the results show that they are ineffective for designing lightweight thermally efficient buildings, especially in passive cooling. For this reasons, the recourse to traditional architecture for the conception of low-environmental impact habitat models with excellent summer thermal comfort levels is needed [9, 10]. Moreover, further studies on reinforcing earthen construction practices using unfired clay bricks are recommended [11].

5. Conclusion

A local building material database for eastern Morocco was developed. The thermal conductivity of most used XPS insulation boards at local level was found to be around 0.03 W/mK. While, the thermal conductivity values of local cement-based products range from 0.65 to 0.7W/mK. It was found also that the hollow bricks are thermally more efficient than their associated solid bricks. The air pores improve the thermal insulation quality of red bricks. The study of pore's position and geometry are encouraged to design high-performant bricks. Thermal dynamic simulations of a typical residential building located in eastern Morocco region using the studied products were carried out. The finding results show that buildings with simple hollow brick walls are poorly insulated, resulting in a huge energy consumption for maintaining the occupant comfort level. The energy needs of simple hollow brick wall building for heating and cooling exceeds the maximum admissible energy consumption by 30%. The use of double hollow brick walls with an air gap (10 cm) can reduce the annual energy consumption by 15% compared the first case. However, the use of 3cm-thick XPS insulation boards in walls instead of air gap are highly recommended.

References

[1] IEA 2009 Energy Policies beyond IEA Countries: Morocco.
[2] CNRST, Appel à Projets "OMRANINNOV" https://www.cnrst.ma/.
[3] Sghiouri H, Charai M, Mezrhab A and Karkri M 2019 Comparison of passive cooling techniques in reducing overheating of clay-straw building in semi-arid climate *Building Simulation* **13** (1) 65–88.
[4] Aketouane Z, Malha M, Bruneau D, Bah A, Michel B, Asbik M and Ansari O 2018 Energy savings potential by integrating Phase Change Material into hollow bricks: The case of Moroccan buildings *Building Simulation* **11** (6) 1109–1122.
[5] He Y 2005 Rapid thermal conductivity measurement with a hot disk sensor. *Thermochimica Acta* **436** (1–2) 122–129.
[6] AMEE, Propriétés thermos-physiques des matériaux locaux de construction au Maroc.
[7] U.S. Department of Energy DOE 2018 *EnergyPlus* 9.0.1.
[8] Charai M, Sghiouri H, Mezrhab A, and Karkri M 2018 Numerical Study of the Impact of Clay-Straw Walls on the Energy Performance of a Residential Building *6th In-ternational Renewable and Sustainable Energy Conference* pp. 1-5: IEEE
[9] Khalili M and Amindeldar S 2014 Traditional solutions in low energy buildings of hot-arid regions of Iran *Sustainable Cities and Society,* **13** 171–181.
[10] Baglioni E, Mecca S, Rovero L and Tonietti U 2012 Traditional building techniques of the drâa valley (Morocco) *DigitAR - Revista Digital de Arqueologia, Arquitectura e Artes.*
[11] Charai M, Sghiouri H, Mezrhab A, Karkri M and El Hammouti K 2020 Comparative study of a clay before and after fired brick-making process. *Materials Today: Proceedings* **31** S103–S108 doi.org/10.1016/j.matpr.2020.06.250.

IOP Conf. Series: Materials Science and Engineering 1138 (2021) 012016 doi:10.1088/1757-899X/1138/1/012016

Exploring the potential for organic and more environmentally friendly plastics in the building industry

G C Chiţonu[1*] and C E Cazacu[1]

[1] Department of Civil Engineering, Faculty of Civil Engineering, Transilvania University of Brasov, Romania,

* gabriela.chitonu@unitbv.ro

Abstract. Modern building techniques, materials and methods open possibilities for rediscovery and new applications of organic and more environmentally friendly solutions in multiple sectors of the building industry. As more scientists and engineers are looking at sustainable resources, the possibilities for managing the potential weaknesses of making more environmentally friendly building and product materials are bringing new materials and methods to the market at a rapid pace. The new biodegradable solutions are looking into natural fibres, or crops and wood. Hemp-based materials are already well re-established in multiple environmentally friendly products in the building industry. Hemp-lime, hemp insulation or hemp plastics are materials with low carbon footprint which are successfully being used in construction, replacing successfully non-renewable building materials. In cars, the BMW i3 has for several years featured hemp and kenaf, and Hunton Nativo wood fibre insulation is another example from the building industry of how renewable climate friendly materials can easily replace non-renewable materials. This article aims to highlight some existing solutions, and possible improvements to products in the building industry.

1. Introduction

In our modern times, most industry is generally designed to achieve economic growth. This is coming at the expense of other vital concerns, human health and environmental, cultural, and social unbalance. The purpose of the industrial development did probably not intend to bring such devastating effects, over the natural environment, as waste, pollution, crude products (defined as not designed for human and ecological health) With overcrowding, space limitations, physical and psychological needs and desires, economic appetency, including patent right consequences, humans have created a rapid competition for getting new, exclusive, convenient, advanced products with material and chemical properties with side effects or unintended consequences which can lead to severe global disasters, including advanced. Many of these negative consequences are the result of a lack of knowledge about Life-cycle assessments. LCA involve cradle-to-grave analyses of production systems and provide comprehensive evaluations of all upstream and downstream energy inputs and multimedia environmental emissions. The "life-cycle" impacts include the extraction of raw materials; the processing, manufacturing, and fabrication of the product; the transportation or distribution of the product to the consumer; the use of the product by the consumer; and the disposal or recovery of the product after its useful life [1, 2].

Humans have over millennia found, utilised, shaped and discarded, including opted out materials and chemicals to increase their chances of survival and prosperity. They have found natural materials that

IOP Conf. Series: Materials Science and Engineering 1138 (2021) 012016 doi:10.1088/1757-899X/1138/1/012016

came with good and bad consequences and opted out of using many that turned out to be severely negative. The last couple of centuries have driven innovation at such a breath-taking speed, with such a wealth of physical and scientific tools, methods, possibilities and materials. Alongside such an instant ability for also uncritical communication, and with globalisation such a historically unique opportunity to drive one class of product and material as the solution to all problems, "petroleum" has received a market share which has quenched most alternatives, even when there have been equally as good or better alternatives.

2. Oil based plastics, problematic and consequences

With global warming facilitating severe weather changes, global thermal contrasts, more frequent storms, rises of sea levels, shifts in seasons, it demands a rethink of the industrial development approach. Incinerating fossil fuels contributes microscopic particles of soot to the environment, causing respiratory and other health problems [3].

Plastic materials over the last 40 - 50 years have become problematic for the natural environment, human habitat and most alarming for the world's oceans, with its long time of decay in nature. Currently, the world plastic production is mainly based on non-renewable resources, and non-biodegradable.

The million tonnes of plastic litter that end up in the oceans every year are one of their most visible and alarming signs of these problems, causing growing public concern.

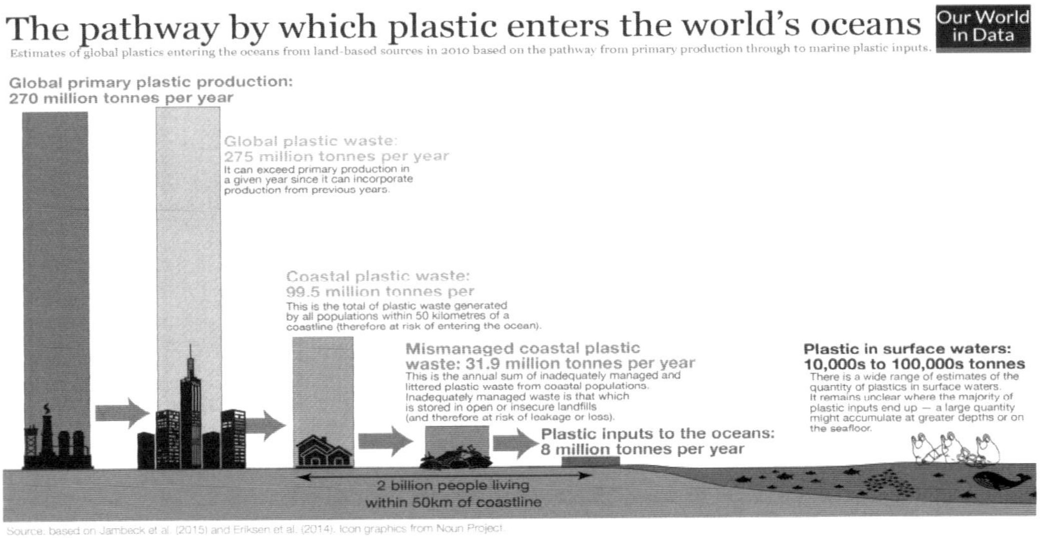

Figure 1. Plastic pollution in the world's oceans [4].

The plastics industry is very important to the European economy, and increasing its sustainability can bring new opportunities for innovation, competitiveness and job creation, in line with the objectives pursued by the renewed EU Industrial Policy Strategy [5].

A smart, innovative and sustainable plastics industry, where design and production fully respect the needs of reuse, repair, and recycling, brings growth and jobs to Europe and helps cut EU's greenhouse gas emissions and dependence on imported fossil fuels [5].

3. Methods of evaluation. Alternatives of biodegradable plastic

On the path of transition to Circular Economy (CE), in a sustainable economic growth, EU is supporting all partners to develop an European plastic industry, based on bio-plastic.

"A new but growing industry, bioplastics can play an important role in this transition. Encompassing a whole family of materials with different properties and applications, bioplastics can be made from renewable resources such as crops and wood, or from waste streams such as the residues of food processing [6].

With the emergence of more sophisticated materials, applications, and products, the global market is already growing by about 20 to 100% per year. By 2021, it is expected that Europe will possess around a quarter of the world's bioplastics production capacity" [6].

European Bioplastics (EUBP), as part of the European Bioeconomy Alliance (EUBA), an informal alliance of 12 leading European organisations is currently representing sectors active in the bioeconomy in Europe, has formulated specific policy asks for the revision of the EU Bioeconomy Strategy to help stimulate the bioeconomy in Europe [7]. Sustainably, efficiently and competitively produced and recycled biomass from the agricultural, forestry and marine sectors, and from waste sources can contribute significantly to Europe's climate commitments as well as to socio economic targets, such as creating more jobs, growth and viable rural areas. It can also help reducing Europe's dependence on imports while increasing the security of supply. In addition, it can enhance biodiversity through sustainable management of land and resources [7].

Despite of some limitations, LCA is a used method to evaluate the bio-plastic products, including the big range of raw natural resources.

LCA thoroughness and accuracy will depend on the availability of data; gathering of data can be problematic; hence a clear understanding of the uncertainty and assumptions is important.

Classic LCA will not determine which product, process, or technology is the most cost effective or top-performing; therefore, LCA needs to be combined with cost analysis, technical evaluation, and social metrics for comprehensive sustainability analysis.

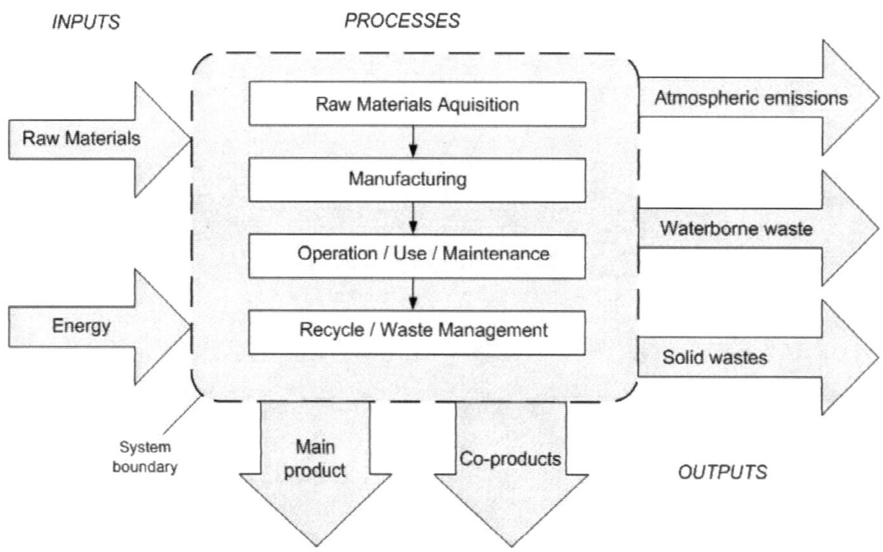

Figure 2. Main stages and typical inflows and outflows considered in lifecycle assessment [2].

The utilization of LCA method can help in the following [8]:

• searching for the most available life cycles, e.g., those with minimal negative impact on environment,

• assuming the decisions in industry, public organizations, or NGOs, which determine direction and priorities in strategic planning, design or design product, or process change,

• choosing important indicators of environmental behavior of organization including measurement and assessment techniques, mainly in connection with the assessment of the state of its environment,

• marketing with the link on formulation of environmental declaration or eco-labeling.

A vast array of plastic products cannot easily, affordably and often actually be replaced with biodegradable alternatives. In applications of extreme stress, rough weather, many chemical, medical and other applications there are few if any more environmentally friendly alternatives to oil-based plastics.

The environmental footprint of these oil-based plastics can be reduced by better managing the materials, by choosing more suitable plastics for the various products and tasks they are designed for, and by recycling more of the plastics. Recycling of plastics is of course a complicated field as there are often uncertainties surrounding the material properties of a plastic mix, which makes it very risky for manufacturers to take the chance on reusing previously molded plastics.

When wanting to help save the environment it is important to delve into the details and understand the challenges associated with recycling. In many cases the cleaning processes, the transport and additives that will be used when recycling plastics and making it ready for being shaped or molded again, can have equally as resource intensive environmental footprints or even worse than using new plastics.

For a wide array of plastic products there are however options, and enormous potential for improvement and more environmentally friendly life cycle footprints by switching to biodegradable products, organic, waste based or even the most easily reusable of petroleum-based plastics. Innovations in similar types of plastics to oil based using as an example AirCarbon with its methane-based production provide more environmentally friendly versions of the types of plastics which it is hard to replace with more traditional plant based or organic alternatives.

AirCarbon has recently been certified as carbon negative by the Carbon Trust [9].

Biodegradable plastics is not with the annual needed volume of material for plastic manufacturing, a solution which can be quickly established without negative consequences along the human and animal food chain as an example. However, by changing the mentality of the designers, and making slight changes to the product specs for many low stress/loads products then resources can be saved in the millions of tonnes of materials + energy annually.

Challenges for companies designing products and manufacturers exists in needing their products to hold up to very high material standards to feel secure about not having to replace products. This is a reasonable way of thinking as many products aren't used in accordance with the intentions, and many users do not read instruction manuals or try to follow them.

Just like BMW, Mercedes, Bentley, Volkswagen and others are making revolutionary changes to the car and manufacturing industries by deliberately seeking to go green by including materials like hemp and kenaf, the building sector still has a huge potential for replacing energy intensive, toxic, polluting and resource intensive products with more sustainable alternatives [10]. Products like spacers, casings, housings, decorative fixtures, ventilation pipes, furniture, kitchen utensils, handles, fans, clothing hooks, bath mats are just a few examples of products that can be made with easily degradable plastics based on organic or biodegradable plastics. Innovative products like QMilch, AirCarbon, algea based bioplastics and many other alternatives can increase the array of possible products and applications, where more sustainable plastics can replace more problematic ones.

4. Hemp industry

There is considerable potential for using environmentally friendly plastics in the building industry.

Hemp plastics is one easy to use solution in this regard, where it can be used in much of the packaging materials for building sites, and for example n products like spacers, baseboards, chair rails, door casings and crown moldings.

Figure 3. Hemp shiv and hemp-lime, building materials.

Figure 4. Hemp plastic.

Figure 5. Hemp pellets.

The hemp industry provides alternatives to pure fossil fuel-based plastics. The industrial hemp plants (which although they are in the Cannabis sativa plant family, are different from Marijuana, as it contains less than 0,3% tetrahydrocannabinol, THC, and will not create a psychoactive effect) contains cellulose fibers and shiv which is the woody core of the hemp plant. Extracting hemp cellulose from the raw industrial hemp material by crushing the stems and separating the cellulose fibers from the shiv, the cellulose can be used to make fabric, paper, concrete, bioplastics and biofuels.

Utilising the hemp for cellulosic ethanol, the hemp plant provides a good alternative to the oil or sugar that is currently used for the majority of fuel.

The hemp shiv is a very porous material with a low bulk density. This makes it a good product for use in insulation both for acoustic and thermal applications [11].

The hemp industry is continuing growing and is proving its efficiency alongside eco-friendly bioplastic, bringing the opportunity to manufacture plastics products that are more earth-friendly. They are also starting to do this at a comparable price to their fossil fuel-based counterparts.

Key Features of Hemp Plastic Bioplastics:

- a near term solution to a customer's sustainability objectives,
- colours - 'earthy' appearance,
- 10% lighter than many other filled plastics
- immediately Positive – Reduce CO_2 and Consume Waste,
- consume a massive and growing industrial waste problem [12].

The products feature significant improvements in CO_2 sequestration, use renewable sustainable materials, and, in some cases, are compostable and are generally designed to be drop-in replacements for conventional fossil fuel-based polymers.

The cost factor related to hemp and kenaf is currently impacted by the availability, but both plants grow very fast, have a wide array of use, and has the potential for significant economic benefits for both farmers and customers. With the highly underemphasized petroleum-sector subsidies being phased out, the costs of the alternatives could start evening out more, but there can be little doubt that the COVID-

19 pandemic effects on the oil prices, and the emergence of electrical alternatives for transport and heating, are not helping the sustainable alternatives. If the oil prices remain low over time, then the environmentally friendly alternatives may need help from support schemes, or fees, taxation, unless one sees the same type of consumer demand initiatives which impacted palm oil in household products in the Nordic countries [13]. Hemp has a significant potential for producing biofuel that can directly fuel the vast majority of current car types [14].

Kenaf is used in engineered wood, fibre boards, insulation, packing material and more for the building industry, while hemp is used in bricks, hempcrete, hemp wood and insulation to mention a few products. A key product for households where there could be good reasons for switching from petroleum-based plastics is growing mats for plants.

5. Conclusions

By better handling our resources, and finding materials that are good enough for their applications, rather than always over shooting for the strongest and deemed safest materials, energy, material, processing and pollution reductions can be made in the millions of tonnes annually. Many sustainable plastic types have advantages over many of the oil-based plastics in less toxicity, less allergies, less negative health consequences, reduced shrinkage when shaping and less environmental consequences when the product wear over time. Products in the building industry have a significant potential for being made from environmentally friendly and/or biodegradable plastics. These are perfectly suited for use in many indoor and limited weathering spaces in buildings, and can bring both considerable emissions and pollutions reductions and improve the indoor climate of buildings.

Changing our habits and the way we design products is possible today, and can bring substantial environmental and economic improvements for tomorrow.

References

[1] Wiley, 2015 *Handbook of Clean Energy Systems*, Jinyue Yan (Editor), Publisher
[2] EPA/600/R-06/060, 2006 *Life Cycle Assessment: Principles and Practice*
[3] McDonough W and Braungart M 2012 *Cradle to cradle: remaking the way to make things.* Distributed in Canada by Douglas & Mclntre LTD, ISBN-13; 978-0-86547-587-9
[4] https://ourworldindata.org/plastic-pollution
[5] EU – An European Strategy for Plastics in a Circular Economy
[6] research*eu, *RESULTS PACK*, Published by The Community Research and Development Information Service (CORDIS) managed by the Publications Office of the European Union 2, rue Mercier 2985 Luxembourg LUXEMBOURG, cordis@publications
[7] https://www.european-bioplastics.org/about-us/mission/More references
[8] Iyyanki V. Muralikrishna and Valli Manickam, 2017 *Environmental Management, Science and Engineering for Industry* Butterworth-Heinemann Publishing
[9] https://www.prnewswire.com/news-releases/newlight-technologies-aircarbon-material-has-been-certified-carbon-negative-by-the-carbon-trust-301138193.html
[10] Deaconu O 2014 *Conditions of the constructions sustainability in natural environments* CIBv 2014 published at Bulletin of the Transilvania University of Braşov
[11] Diquelou Y, Gourlay E, Arnaud L and Kurek B 2015 *Impact of hemp shiv on cement setting and hardening: Influence of the extracted components from the aggregates and study of the interfaces with the inorganic matrix* (Cement and Concrete Composites) 55:112-121
[12] https://www.hempplastic.com/
[13] https://news.mongabay.com/2012/07/campaign-cuts-norways-palm-oil-consumption-64/
[14] Sankpal S and Naikwade P 2013 *Important Bio-fuel Crops: Advantages and Disadvantages* International Journal of Scientific and Engineering Research 4(12):1-5

IOP Conf. Series: Materials Science and Engineering 1138 (2021) 012017 doi:10.1088/1757-899X/1138/1/012017

Considerations on the efficient functioning of the urban water pumping stations

V Cojanu[1*] and E Helerea[1]

[1] Department of Electrical Engineering and Applied Physics, Transilvania University of Brasov, Romania

* vlad.cojanu@unitbv.ro

Abstract. Current urban water systems aim to ensure hydraulic and water quality parameters for all consumers connected to the water distribution network, with the highest possible efficiency. This corresponds to reduce primary energy consumption and a responsible control of drinking water sources in order to supply water to urban areas. Most of the time, a low efficiency of water supply systems is due to the following factors: inefficient operation, poor maintenance, incorrect sizing and unmetered water consumption. This article analyzes the efficiency of the solutions applied in the pumping station as a component part of the water supply systems. The case study focuses on the operation of pumps in water supply systems and presents the analysis of the relationship between water consumption and electricity of a pumping group in an urban water supply system. This paper is a first step in order to obtain global parameters for energy optimization of the urban water system.

1. Introduction

The term of drinking water include water intended for human consumption, which can be: water in its natural or treated form, used for drinking, food preparation or other household purposes or water used as a source in the food industry.

Council Directive 98/83/EC regulates water quality parameters for all the member states of European Union and, at national level Law 458/2002 in actualize form is applied. The water quality indicators that are monitored are: physical indicators, chemical indicators, biological indicators, bacteriological indicators.

The term Urban Water System (UWS) defines all the systems that serve a city in terms of water. UWS consists of a number of subsystems (figure 1):

- Urban Water Supply System, which have the role of capturing and transporting water from the source to the treatment area;
- Urban Water Purification System, which have the role of treating water for drinking;
- Urban Distribution System, having the role of supplying consumers with water, it is often the largest and most complex component of UWS;
- Urban Water Draining System, which has the role of collecting waste water and treating it;
- Urban River System, which is composed of all the drainage basins that flow through an urban area [1].
- Our research has focused on the Urban Water Supply System and Urban Distribution System, where the main energy consumers are pumps.

2. Research methodology on pump efficiency

This research was made by consulting various online data-bases (Science Direct, IEEE, Springer Link, Google Scholar and other on-line sources) using a predefine search algorithm. For an increased accuracy the scientific literature was filter based on research title, keywords, year of publication and subject areas to determine the relevancy of the sampled research.

Water Energy Nexus (WEN) concept is the best way to understand the relationship between water and energy. This concept is an important factor in the present urban development and for the future developments. In the last three years WEN has been studied in order to understand and to increase the UWS global efficiency.

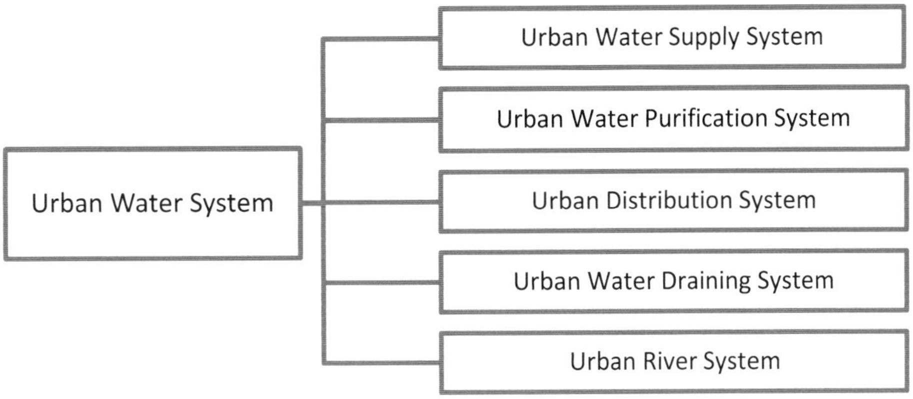

Figure 1. Subsystems of Urban Water System.

According to Lam et al. [2] the most important consumers of the UWS are the pumping stations and therefore it is an important task to increase their energy efficiency. The study was conducted on a population of around 170 million persons which live in 30 major cities.

Blinco et al. [3] propose an optimization strategy based on dynamic- level-feedback-control approach to increase the efficiency of a studied tow stage pumping system. A large amount of energy is being used for water pumping in UWS. The energy cost for water suppliers in a municipal area is 65% of their total operating costs. A 7% energy reduction cand be obtained by optimizing the driven energy and driven cost for a tow stage pumping system with uncertain water demand for a better reliability, durability and economy.

Another approach on pump efficiency can be made by controlling the pumps speed to be closer as possible to the best efficiency point (BEP). The ideal case is when the pressure demand and water flow are delivered with the minimum energy consumption. The manufacturers are providing the efficiency characteristic and the performance of each pump type [4].

Supervisory Control and Data Acquisition (SCADA) are complex systems that help control an UWS. These systems are complex and need constant maintenance and investments. Salomons et al. [5] present a simple and practical methodology for real time pumping scheduling without computation hardware. This methodology can be deployed on the pumping stations standard handwear for reducing the operation cost and the energy consumption.

Considering the rewired studies, this was the motivation to monitor the electric energy consumption of pump systems and to establish the ways for improving the efficiency of this system.

The case study was conducted using gathered data from a part of Brasov's UWS that cover an area of 0.4 km^2 in which the drinking water is being pumped to the consumers. We propose to manage the measurement of the hourly electrical energy consumption of the pumping station for a period of 30 days, for 1 January to 31 January 2020, and to analyze the pumps consumption, pump efficiency and system reliability.

2.1. Centrifugal pumps efficiency

Hydraulic machines are technical systems that transform mechanical energy and hydraulic energy in both directions depending on the application. Hydrodynamic pumps, such as centrifugal pumps, are machines in which the mechanical energy is converted in hydraulic energy, using an impeller which interacts with the working fluid. The energy received by the fluid from the impeller is in the form of kinetic energy and potential pressure energy [6].

Centrifugal pumps are widespread in areas such as: water industry, sewage, agriculture industry and petroleum industry. The main advantages of this constructive type are: reliability, simple engineering, ability to transport high flows and low maintenance.

The mathematical relation mechanical power which converted in hydraulic power. ns (1) and (2) define the hydraulic energy E_h and power P_h for a centrifugal pump:

$$E_h = \rho \cdot g \cdot V \cdot H \tag{1}$$

$$P_h = \rho \cdot g \cdot Q \cdot H \tag{2}$$

where: ρ- fluid density, g- gravitational acceleration, V- water volume, H- pumped head in m, and Q- water flow in m^3/s.

If the power needed for supply the electric motor of the pump is P_e and the efficiency of the motor in η_e, a relationship for the efficiency of the pump results:

$$\eta_p = \frac{P_h}{P_e} = \frac{P_h}{P_m} \cdot \frac{P_m}{P_e} = \eta_h \cdot \eta_e \tag{3}$$

where η_p is the pump efficiency and P_m is the mechanical power, which is converted in hydraulic power.

There are empirical relations which describe the average pump efficiency in relation with fluid flow Q, as those proposed by Araceli Martin-Candilejo et al. in [7]:

$$\eta_p = 0.1286 \cdot ln(2.047 \cdot ln\, Q - 1{,}7951) + 0.5471 \tag{4}$$

where: Q is fluid flow in l/s.

2.2. *Ways for pump efficiency optimization*

One important step is to increase the efficiency of the motor. Nowadays, electric motors used for drive the pumps are of efficiency series E3 even E4, of high efficiency.

Another important step is to optimize the operational schedule. The most frequently method that is been used to operate the pumping station is by using the feedback control strategies such as pressure or water level monitoring. Using a water level monitoring system, the pumps are being turn on/off according to a predefined water level in the storage area.

Another method that is commonly used in pumping stations operation is by scheduling the functioning program. Due to the constantly changing water behavior of the consumers and the complexity of the system, the scheduling control approaches is more challenging [8].

3. Case study

The case study was conducted using gathered data from a part of Brasov's UWS that cover an area of 0.4 km^2 in which the drinking water is being pumped to the consumers.

This part of the city was manly an industrial area, but in the last five years the area is being constantly changing and evolving, a lot of new buildings are being constructed.

The pumping station is equipped with multistage centrifugal pumps with a nominal flow of 64 m^3/h and a maximum flow of 85 m^3/h. The maximum operating pressure is 16 bar.

The current system was designed for a required flow Qr = 193 m^3/h, at a pressure of P=4 bar. As urban areas are constantly evolving, the system was design with the possibility to add two more pumps to ensure water flow in the future.

3.1. Pump system

The studied pumps are vertical multistage centrifugal type with the suction side and the pressure side in line. These pumps have a cast iron base with three impellers and tow reduced diameter impellers made of stainless still. The net weight of a pump is 179 kg.

For a safe operation is recommended to not exceed the Ambiental temperature of 60°C. Every pump is driven by a three-phase 2 poles motor rated at 28 A, 380V. The starting current can vary between 660% and 780%. In table 1, the electrical characteristics according to the producer are shown.

According to the producer η_p is between 60 - 80% and is directly influenced by Q and H. BEP is obtain at $Q_{BEP} = 65.6$ and $H_{BEP} = 55.5$. The pump performance curves are represented in figure 2.

Table 1. Pumps - electrical motor data [9].

Rated power:	15 kW
Main frequency:	50 Hz
Power factor:	0.89-0.87
Number of poles:	2
Rated speed:	2930 - 2950 rpm
Rated full-load torque:	49 Nm
Efficiency according to IE3:	91.9%

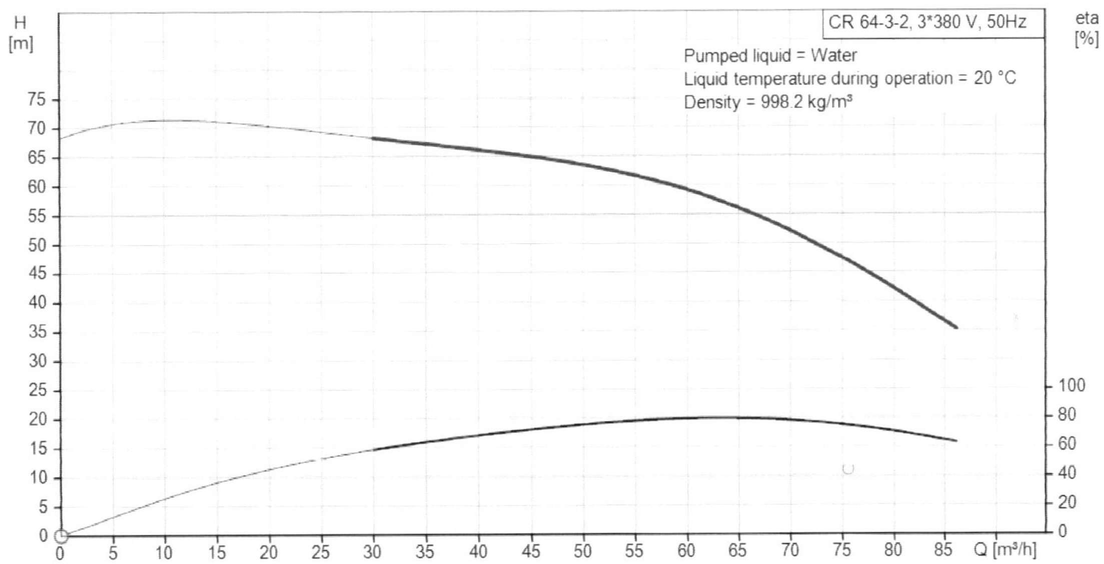

Figure 2. Pump performance curves [10].

3.2. Measuring data

We manage to measure the hourly electrical energy consumption of the pumping station for a period of 30 days, for 1 January to 31 January 2020. A total of 744 values were measured according to a predefine interval of 60 minutes starting with 0 hours of each day.

Figure 3 represents the hourly electrical energy consumption of the studied pumps that is directly related to the water consumption.

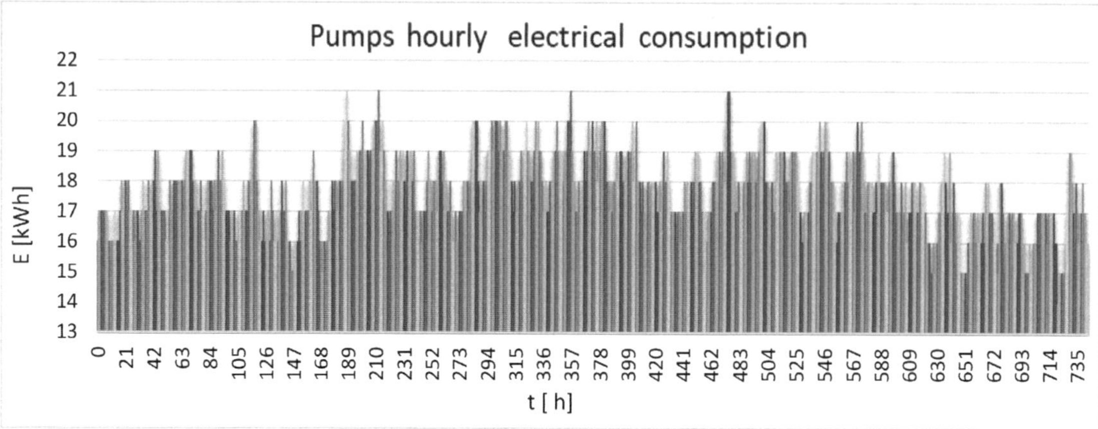

Figure 3. Hourly electrical consumption.

Figure 4 represents the daily electrical energy consumption of the studied pumps that is also directly related to the water consumption.

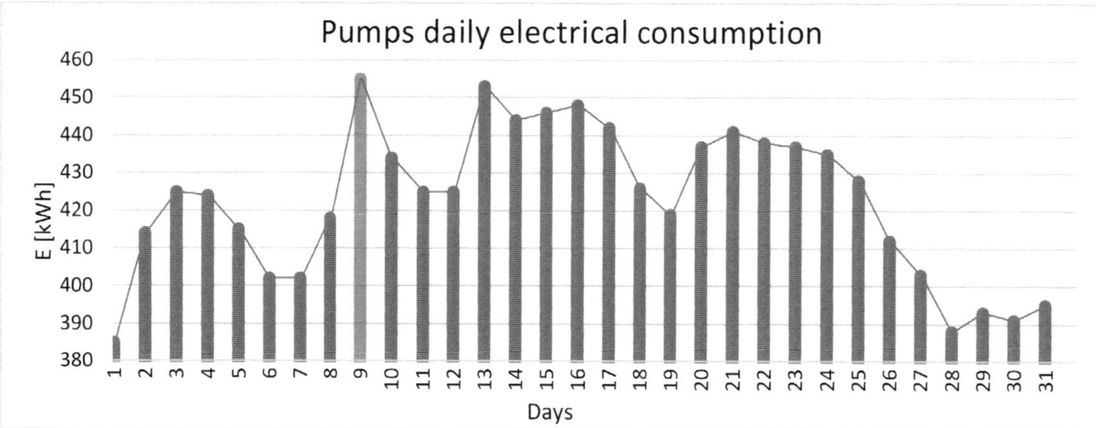

Figure 4. Daily electrical consumption

3.3. Data proceeding

The analysis of hourly measurement data (figure 1) has shown a maximum hourly consumption was E_{max} = 21 kWh/h. This value was reached on 08.01, 09.01, 15.01 and 20.01. On 20.01.2020, the maximum hourly consumption for both days 21 and 22 was also reached. These measured values of consumption peaks are obtained in the second part of the day.

The minimum electric energy consumption value of E_{min} = 14 kWh/h was recorded on 28.01 and 31.01. Unlike the peak consumption which is obtained in the evening, the minimum consumption is realized in the morning.

The average hourly consumption for January 2020 was E_{av} = 17.6 kWh/h.

Analyzing the daily electrical consumption (figure 4), we manage to identify the minimum and maximum consumption. The daily peak consumption of E_{max} = 455 kWh/day was measured in 09.01, the same day with the maximum hourly consumption.

The minimum consumption was recorded in 01.01, and was E_{min} = 385 kWh/day.

For establish the correlation between the electric energy consumption and water flow, the relation (4) is applied. Figure 5 represents the pumps efficiency for the maximum and minimum daily consumption.

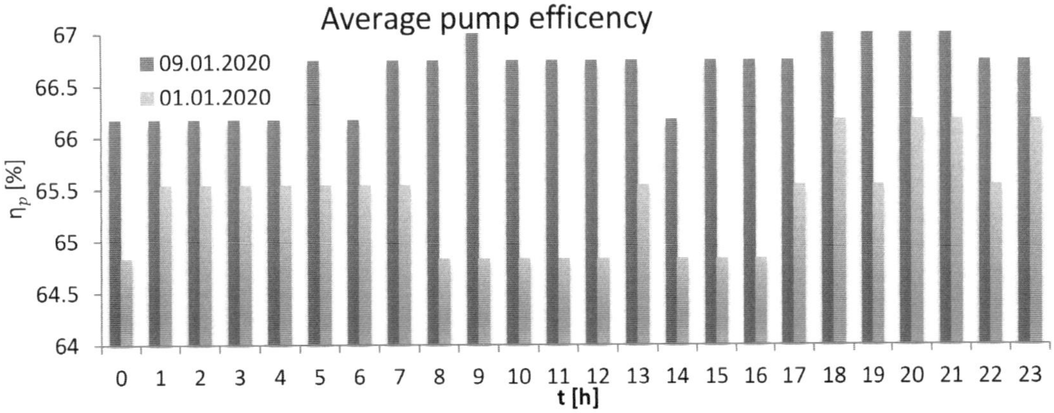

Figure. 5 Pumps average efficiency for 01.01.2020 and 09.01.2020.

The obtained overall efficiency of the pumps was between $\eta_{pmin} = 64.8\%$ and $\eta_{pmax} = 67.73\%$. The average efficiency for the maximum consumption day (09.01.2020) was $\eta_{pav} = 66.7\%$.

According to performance curves at nominal point the global efficiency is around 80% [10] but we obtain an average efficiency of 66.7% which means the pumps are working far from BEP.

4. Conclusions

During the analyzed period, water consumption was $V = 20743$ m^3 and the electrical energy consumption was $E = 13116$ kWh, which correspond to a specific energy consumption: $E_s = 0.63$ kWh/m^3. This is an important indicator of the efficiency and will be monitored in the future research to find tools and methods of its reduction.

Following the analysis of hourly consumption, it was observed that the consumption peaks are in the second part of the day, between 21.00-22.00, and the minimum hourly consumptions in the morning, in period 3.00-5.00. This type of water consumption behavior is characterizing for domestic consumers, which means the studied area served is manly a residential area.

Measures proposed for increasing the pumps efficiency should be: operate the pumps according to a well-established schedule, so that the motors have the maximum efficiency; decrease the hydraulic losses from the intake and discharge pipes; regular maintenance, and worn parts replacement.

References:

[1] Ahmad S, Jia H, Chen Z, Li Q and Xu C 2020 Water-energy nexus and energy efficiency: A systematic analysis of urban water systems *Renew Sustain Energy Rev*

[2] Lam KL, Kenway SJ and Lant PA 2017 Energy use for water provision in cities. *J Clean Prod* **143**:699–709

[3] Blinco L J, Simpson A R, Lambert M F, Auricht C A, Hurr N E, Tiggemann S M, and al. 2014 Genetic algorithm optimization of operational costs and greenhouse gas emissions for water distribution systems. *Procedia Eng* **89**:509–16

[4] Koor M, Vassiljev A and Koppel T 2014 Optimal pump count prediction algorithm for identical pumps working in parallel mode. *Procedia Eng* **70**:951–8

[5] Salomons E and Housh M 2020 Practical real-time optimization for energy efficient water distribution systems operation. J *Clean Prod* **275**:124148

[6] Crăciun O M, Țârulescu R and Ungureanu V B 2012 Mașini și aparate fluidice. Brașov: Editura Universitățoo Transilvania din Brașov

[7] Martin-Candilejo A, Santillán D and Garrote L. 2020 Pump efficiency analysis for proper energy assessment in optimization of water supply systems. *Water (Switzerland)* **12**

[8] Chen W, Tao T, Zhou A, Zhang L, Liao L, Wu X, and al. 2021 Genetic optimization toward operation of water intake-supply pump stations system. *J Clean Prod* **279**:123573

[9] *** Grundfos - MG160MD n.d. https://product-selection.grundfos.com/products/mg-ml/mg-ml-160/mg160md-85U17526?productnumber=85U17526&custid=GMA&tab=variant-specifications (accessed December 2, 2020)
[10] *** Grundfos - CR 64-3-2 A-F-A-E-HQQE n.d. https://product-selection.grundfos.com/products/cr-cre-cri-crie-crn-crne-crt-crte/cr/cr-64-3-2-96123487?productnumber=96123487&custid=GMA&tab=variant-curves&pumpsystemid=1155343516 (accessed December 4, 2020)

IOP Conf. Series: Materials Science and Engineering 1138 (2021) 012018 doi:10.1088/1757-899X/1138/1/012018

Conversion of industrial liquid construction waste with plasma on hydrogen

I Cucos[1*], D N Isopescu[2], I Antonescu[2], C I Alecu[3], V Caunii[2], V C Cucos[1] and M O Agavriloaie[1]

[1] Faculty of Hydrotechnics, Geodesy and Environmental Engineering, "Gheorghe Asachi" Technical University of Iasi, Romania.
[2] Faculty of Civil Engineering and Buildings Services "Gheorghe Asachi" Technical University of Iasi, Romania.
[3] "Gheorghe Zane" Institute for Economic & Social Research, Romanian Academy, Iasi

* icucos@tuiasi.ro

Abstract. This paper presents a modern system for conversion with plasma with hydrogen of industrial liquid waste which has implemented a control system through thermoregulatory with PID control laws. Modeling and simulation of thermochemical plasma process performed using specialized software Matlab & Simulink, Ansys allows the optimum control of the temperatures and chemical reactions of the plasma reactor on hydrogen, allowing the conversion of hazardous industrial liquid waste into electrical and thermal energy and useful raw materials that can be reused in the production process.

1. Introduction

The main objective is to design a plasma eco-innovative technology to optimize syngas by controlling the plasma conversion of liquid waste according to the results obtained from modeling and simulation of thermochemical plasma processing. The management of plasma hydrogen reactor is controlling the temperature of the reactor so that it can evolve as a profile imposed by technology, designed based on the information on the type, chemical composition, viscosity, quantity and density of dangerous industrial liquids.

The general principles of waste management are [1-4]:

- waste prevention targets to reduce emissions,
- reduction of hazardous substances in material streams and increase resource efficiency,
- preparation for reuse involves checking, cleaning or recovery by which products or components of products that have become waste are prepared for reuse,
- energy recovery from waste and other recovery activities.

2. The principal technologies for treating of dangerous industrial liquids.

2.1. Comparative analysis of the technologies used for the treatment of industrial liquid waste

The following figure presents a comparative analysis of the main methods used for the treatment of industrial liquid waste with the highlighting of technical and economic particularities [1-8].

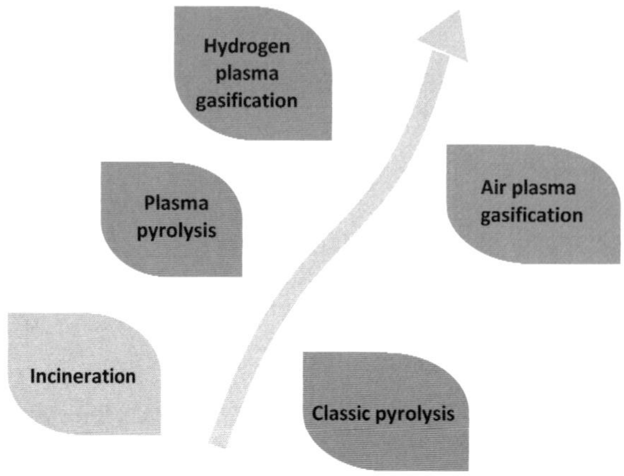

Figure 1. The increase in the efficiency of the technological process

a) Incineration [1, 2]:
- an auxiliary fuel is used to start the process, the combustion temperature is reduced;
- gases with a low calorific value, a different chemical composition and a low degree of purity are produced;
- the process has a negative impact on the environment;
- the amount of inert waste is high;
- the area occupied by the installation is large.

b) Clasic pyrolysis [3-5]:
- the process temperature is high;
- gases with a low calorific value, a different chemical composition and a low degree of purity are produced;
- the process has a negative impact on the environment;
- the amount of inert waste is high.

c) Plasma pyroysis [3-5]:
- the temperature in the reactor is high above 2000°C;
- a small amount of syngas used in cogeneration equipment is produced;
- the amount of inert waste is more than 15%.

d) Air plasma gasification [8-10]:
- the temperature in the reactor is high over 4000°C;
- high-purity, high-calorie syngas is produced and used for cogeneration equipment for energy production;
- the gases emitted into the atmosphere have a minimal impact on the environment;
- the amount of waste is reduced approx. 5 - 10%;
- the installations are large, require expensive consumables, the carbon footprint of the equipment is reduced.

e) Hydrogen plasma gasification [10-14]:
- the temperature in the reactor is very high above 10000°C;
- high-purity, high-calorie syngas is produced and used for cogeneration equipment for energy production;
- by controlling the chemical reactions in the reactor, raw materials can be obtained, the gases emitted into the atmosphere have a minimal impact on the environment;
- the amount of waste is reduced approx. 5%;

IOP Conf. Series: Materials Science and Engineering 1138 (2021) 012018 doi:10.1088/1757-899X/1138/1/012018

- the installations are compact, small in size, fully automated, the carbon footprint of the equipment is reduced.

The following figure shows the system for conversion with hydrogen plasma for hazardous recyclable liquid waste, the authors have redesigned plasma torches on hydrogen and system for controlling the variable power sources.

Figure 2. The system for conversion with plasma with hydrogen of industrial hazardous recyclable liquid waste

Designing a system for automatic control of a reactor temperatures involves the following steps:

- mathematical modeling for the prediction of the technological parameters of the plasma – chemical gasification unit;
- mathematical characterization of the behavior of the controlled plasma hydrogen reactor and acting on operating mode, setting goals adjustment determined by the type of reactor process management, external signals and nature awareness of the mathematical model of the reactor, establishing the adjustment of the objectives determined by the management of the process in the reactor and the external disturbing signals;
- establishing the influence of certain technological factors specific to the plasma conversion technology of municipal waste;
- modeling the ordering process of the installation, establishing the agreement parameters for thermoregulators (classical regulation, based on fuzzy neural networks);
- choice of method design involves optimizing parameter values agree, establish criteria for the selection and award regulators and determining optimal control algorithms, identification experimental model driven process, identification is performed online in safe working condition without raw material inside the reactor, as input data becomes available through measurement;
- simulating the control structure of the plasma hydrogen reactor using the Matlab & Simulink package;
- test design and analysis of algorithms, possibility of achieving implementation, the optimal choice of equipment that ensures a precise implementation of the algorithm of management;
- finite element analysis of the evolution in time of the temperature and heat flow of the raw material subjected to heat conducted in the reactor with thermo-regulator controlled PID algorithm;
- experimental identification of the driven process model, the identification is performed online in the normal working conditions of the installation, as the input data become available by measurement;

- implementation on the plasma - chemical gasification unit of the simulated control structures and following the evolution in time of the monitored parameters.

The advanced temperature control of the hydrogen plasma reactor has been achieved with the help of thermoregulators that have implemented algorithms of regulation based on fuzzy neural networks that allow the acquisition of data and real-time decisions in conditions of risk and uncertainty that characterize the conversion with plasma of municipal waste.

2.2. Design of multivariable control system from plasma hydrogen reactor

Simulation of the control process with the Simulink software package (PID and one-step predictive) and optimization of parameters for process management. Automatic control algorithms for temperature control in the plasma - chemical gasification unit, the main problem of the management of the plasma - chemical gasification unit is the temperature control so that it evolves according to the profile imposed by the technologist [15-18].

For the implementation of one-step predictive control, the non-interactive open circuit decoupling structure of figure 3.

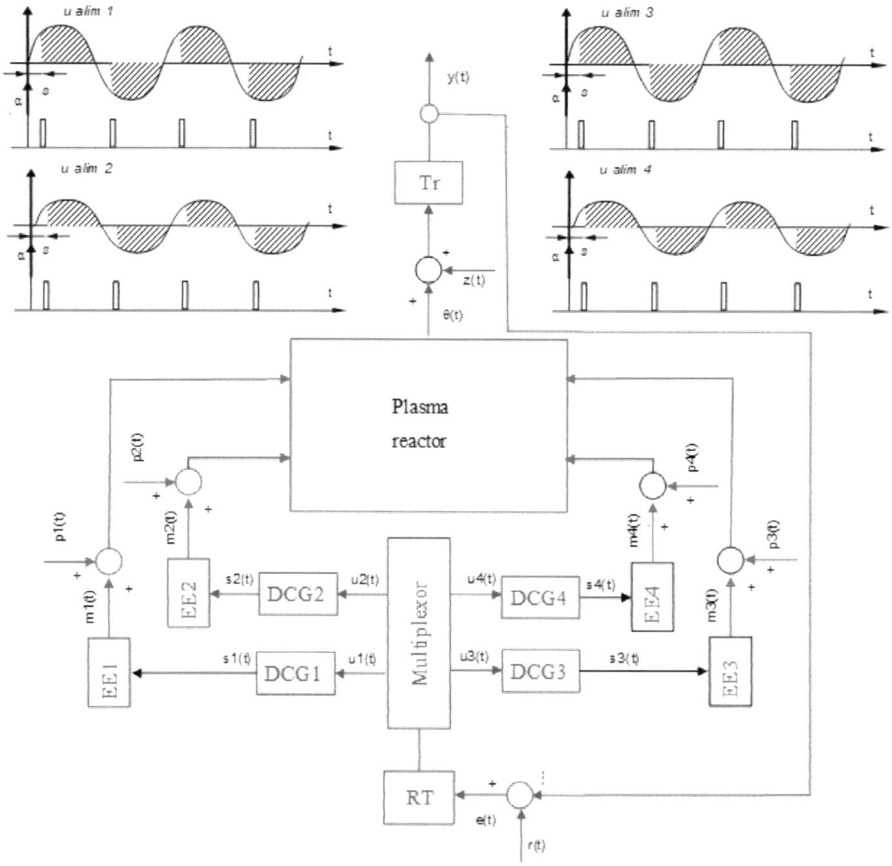

Figure 3. Non interactive multivariable adjustment of a reactor with two temperature zones

The control of the plasma-chemical gasification unit must take into account the following aspects:

- the regulated temperature profile of plasma hydrogen reactor are machines with very long delay times, errors due to adjustment must be 0;
- the quality of the thermal process is given by the accuracy of the control system.

Thermal processes can be accurately modeled with transfer functions for first-order elements with dead time. PID algorithms are sometimes successfully implemented in the control of such reactors but downtime has unfavorable effects in regulation, the best regulation can be achieved using a one-step predictive algorithm [17-19].

Implementing the plasma hydrogen reactor control structures and tracking simulated time evolution of track parameters, validation solution by analyzing the performance of the entire control system implemented allows the reactor if necessary, redesign or adjustment parameters for the operating agreement.

Modeling and simulation of the industrial operation of the hydrogen plasma reactor controlled by termoregulators with PID algorithm one step ahead was realized with the help of the Matlab & Simulink.

The experimental tests were performed on the plasma-chemical gasification unit shown In figure 3 modified in the sense that the two heating zones are supplied separately with electricity through a system based on thermoregulators controlled with classical, PID and one-step predictive algorithms.

The sampling of the data from the reactor (temperature measured by optical pyrometers located in the jacket) is performed with a data acquisition plate mounted in the computer that controls the plasma - chemical gasification unit, this information is processed by the mathematical model to determine the agreement parameters of thermoregulators. which adjusts the plasma supply voltage separately for each area of the reactor.

The sampling period used by the control system algorithms is 5 sec equal to the thermal inertia of the optical pyrometers used to measure the temperature in the installation. The actual results from the installation obtained from the plasma control were plotted with a sampling period of 10 sec.

Real-time implementation of the temperature regulation structure with one-step predictive algorithm. To improve the performance of the dead time control system used to control the heat treatment reactor, we used in the following experiment thermoregulators controlled with one-step predictive type control algorithm.

2. 3. Design of main components of the plasma reactor

Plasma torches on hydrogen have been redesigned so that fluid flow is swirling to uniformly blend spray liquids and compressed air resulting a homogeneous flame with high temperature and very high dispersion, figure 4 is a part of the modeling and simulation performed in Ansys software (figure 4, 5, 6) that have been validated experimentally on the existing plant in the research laboratory.

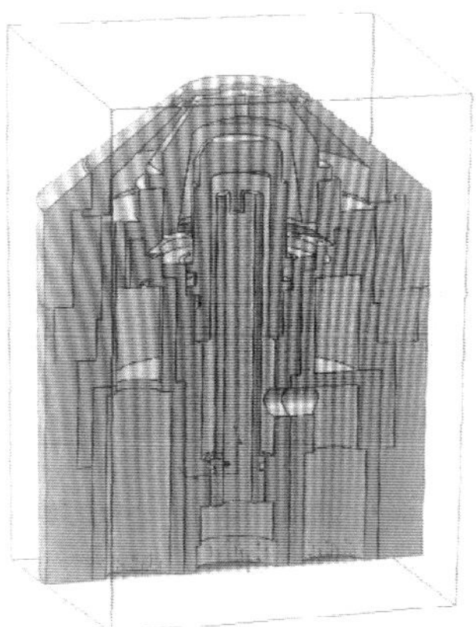

a) 3D drawing of plasma on hydrogen

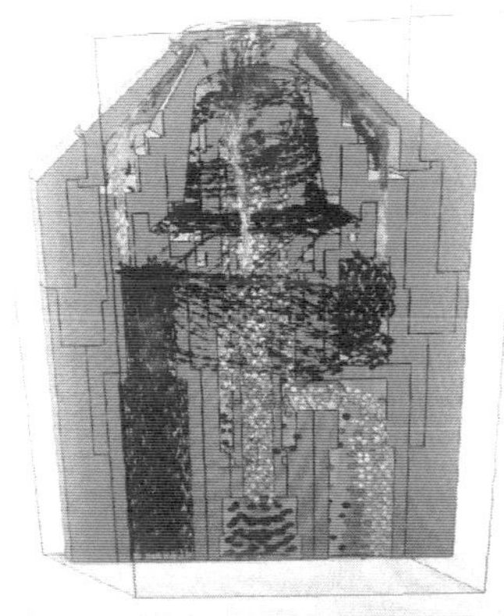

b) Fluid flow simulation

Figure 4. Plasma torches on hydrogen, nozzle diameter 2,5 mm

The temperature generated by the plasma on hydrogen varies over 4000 - 5000 °C for this reason the cooling is very important, the agents used for cooling the plasma are air or water, the main component of the cooling equipment is the air or water turbine [20, 21].

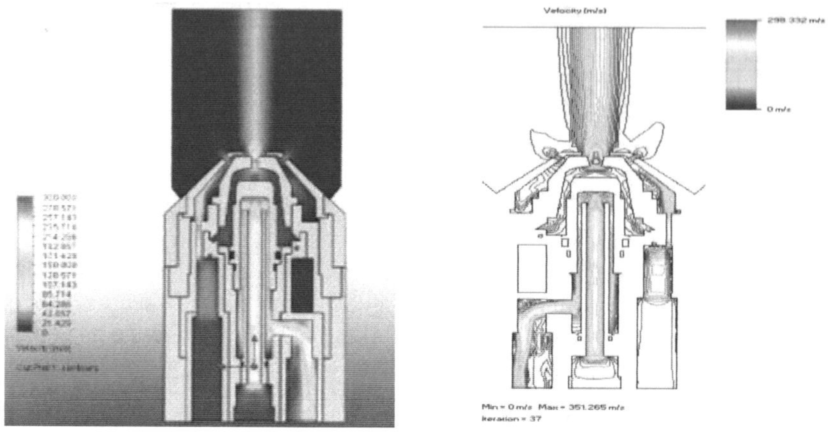

Figure 5. Modeling of fluid flow and thermal analysis of plasma on hydrogen nozzle diameter D = 1.5 mm

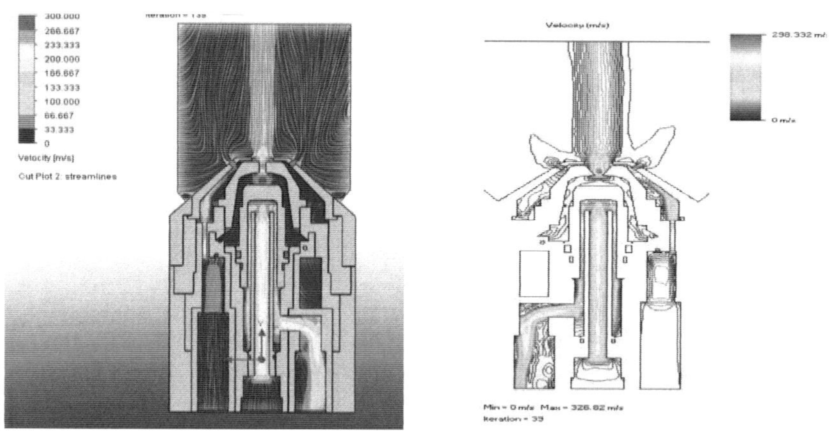

Figure 6. Modeling of fluid flow and thermal analysis of plasma on hydrogen nozzle diameter D = 2.5 mm

3. Conclusions

The main objective is the design of an eco-innovative plasma-based technology for the recovery of waste as a renewable energy source, improvement of the technical and economic performance of the plasma installation for the waste processing by upgrading some components of the plant (plasmas, adaptive thermal conduction processes from the plasma reactor enclosure, plasma gas purification system) and optimization of the syngas production technology by controlling the plasma conversion of waste depending on the results obtained from the modeling and simulation of thermo-chemical processes of plasma waste processing with the help of specialized software.

This paper presents a modern system for conversion with plasma with hydrogen of industrial liquid waste which has implemented a control system through thermoregulatory with PID control laws. Modeling and simulation of thermochemical plasma process performed using specialized software Solidworks, Matlab & Simulink, Ansys allows the optimum control of the temperatures and chemical reactions of the plasma reactor on hydrogen, allowing the conversion of hazardous industrial liquid waste into electrical and thermal energy and useful raw materials that can be reused in the production process.

Acknowledgements
This paper was elaborated with the support of the “Eco-innovative Products and Technologies for Energy Efficiency in Constructions – EFECON” research grant, project ID P_40_295/105524, Program co-financed by the European Regional Development Fund through Operational Program Competitiveness 2014-2020.

References

[1] Begum S, Rasul M and Akbar D 2014 A Numerical Investigation of Municipal Solid Waste Gasification Using Aspen Plus, *Procedia Engineering* **90**, pp. 710-717, ISSN 1877-7058

[2] Brunner P and Rechberger H 2015 Waste to energy – key element for sustainable waste management, Waste Management **37**, pp. 3-12, ISSN 0956-053X

[3] Couto N, Silva V, Monteiro E and Rouboa A, 2015 Assessment of municipal solid wastes gasification in a semi-industrial gasifier using syngas quality indices, Energy **93** 1 pp. 864-873, ISSN 0360-5442

[4] Consonni S and Viganò F 2012 Waste gasification vs conventional Waste-To-Energy: A comparative evaluation of two commercial technologies, *Waste Management* 32 4, pp. 653-666, ISSN 0956-053X

[5] Gomez E, Amutha R, Cheeseman D, Deegan C R, Wise D and Boccaccini M 2009 Thermal Plasma Technology for the Treatment of Wastes: A Critical Review, *Journal of Hazardous Materials* **161**, pp. 614-626

[6] Hofbauer H and Materazzi M 2019 Waste gasification processes for SNG production, Editors: Massimiliano Materazzi, Pier Ugo Foscolo, Substitute Natural Gas from Waste, *Academic Press*, pp. 105-160, ISBN 9780128155547

[7] Lombardi L, Carnevale E and Corti A 2012 Analysis of energy recovery potential using innovative technologies of waste gasification, Waste Management **32** 4, pp. 640-652, ISSN 0956-053X

[8] Martínez J D, Puy N, Murillo R, García T, Navarro M V and Mastral A M 2013 Waste tyre pyrolysis – A review, *Renewable and Sustainable Energy Reviews* **23**, pp. 179-213, ISSN 1364-0321

[9] Lombardi L, Carnevale E and Corti A 2015 A review of technologies and performances of thermal treatment systems for energy recovery from waste, *Waste Management* **37**, pp. 26-44, ISSN 0956-053X

[10] Mukherjee C, Denney J, Mbonimpa EG, Slagley J and Bhowmik R, 2020 A review on municipal solid waste-to-energy trends in the USA, *Renewable and Sustainable Energy Reviews* **119**, 109512, ISSN 1364-0321

[11] Moustakas K, Xydis G, Malamis S, Haralambous K J and Loizidou M 2008, Analysis of Results from the Operation of a Pilot Plasma Gasification/Vitrification Unit for Optimizing its Performance, *Journal of Hazardous Materials* **151**, pp. 473-480

[12] Ramos A, Teixeira C A, Rouboa A, 2018 Assessment study of an advanced gasification strategy at low temperature for syngas generation, *International Journal of Hydrogen Energy* **43** 21, pp. 10155-10166, ISSN 0360-3199

[13] Sharma A, Pareek V, Zhang D, 2018 Biomass pyrolysis — A review of modelling, process parameters and catalytic studies, *Renewable and Sustainable Energy Reviews* **50** pp. 1081-1096 ISSN 1364-0321

[14] Sogancioglu M, Ahmetli G, Yel E 2017 A Comparative Study on Waste Plastics Pyrolysis Liquid Products Quantity and Energy Recovery Potential, *Energy Procedia* **118**, pp. 221-226, ISSN 1876-6102

[15] Avram P, Imbrea M S, Istrate B, Strugaru S I, Cucos I, Axinte C and Munteanu C 2014 Wear Resistance and XRD Analyses of CMoCuNiCrSiBO Coatings Obtained by Thermal Deposition on OLC45 Substrate, *Applied Mechanics and Materials* **659**, pp. 10-15 Trans Tech Publications, Switzerland

[16] Avram P, Imbrea M S, Istrate B, Strugaru S I, Cucos I, Axinte C and Munteanu C 2014 Wear Resistance and XRD Analyses of CNiCrSiBO Coatings Obtained by Thermal Deposition on OLC45 Substrate, *Applied Mechanics and Materials* **659**, pp. 16-21, Trans Tech Publications, Switzerland

[17] Cucos I and Munteanu C 2015 The experimental researches of advanced systems for managing thermosensitive heat treatments, *Journal of Optoelectronics and Advanced Materials*, **17** 3-4, pp. 409-420

[18] Cucos I, Munteanu C and Mirea C 2015 The electric furnaces for heat treatment with PID predictiv system, *Applied Mechanics and Materials*, **809-810**, pp. 778-783, Trans Tech Publications, Switzerland

[19] Mazzoni L, Almazrouei M, Ghenai C, Janajreh I 2017 A comparison of energy recovery from MSW through plasma gasification and entrained flow gasification, *Energy Procedia* **142**, pp. 3480-3485, ISSN 1876-6102

[20] Amjed A, 2013 Thermal analysis of cooling effect on gas turbine blade, *International Journal of Research in Engineering and Technology*, ISSN 2319 -1163

[21] Sunshil S, 2012 Review of heat transfer augmentation for cooling of turbine blade tip by geometrical modification to the surfaces of blades, *International Journal of Research in Engineering and Technology*, ISSN 2319-1163

IOP Conf. Series: Materials Science and Engineering 1138 (2021) 012019 doi:10.1088/1757-899X/1138/1/012019

Quantitative fire risk assessment for an office building

R Dârmon[1]

[1] Technical University of Cluj-Napoca, Romania

* ruxandra.darmon@ccm.utcluj.ro

Abstract. In Romania, the risk of a fire occurring within a building is determined in a very conservative manner, as a function of the building occupancy type and the thermal load contained within each enclosure. The result is a qualitative expression of low, moderate or high fire risk. On the other hand, the fire is a serious hazard which has to be taken into account within the decision-making process by any organisation for the elaboration of their risk assessments and the risk management strategies. Nevertheless, the one carrying out a fire risk assessment has a various range of methods and techniques to apply, [1, 2] since the legislation does not specifically require a certain approach. This article is an example study of the application of Bayesian statistics to determine the probability of a fire occurring within an office building, due to an electrical fault.

1. Introduction

Once, with the inclusion in the European Union, Romania had to adapt and implement a series of design regulations and new standards. The Eurocodes, for the buildings and the construction products design, are currently in use. Nonetheless, some other European regulations [1] have only been translated and kept idle due to the lack of legal infrastructure that may prove it applicable. Although, there are several methodologies for assessing the fire risk, there are not enforced any specific requirements concerning the methods and technics to be used for the risk analysis, nor for the verification and certification of a risk analysis. Moreover, the third-party peer review for the fire risk analysis is not deem necessary or required by any of the regulations.

One commonly used approach for assessing the fire risk, in Romania, is the cumbersome method used in Switzerland and France [2], which is highly prescriptive and it is based on several coefficients which have been derived on certain assumptions that may not be actual nowadays and are not generally valid.

In general, a fire risk assessment is required for complex buildings of special importance as hospitals, hotels, shops and malls, offices, museums, industrial buildings etc. The occurrence of a fire in such buildings will result in many life losses or considerable material losses. Most of these buildings are unique and form very complex systems which need to be assessed in a holistic manner, therefore a probabilistic risk assessment approach would be more appropriate instead of the implicit fire risk analysis.

In practice, especially when it comes to businesses and organizations, there is a permanent need of a decision-making process and the decisions have to be made, despite the lack of knowledge or data. Bayesian statistics may offer a useful interpretation of the uncertainty and it proves to be reliable in such situations, when the number of identical structures is very limited and the operation and the loading conditions are different for each kind of structure [3].

The present paper is a study case of Bayes theorem practical application in order to estimate what is the probability that a faulty electrical appliance may be the cause of a fire in an office building in Romania.

2. Model framework

For the decision-making process, it is important to have reliable data as a starting point of a risk analysis. Assessing the initial probabilities, it is often a difficult task, given that statistical data are scarce and not always available.

2.1. Statistical data

In order to review the most common causes for fires in Romania, one forthcoming way is to use the statistics [4] available on the General Inspectorate for Emergencies Situations website. In 2018, according to the forementioned study, a total number of 28468 fires have been reported over the whole country. Electrical faults, short-circuits or flawed electrical appliances accounted for the initiation of almost a third of the total number of fires, as it can be seen in the chart, below.

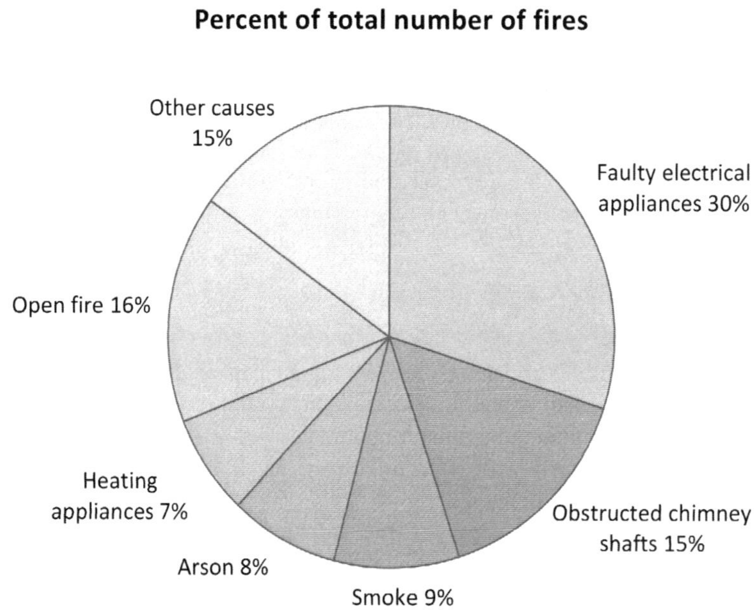

Figure 1. The fire sources distribution in 2018 [4].

From figure 1, results that 30% of the total number of fires, in Romania, occurred due to an electrical fault. In order to carry out a probabilistic fire risk assessment, instead of assuming a probability of 0,3 for electrical fires, one should look into the proportion of the office buildings to the total amount of buildings, since the 30% represents an accumulative proportion of all the fires. Thus, a useful website for statistical resources is that owned by the National Institute for Statistics [5]. Analyzing the data found in a study [6] about the building stock breakdown published by Buildings Performance institute Europe (BPIE), it is revealed that the percentage of non-residential buildings, in Romania one can track down the percentage of the office buildings in Romania, shown in Figure 2, which is about 2.2% of the total.

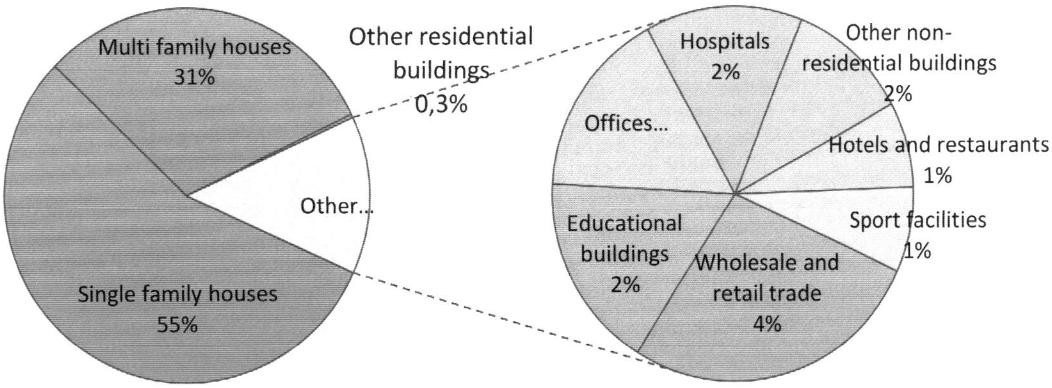

Figure 2. The breakdown of the building stock by building type [6].

2.1.1. Conditional probabilities.
When there are two independent events, as a fire occurring into an office building, O, and an electrical fault, E, the probability of both events happening in the same time given that the latter event has occurred is given by the proportion of OE to all of E. Similarly, the probability of both events happening, giving that the former event, O, has occurred is given by the proportion EO to all O, shown graphically in Figure , where:
P(E) is the prior probability that the fire occurs due to electrical fault = the cause,
P(O) is the probability to have a fire in an office building = the event,
P(E|O) is the probability that a fire occurred due to an electrical fault, given that it started in an office building,
P(O|E) is the probability that the fire occurred inside an office building, given that it has been an electrical fault.

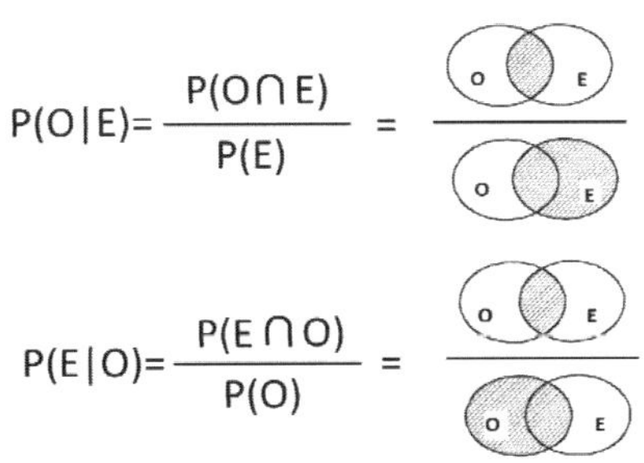

Figure 3. Conditional probabilities.

$$(\text{Cause}|\text{Event}) = \frac{P(\text{Cause}) \cdot P(\text{Event}|\text{Cause})}{P(\text{Event})}$$

$$\Downarrow$$

$$P(E|O) = \frac{P(E) \cdot P(O|E)}{P(E)}$$

Figure 4. Bayes law for two events.

Conditional probabilities are the basis for updating the probability estimates based on new information, evidence or knowledge and experience. Bayes law, shown above, in figure 4, describes the probability of an event based on the prior knowledge i.e. that 30% of the fires occur due to electrical faults, P(E). If the conditional probability P(O|E) is known, or can be estimated, Bayes law can be used to determine the reverse probability P(E|O), called *the posterior* probability.

2.1.2. Bayes law describes the probability of an event based on the prior knowledge of the conditions that might be related to that event. If the conditional probability P(O|E) is known, Bayes law can be used to determine the posterior probability P(E|O).

2.2. *Risk assessment*

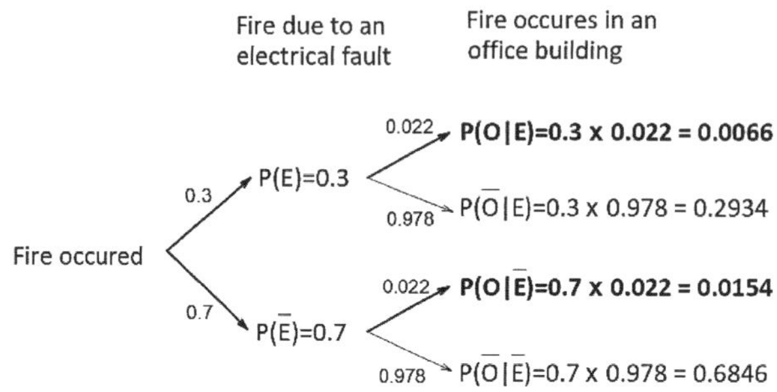

Figure 5. Event tree for the fire occurrence

The total probability to have a fire in an office building, P(O), can be expressed as the overall chance of the event O happening, i.e. a weighted combination of its probability given that the event E occurs and the probability of O given that the event non-E happens. [7] Hence, P(O), calculated from the event tree shown in figure 5, is replaced in Bayes law resulting a probability that the fire has started from an electrical fault, given that there is an office building, P(E|O) equal to 0.155.

3. Discussions and results

Assessing the initial probabilities, it is often a difficult task, given that statistical data are scarce and not always available. Bayes law is a powerful tool which enable updating the prior belief as the new information is revealed. In this study, the prior probability of having a fire due to electrical faults, based on statistical data, was initially 0.3. After taking into account the proportion of the office buildings to the total amount of the existing buildings in Romania, the same probability has been calculated as 0.155, decreased by a half. The more accurate are the statistical data, the better results can be obtained. This method is very useful to apply within the whole decision making process involved in the risk management, providing a dynamic tool for the risk assessment.

4. Conclusions

The conditional probability is useful to estimate the probability of an event (the fire starting into an office building) given that one of its possible causes (electrical faulty appliances) has occurred and the Bayes law is useful to find the probability that one of the possible causes of an event, given that the event has happen.

References

[1] SR EN 31010:2010 2011 Risk management. Risk management techniques, *Romanian Society for Standards, ASRO*

[2] Evaluation du Risque d'Incendie, Methode de Calcul 1984 *Societe Suisse des Ingenieurs et des Architectes*

[3] Faber M H, 2012 Statistics and Probability Theory In Pursuit of Engineering Decision Support, *Dordrecht Heidelberg London* New York: Springer

[4] Analiza statistică privind acţiunile de prevenire, pregătire şi răspuns în situaţii de urgenţă 2018 IGSU, https://www.igsu.ro/Resources/COJ/Rapoarte Studii/65201914407%20PM.pdf

[5] Institutul National de Statistica, Fondul de locuite 2020 Institutul National de Statistica https://insse.ro/cms/ro/content/fondul-de-locuin%C5%A3e-3

[6] B. P. I. E. (BPIE) 2014 Renovating Romania. A Strategy for the Energy Renovation of Romania's Building Stoc http://bpie.eu/wp-content/uploads/2015/10/Renovating-Romania_EN-final.pdf

[7] Kaplan M and Kaplan E 2007 Chances are...Adventures in probability London Penguin Books Ltd p. 185

IOP Conf. Series: Materials Science and Engineering 1138 (2021) 012020 doi:10.1088/1757-899X/1138/1/012020

Aspects regarding the extension and consolidation of a house located on a sloping plot of land

O Deaconu[1]*

[1] Faculty of Civil Engineering, Transilvania University of Brasov, Romania

* ovideaconu@unitbv.ro

Abstract. When the land available for a construction is in slope, with a high degree, there can be a series of problems related to geotechnical data and execution difficulties, considering the foundations and the structure itself. The grade of difficulty is considerably increased when an extension or attic is intended to be design and build. The paper is presenting some considerations to this problem, taken into account a study case – extension of an existing structure for a family house.

1. Introduction

There is a continuous expansion of the urbanization in Romanian cities. The lack of buildable plots is partially solved by extensions of the existing buildings, which provides benefits in terms of solving the needs for housing, while preserving the urban habitat for the local population.

This paper is presenting some considerations upon the extension of the existing house build on a sloped plot. There is presented a study case for an extension with an attic for an existing construction located in Săcele city (Brașov county), on a ground with slope ~ 35 degrees (figure 1, 2). Through the design theme, the owner wants to transform a family house into a two apartments house. It is also desired that the ground floor does not change its current functions or dimensions. The project respected this design theme, so that the same wall partitioning was kept on the ground floor, the extension area having more of an access role to the attic.

The conditions of the building location, in a landscape specific to mountains areas, are generally challenging in terms of structural solutions [1].

Considering the study case, in order to accomplish a functional building extension with an attic, several infrastructure problems have been encountered: foundations with insufficient depth and width, the nature of foundation soil, problems in rainwater drainage, natural slope of the land and problems related to existing structure: lack of columns confinement for the load-bearing brick walls, the lack of the perimetral belt at the level of the reinforced concrete slab above the ground floor.

The article presents solutions for identified problems in the existing construction and for details for the extension and attic. There is also taken into account the increasing of the energy efficiency of the house, in the end comparing the existing and the designed situation [2].

2. Description of design and execution difficulties

According to the design theme, the owner wants to build an extension and an attic for an existing house. The existing height regime is ground floor and will become a ground floor with an attic, increasing the total height of the construction. As presented in figures 1 and 2, there are two retaining reinforced concrete walls: one located in the area of the house and one in the back of it (which is approximately at the level of the ground floor of the existing house) [3]. Also, in figures 3 and 4 there are presented the facades of the existing building.

Figure 1. Left side facade, first support wall.

Figure 2. The second support wall.

Figure 3. Existing main facade.

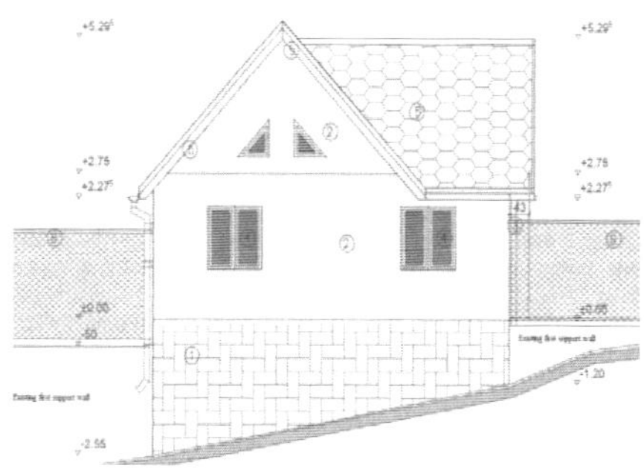

Figure 4. Existing left facade.

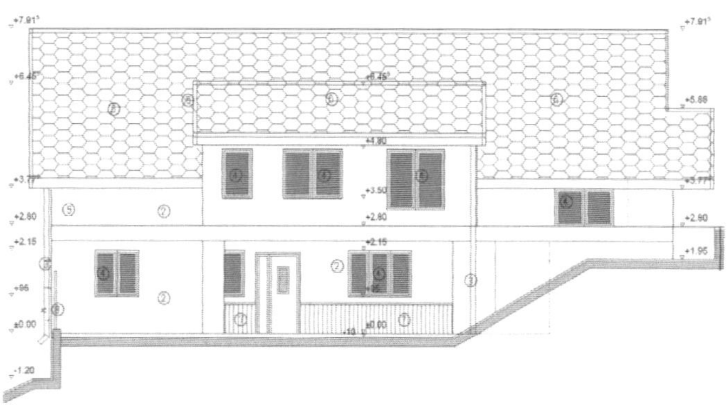

Figure 5. Proposed main facade.

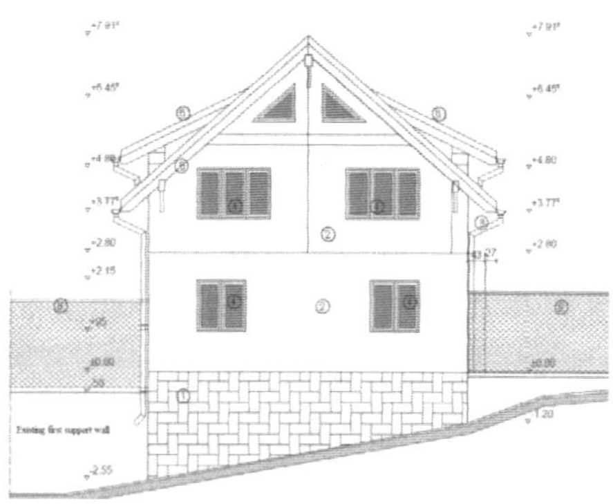

Figure 6. Proposed left facade.

IOP Conf. Series: Materials Science and Engineering 1138 (2021) 012020 doi:10.1088/1757-899X/1138/1/012020

The interventions on the existing structure are presented in figures 5, 6 and 7. As one can see, there is an extension of the existing house from axis 2 to 8 and from axis A to F (figure 7). According to the foundations plan the difference level of the natural terrain is significant, about approximately 5.50 m (figure 7). Considering the access road level, the difference increases up to 7.00 m. For this reason, the existing construction was provided with foundation system in steps in order to undertake the height differences. Extension is only possible towards up of the sloping land.

According to the geotechnical study, the existing construction was provided with continuous foundations in steps, but for the most part they do not have a width corresponding to the conventional pressure, and in the lateral areas (axes 1 and 5) the foundations do not respect the minimum frost depth corresponding to Brasov of -1.10 m [4, 16].

The conventional pressure indicated by the geotechnical study was 240 kPa. Due to these inconveniences described above but also due to the need to have the same foundation level, a part of the foundations of the existing building required sub-foundations [5].

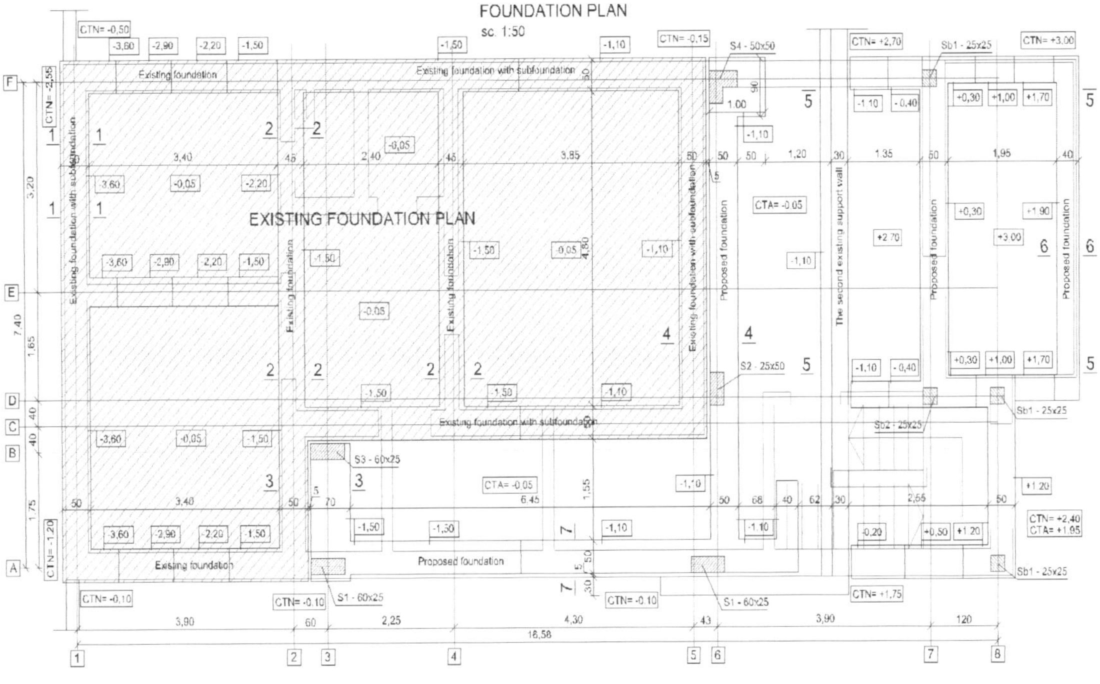

Figure 7. Foundation plan.

Between the existing construction and the whole extension area, it was designed a seismic gap on the entire height of the structure [6]. In this way, the existing construction will be completely decoupled from the extension. In order for none of the areas to negatively influence the other, in the adjacent area of the two buildings the foundation depth will be at the same level. Figure 9 presents some of the foundation details and in figure 8 a cross section of the foundation is shown, revealing the difference of 3 m – difference between -1.10 m of the existing foundation and +1.90, the level of the extension foundation. The beams G1 and G2 are embedded on the left side directly in the foundation [7].

According to laboratory tests, and the design calculations, the interior and exterior walls of the existing construction can withstand the load difference between those of the existing bridge compared to the new demands of the attic loads [8].

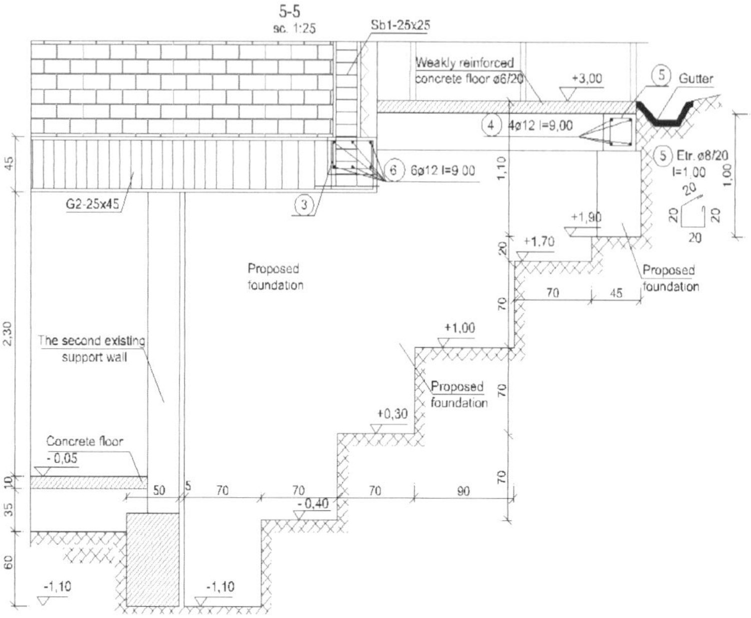

Figure 8. Detail 5 of the foundation.

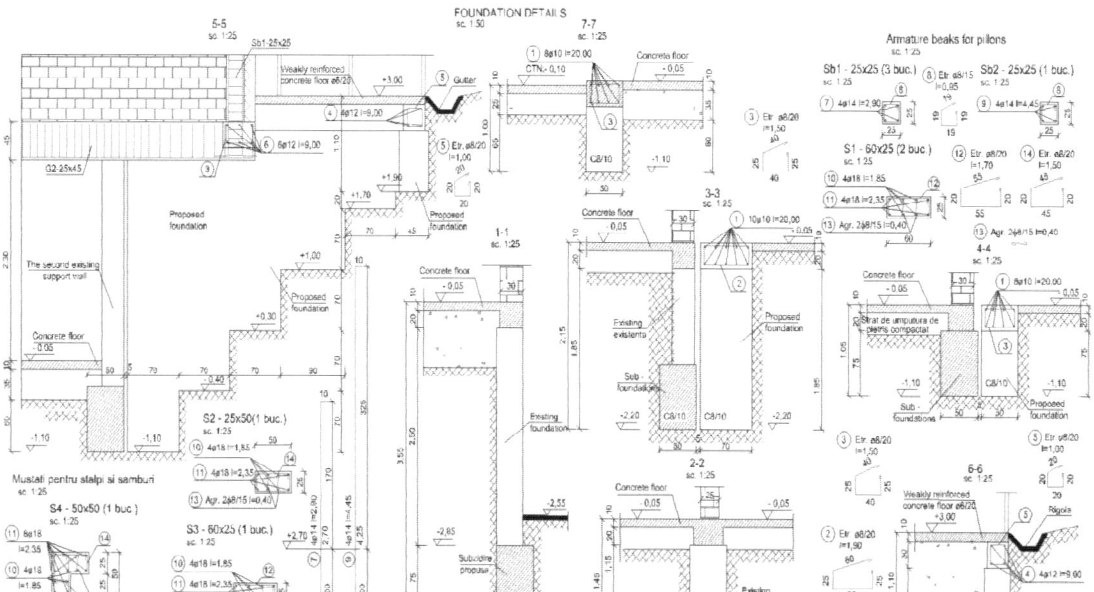

Figure 9. Foundation details.

The seismic gap at the level of the roof was made by doubling the wooden support elements. This mode of execution also allows independent movement at the roof level. Fortunately, in axis C, between axes 2 and 6, the architect provided a skylight who changes the slope of the roof. The roof tiles are made of steel sheet. Under these conditions, the steel sheet will be cut at the slope change, and the bottom sheet can move under the protection of the sheet above.

IOP Conf. Series: Materials Science and Engineering 1138 (2021) 012020 doi:10.1088/1757-899X/1138/1/012020

A reinforced concrete belt was designed under all the external walls and under a part of the structural walls of the attic in order to withstand the seismic loads – ensuring a rigid slab. The columns provided in the attic, over the existing building area, are also embedded in this belt.

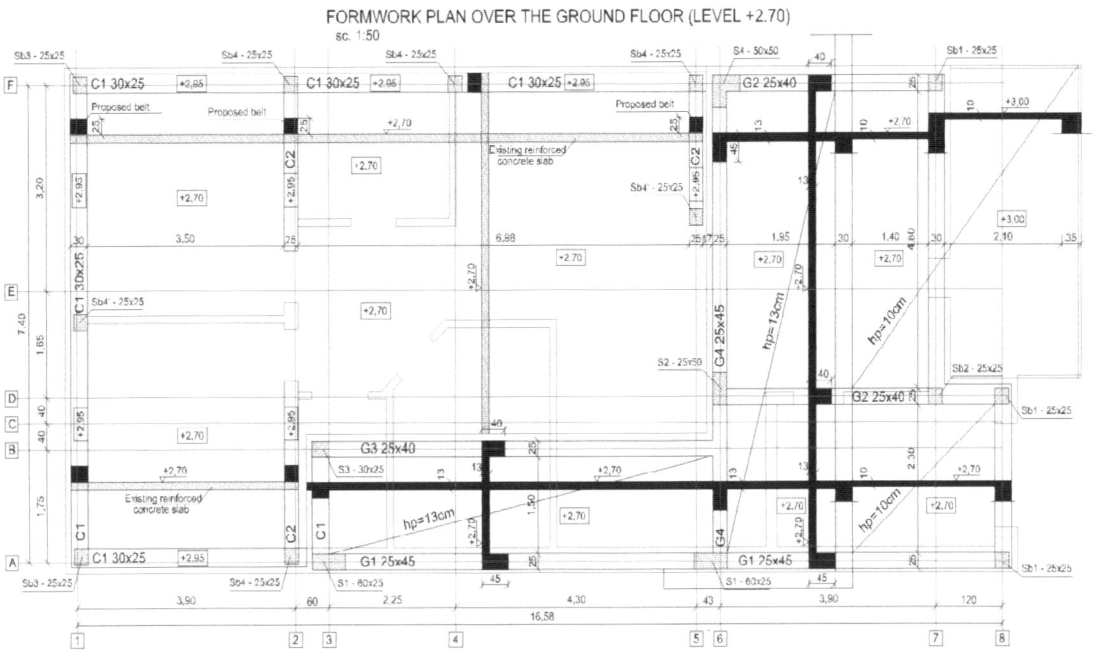

Figure 10. Formwork plan over the ground floor (level +2.70).

Considering the execution, the pouring of the concrete was a challenge due to the fact that the road access nearby the construction had to be blocked. To enter in the yard of the house you have to make a sharp turn, impossible for a big vehicle. Thus it was needed to pour the concrete into gutters, supported on the formwork, to the edge of the terrace and the extension area, directly from the road area. Due to the traffic conditions, pouring of the new foundations, half of the sub-foundations and the belt over the attic was erected in a single stage. In the second stage it was erected the other half of the sub-foundations, the pillars supporting the extension and the pillars over the old construction. In the third stage, the floor of the extension was made. In the last stage, the pillars from the attic in the extension area and the belts and beams under the roof were made [7].

3. Results and Discussion

The first investigated problem [9] was the one related to the presence of a small damage at the level of the existing construction foundation. In addition to the access road to the property there is a water channel protected with precast concrete elements. This channel collects excess water from a swamp area (ponds, ravines, meanders, etc.) located upstream of the construction site. Due to this channel as well as the presence of a rainwater collection pipe on the property, this pipe being embedded in the first retaining wall, a slight degradation of the foundation of the existing building was identified. For this reason, a consolidation was recommended to stabilize the land in the area together with the necessary works for the attic and the extension. The owner considered these additional works too expensive. Due to the fact that the degradations in the foundations are minor without the appearance of dislocations or rupture of the foundation continuity, the consolidation of the land will be done at a later stage when measures will be taken to consolidate the entire access road [7, 10].

The second investigated problem was related to the seismic gap. Based on the geotechnical study, as well as the laboratory tests performed on the existing construction, the following were found: areas of the continuous foundation is embedded above the frost depth [4, 11], or do not have a sufficient width, lack of reinforced concrete pillars for masonry confinement, both interior and exterior structural walls are made of vaults with natural aggregates, lack of waterproofing, wrong location of drains. In order not to affect the resistance structure of the initial construction, the decision was made that the extension be completely disconnected from the existing structure. In this way, a seismic gap of 5 cm at the level of the foundation and of 10 cm at the level of the structure was provided, a gap that will separate the entire structure into two building structures [6]. The width of the seismic gap was calculated so that from the most unfavorable combination of lateral displacements, including those resulting from the general torsion, the two bodies do not touch each other. The approximation of the calculated maximum displacements was 7.32 cm, so below the dimension of the gap provided in the superstructure [7]. The seismic gap is provided on the entire height of the two constructions. At the level of the frame, on the area between axis 2 to 6, it was necessary to double the frame support elements. At the level of the attic, the architect provided a skylight necessary both for a good natural lighting and for increasing the useful interior surface. Following discussions with the architect, the skylight was made in the form of changing the main slope of the roof and on the entire area of internal longitudinal overlap of the two bodies, i.e. on axis B, C between axis 2 and 6. The change point of the slope was chosen between the axis B and C, which facilitates the sliding of the roofing surfaces one below the other [12].

Between the existing construction and the second retaining wall there is a passage corridor, between the courtyard in front of the house and the land behind the house, which the owner wanted to keep. For this reason the second retaining wall had to be kept in its original form. The extension was necessary, to support, on the continuous foundations located behind the retaining wall. In order not to introduce additional forces from pushing the ground due to the loads from the extension, the foundation for the extension was made in minimum steps of length, the starting quota of the foundation coinciding with that of the retaining wall. Above the passage corridor, the floor above the ground floor of the extension was provided, thus forming a kind of passage tunnel. Normally the ends of the tunnel should have been blocked with the passage of the continuous foundation. The solution of this problem consisted in the support of the main beams G2-25x45, G2-25x45 on the left on the pillars S1-60x25, S2-25x50 respectively S4-50x50, and on the right directly on the simple foundation block, the beams being connected to the foundation belt from the beginning (figure 8) [13].

The wooden structural elements of the frame over the initial construction, show some damage, respectively most of them have smaller dimensions than the structural elements of the future frame, it was proposed that this wood should not be reused. It was also proposed that the structure of the frame for the entire attic be made of ferrocement. According to economic calculations, this solution would have been cheaper. The beneficiary opted for the classic solution with wooden elements [3].

Besides the structural consolidation on the initial building, a thermal rehabilitation was proposed to increase the energy performance. For the final energy performance of the entire building, the technical characteristics of the building and installations, the location of the building in relation to external climatic factors, sun exposure, the influence of neighboring buildings, own energy production sources and the internal climate of the building were taken into account. It was proposed to mount on the roof several solar panels for the production of electricity and heat. They were also provided for thermal insulation with 15 cm thick polystyrene and the installation of triple-layered double-glazed windows [14 - 16].

4. Conclusions

The design of the extension and attic for the construction analyzed in this paper, raised significant series of difficulties in terms of consolidating existing foundations, joining the new extension area, founding on an area with high slope, difficult access of construction works due to the narrow area and on the slope.

There are presented several details regarding a possible structural solution for extension of an existing house taken into account the seismic gaps and the foundation issues, considering the different foundation levels.

Currently, construction work is almost complete. The owner was asked to make a periodically current inspection (at intervals of one year) for the entire building, both to verify the behavior over time of the consolidations made to the existent construction and at the new part of the extension. If it is proved that the executed works will not show damages, the adopted solutions can be used for other similar constructions.

References

[1] Chițonu C 1991 The Traditional Habitat of Bran village, Romania. Sustainable development in rural landscape. Paripex *Indian Journal of Researh* 3 10 ISSN-2250-1991

[2] Cazacu Ch and Chițonu C 2018 Ecological criteria considering the connections between the construction, building-site and the spatial planning development. 3rd China-Romania Science and Technology Seminar (CRSTS 2018) *IOP Conf. Series: Materials Sciences and Engineering* **399**, doi:10.1088/1757-899X/1/2006

[3] Deaconu O 2016 Considerations regarding using concrete with dispersed reinforcement and ferrocement *CIBv 2016* Bulletin of the Transilvania University of Brașov

[4] Deaconu O, Dimulescu Al and Litra E 2004 Propuneri de interventii pentru diverse defecte ale betonului *Sesiunea stiintifica Constructii - Instalatii CIBv 2004*, Universitatea Transilvania Brasov

[5] Dimulescu Al, Litra E and Deaconu O 2004 Consolidarea fundatiilor continue din beton armat *Sesiunea stiintifica Constructii - Instalatii CIBv 2004*, ed. Universitatii Transilvania Brasov

[6] P100-1/2013 Cod de proiectare seismică (Seismic design code)

[7] Deaconu O 2014 Conditions of the constructions sustainability in natural environments. *CIBv 2014 Bulletin of the Transilvania University of Brașov*

[8] Eurocode 1 Actions on structures

[9] Chițonu C 2015 Building In Line With Tradition *Bulletin of the Transilvania University of Brașov , CIBv 2015* **8** (57) 1

[10] Popaescu A and Deaconu O 2002 Durabilitatea structurilor din beton *Sesiunea stiintifica aniversara Constructii - Instalatii Brasov CIBv Universitatea Transilvania Brasov*

[11] STAS 6054-77 Adancimi maxime de inghet (Maximum depths of frost)

[12] Hobjilă V and Luca M 2000 Complementary instructions to the design and expertise of some types of hydrotechnical constructions Iași Cermi Press p 205, ISBN: 973-8000-28-6

[13] Eurocode 2 Design of concrete structures

[14] Kuzmin A 2011 Quality of life and quality of the spatial environment - social standards and standards in urban planning, architecture, construction, *Urban Development* **4**, pp. 16-19

[15] Skobeleva E 2015 Analysis of proposals for assessing the microclimate environmentally safe and comfortable urban environment, *Construction and reconstruction* **4**, pp. 123-132

[16] SR EN 13164 Thermo-insulation products for buildings. Products made of extruded polystyrene foam

IOP Conf. Series: Materials Science and Engineering 1138 (2021) 012021 doi:10.1088/1757-899X/1138/1/012021

Insights from the management of a large research infrastructure in Romania: National Network for the Seismic Monitoring and Protection of Building Stock at NIRD URBAN-INCERC

C S Dragomir[1,3], I G Craifaleanu[1,2*], V Meiţă[1], E S Georgescu[1], D Dobre[1,2], M Sandu[1] and A Cişmelaru[1]

[1] National Institute for Research and Development in Construction, Urban Planning and Sustainable Spatial Development "URBAN-INCERC", Romania
[2] Technical University of Civil Engineering Bucharest, Romania
[3] University of Agronomic Sciences and Veterinary Medicine of Bucharest, Romania

* i.craifaleanu@gmail.com

Abstract. The National Network for the Seismic Monitoring and Protection of Building Stock is a key infrastructure of the National Institute for Research and Development in Construction, Urban Planning and Sustainable Spatial Development "URBAN-INCERC" in Romania. This infrastructure includes at present, among other research facilities, a strong-motion network of 64 permanent seismic stations, to which other temporary stations are added for various short-term vibration monitoring projects. Since its establishment in 1967, the seismic network has played an essential role in recording the earthquakes that affect Romania. Today, the seismic stations are distributed all over the country, according to its characteristic seismicity patterns. Most of them use state-of-the-art recording and communication equipment, being permanently connected online to the Data Center in Bucharest. The management of the network implies conducting a complex set of activities, centered on the recording, processing, analysis and interpretation of natural- and human-induced vibrations. All of these are seconded, in parallel, by various maintenance, reporting, logging, communication and dissemination tasks. The logistic and financial management also plays an important role. A rational and efficient administration of the seismic network cannot be acquired without a rigorous organization of data flows pertaining to the various activities performed. The paper provides insights from the organization of the network, highlighting the intercorrelation between the hierarchical levels of its functionality and the logical course of operation. A brief review of the challenges and solutions gradually developed and implemented by the members of the network team, in order to provide the required functionality, is presented.

1. Introduction

In seismic-prone areas, ground motion monitoring is a crucial activity for acquiring data on the local and regional earthquake hazard. Starting with the installation of separate instruments in a relatively small number of locations, ground motion monitoring has evolved to the deployment of several stations, forming a seismic network, in which collected data is joined and processed together [1]. The progress of seismic networks occurred gradually, as technological advances allowed for higher precision sensors, real-time communication systems, automated data processing and advanced storage solutions. The types of recording instruments have also diversified, depending on the type of monitored ground motions and on the scale at which the network is deployed. Today, there are several local or national networks, but also global networks, that record earthquakes from the whole world, as the Global Seismographic Network [2] or the International Federation of Digital Seismograph Networks, FDSN [3].

A large number of seismic networks are nowadays installed in most earthquake-prone areas, covering large surfaces and implying the management and maintenance of tenths or hundreds of instruments and of the associated communication lines and equipment. The obtained ground motion records, often totaling huge amounts of data, have to be undergo various processing phases, which are required for their correct analysis and interpretation. These operations require specialized software, high-end computers, dedicated storage capacities and qualified personnel. Once the processing completed, data classification and backup has to be done, for easy retrieval and distribution. The entire management of a seismic network requires thus the coordination of several interconnected activities, which concur to the proper functioning of the entire system.

Even though the technical solutions used and the general organization of existing seismic networks in various countries have at present many things in common and tend to become standardized, there are however several aspects that depend on specific conditions and on the historical evolution of each network. The paper presents some insights on the organization and management of the seismic network of the National Institute for Research and Development in Construction, Urban Planning and Sustainable Spatial Development "URBAN-INCERC" (NIRD URBAN-INCERC) in Romania.

2. The seismic network of NIRD URBAN-INCERC

The seismic network of NIRD URBAN-INCERC has been in continuous operation since its establishment in 1967 [4]. Over the years, the network contributed with essential seismic records to the characterization of the seismicity of the country. The most important record was the unique accelerogram of the catastrophic $M = 7.2$ earthquake that occurred in the Vrancea (Romania) seismic zone on March 4, 1977. The analysis of the unusual features revealed by this record and the general lessons learned from the earthquake led to a radical change of the Romanian seismic design code, which was significantly revised after the earthquake.

During the more than five decades of operation, the network went through several transformations, evolving both technologically and in terms of area coverage. Today, together with other related research facilities, the seismic network is part the National Network for the Seismic Monitoring and Protection of Building Stock (IOSIN), a key infrastructure of NIRD URBAN-INCERC [5]. It operates 64 permanent accelerometric stations. Most stations use state-of-the-art recording and communication equipment, a large number of them being permanently connected online to the Data Center located at the INCERC Bucharest Branch of the institute. The stations are installed in various locations across the country (figure 1), the sites being selected according to the particular characteristics of the seismicity of Romania and taking into account the various seismogenic sources.

The management of the seismic network implies conducting a complex set of activities, centered on the recording, processing, analysis and interpretation of natural- and human-induced vibrations. All of these are seconded, in parallel, by various maintenance, reporting, logging, communication and dissemination tasks. The logistic and financial management also plays an important role. A rational and efficient administration of the seismic network cannot be acquired without a rigorous organization of data flows pertaining to the various conducted activities.

The seismic network is managed based on a set of customized hierarchical procedures, which has been gradually developed by the members of the network team, during a work spanning over several years. The procedures have to be periodically updated, in order to respond to the contemporary needs generated by the specific layout, configuration and functionality requirements.

IOP Conf. Series: Materials Science and Engineering 1138 (2021) 012021 doi:10.1088/1757-899X/1138/1/012021

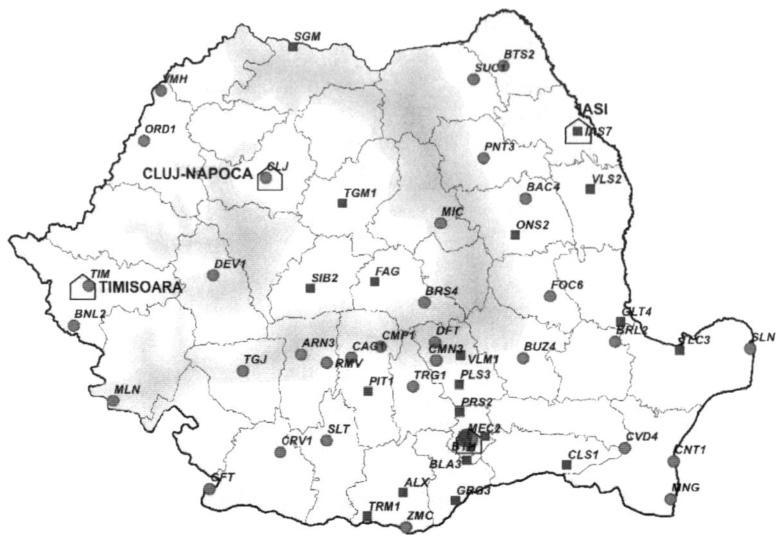

Figure 1. The National Seismic Network of NIRD URBAN-INCERC, Romania, and the locations of the regional research facilities of the National Network for the Seismic Monitoring and Protection of Building Stock (Bucharest, Cluj-Napoca, Iaşi and Timişoara).

3. Seismic network management data flows

The current structure of data flows [6] within the seismic network of NIRD URBAN-INCERC is shown in figure 2.

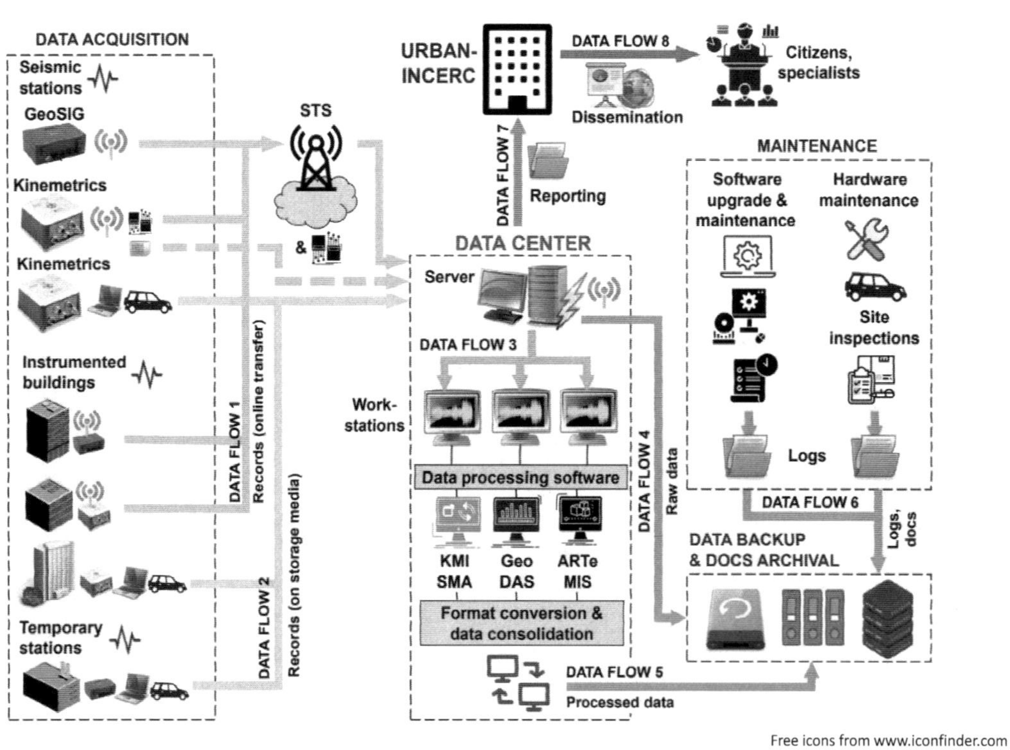

Figure 2. Current structure of the data flows within the NIRD URBAN-INCERC seismic network.

3.1. *Data acquisition*

The main data provider is the data acquisition equipment, consisting of strong motion accelerometers. There are three types of seismic stations in the network, according to their location and to the position of the instruments:

a) stations installed in free-field or in small buildings - the latter similar to the Strong-Motion Reference Station: Small Building (SMRS-SB) in the ANSS classification [7];
b) stations installed in higher buildings, on the base floor of the building, similar to the Strong-Motion Building Reference Station: Densely Urbanized Area (SMBRS-DU) in the ANSS classification [7];
c) instrumented buildings.

According to the way in which data is transmitted to the seismic network data center, the stations in the seismic network of NIRD URBAN-INCERC are either connected online or without a connection. In the second case, recorded data is retrieved by site inspections. Most of the online stations use the dedicated network of the Special Telecommunications Service (STS); however, a small number are connected through commercial mobile phone networks.

Based on the above, two distinct data flows are shown in figure 2: Data Flow 1 (yellow), for online stations (STS network with continuous line and mobile data operators with dotted line), and Data Flow 2 (light blue), for the other stations. Given that the seismic network uses Kinemetrics and GeoSIG accelerometers, the type of connection for each of the two equipment providers is shown in the figure.

Temporary stations are also used occasionally, for various vibration monitoring projects. These do not have generally an online connection; thus, the acquired data is transferred to the Data Center through Data Flow 2.

3.2. *Data center operations*

Once the recorded accelerograms have reached the Data Center, processing, conversion and consolidation of the data have to be done. Given the different types of instruments in the network, data is provided in various formats. The first processing steps are carried out using specific software from each equipment provider, i.e. SMA for Kinemetrics [8] and GeoDAS for GeoSIG [9].

For the structural health monitoring of instrumented buildings, the SVIBS ARTeMIS software is used [10-13] (figure 3).

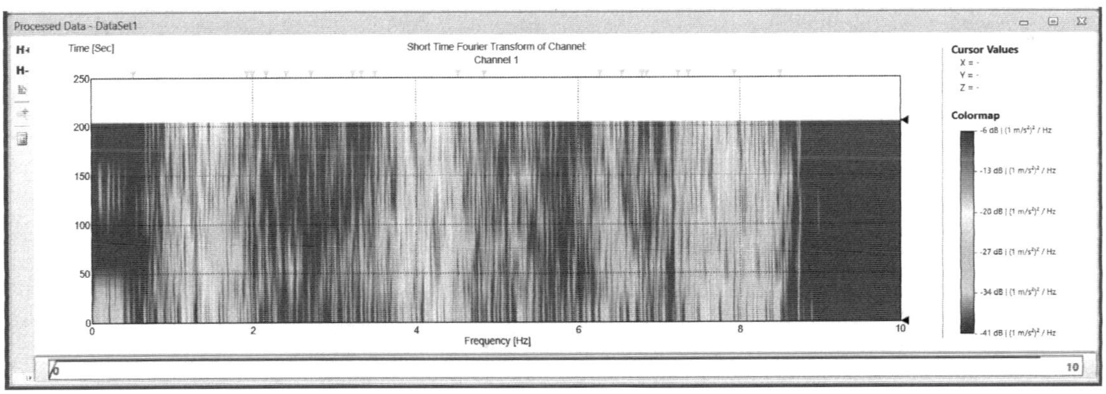

Figure 3. Use of records obtained on instrumented buildings in SVIBS ARTeMIS. Case study on the structural health monitoring of a residential building in Bucharest.

To use the processing results in subsequent research and analyses, data is converted in general formats and consolidated in a unified form. All the above operations address data pertaining to Data Flow 3 (marked with light green arrows in figure 2).

3.3. Data backup and document archival
Raw data (Data Flow 4), as well as processed data (Data Flow 5), are backed up to dedicated storage media (lavender arrows in figure 2). Data is organized primarily on a chronological basis. In addition, a codification system is used, to allow quick data retrieval for specific locations. Thus, each station is identified by a unique three-letter code, a system that is compatible with other internationally and nationally used approaches. To this code, a one-digit number is added in the case of more stations existing at the same location. Strong-motion records with magnitude larger than 5, envisaged for inclusion in the URBAN-INCERC reference database [14], are identified by a code created through the concatenation of the year of the earthquake, the reference number of the seismic event and the station code. Given the huge amount of data, this is a critical step, both in terms of data organization and in terms of ensuring redundant and reliable storage.

3.4. Maintenance and logistics
To support the main operations, various additional activities are required. These address data center software upgrade and maintenance, as well as site inspections for equipment check, repair and replacement. The activities require rigorous management and planning, as well as reliable logistics. Moreover, all operations should be logged and archived, to provide traceability and to comply with the internal procedures and the generally applicable legal framework. Various other activities, as those related to procurement, accounting and general management are here involved. The data associated to the entire set of the above activities forms Data Flow 6 in figure 2 (magenta arrows).

3.5. Reporting and dissemination
Periodic reports on various aspects of seismic network activity are compiled and sent to the management staff and superior hierarchic departments, according to internal procedures. The communicated data is shown in figure 2 as Data Flow 7 (blue arrow).

A special data flow is dedicated to data and information communication and dissemination. After each significant earthquake, seismic reports are prepared, including ground acceleration data, maps and comments on the characteristics of the event. Moreover, given that the seismic network is a central infrastructure of a research institute, acquired ground motion data is directly used within studies performed in the frameworks of national and international research projects. Results are presented in scientific events and published in conference proceedings and journals. While these types of information are disseminated to academia, professionals or authorities on a regular basis, other specific actions are dedicated to raising seismic risk awareness and to the education for seismic risk mitigation of citizens, decision factors, business owners and other audience categories. Several activities of the latter type were carried out during recent years by the European Center for Buildings Rehabilitation (ECBR) in URBAN-INCERC as, for instance, those in the project "Seismic risk preparedness and disaster risk reduction – DRR training for vulnerable groups of populations, school students and volunteers in neighborhoods in Romania" [15].

These activities are reflected in Data Flow 8 (red arrow), shown in figure 2.

4. Implementation and perspectives
To ensure the efficient operation of the network infrastructure and a proper data flow, various functional issues have to be tackled with. These can arise for every type of data flow presented in the previous section and specific solutions apply.

Some of the most frequent issues occur with regard to recording and communication equipment, as this requires permanent maintenance and repair and, if needed, updating. This involves not only technical operations, but also mobilizes some activities associated to other data flows, such as procurement, reporting and logging, as shown in Sections 3.4 and 3.5 above. If these operations are not carried on in a timely manner, equipment malfunction can lead to the irremediable loss of data in case of a seismic event. For this reason, ensuring the functionality of Data Flows 1 and 2 in figure 2 is a central point, as it impacts the functionality of the entire network and its main goal of providing as much as possible quality data from seismic stations and instrumented buildings. It is worth mentioning that, if communication issues occur during the transmission of recorded data from online instruments to the server, data will not be lost, given that it is stored locally, in the instrument. Consequently, data can be downloaded once the connection is re-established or, in the worst case, it can be recovered by a site inspection. The redundancy of data recording is thus ensured. In addition, raw and processed data is backed up systematically on reliable media, within the Data Center (see Data Flows 4 and 5 in figure 2).

The activities in the Data Center involve complex processing (Data Flow 3 in figure 2) and many of them are performed, at present, by human operators. Not only the data comes in various formats, given the different types of instruments that provide it, but, in addition, different goals are associated to free-field and SMRS-SB recordings and to structural health monitoring, respectively. One of the ways of improving the efficiency of Data Center operations is to increase, as much as possible, the automation of all processes.

The functionality of the network also relies, in a large measure, on the proper coordination of all activities and on a systematic organization of underlying operations, including logging, reporting and document archival for traceability. Collaboration is a key word for the success of the entire process.

The gradual increase of the number of stations and their rational spatial distribution, according to the specific of the country's seismicity, are among the main objectives for future development, aiming to achieve an improved seismic instrumentation of the whole territory of the country, by coordination also with other Romanian networks [13]. The extension of building instrumentation, in terms of building number, significance and covered typologies are another important objective. Finally, no less important are the focus on the quality of seismic records, the upgrading of the equipment and the implementation of a higher degree of automation for all network management operations.

5. Conclusions

The acquisition, processing and interpretation of strong ground motion and vibration data are among the main activities conducted within the National Network for the Seismic Monitoring and Protection of Building Stock at NIRD URBAN-INCERC. The efficient management of the very large amounts of data generated within these specific activities is a key condition of a systematic and organized process, in which an effective valorization of these data can be achieved. Beyond the intrinsic value of the acquired information, this serves the major purpose of saving lives and of minimizing losses in future earthquakes, contributing as well to providing a safe and sustainable built environment.

Acknowledgments

The paper refers results obtained at NIRD URBAN-INCERC within the projects PN 18-35 01 01 “Research for the development of an instrumental detection system for structural damage caused by seismic or non-seismic sources” (CONCRET) and PN 19-33 01 01 “Research on the implementation of an integrated system for ensuring the security of the constructed space, with semi-automatic generation of PGA maps provided by seismic actions or other vibratory sources and quick evaluation of vulnerability of instrumented buildings” (ECOSMARTCONS), funded in the framework of the “Nucleu” Programme by the Romanian Ministry of Research and Innovation. Also, the paper references the project “Seismic risk preparedness and disaster risk reduction – DRR training for vulnerable groups of populations, school students and volunteers in neighborhoods in Romania”, carried on within the European Centre for Buildings Rehabilitation in NIRD URBAN-INCERC, funded by the Council of Europe through the EUR-OPA Major Hazards Agreement, Grant Agreement with ECBR GA/2019/16. FIMS PO No. 623192. This funding is gratefully acknowledged.

References

[1] Havskov J and Alguacil G 2016 Seismic networks *Instrumentation in earthquake seismology* (Springer Intl. Publishing Switzerland) doi.org/10.1007/978-3-319-21314-9_8
[2] Global Seismographic Network IRIS 2015 https://www.iris.edu/hq/programs/gsn

[3] FDSN: About the FDSN 2020 https://www.fdsn.org/about/
[4] Craifaleanu I-G, Borcia I S and Prăun I C 2011 Strong-motion networks in Romania and their efficient use in the structural engineering applications. *Earthquake data in engineering seismology. predictive models, data management and networks*, ed S Akkar, P Gülkan and T van Eck (Springer Netherlands) chapter 17 pp 247-259
[5] Dragomir C S, Meiţă V, Georgescu E S, Craifaleanu I-G and Dobre D 2019 The Romanian URBAN-INCERC seismic strong-motion recording network as a tool for reducing earthquake disasters. *UNESCO International Platform for Reducing Earthquake Disasters (IPRED) 11th Annual Session - International Workshop on "Build Resilient Society with Scientific Knowledge and Experience on Seismology and Earthquake Engineering"* (Bucharest: Technical University of Civil Engineering Bucharest)
[6] Dragomir C S, Craifaleanu I-G and Meiţă V 2020 Innovative concept for the integrated management, structuring and processing of data flows within the National Network for the Seismic Monitoring and Protection of Building Stock at NIRD URBAN-INCERC, Romania. *Creative Invention Contest* – CIC 2020 (Seoul)
[7] COSMOS 2001 *Guidelines for installation of Advanced National Seismic System strong-motion reference stations* (Richmond: Consortium of Organizations for Strong-Motion Observation Systems) https://www.strongmotion.org/archive/publications/guidelines/Guidelines_ANSS.pdf
[8] Kinemetrics *Strong Motion Analyst datasheet* https://kinemetrics.com/wp-content/uploads/2017/04/datasheet-software-strong-motion-analyst-sma.pdf
[9] GeoDAS *GeoSIG Software* https://www.geosig.com/GeoDAS-id12565.aspx
[10] Structural Vibration Solutions *ARTeMIS Modal* https://svibs.com/
[11] Dragomir C S, Craifaleanu I-G, Dobre D and Georgescu E S 2019 Prospective studies for the implementation of a remote access earthquake damage detection system for high-rise buildings in Romania. *IOP Conference Series: Earth and Environmental Science* **221(1)** 012036
[12] Dragomir C S, Dobre D, Craifaleanu I-G and Georgescu E S 2020 An integrated national system for assuring the quick evaluation of the vulnerability of all instrumented buildings after an earthquake. Recent developments *Scientific Papers. Series E. Land Reclamation, Earth Observation & Surveying, Environmental Engineering* **IX** 94
[13] Mărmureanu A, Ionescu C, Grecu B, Toma-Danila D, Tiganescu A, Neagoe C, Toader V, Craifaleanu I-G, Dragomir C S, Meiţă V, et al 2021 From National to Transnational Seismic Monitoring Products and Services in the Republic of Bulgaria, Republic of Moldova, Romania, and Ukraine, *Seismol. Res. Lett.* **92** 2 doi: 10.1785/0220200393 (pre-issue publication)
[14] Borcia I S, Craifaleanu I G, Calarasu E A, Tanase N F and Praun I C 2015 SM-ROM-GL (Strong Motion ROMania Ground Level) database *Construcţii* **16(1)** 3
[15] Georgescu E S 2019 ECBR – European Centre for Buildings Rehabilitation report on ECBR activities in 2019. Proposal of ECBR project in 2020-2021 Presentation at Joint Meeting 2019 https://rm.coe.int/ecbr-romania-presentation-joint-meeting-2019/168098b48f

IOP Conf. Series: Materials Science and Engineering 1138 (2021) 012022 doi:10.1088/1757-899X/1138/1/012022

Using E-waste in asphalt mixtures – a laboratory investigation

M L Dragomir[1*], R D Cadar[1] and R M Boitor[1]

[1] Transport Systems Research Group, Technical University of Cluj-Napoca, Romania

* mihai.dragomir@cfdp.utcluj.ro

Abstract. Incorporating electronic waste (E-waste) in a new asphalt mixture is a subjected connected to the recycling process of PCB (Printed Circuit Boards), in which PCB waste is turned into a powder. The aim of the paper is to present the performance of the new asphalt pavement obtained by using PCB waste to substitute lime filler in a share of 100%. First, laboratory analyses are performed to determine the composition of PCB. Then, the feasibility of using PCB waste in the asphalt pavement industry is analysed in a series of laboratory investigations conducted on asphalt concrete BA16 rul 50/70 (AC16) that is a standard asphalt mixture in Romania. The following tests are performed on the new material: cyclic compression-dynamic flow, rutting test, and Marshall Stability and flow. The conclusion after the first experimental stage of this research is that the mixture containing PCB is softer than the control/witness mixture containing lime filler. Further investigations regarding the use of E-waste in road asphalt pavements will be conducted to find a way to reduce the use of new raw materials in the road construction industry.

1. Introduction

Road safety and road maintenance are a permanent concern for road administrations and finding alternatives to reduce the use of new raw materials in the road construction industry has led to creating and testing new pavements that incorporate different type of waste.

The idea of incorporating waste as a substitute for filler is not a new one. Various types of materials that have a mineral origin, such as brick dust, fly ash, hydrated lime dust, recycled concrete waste, aggregate dust and stone dust were used to substitute lime filler in asphalt mixtures so far [1].

Others deal with the recovery of metals and non-metals from electronic waste by the physical and chemical recycling process, used the so-called E-Waste [2].

The new material that is used to substitute filler in a share of 100% in the asphalt mixture in the research presented in the paper is a powder obtained by recycling electronic waste - Printed Circuit Boards. This raw material is used to support and electrically connect electronic components that have attached Cu sheets glued over a non-conductive support.

According to [3], PCB assemblies contain a green/yellow board and ECs (i.e. resistors, relays, capacitors and integrated circuits (chips) attached to it. PCBs are a mixture of phenolic/cellulose paper (yellow and low grade) or epoxy (green and high grade) resins, woven glass fiber and multiple kinds of metals (Cu, Sn, Pb etc.). The basic structure of the PCBs is the Cu-clad laminate consisting of fiber glass-reinforced epoxy resin and a number of metallic materials including precious metals. The concentration of precious metals especially Au, Ag, Pd and Pt is much higher than their respective primary resources, making waste PCBs an economically attractive material.

Additionally, PCBs also contain different hazardous elements including heavy metals (Cr, Hg, Cd etc.) rare earth elements (Ta, Ga etc.) and flame-retardants (Brand Cl) that pose grave danger to the eco-system during conventional waste treatment such as land filling and incineration.

PCB composition investigated previously by [4] is presented in table 1.

Table 1. PCB material composition.

Metals About 40%	%	Ceramics About 30%	%	Plastics About 30%	%
Cu	6-27	SiO_2	15-30	PE	10-16
Fe	1.2-8.0	Al_2O_3	6.0-9.4	PP	4.8
Al	2.0-7.2	Alkali-earthoxides	6	PS	4.8
Sn	1.0-5.6	Titanates-micas	3.0	Epoxy	4.8
Pb	1.0-4.2			PVC	2.4
Ni	0.3-5.4			PTPE	2.4
Zn	0.2-2.2			Nylon	0.9
Sb	0.1-0.4				
Au(ppm)	250 - 2050				
Ag(ppm)	110-4500				
Pd(ppm)	50-4000				
Pt(ppm)	5-30				
Co(ppm)	1-4000				

Other research papers [5-7] consider the use of the brake pad waste in new road pavements due to the waste composition including metal powder, graphite, rubber and fibers. Furthermore, [8] highlights the advantages in using fibers in asphalt concrete.

There are more waste materials that can also be reused to reduce the consumption of raw natural resources, providing opportunities to make new progresses in the building materials industry such as steel slag, rubber particles, graphite powder, PET and so on [9-13].

The research presented in the paper focuses on substituting the lime filler in a standard asphalt mixture used in Romania with PCB in a share of 100%.

The results of the laboratory investigation conducted on the new asphalt mixture shows that further investigations are needed in order to obtain a more proper asphalt mixture for road pavements. Different percentages of PCB will be used to obtain and to test a new material that can be used with great success considering the standards and regulations in Romania. In this way, by reusing PCB waste a great help will be given to reducing the use of new raw materials in the road construction industry.

The sections of the paper are Introduction, Experimental study, Results and discussion, Conclusion.

2. Experimental study

2.1. PCB waste description

The first step in analysing the waste is to use X-ray diffraction to show the composition of PCB (see Figure 1). X ray radiography shows that the waste looks like the PCB described by [4] or in a different manner. The average for the three specters is shown in table 2.

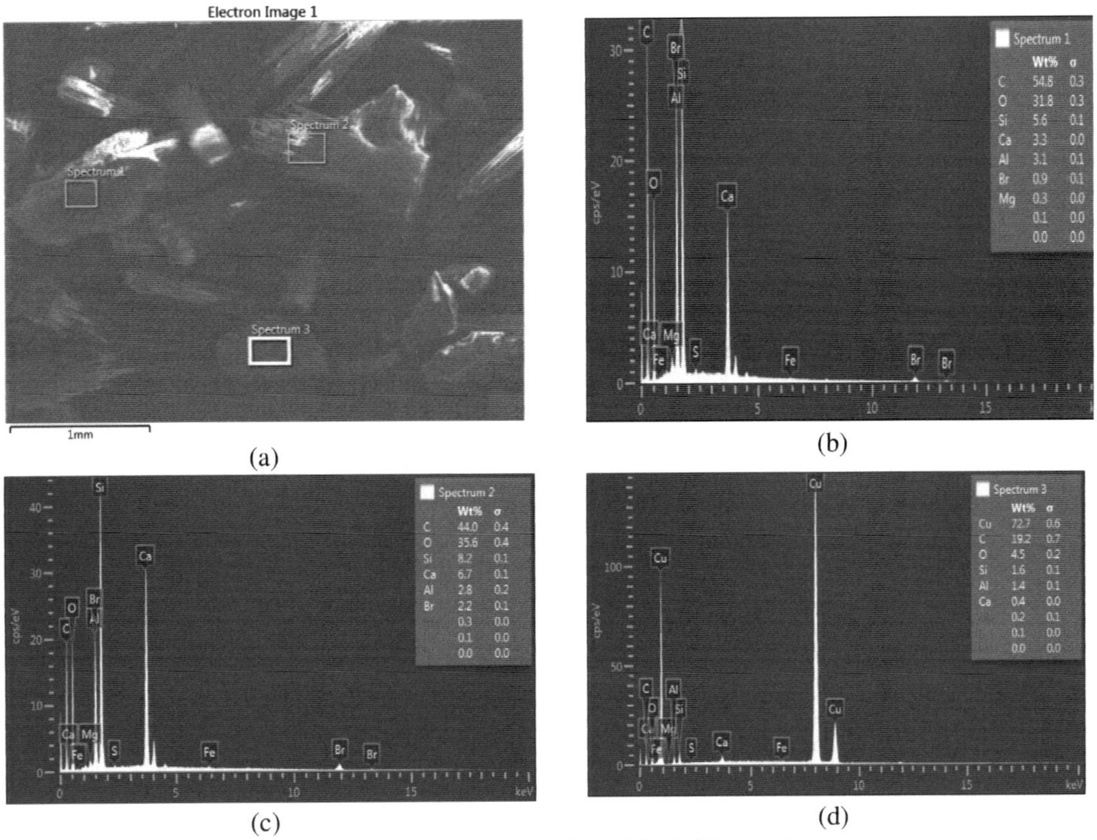

Figure 1. X-ray radiography of the PCB sample:
(a) Electron image; (b) Spectrum 1; (c) Spectrum 2; (d) Spectrum 3.

According to the results of the laboratory test, the sample of PCB waste contains a large amount of non-metals. This leads to the simple conclusion that there are different types of PCBs made from different materials. In this case, the results of the analyses show that Carbon represents up to 72.7% from the waste mass.

Table 2. PCB material composition.

Non-metals About 93.40%	%	Metals About 6.60%	%
C	44-72.7	Ca	1.6-6.7
O	19.2-35.6	Al	1.4-3.1
Si	4.5-8.2	Mg	0.2-0.3
Br	0.4-2.2	Fe	0.00
S	0.1		

The granularity of the W-waste is compared with the lime filler one (see table 3). Therefore, the research question is whether PCB can substitute the filler in the asphalt mixture in a share of 100%.

2.2. Asphalt mixture design

The asphalt mixture design was made according to European standards (EN), SR EN 12697 [16] and to the Romanian standard AND 605-2016 [17]. A mixture for the final course layer – top layer – that is the most exposed to traffic and climatic changes from all road structures interlayers was prepared.

The name of the layer according to AND 605-2016 is BA 16 rul 50/70.

Table 3. PCB physical characteristics versus lime filler.

Laboratory tests	PCB	Filler	Standards
Granularity [mm]	%	%	
1	99.3	100	SR EN 933-10:2009 [14]
0.63	90	100	
0.25	62.7	99.8	
0.125	53.2	93.9	
0.063	44.1	79.9	
0.02	17.2	18.9	
Humidity			
Max.2%	1.44	0.12	STAS 539/79 [15]
Apparent density	0.48	0.54	
0.5-0.8g/cm^3			
Hydrophilicity coefficient Max.1cm^3	0.78	0.28	STAS 539/79
Voids content range0.3-0.5%	0.82	0.41	STAS 539/79

BA 16 rul 50/70 means: the maximum size of the aggregates is 16mm, rul- came from the wearing layer (the final one) and 50/70 is the binder/bitumen type. Bitumen 50/70 is the most common asphalt binder used in Romania.

First tested hypothesis is that PCB substitutes Filler in a share of 100%. The goal was to make an eco-mixture that can respond well to all the laboratory tests and can be successfully used in the road construction industry for roads or highways. The new eco-mixture results from using this waste instead of classical lime filler are presented below.

The mixture recipe is described in table 4. The results of testing the mixtures with PBC and with filler are presented in table 5.

Table 4. BA16 rul 50/70 recipe.

Material-granular and binder	%
0-4mm	42.48
4-8mm	23.6
8-16mm	19.82
Lime filler – PCB	8.50
Bitumen	5.60

Table 5. Primary results of laboratory tests.

Lab. test	BA16 Rul50/70 PCB	BA16 Rul50/70 Filler	Standards
Marshall Stability	10.7	11.3	SR EN 12697-34:2012
Flow	8.7	3.3	SR EN 12697-34:2012
Density	2129	2399	SR EN 12697-6:2012
Water absorption	7.3	2.4	AND605:2016
Void content	8.5	2.7	SR EN 12697-5:2012
Void content	8.5	2.7	SR EN 12697-5:2012

IOP Conf. Series: Materials Science and Engineering 1138 (2021) 012022 doi:10.1088/1757-899X/1138/1/012022

2.3. *Dynamic results*

The first laboratory investigations regard the behaviour of both asphalt mixtures. The standard one using filler and the new one that has instead of filler the PCB waste. According to the Romania Standard AND 605:2016 [17], for the asphalt mixture design is very important that all bituminous mixtures respond to dynamic actions too: dynamic flow-cyclic compression and rutting.

BA 16 rul 50/70filler samples have the following characteristics: 101.1 mm diameter, 63.2 mm high. The cyclic load was set up at 300 kPa, temperature 50°C, frequency 3Hz, lateral pressing 150kPa and 10.000pulses.

BA16 rul 50/70 PCB samples have the following characteristics: 100.2 mm diameter, 63.6 mm high. The cyclic load is set up at 300 kPa, temperature 50°C, frequency 3 Hz, lateral pressing 150 kPa and 10.000pulses.

The results were compiled to describe the mixtures:

- BA16 rul 50/70 with Filler in the able 6 and Figure 2.
- BA16 rul 50/70 with PCB in table 7 and Figure 3.

Table 6. Dynamic flow-Cyclic compression for BA16 rul 50/70 with Filler.

	Determination	MU	Result	Standard method
1	Deformation speed at 50°C at 10000 pulses	μm/m/cycle	0.30	SR EN 12697-25:2016
2	Deformation at 50°C at 10000 pulses	μm/m	17602	SR EN 12697-25:2016

Table 7. Dynamic flow-Cyclic compression for BA16 rul 50/70 with PCB.

	Determination	MU	Result	Standard method
1	Deformation speed at 50°C at 10000 pulses	μm/m/cycle	0.51	SR EN 12697-25:2016
2	Deformation at 50°C at 10000 pulses	μm/m	55718	SR EN 12697-25:2016

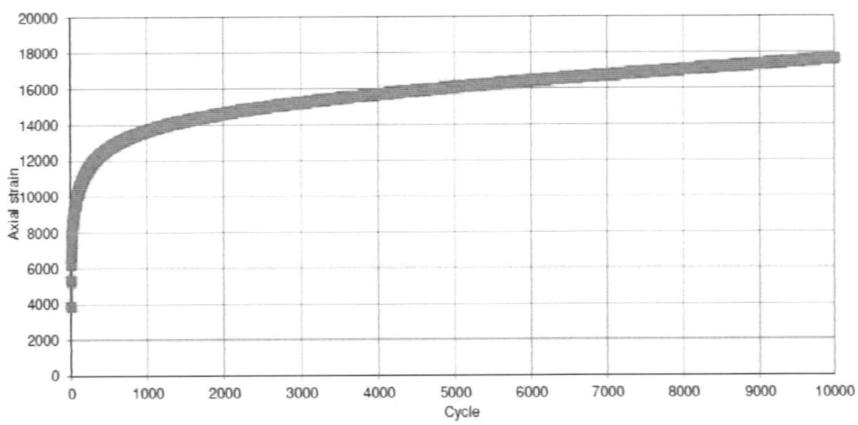

Figure 2. Dynamic flow-Cyclic compression for BA16 rul 50/70 with Filler.

Rutting behaviour was tested according to SR EN 12697:22, B procedure, at temperature 60°C, 40mm thickness of the slab, number of vehicles over 6000.

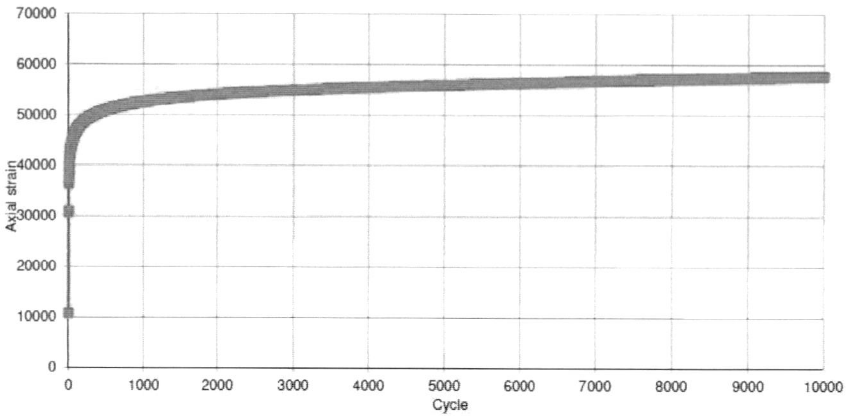

Figure 3. Dynamic flow-Cyclic compression for BA16 rul 50/70 with PCB.

The results of the rutting behaviour tests for the two asphalt mixtures are presented in table 8 for BA16 rul 50/70 with Filler and in table 9 for BA16 rul 50/70 with PCB.

Table 8. Rutting behaviour for BA16 rul 50/70 with Filler.

	Test	m.u	Result	Standard-method
1	Deformation speed WTS_{AIR}	mm/10^3cycles	0.16	SR EN 12697-22+A1:2007
2	Proportional Rut depth PRD_{AIR}	%	4.6	SR EN 12697-22+A1:2007
3	Rut depth RD_{AIR}	mm	1.82	SR EN 12697-22+A1:2007

Table 9. Rutting behaviour for BA16 rul 50/70 with PCB.

	Test	m.u	Result	Standard-method
1	Deformation speed WTS_{AIR}	mm/10^3cycles	0.35	SR EN 12697-22+A1:2007
2	Proportional Rut depth PRD_{AIR}	%	14.1	SR EN 12697-22+A1:2007
3	Rut depth RD_{AIR}	mm	5.63	SR EN 12697-22+A1:2007

3. Results and discussion

As a result of testing the new road asphalt mixture with 100% PCB E-waste replacing the lime filler some conclusions can be drawn.

The basic experimental set of tests that were conducted, namely Marshall flow and stability, density, water absorption and void content, show that both asphalt mixture possess good properties.

Not the same thing can be said about the results of the dynamic tests, where the differences between the two asphalt mixtures are significant.

The difference for the dynamic flow is: 17602 μm/m for the mixture containing filler compared to 55718 μm/m for the mixture containing PCB. Therefore, the dynamic flow modulus is 3.16 times higher for the mixture containing PCB which cannot be admitted in practice.

In terms of rutting, the difference between the two asphalt mixtures is similar to that of the dynamic flow. Rut depth is 3.09 higher for the mixture containing PCB.

4. Conclusions

The results of the first experimental stage, show that according to the Romanian Standard AND 605:2016, the eco-mixture using PCB E-waste to replace the filler in a share of 100% cannot be used as an asphalt pavement for the final course layer – top layer - to successfully support vehicle traffic because the deformations are significant.

Therefore, the new eco-mixture is not proper to use as asphalt pavement on roads.

However, it is proper to use on pedestrian or bicycle dedicated areas, where the asphalt mixture has a different role.

Exploring the good interaction between carbon and asphalt binder, a second step of the research will be carried out considering different ways to use this waste.

It is of great importance to continue this research to find some ways to help the road construction industry to face the great challenges concerning the diminishing of the raw material deposits and increasing costs with materials, as well as to help the environmental protection dealing with the increasing waste generation and reduced recycling rate.

E-waste recycling is one step further to a cleaner environment. PCB waste can be used in road materials but more hypotheses will be tested in order to find the proper usage in the asphalt pavement for the final course layer – top layer - to successfully support vehicle traffic.

Acknowledgments
In memory of PhD Eng. Andrei Boboc, who has inspired the research in this paper.

Authors wish to acknowledge: Technical University of Cluj-Napoca for the financial and logistic support in obtaining these results, Physics National Institute of Cluj-Napoca for the PCB Xray examination and TPA Cluj-Napoca for the laboratory testing and results. TPA is a private society for innovation and quality on laboratory examination, member of Strabag group.

References

[1] Kuity A, Jayaprakasan S and Das A 2014 Laboratory investigation on volume proportioning scheme of mineral fillers in asphalt mixture *Constr Build Mater.* **68** 637-43
[2] Kaya M 2016 Recovery of metals and nonmetals from electronic waste by physical and chemical recycling processes *Waste manage.* **57** 64-90
[3] Li J, Shrivastava P, Gao Z and Zhang HC 2004 Printed circuit board recycling: a state-of-the-art survey *IEEE Trans. Electron. Packag. Manuf.* **27** 33-42
[4] Duan H, Hou K, Li J and Zhu X 2011 Examining the technology acceptance for dismantling of waste printed circuit boards in light of recycling and environmental concerns *J. Environ. Manage.* **92** 392-9
[5] Hu X, Wang N, Pan P and Bai T 2017 Performance evaluation of asphalt mixture using brake pad waste as mineral filler *Constr Build Mater.* **138** 410-7
[6] Pan P, Wu S, Xiao Y, Wang P and Liu X 2014 Influence of graphite on the thermal characteristics and anti-ageing properties of asphalt binder *Constr Build Mater.* **68** 220-6
[7] Sugozu I, Can I, Oner C and Bagirov H 2015 Friction behavior of granite powder added brake pads *Mater. Test.* **57** 685-9
[8] Wang H, Yang J, Liao H and Chen X 2016 Electrical and mechanical properties of asphalt concrete containing conductive fibers and fillers *Constr Build Mater.* **122** 184-90
[9] Răcănel C and Burlacu A *Int. Multidisciplinary Sci. GeoConf.: SGEM* „Surveying Geology & Mining Ecology Management (SGEM) 2016 Sofia" vol 2 pp 33-40
[10] Popescu D, Burlacu A and Răcănel C *Int. Multidisciplinary Sci. GeoConf.: SGEM* „Surveying Geology & Mining Ecology Management (SGEM) 2016 Sofia" vol 3 pp 585-591
[11] Răcănel C and Burlacu A *Int. Multidisciplinary Sci. GeoConf.: SGEM* „Surveying Geology & Mining Ecology Management (SGEM) 2015 Sofia" vol 1 pp 795-802
[12] Răcănel C and Burlacu A *Proc. of the 3rd Int. Conf. on Road and Rail Infrastructure - CETRA, Road and Rail Infrastructure III*, „University of Zagreb 2014 Spilt Dalmatia" pp 395-400
[13] Popescu D and Burlacu A 2017 Considerations on the benefits of using recyclable materials for road construction *Rom. J. Transp. Infrastruct.* **6** 43-53
[14] SR EN 933-10:2009 Încercări pentru determinarea caracteristicilor geometrice ale agregatelor.

Partea 10: Evaluarea părţilor fine. Determinarea granulozităţii filerului (cernere în curent de aer) (Romania: Technical standard)

[15] STAS 539/79 Filer de calcar, filer de cretă şi filer de var stins în pulbere (Romania: Technical standard)

[16] SR EN 12697-5:2012 Mixturi asfaltice. Metode de încercare. Partea 5: Determinarea densității maxime (Romania: Technical standard)

[17] AND 605-2016 2016 Normativ privind mixturile asfaltice executate la cald. Conditii tehnice de proiectare, preparare si punere in opera a mixturilor asfaltice (Romania: Technical standard)

IOP Conf. Series: Materials Science and Engineering 1138 (2021) 012023 doi:10.1088/1757-899X/1138/1/012023

The influence of materials on the behaviour of joints with multiple bar connections

P Heinemann[1*], D N Isopescu[1] and S G Maxineasa[1]

[1] "Gheorghe Asachi" Technical University of Iasi, Faculty of Civil Engineering and Building Services, Department of Civil and Industrial Engineering, Romania

* patrick.heinemann@tuiasi.ro

Abstract. Joints made out of tubular steel members are often used in industrial constructions, like offshore, trusses or tree-shaped columns. The joint area is the weakest point in a truss structure. Beside the normal static behaviour many non-linearities due to the geometry or to the welding process have to be taken into account. Steel hollow sections are often used. Much research was made on welded nodes so far. To increase the resistance of a joint it is an option to vary the material choice. Many variations were tested in the past. It is possible to change the overall material of a joint or using composite structures, like a concrete filled column. Or the steel of the welding line can get an additional material component. Another aspect is to reinforce the joint for example by a stiffened ring. In this paper a state-of-the-art review of the influence of different materials to the resistance of welded nodes is given.

1. Introduction

For industrial buildings or extravagant structures, the main focus is on their stability and resistance. At the same time, it must be a very low weight to reach the requirements to the height or length, for example at widespan constructions. This is the reason why trusses, especially together with hollow profiles are often used. Rectangular (RHS) or square hollow sections (SHS) can be found in truss structures, bridges and high-rise buildings. Mostly for offshore platforms circular hollow sections (CHS) are used. To calculate the nodes of these structures is the more complex part. There are different types of joints. Figure 1 shows the types which are listed in the standards. The most common type in industry is the K-joint [1].

In tubular joints where two pipes have different diameters the lager pipe is called chord and the smaller one brace. The angle between brace and chord is called inclination angle. The joint area with a small inclination angle is called brace crown heel, with large brace crown toe.

A great part of welding research is about the microstructure of the welding lines, which is influenced by the welding heat/procedure. Younise et al. [3] did tests and determined the influence of material heterogeneity. He calculated the critical void volume fraction und used a mesh size of 0.2 - 0.5 mm. Baniari et al. [4] explained a non-destructive method of analysing the residual stress, called X-ray. There are three types of residual stresses: Macrostresses, Microstresses and stresses without specific name. By method of X-Ray stress can be determined by the measurement of the distance between crystallographic planes.

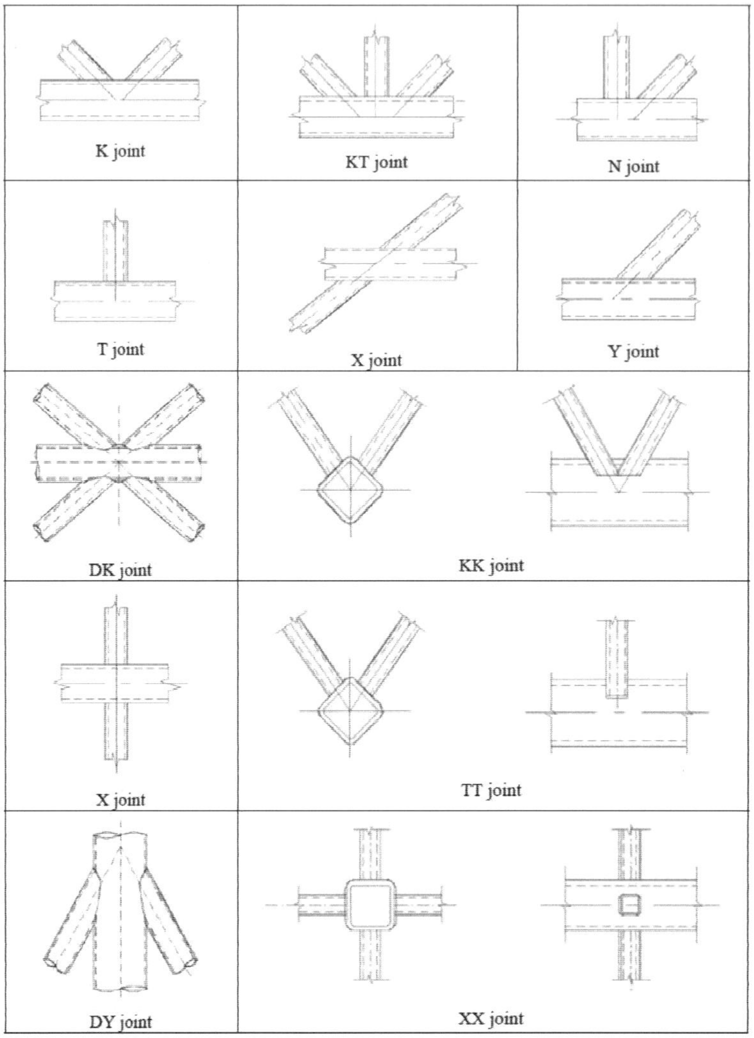

Figure 1. Joint types [2].

To increase the resistance of a above described joint it is an option to chance the material choice. There are many variations to change the overall material, like stainless steel, timber, cold-formed steel or concrete filled hollow sections. Another option is to vary the welding line material, or to add additives to the welding material. Examples for the physical variation are reinforced structures, like stiffened rings or rib-plates. In the following paper a state-of-the-art review of the described variations is given.

2. Analysis of joints made of different materials

2.1. Stainless Steel

For industry purpose stainless steel is an underpart. But due to its resistance to corrosion much research is done to that material.

Buchanan et al. [5] did research about stainless steel CHS beam columns under axial loading and bending moments. Buchanan explained, that the advantages of stainless steel are the high ductility, stiffness, strength, corrosion resistance, recyclability and aesthetics. The most common types of stainless-steel constructions are austenitic, duplex and ferritic. The austenitic grades are most prevalent, duplex offers improved mechanical and corrosion resistance and ferritic have the lowest costs. The most

common profile types are the CHS profiles, including high torsional resistance. There is a limited data for stainless steel CHS columns or beams. It was concluded, that the current design code is conservative regarding the described material for members with a high ratio of bending moment to axial load, while it is slightly unconservative for members with a low ratio of axial loading to bending moment.
Gardener [6] explained, that stainless steel has got a number of established benefits over carbon steel, but also higher costs.

Ashraf et al. [7] did research about stainless steel CHS columns. His approach does not require a classification of cross-sections, but it is based on a continuous relationship between the slenderness of a cross-section and its deformation capacity. The concept is extended further to determine flexural buckling resistance for long columns based on Perry type column curves. It was said, like in the first paragraph, that the Eurocode is highly conservative regarding the resistance of the CHS stub columns. In case of long columns, the Eurocode guidelines result in overprediction of the considered test results.

Feng et al. [8] did research about hybrid CHS to SHS tubular joints under axial compression. Experimental tests were done to 4 T-joint, 5 Y-joints and 9 X-joints made of stainless steel. IT was found out, that the current design rules are not appropriate for these types of joints. The FEM model was built out of 4-noded 10 x 10 mm shell elements. This is also valid for the welding line, which is not as exact as solid elements. But in this case, it is comprehensible, because mainly the study is about the difference of the material in the braces and chord. It is not about the welding line. In opposite to the author's numerical studies, Feng calculated the load capacity by the deformation limit criteria, invented by Lu et al. [9]. Next to the introduction to the four failure modes, Gao and Gho [10] also did studies about the geometric parameters. Most results are depending to the hybrid material, they are not described to that point, but the results are about the decreasing load capacities for T/Y- and X-joints with an increasing inclination angle. Feng concludes that four failure modes can be occur. The strength of the joints is affected by geometric parameters, especially by the β value. The methods to calculate the load-carrying capacity by the design codes and guides are close to the FEA results, but on the conservative side.

Zamzami et al. [11] explained that by comparing the strength of the welded and non-welded components, made of the same material, that there is a significant reduction in the fatigue strength of the welded components. Azari-Dodaran et al. [12] stated it more precisely and concluded a difference of nearly 10% in the ultimate strength.

Sometimes because of weight reduction the industry uses different materials. Zamzami did some research on hybrid joints, where the chord is made of steel and the braces are made of aluminium.

A very big issue is the resistance of the welding line itself. There are three zones in a welding area with different resistance, including the welding line. Mostly the welding steel is combined with some chemical additions.

2.2. High and low strength steel

The less expensive alternative to stainless steel or aluminium is to vary the hardness grade of the steel. In the following paragraph the main research results to the high and low strength steels are presented.

Xin et al. [13] did research about welded rectangular hollow section joints made out of high strength steel. It has got many advantages, but it is also expensive and only relevant for special constructions. It is used especially for truss girders of long spans used in sport arenas and bridges [14, 15]. Xin explains, that the steel grade is included in the books of the **C**ommittee for **I**nternational **D**evelopment and **E**ducation on **C**onstruction of **T**ubular structures (CIDECT) up to S460 and in the EC3 up to 700 MPa. The secondary bending moment is strongly dependent on the geometry and material properties [16]. Especially for high strength steel the secondary bending stress should be analysed in detail. Xin concludes that K-joints made out of HSS have got a higher ultimate load resistance of 20 - 142% (depending on the steel grade) compared to S355. The ultimate load resistance increases with increasing of the yield strength. Systems with a larger β-value result in larger secondary bending stresses. Joints with a lower gap size create a larger secondary bending stress.

Swoo-Heon Lee et al. [17] did numerical and experimental research on CHS to plate joints. For his FEA simulation he used shell and solid mesh elements. The element size was in the range of 30, 20 or 10 mm for the coarser mesh. For the refined area a size with a factor of 0.5 was implemented. Also, a comparison between the different methods is expressed, which are quite similar. First there is the NJS (Normalized Joint Strength) ($R_{3\%D}/F_y t^2$), secondly the AISC (American Institute of Steel Construction) ($R_n/R_{3\%D}$) of 2016, thirdly the Eurocode ($N_{1,Rd}/R_{3\%D}$) of 2005 and the last is the ISO (International Organisation of Standardization) ($F_1/R_{3\%D}$) of 2013.

It was concluded that the axial tension chord stress acting on ISO 14346 is similar at the yield strength of 460 MPa, but AISC and Eurocode 3, which do not consider the strength reduction effect are similar with increasing yield strength. Lee explained that by using high-strength steel under axial tension chord stress, the strength reduction effect can be neglected.

Islamovic et al. [18] did experimental tests about welded joints induced by bending moments. Steel plates with different conditions were analysed. The plates had butt welding lines. It was concluded, that for a bending angle of 120° no cracks appear in case of the steel S.0361.

Mishra [19] collected in his presentation many aspects about welded joints, including calculation methods. According to filled welds, advantages and disadvantages were listed. In summary the main advantages are: Filled welds are easy to prepare, they can be formed between two dissimilar metals, accommodation of different thicknesses and thin material such as diaphragms and foils can also be used. In contrast there are disadvantages like the chance of lower tensile strength, it is less rigid than the base material, overlaps may be undesirable for mechanical or aesthetic reasons, Micro-cracks and cavity deflects may occur or corrosion and fatigue cracking may occur. The butt-welds have different characteristics. It is the simplest form of welding, it does not require cutting the material, two metals are joined by simple placing their ends together and if the thickness is smaller than 5 mm bevelling is not necessary.

Al Rashed [20] did a presentation about the tensile strength of welded joints. Information about the hardness in different zones in the microstructure were already given. Especially a focus was given to the heat-affected zone. Different experimental bending test were made to see the difference of the absolute strength.

Shakhmatov and Usmanova [21] did research about the strength of welded joints under quasi brittle fracture. It was explained, that the strength of welds with structural faulty welds depends on the mechanical characteristics of the material, the length of the faulty weld, as well as on the radius at its top. The critical opening of the concentrator depends on the radius at its top.

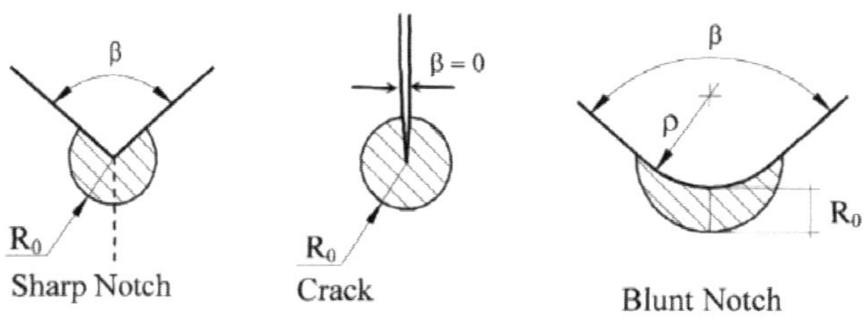

Figure 2. Notch geometries [22].

Remes et al. [22] did numerical and experimental tests about the fatigue strength of welded joints. Especially the different weld geometry and the plate thickness were focused. It was distinguished between normal and high-performing welds, which are characterised by a benefit of fatigue strength, e.g. due to weld notch geometry. The three kinds of notches are presented in figure 2. Beside this, different methods for calculating fatigue life were presented. Examples are Linear Elastic Fracture Mechanics (LEFM), Effective Notch Stress approach (ENS), Strain-based Crack Growth (SCG) or the averaged Strain Energy Density approach (SED). It was concluded, that a simplified modelling of the weld notch effect can cause significant uncertainties in the fatigue strength of high-performing welds.

2.3. Cold formed steel

The influence of the process of cold forming steel is explained in the following paragraph. A focus is given to the imperfection of the hollow section members, which is automatically generated by the process.

Mitsui et al. [23] did research about the effect of initial imperfection on elasto-plastic behaviour of SHS columns under axial force with bending moment. It was explained, that due to the cold forming process and geometrical initial imperfections, residual stresses can occur. In the numerical model, on one hand global imperfection, in form of initial bow imperfection was set, on the other hand there are local imperfections, which set as lack of flatness in the plate. Beside this, there is a distribution of the residual stresses in the thickness direction as material initial imperfection. The imperfections are given with equation (1) to equation (3). A four-node shell model was used (figure 3).

$$u_{y,G} = D_L L sin\frac{\pi}{L}z \tag{1}$$

$$u_{x,L} = \pm\frac{D_B B}{2}\left(cos\frac{2\pi}{B}y - 1\right)sin\frac{\pi}{B}z \tag{2}$$

$$u_{y,L} = \pm\frac{D_B B}{2}\left(cos\frac{2\pi}{B}x - 1\right)sin\frac{\pi}{B}z \tag{3}$$

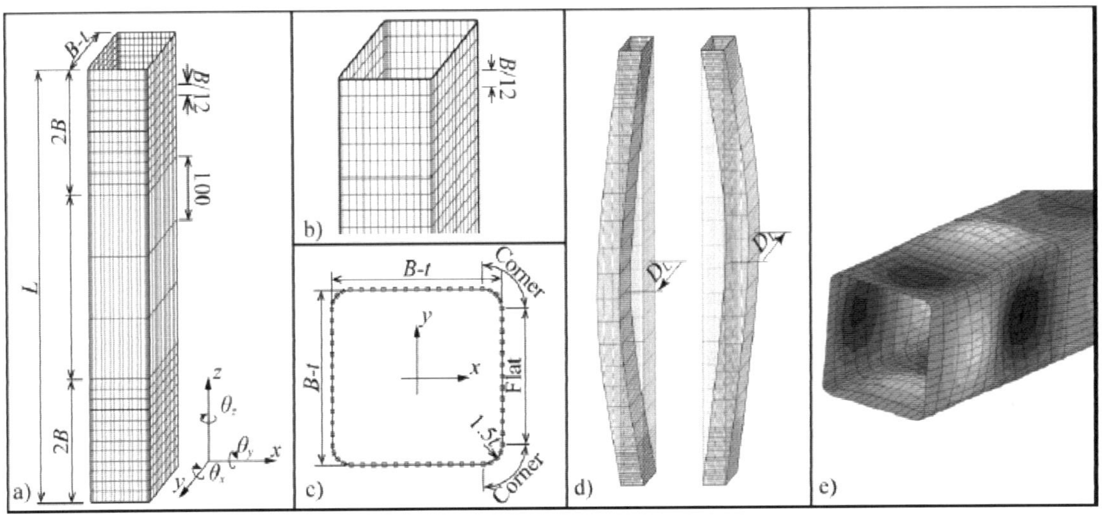

Figure 3. Analysis model [23].

It was explained, that in this case, the analysis model with individual initial local imperfection, regardless of the collapse mode, the maximum bending moment and the plastic deformation capacity decrease with the increment of the amplitude of the initial local imperfection. Beside this, it was concluded, that the magnitude of the residual stress which changes in the thickness direction has hardly an effect on performance. The amplitude of the residual stress which does not change in the thickness direction has an impact on performance. The amplitude of the local imperfections is the most influential imperfection on the deformation capacity.

2.4. Timber

Timber is a lightweight and ecological material. Especially for building constructions, which should be integrated in nature setting, timber structures are often chosen. Because of the heterogeneity of the microstructure and the difficultness of connection, often a hybrid technique is used. The wooden chords and braces are axial compression or tensional members, while the joint is made out of steel, which can be seen in figures 4 and 5. However due to the normal forces these constructions are comparable to truss structures.

A. Lokaj and Klajmonova [24] did tests on timber joints. The influence of the angle of the joint is analysed. Test with an inclination angle of 0, 60 and 90 degrees were done. Figure 6 shows a comparison between test records of these three different angles.

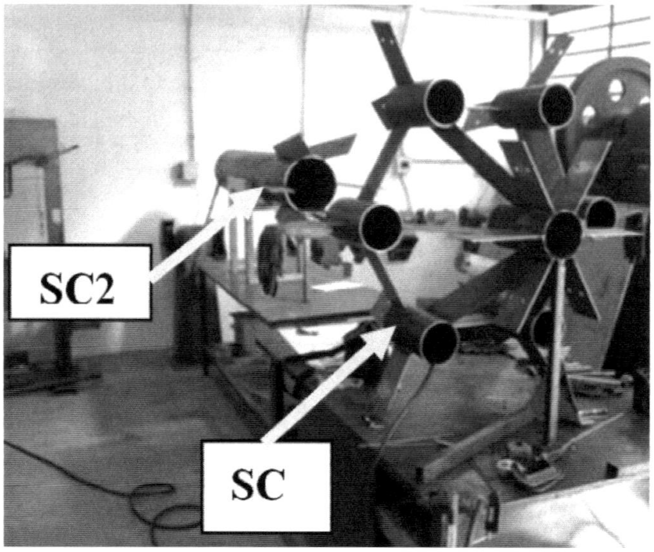

Figure 4. Steel joint [25].

Figure 5. Connection to timber members [25].

Goncalves et al. [26] did research to tree-shaped timber structures. Especially the six faces "diamond" hexagonal columns (DC) were analysed. The columns are constructed out of 15 cm diameter steel connections compared with 16 cm diameter C30/C40 timber dowels. It is designed in the way, the wood parts acted like compression members, without having bending moments. The resultant model can be seen in figure 7. There is only one support point in the soil, which defines the Tree-shaped timber stayed structural system (SETAM), which builds the opposite to the SETA – Tree-shaped timber structural system, having three support points in the soil. An advantage is the light structural weight, which can be a benefit to the application area of centres, malls hangars and so on. Beside this there is only a minimal environmental impact due to the natural architectonical shape.

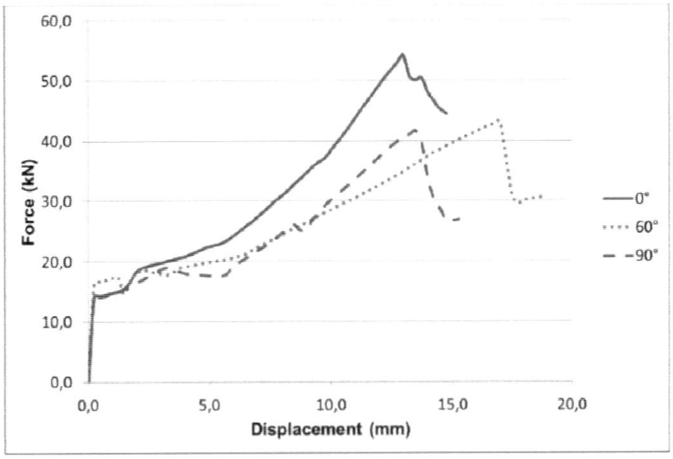

Comparison of test records of round timber samples in tension at a different angles

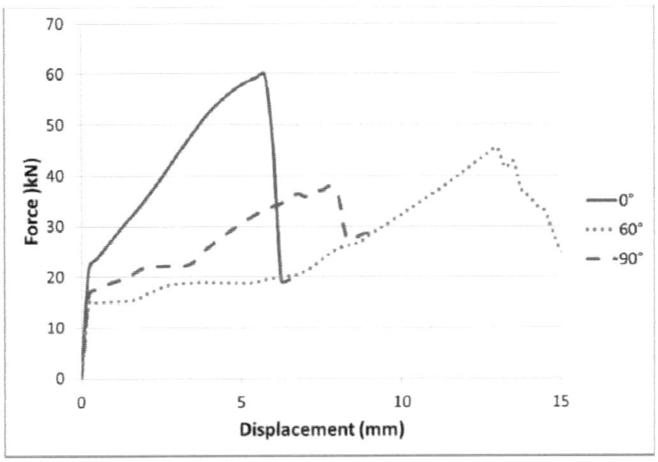

Comparison of test records of squared timber samples in tension at a different angles

Figure 6. Comparison of the influence of the joint angle [24].

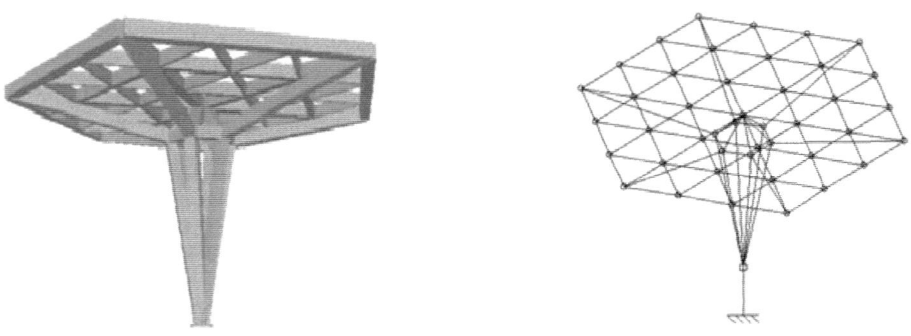

(a) (b)

Figure 7. Hexagonal column [25].

2.5. Carbon Fiber Reinforced Polymer (CFRP) and Concrete Filled Steel Tubular (CFST)

Carbon Fiber reinforced polymer is a composite material made out of a matrix and a reinforcement. For industrial purpose this kind of material is too expensive for an area-wide use, but a lot of research is done to that topic. It is a very low-weight and simultaneously a high strength base material.

The method of concrete filled tubular steel is mainly used for large diameter hollow section geometries. Inside of the steel tubular profile, reinforcement together with concrete is set. This is done to reach a higher stability and resistance to the column.

Ammari and Narmashiri [27] did research about SHS steel columns, strengthened by CFRP. The method of Fiber Reinforced Polymer (FRP) or the Carbon Fiber Reinforced Polymer (CFRP) method is a very seldom, but effective method to reinforce columns to get a higher strength, corrosion resistance and easy application. Jiao and Zhao [28] concluded a 25% - 76% higher tensile strength by using this method.

Dong et al. [29] concluded that CFRP warping on hollow and filled circular columns had better results than square sections. Ammari and Narmashiri simulated the model by the use of shell elements. He concluded that the horizontal deficiency in steel columns indicated significant local buckling, which causes stress concentration at damaged areas. The horizontal deficiency in the middle of columns caused the most critical of middle and corner elements. Carbon fiber had a significant effect on lateral deformation and local buckling.

Shahraki [30] did experimental and numerical investigation of strengthened SHS steel columns under axial compressive loads. The advantages of FRP reinforced columns were focused, which are the high strength-to-weight ratio and the optimal corrosion resistance.

S. Saleh et al. [31] did research about K-joints. Two design types were compared. On one hand there are CHS and on the other hand CFST (concrete filled steel tubular) joints. Because of the material choice for the thesis, only the first type is relevant. A main focus in their study should be the effect of corrosion, which was implemented by thickness reduction. This method is an approach, because there is no numerical model, which is able to replicate the corrosion effect [31].

Saleh occurs different failure modes for the CHS K-joint in the compression load case. On one hand there is the failure by punching into the chord and on the other hand there is the failure by punching in the chord combined with local buckling. For the tension load case, there is the punching out of the chord mode. Saleh concludes, that corrosion has only a moderate influence on the behaviour of the joint.

In his experimental research Garifullin [32] did tests with brace-to-chord width ratios of 0.6 to 0.8. Beside this the material S420, S500 and S700 together with the welding types a6, a10 filled welds and 1/2 v butt welds were analysed. It was concluded, that the filled welds increase the brace to chord diameter ratio β to a value of 0.85 and even greater. The ultimate resistance exceeded the plastic resistance of a steel joint by 16% average for all the joints. For joints with a higher β the difference was smaller. So, designing joints by plastic resistance is the more conservative way.

3. The influence of the material on the behaviour of a joint

In paragraph 2 several options are presented according material change for joints. In a numerical study the influence on von-Mises stress, strain and deformation distributions of the different materials are analysed. Overall, six different materials, representing each presented material group, is studied. The exception is timber (2.4), because of the different joint type and it is not comparable to the other materials. The materials are: Aluminium, stainless steel, steel with a high ratio of Carbon, steel with a low ratio of Carbon, Carbon and S 235 as common steel grade. Even it is not realistic, the composite material Carbon with its non-linear material parameters is set in the same way, like the others. The material parameters of the different categories are shown in table 1.

To see the differences in geometric purposes, a circular hollow-section Y- and a K-joint with inclination angles of 60° is simulated. The chord's profile is a CHS DN 50 with a length of the chord of 350 mm. The braces are CHS DN 40 profiles with a length of 175 mm. Figure 8 shows the geometry of the K-joint. It is meshed by quadratic tetrahedral elements with a 10 mm coarser and a 1.3 mm refined mesh. For the simulation the numerical software Ansys [33] is used.

Table 1. Material parameters.

Material	S 235	Aluminium	Stainless Steel	High Carbon Steel	Low Carbon Steel	Carbon
Density [kg/m³]	7850	2770	7750	7850	7850	1490
Tensile strength [N/mm²]	360	310	586	1070	365	2231
E-Modulus [N/mm²]	210.000	71.000	193.000	212.000	210.000	121.000
ν	0.3	0.33	0.31	0.29	0.29	0.27
Bulk-Modulus [N/mm²]	175.000	69.608	169.300	168.250	166.666	--
Shear-Modulus [N/mm²]	80.769	26.692	73.664	82.171.	81.395	60.000

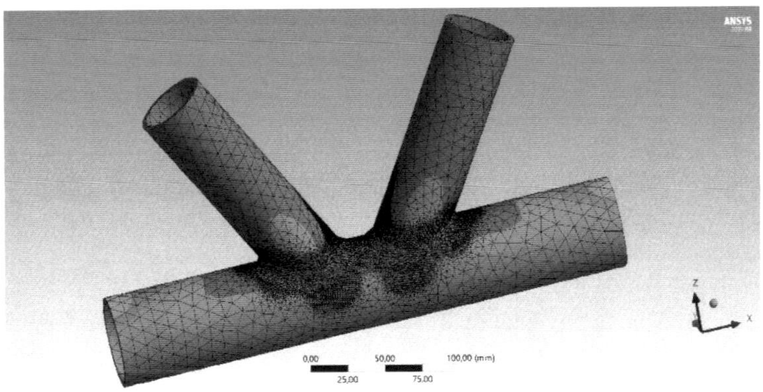

Figure 8. Geometry and mesh of the K-joint.

An axial force of 20.0 kN is affected to the braces' ends. The chord is set as pinned. The 3 mm fillet weld is made out of the same material, like the members, even it is unrealistic in case of the composite material Carbon.

The figures 9 to 11 visualise the results for the stress, deformation and strain for both joint types. The distributions of the Y- and K-joint have got a similar trend. There is only a larger difference in case of Aluminium. In general, there is a gap of around 40 N/mm² in stress and a difference of smaller than 1% in stain and deformation for the steel types. The deformation of the Aluminium is significantly higher than for the other steel types. A large peak arises in all three categories for the non-linear material carbon. The stresses are larger by a factor of around 3.

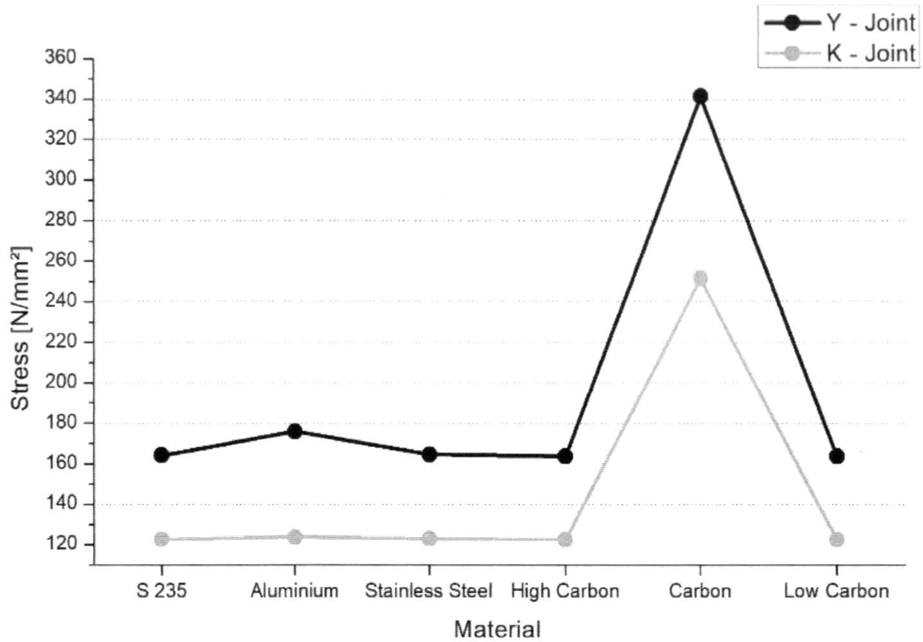

Figure 9. Stress results.

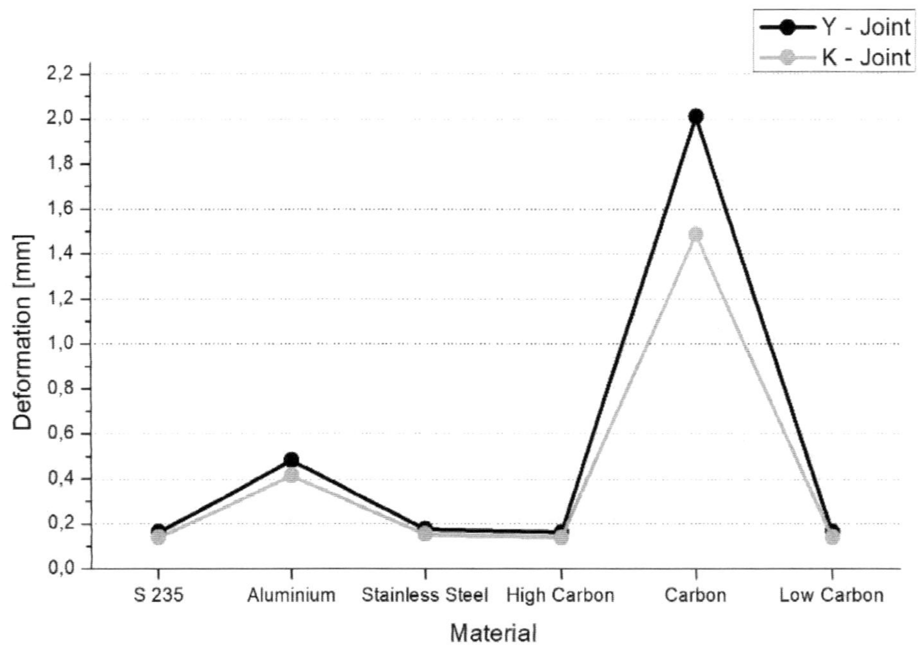

Figure 10. Deformation results.

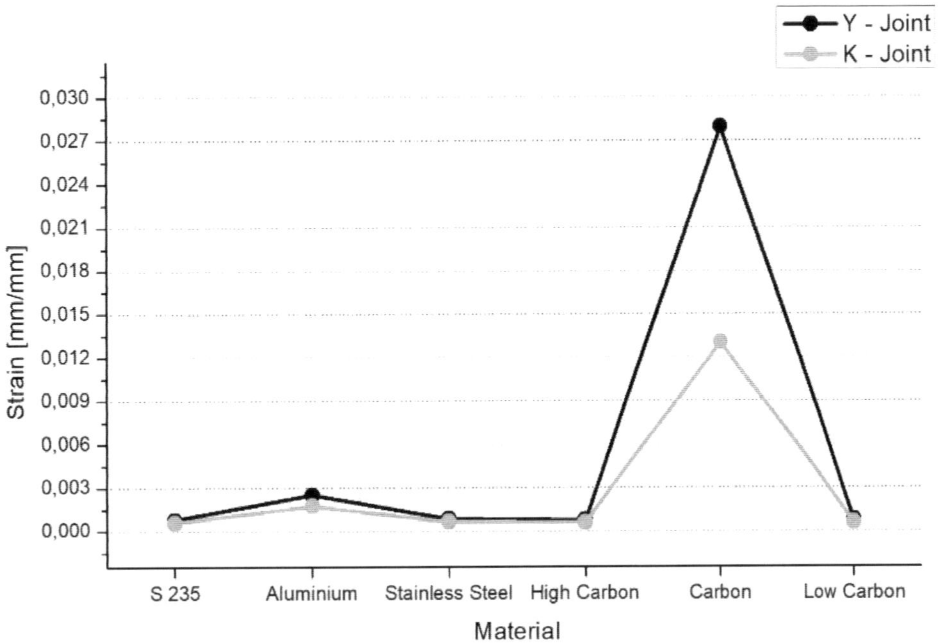

Figure 11. Strain results.

4. Conclusions

This paper summarises the different options regarding the material choice for structural joints. Connections made out of high-, low-strength steel, timber, CFRP, CFST, aluminium, cold formed and stainless steel were introduced. Each material has got advantages and disadvantages, but in common it has increased the resistance and stability of the normal used industrial steel. The timber structures are presented as an alternative to the steel structures. Much research is done to the high-performance materials to realise a slenderer construction. Some of the described materials, especially the Carbon Fiber Reinforced Polymer is just too expensive for the area-wide use.

A numerical study to a Y- and K-joint made out of different materials, which were presented in this paper was done. The results of the von-Mises stresses, deformation and strain are nearly superimposable for the S 235, stainless, high and low carbon steel. The Aluminium has not got the high resistance like the steel grades.

The Carbon has got much more higher stresses, strain and deformations. A reason is, that the theoretical model according the connection of one or two members by using a fillet weld is not applicable for a material like Carbon, which is produced out of a high number of fibres.

References

[1] Espinosa J M, Schafhirt S, Muskulus and M. Mesh Parametric 2017 Study for Fatigue Assessment of tubular K-joints using Numerical Methods. *Energy Procedia* **137** 514-522

[2] DIN EN 1993-1-8, Eurocode 3 2010-12 Bemessung und Konstruktion von Stahlbauten – Bemessung von Anschlüssen. *Deutsches Institut für Normung*

[3] Younise B, Rakin M, Gubeljak N, Medo B and Sedmak A 2016 Numerical prediction of ductile fracture resistance of welded joint zones. *Procedia Structural Integrity* **2** 753-760

[4] Baniari V, Blatnicka M, Sajgalik M, Vasko M and Saga M 2017 Measurement and numerical analyses of residual stress distribution near weld joint. *Procedia Engineering* **192** 22-27

[5] Buchanan C, Zhao O, Real E and Gardner L 2020 Cold-formed stainless steel CHS beam-columns- testing, simulation and design. *Engineering Structures* **2020**

[6] Gardener L 2005 The Use of stainless steel in structures. *Progress In structural Engineering and Materials* **7(2)** 45-55

[7] Ashraf M, Gardner L and Nethercot D A 2008 Resistance of stainless steel CHS columns based on cross-section deformation capacity. *Journal of Constructional Steel Research*
[8] Feng R, Huang Z, Chen Z, Roy K, Chen B and Lim J B P 2020 Finite-element analysis and design of stainless steel CHS-to-SHS hybrid tubular joints under axial compression. *Thin-Walled Structures*
[9] Lu H L, Winkel G D, Yu Y and Wardenier J 1994 Deformation limit for the ultimate strength of hollow section joints. *Proceedings of the Sixth International Symposium on Tubular Structures* 341-349
[10] Feng R, Huang Z, Chen Z, Roy K, Chen B and Lim J B P 2020 Finite-element analysis and design of stainless steel CHS-to-SHS hybrid tubular joints under axial compression. *Thin-Walled Structures*
[11] Zamzami I A, Davison B and Susmel L. 2019 Experimental fatigue curves to perform the fatigue assessment of aluminium-to-steel thin welded joints. *Procedia Structural Integrity* **18** 255-261
[12] Azari-Dodaran N and Ahmadi H 2019 Static behaviour of offshore two-planar tubular KT-joints under axial loading at fire-induced elevated temperatures. *Journal of Ocean Engineering and Science*; S2468-0133 **19** 30079-8
[13] Xin H, Kisoensingh P and Veljkovic M 2019 Welded Joints of Rectangular Hollow Section Made of High Strength Steel (HSS). *17th International Symposium on Tubular Structures*
[14] Bijlaard F, Veljkovic M, Shi G, Qiang X 2018 Editoral: Implementation of high-strength, high performance steel structures. *Steel construction* **11 (4)** 247-248
[15] Veljkovic M, Johansson B 2004 Design of hybrid steel grinders. *Journal of Constructional Steel research* **60** (3-5) 535-547
[16] Bjork T, Tuominen N P and Lahde T 2015. Effect of the secondary bending moment on K-joints capacity. *Proceedings of 14th International symposium on tubular structures*
[17] Lee S-H, Shin K-J, K A-Y, Lee H-D 2019 Numerical Study on the Deformation Behaviour of longitudinal Plate-to-High-Strength Circular Hollow-Section X-Joints under Axial Load. *Applied Sciences* **9** 3999
[18] Islamovic F, Muratovic P, Gaco D and Kulenovic F 2009 Bend testing of the welded joints. *7th International Scientific Conference on Production Engineering*
[19] Mishra A 2020 Mechanics of Welded Joints. *Presentation*
[20] Al Rashed M 2019 Investigation of Welded Joints. *Presentation*
[21] Shakhmatov M V and Usmanova E A 2020 Strength of welded joints under quasi-brittle fracture. Journal of Physics: *Conference Series* **1441** 012114
[22] Remes H, Gallo P, Jelovica J, Romanoff J and Lehto P 2020 Fatigue strength modelling of high-performing welded joints. *International Journal of fatigue* **135** 105555
[23] Mitsi K, Sato A, Aoyama M 2017 Effect of initial imperfection on elasto-plastic behaviour of SHS columns under axial force with bending moment. *Eurosteel*
[24] Lokaj A and Klajmonova K 2016 Comparison of round timber and squared timber joints capacity in tension at different angles to the grain. *Procedia Engineering* **142** 357-363
[25] Karamanos S A, Romeijn A and Wardenier J 2000 On the fatigue Design of K-Joint Tubular Girders. *International Journal of Offshore and Polar Engineering* **Vol. 10**
[26] Goncalves D, Varanda L D, Christoforo A L, Lahr F A R 2014 Tree-shaped timber Structural System: An Ecological Way of Designing Spatial Structure. *Journal of Civil Engineering Research* **4(1)** 1-7
[27] Ammari A S and Narmashiri K 2020 Effects of horizontal deficiency location on the structural behaviours of steel SHS short columns strengthened using CFRP. *Mechanics of advanced composite Structures* **7** 39-47
[28] Jiao H and Zhao X L 2004 CFRP strengthened butt welded very high strength (VHS) circular steel tubes. *Thin wall Structures* **42(7)** 963-78
[29] Dong J F, Wang Q Y and Chen X 2013 Structural behaviour of recycled aggregate concrete filled steel tube columns strengthened by CFRP. *Engineering Structures* **48** 532-542

IOP Conf. Series: Materials Science and Engineering 1138 (2021) 012023 doi:10.1088/1757-899X/1138/1/012023

[30] Shahraki M, Sohrabi M R, Azizyan G R and Narmashiri K 2018 Experimental and numerical investigation of strengthened deficient steel SHS columns under axial compressive loads. *Structural Engineering and Mechanics* **67** No. 2 207-2017
[31] Saleh S, Saqif M A and Ramiz F 2019 Comparison of behaviour between hollow and composite K-joints under sustained loading and corrosion. *Materials Science and Engineering* **513** 012039
[32] Garifullin M 2018 Experimental moment resistance of rectangular hollow T joints. *MATEC Web of Conferences* **245** 08003
[33] ANSYS Inc. Products 2019 R3 © 2006-2019

Thermo-energetic assessment of industrial buildings

I E Iancu[1] and L Moga[1*]

[1] Faculty of Civil Engineering, Technical University of Cluj-Napoca, Romania

* ligia.moga@ccm.utcluj.ro

Abstract. The present paper is focused on the assessment of industrial buildings from their thermal and energy performance perspective, driven by the need to reduce energy consumption and CO_2 emissions on a worldwide level. A state of the art will be elaborated regarding the environmental impact of buildings that led to an increased pressure on a more energy efficient design. The focus will afterwards turn on the constructive solutions of industrial buildings for a better understanding of their elements and materials used. Aspects regarding the national and international stock of industrial and logistics spaces, demand, price range and future evolution of this sector, are also presented. Description and comparison is conducted on the building envelope thermal performance and its impact on the energy performance of buildings. Several conclusions will be drawn based on the obtained energy indexes and future research paths will be suggested.

1. Introduction

Constructions occupy a significant percentage in the process of environmental deterioration in all stages of life, from design (i.e. design of life expectancy), production of construction materials, transport and commissioning, operation and maintenance, to demolition and recycling [1]. In 2014, the local cement market was valued at about 500-600 million euros [2]. Cement, the basic material in the construction industry is polluting through the production process, generating a very large amount of CO_2.

Construction is a sector that, by its nature, has a major impact on the environment. Globally, the cumulative impact of construction processes has increased exponentially due to the accelerated development of the urban environment. It is reported that the operation of buildings (maintenance, occupation) contributes to a third of global greenhouse gas emissions and more than 40% of the use of energy resources globally [3]. Population growth and migration to large urban centres is the premise of the need to develop new homes, shopping centres, industrial buildings, etc.

Globally, there is a concern for the field of metal constructions due to certain advantages that the adoption of such a solution can bring to the realization of the structure of a building. There are two usual methods of approaching a structure, each with its specific advantages and disadvantages, i.e. reinforced concrete and / or prestressed concrete structure and metallic structure.

2. Warehouse - characteristics and envelope elements

A warehouse is defined as a very large hall, covered, with one or more openings, used for industrial purposes, as storage, as a workshop in factories, as a hall for sporting and cultural events, for exhibitions, etc. The industrial warehouse is in addition, a building that houses technological processes.

Types of halls:

- depending on the material: Metal halls; Concrete halls (monolith or prefabricated elements); Masonry halls; Mixed halls;
- depending on the height regime: Single storey hall; Multi-storey halls.

Industrial halls can be made with one or more openings (of equal or variable sizes) and can be provided with or without overhead cranes, depending on the housed technological processes.

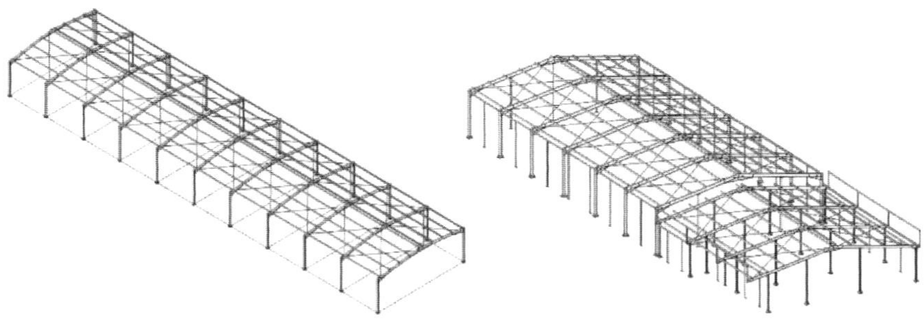

(a) (b)

Figure 1. Hall with one opening (a), hall with several openings (b) [5].

Usually, the halls have openings from 12 m up to 36 m transversely and openings from 5 m up to 12 m longitudinally. At very large dimensions, as well as at irregular shapes, joints are provided for the fragmentation / separation in the plan of the construction. These joints can be: expansion joints to combat temperature variations; seismic joints to prevent degradation of the construction due to the action of the earthquake; settlement joints to counteract any differences in the foundation ground, in order to avoid differentiated settlements of the construction, which can lead to a major degradation of the building.

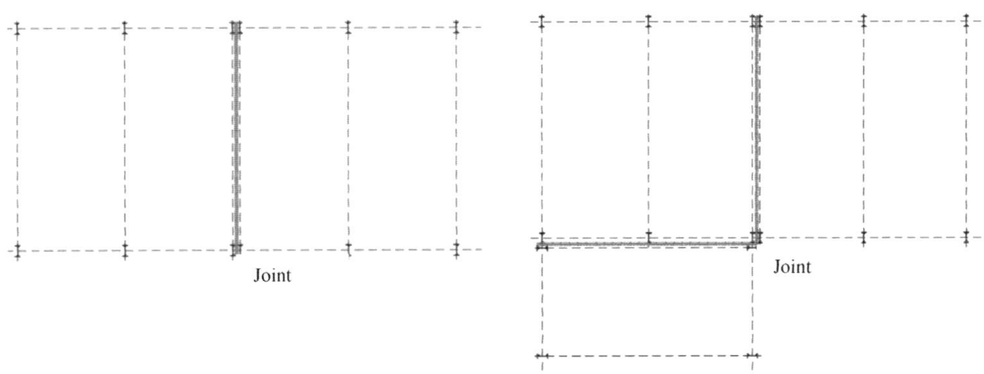

Figure 2. Hall separated with joint.

The envelope of the hall type constructions consists mainly of sandwich panels. It is the first choice of locking systems because it brings speed and efficiency in reducing execution time. They are made of two sheets of steel, between which there is a thermal insulation layer. They are fixed on a skeleton made of metal profiles (steel), wooden profiles or on metal inserts (in concrete) and are joined longitudinally - through the profiled joint system - and transversely - by partial cutting of the lower metal plate and the thermal insulation layer, overlapping the panels and fixing with self-tapping screws[6]. The most widely used solutions for industrial constructions are [13]:

For the roof:

- prefabricated sandwich panels with polyurethane foam PUR (or PIR)
- prefabricated sandwich panels with mineral wool
- panel + mineral wool + waterproofing membrane (made at construction site)
- panel + glass wool + metal panel (made at construction site)

For the walls:

- prefabricated sandwich panels with PUR polyurethane foam (or PIR)
- prefabricated sandwich panels with mineral wool
- panel + glass wool + metal panel (made at site)
- structural boxes + cotton wool (glass or mineral) + outdoor metal panel (which can be of several types depending on the chosen architecture)

Available products on the market consist of sandwich panels with cores of expanded polystyrene (EPS), mineral wool and polyurethane foam (PIR, or polyisocyanurate) [14].

Each material used as the core of the sandwich panel vary in their thermal insulating performance, sound insulating performance, reaction to fire and weight.

- Polyurethane foam core panels [15]:

- are lightweight prefabricated elements, sandwich panel type;

- are made in a "tied" system, with total adhesion between the components, in continuous technological process, by injecting under computer control rigid polyurethane flame retardant foam expanded between 2 plates of metal (zincate steel, pre painted aluminium or copper);

- polyurethane foam used – free of fully halogenated fluoride-chlorine hydrocarbons (CFCs), is made of polio and isocyanate, with expansion agent;

-fastening shall be carried out on a skeleton made of metal profiles (steel) of a thickness greater than 1,5 mm, wooden profiles with a minimum section of 60x80 mm, or on metal inserts (in concrete) of a width of at least 60 m and a thickness of at least 1,5 mm;

-sandwich panels are joined longitudinally by the profiled joint system, and in a transverse direction by partially cutting the lower metal plate and the insulating layer, overlapping the panels and fixing with self-tapping screws.

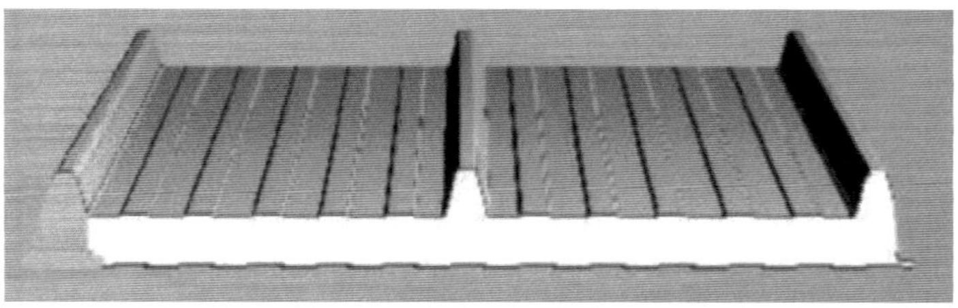

Figure 3. Polyurethane foam core panel [15].

Thickness varies from 30mm up to 200mm for standard panels and have a thermal transmittance averaging from as high as 0.70 $W/m^2 \cdot K$ to as low as 0.15$W/m^2 \cdot K$.

- Basaltic mineral wool thermal insulation panels [15]:

- are lightweight prefabricated layered envelope elements;

- are made in a "tied" system, with total adhesion between the components, in continuous technological process, by bonding panels of basaltic mineral wool of different thicknesses between 2 plates of metal (zincate steel, pre painted aluminium or stainless steel);

- mineral wool is of basaltic type, with a density of 100 kg/m;

- fire resistance is between 30 and 180 minutes, depending on the type of panel, its thickness and mounting position;

- fastening shall be carried out on a skeleton made of metal profiles (steel) of a thickness greater than 1,5 mm, wooden profiles with a minimum section of 60x80 mm, or on metal inserts (in concrete) of a width of at least 60 m and a thickness of at least 1,5 mm;
- sandwich panels are joined longitudinally by the profiled joint system, and in a transverse direction by partially cutting the lower metal plate and the insulating layer, overlapping the panels and fixing with self-tapping screws;

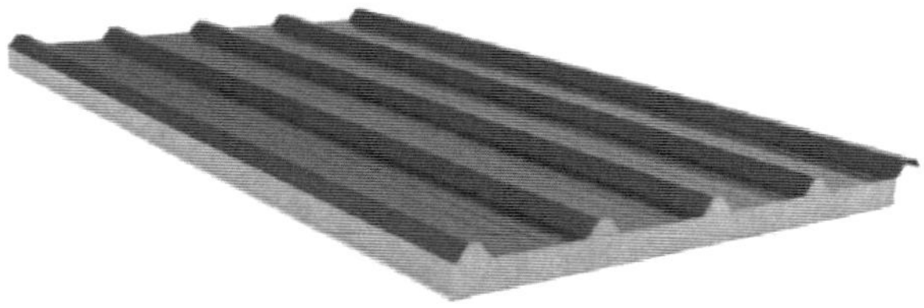

Figure 4. Basaltic mineral wool thermal insulation panels [16].

Thickness varies from 50mm up to 200mm for standard panels and have a thermal transmittance averaging from as high as 0.78 $W/m^2 \cdot K$ to as low as 0.20$W/m^2 \cdot K$.

- EPS thermal insulation panels (expanded polystyrene)

The expanded polystyrene core of the sandwich panels (EPS) provides a high density, effectively covering a wide range of areas of applicability: industrial construction, commercial, residential, halls and warehouses (including refrigerators), container constructions and agro-zoo technical buildings.

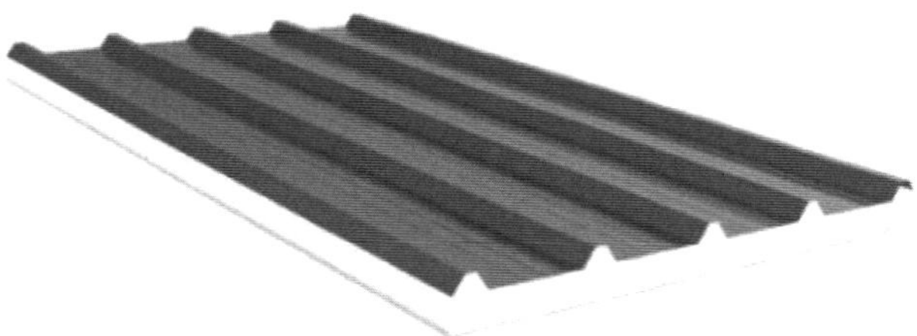

Figure 5. EPS thermal insulation panels [16].

Thickness varies from 50mm up to 200mm for standard panels and have a thermal transmittance averaging from as high as 0.72 $W/m^2 \cdot K$ to as low as 0.18$W/m^2 \cdot K$.

The most efficient material is thus the polyurethane foam, and even more so, isophenic rigid foam which is similar with polyurethane foam except it uses a different reaction and produces a higher concentration of methylene diphenyl diisocyanate.

Access inside warehouses is done through industrial doors, which are special technical solutions whose functionality and characteristics cannot be supplemented by conventional closures. Industrial door systems perform several unique functions at the same time as safe and long-term use. When we talk about industrial doors, we think of solutions like stainless steel sectional doors, glazed sectional doors, industrial sectional doors, industrial rolls and industrial sliding doors through the ability to close large gaps. Depending on the opening system [18], there are:

- Standard sliding system - Panels rise below the ceiling, parallel to it
- Bottom sliding system - for sectional doors with surfaces up to 20 sqm
- High sliding system - for sectional doors up to 40 sqm
- Vertical sliding system – where high heights are available
- Angular sliding system – for sectional doors with surfaces up to 20 sqm

Thermal insulation coefficient (U value) of industrial doors range from 0.5W/m²·K for full sectional doors [17] to 1.2 W/m²·K[19]; 1.5 W/m²·K to 3.4 W/m²·K for glazed sectional doors [19]; 1.6 W/m²·K to 3.6 W/m²·K for aluminium glazed doors [19]. Another solution for glazed doors is in the form of acrylic glass instead of standard safety glass. These provide a U value ranging from 1.52 W/m²·K to 4.46 W/m²·K[18] depending on the number of layers and surface used.

Aside from glazed panels, natural illumination and ventilation in industrial halls can be done through skylights. Optimal skylight/roof surface ratio is between 2.5-15% for sufficient daytime lighting without the need for artificial lighting[20]; and also, for natural ventilation [21]. Ventilation made by the hatches is more efficient than that carried out by side windows [22]. The thermal transmission identified for fiberglass reinforced polyester skylight, protected at surface level with a gel layer or polyester film and an additional 10 mm polycarbonate sheet is 2.5W/m²K [23]. Comparatively, polycarbonate cell panels with 5 cents, have a thermal transmission coefficient of 2.65W/m²·K for a thickness of 12mm and 2.02W/m²·K for a thickness of 16mm [24].

3. Industrial stock market

The building stock of industrial and logistics spaces nationwide has grown rapidly over the past five years, and analysts estimate that growth will continue. In 2019, the share of industrial spaces shown in the next picture, according to business magazine [25], outlines the current state of the industrial construction sector, with Bucharest and Timisoara leading the field.

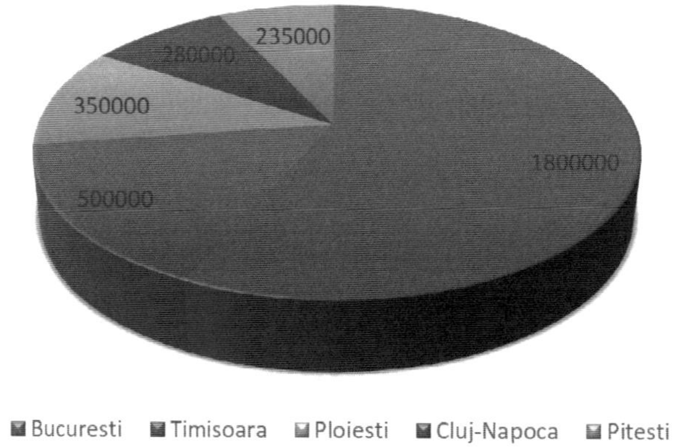

Figure 6. Building stock of industrial and logistics spaces in square meters.

Analysts estimate an increase of at least another 1 million square meters in 2020. The demand for logistics space has been directly influenced by the increased volume of online commerce. The evolution of these logistic spaces was captured by the consulting company Cushman & Wakefield Echinoxduring 5 years, 2015 - 2020, for capitals in Central and Eastern Europe, in table 1.

Table 1. Evolution of the building stock of logistics spaces in the capitals of Central and Eastern Europe in the last 5 years [7].

City	Building stock of logistics spaces (2015) [thousand sqm]	Building stock of logistics spaces (1st trim. 2020) [thousand sqm]	Evolution [%]
Bratislava	852	1308	54%
București	959	2039	113%
Budapesta	1856	2287	23%
Praga	1862	3162	70%
Varsovia	2761	4320	56%

IOP Conf. Series: Materials Science and Engineering 1138 (2021) 012024 doi:10.1088/1757-899X/1138/1/012024

It is expected to attract new investments on the markets of Cluj Napoca, Timisoara, Ploiesti and Bucharest, in the context of shortening the supply chain and encouraging local production, those being important centers of industrial development. Thus, in an analysis of the first half of 2020 made by IBC Focus, 528 industrial sites were identified in the country, with their distribution in figure 7 [8].

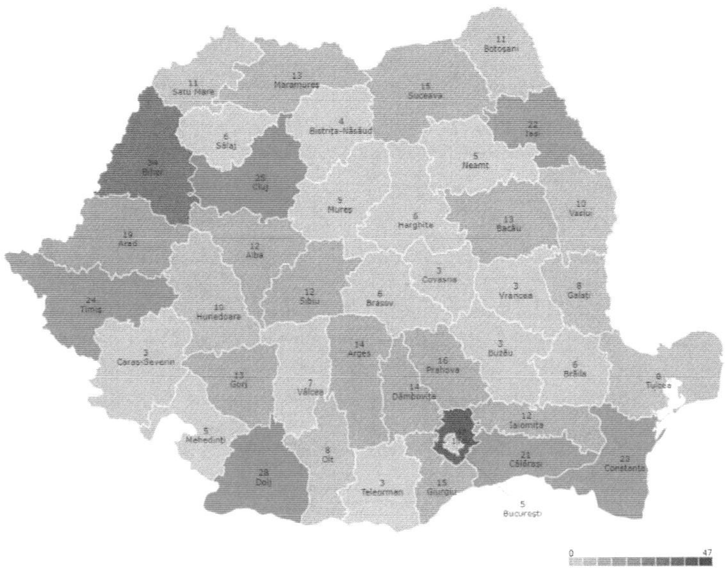

Figure 7. Numerical distribution of construction sites by counties [8].

Analysing the local market in the above-mentioned cities and the metropolitan area, it is possible to elaborate a synthesis of the available commercial and industrial spaces, surface, location, price per sqm. (figure 8).

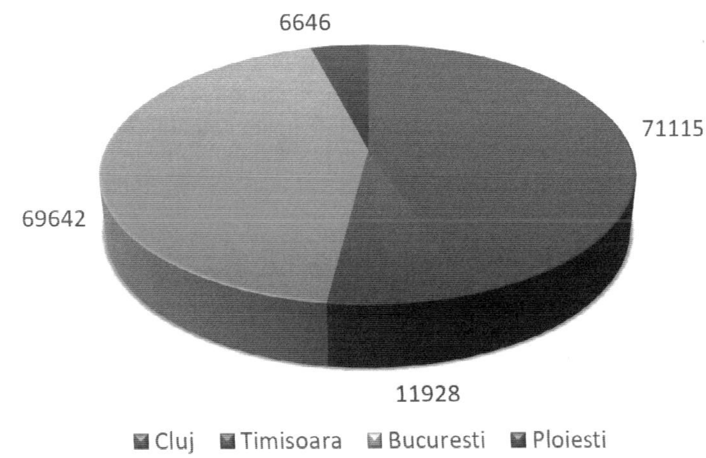

Figure 8. Area available, June 2020.

Included in the sample, only the surfaces representing industrial / commercial spaces of hall type buildings were considered, and the validity of the offers was limited to June 2020. Most spaces have a small surface area, below 1000 m^2. The average surface of the spaces is useful in establishing the optimal solutions for streamlining the hygrothermal behaviour. In Europe, the industrial real estate market is constantly expanding [9]. There is a low supply compared to an excessive demand, and this

IOP Conf. Series: Materials Science and Engineering 1138 (2021) 012024 doi:10.1088/1757-899X/1138/1/012024

is reflected in the increase of rents and increasing returns for investors. E-commerce seems to be the catalyst for much of the increase in demand. In terms of investments in logistics and industry, 2019 has so far continued the recent trend of seeing a growing interest in this sector from investors. Active investors in this sector are increasingly coming from Asia-Pacific and North America, including from sovereign wealth funds and institutional backers.

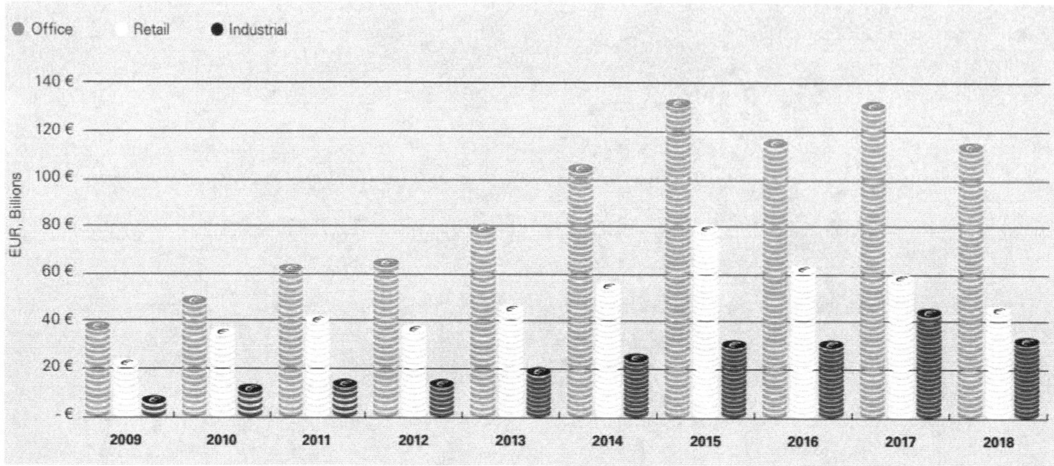

Figure 9. Volume of investments by sectors, source "Real Capital Analytics", taken over [9].

Although there are still ample opportunities in Western Europe, especially in facilities close to large population centers, many in this sector see the most untapped potential in the so-called markets of Poland, the Czech Republic, Slovakia, Hungary, Romania and Bulgaria.

Colliers International predicts that there will be a requirement for more than one million square meters of new logistics and distribution space in these markets, cumulated by years, between 2018 and 2022 [9] and this equates to about 5 percent of current stock. This level of requirements would absorb all current construction activity in the industrial sector in the Czech Republic, Slovakia and Hungary and almost all activity in Romania.

Developers are beginning to respond to this new increase in demand in the Central and Eastern European region. According to figures from Cushman & Wakefield, 2017 saw a total supply of industrial stocks, surpassing the previous peak years of 2007 and 2008.

Growth in Poland has been particularly dramatic, driven by closer proximity to German markets and closer integration with EU supply chains. The total supply of industrial space in Poland in 2017 was almost double compared to 2016 (2.3 million square meters compared to 1.2 million square meters). The Polish market has also been stimulated by a wave of major projects by large occupants, including Amazon, H&M and Zalando [9].

4. Energy use and carbon emissions of industrial constructions

The construction of new buildings has a direct impact on the environment, due to the use of material resources, energy and water, as well as due to the generation of greenhouse gas emissions. Greenhouse gas emissions result either directly from the burning of fossil fuels in the construction site or in the transport of materials, or indirectly from the production of materials and electricity used during the construction and operation phase of the building until its demolition. Therefore, energy consumption and emissions are highlighted directly during the construction, renovation, operation and demolition phases of the building and indirectly, by the use of energy and emission generation associated with the production of each material that makes up the building.

An LCA-type analysis for an industrial unit of 6733 sqm [10] outlines the built-in carbon and energy for a life expectancy of 50 years and a total land area of about 21400 m^2. The building consists of two functional floors: the ground floor (5847.15 m^2) in which equipment and materials stocks are located and the first floor (885 m^2) for offices, dining rooms, bathrooms and recreation. The industrial building also includes an underground gallery with installed cables and pipes, which are necessary for proper operation, cumulating6733 m^2 built area with 13.60 m high, summing a volume of 68107 m^3.

For life cycle inventory analysis (LCI), primary data on electricity, water and fuel consumption measured directly on site were collected. Monthly visits were made to collect this data during construction work over 160 days [10]. The analysis and calculation of embedded energy and carbon emissions were carried out based on these data and on the basis of the technical data sheets of the materials used. As a result, table 2 shows the amount of carbon and energy embedded associated with electricity, water and fuel used in the construction of this industrial building and table 3 shows the energy and carbon embedded associated with the materials used in the structure and architectural elements.

Table 2. Carbon and energy embedded in the use of electricity, water and diesel in the construction of the industrial building, adapted from [10].

	Quantities collected in 160 days	Carbon embedded kgCO2-eq	Embedded energy MJ
Electricity	21200 kWh	8162	76320
Water	780 m^3	608.4	7176
Fuel	29400 l	77296.13	1134840
Total		*86066.53*	*1218336*
Per m^2 built		*12.78*	*180.95*

Table 3. Energy and carbon embedded in the building materials used [10].

Structure	Amounts [kg]	Carbon embedded [$kgCO_2$-eq]	Embedded energy [MJ]
Reinforced concrete	2985399.75	665932.36	6897809.75
Metal works	37579.3	525278.14	6528692.81
Bituminous emulsions	1585.18	792.59	80844.38
Pavement	80844.38	353771.39	3491701.63
Total structure		1545774.48	16999048.57
Per m^2 built		229.58	2524.74
Architecture	**Amounts [kg]**	**Carbon embedded [kgCO2-eq]**	**Embedded energy [MJ]**
Masonry	648345.98	89776.43	842367.62
Rooftop	142514.44	510262.81	6935648.6
Wall finishes	119227.09	739861.46	5982491.44
Floor finishes	77632.18	337602.41	293366.56
Ceiling finishes	7188.68	114889.09	778895.22
Total architecture		1792392.2	14832769.44
Per m^2 built		266.21	2203.00

One of the most significant categories of emissions in the supply chain, apart from transport, is storage, including material handling in logistic buildings. Carbon emissions from material handling, including warehouses and sorting units, are significant and account for 13% of the overall supply chain emissions [12]. As deposit facilities are an essential element in global supply chains, it is inevitable to take into account carbon emissions from storage and handling operations for full lifecycle assessment and for the identification of effective decarbonisation measures in general logistics activities to achieve a sustainable long-term practice. The resulting emissions are generated from both total energy consumption and private consumer, assuming constant use of the units for 50 hours per week [11].

Table 4. Primary energy sources and emissions [11]

Energy sources	CO_2emissions [$kgCO_2/kWh$]
Electricity	0.793
Gas	0.605
Fuel	0.835
Propane	0.234

In view of the parameters described above, the annual energy consumption for a typical warehouse ranges from 1025 MWh to a maximum of 1256 MWh depending on the storage technology adopted. This leads to annual emissions between 888 and 1087 tCO_2 [11].

5. Conclusions

In light of the rapid development of the industrial sector (presented in section 3), cumulated with the energy use of warehouse type buildings, design directions can be anticipated to reduce the environmental impact that these buildings have.

The energy and carbon embedded in the construction of an industrial building is an important criterion in the choice of materials and method of execution. Materials are predominant in embedded energy and carbon emissions. In this respect, the processing of materials such as metals and concrete contributes the most, as their production involves more energy and therefore generates more carbon emissions. On the contrary, natural or less processed materials, such as soil, stone and wood, are generally low in carbon embedded. Based on this and subsequent developments, it will be possible to identify alternative low-impact building materials that can reduce built-in energy and embedded carbon, helping to increase the sustainability of construction.

Storage-related emissions reduction should be adopted taking into account energy-efficient lighting and advanced heating, ventilation and air conditioning systems, while reducing heat loss due to transmission and ventilation. Designing optimal skylights ratio and lighting control and ventilation (through ridge openings) can provide a reduction in energy used for lighting up to 70% [21] and a drop of up to 50% in CO_2 emissions [21] (compared to no lighting control and ventilation).

The increase in demand leads to future developments of industrial buildings all around the world. To stop climate change and design environmentally friendly and sustainable buildings, engineers must come up with better design for building envelope components by also considering the energy and climate regulations for the nearly Zero Energy Buildings. In the same time, they must conduct LCA analyses that lead to a better choice of materials and design. Future research is needed in the direction of sustainable envelope systems that have at least the same hygrothermal properties as classic solutions, if not better, but significantly lower embodied energy and carbon.

References

[1] Georgescu D, Apostu A and Cosma C 2009 Reinforced concrete buildings with reduced impact on the environment and health, Matrixrom
[2] http://ro.wikipedia.org
[3] United Nations Environmental Programme 2009 Buildings and Climate Change Sustainable Buildings & Climate Initiative
[4] Câmpian C 2018 Structural sustainability University Course, Technical University of Cluj-Napoca
[5] Damaschin A F 2019 Metal structures and halls. Metallic structures https://structuri-metalice.com
[6] Built space 2019 How to choose the right sandwich panels for halls https://spatiulconstruit.ro

[7] Lefter A 2020 The market of industrial and logistics spaces in the Capital area has doubled in the last five years. Development continues in 2020, fuelled by online retailers and food retailers, accessed at https://cpi.imobiliare.ro
[8] Ghiorghiu M 2020 528 investments in the Industrial field Accessed at https://infohale.ro/iunie-528-investitii-in-domeniul-industrial/
[9] Panek P 2019 European logistics and warehousing: The outlook for 2019 accessed at http://www.whitecase.com/publications/insight
[10] Rodrigues V, MartinsA A, Nunes M I, Quintas A, Mata T and Caetano N S 2018 LCA of constructing an industrial building: focus on embodied carbon and energy *Energy Procedia* **153** p 420–425
[11] Ries J M, Grosse E H and Fichtinger J 2017 Environmental impact of warehousing: a scenario analysis for the United States, *Inter J. of Prod. Res*. 55:21 6485-6499
[12] World Economic Forum 2020Supply Chain DecarbonisationAccessed at http://weforum.org/
[13] ITKA Industrial Buildings, accessed https://www.itkabuildings.ro
[14] Balex Metal 2019 Sandwich panels – what should you know? accessed at http://balex.eu/en/knowledge-base/news
[15] Mathaus 2020 Thermal-insulating panel system, accessed at http://mathaus.ro/
[16] Metigla 2019 Production of materials for metal closures, accessed at https://metigla.ro
[17] https://smilo.ro
[18] https://usi-tehnice.ro
[19] https://hormann.ro
[20] Lapisa R, Karudin A and Martias M et al. 2020 Effect of skylight–roof ratio on warehouse building energy balance and thermal–visual comfort in hot-humid climate area *Asian J. Civ. Eng.* **21** 915–923
[21] Wang X, Kendrick C, Ogden R, Walliman N and Baiche B 2013 A case study on energy consumption and overheating for a UK industrial *Appl. Ener.* **104** 337-344
[22] Kadra 2020 Versatile natural ventilation solutions for health and energy saving, accessed at https://kadra.ro
[23] http://www.astron.biz/en/
[24] http://panourihale.ro
[25] http://www.businessmagazin.ro

Influence of the input frequency for soundness evaluation of concrete structure based on waveform analysis

K Ikebata[1*], Y Kobayashi[1], K Oda[1] and K Nakamura[1]

[1] Dept. of Civil Engineering, Coll. of Sci. and Tech., Nihon University, Tokyo, Japan

* csku19001@g.nihon-u.ac.jp

Abstract. In the soundness evaluation method based on waveform analysis in time domain, the response waveforms that are measured while the target is in sound condition is indispensable. Therefore, it is expected that this method is for the evaluation of precast concrete products because the state of the products is almost identical and it is able to measure the response waveforms before using the products. On the other hand, although there are difficulties to apply this method for the evaluation of the existing structures because the response waveform is not measured before using, it is expected that this method can be applied to the evaluation of the existing structures by making the response waveform on the basis of the numerical simulation. The computational conditions should be set properly based on the actual measurement conditions for the numerical simulation. Among the computational conditions, the input frequency is determined based on the diameter of the steel ball and contact time if the excitation is given by making an impact with the steel ball by using the theory proposed by Sansalone et al. However, Kubo et al. reported that the measured contact time is different from the theoretical value and the difference reaches maximum twice. The influence of the difference would not be negligible for the soundness evaluation and should be considered carefully to perform the soundness evaluation accurately. Therefore, in this study, the response waveforms are computed on a numerical model with excitations of various input frequencies, and the influence of the input frequency to the response waveform is discussed by comparing the computed response waveforms.

1. Introduction

In Japan, the quality and performance of structures are managed based on the results of testing that are performed during the construction and the service period of the structures to make the structures safe and longer life. In this background, non-destructive testing is frequency used because it gives information of a structure without any damage.

The impact elastic wave method is one of the non-destructive testing, for detecting the defects in the concrete structures. This method uses elastic waves that are excited by using a hammer or a steel ball as an input device. The frequency range of the elastic wave is relatively low, and it is suitable for the investigation of concrete structures due to the low attenuation of the input wave and long wavelength. However, it is known the conventional soundness evaluation method cannot be applied if the target is not a plate.

Therefore, in order to evaluate the position and size of internal defects regardless of the shape of the target structures, the authors propose a new soundness evaluation method that detects internal defects on the basis of change of the response waveform in time domain. On the other hand, there are difficulties to apply this method for the evaluation of the existing structures. Because the response waveform is not measured generally at the initial state, it is impossible to compute the difference value for the structures. However, it is expected that this method can be applied to existing structures if the response waveforms can be obtained by numerical simulations. In the generation of response waveforms by the numerical simulation, the parameters for the simulation should be set properly to simulate the measurement conditions.

The input frequency in the impact elastic method is determined by calculating the contact time from the diameter of the steel ball using the theoretical formula proposed by Sansalone et al. [1]. However, Kubo et al. reported that the measured contact time is different from the theoretical value, and the difference reaches twice at a maximum [2]. The influence of this difference would not be negligible, and it is influence should be revealed.

Therefore, in this study, the response waveforms are computed on a numerical model with excitations of various input frequencies, and the influence of the input frequency to the response waveform is discussed.

2. Soundness evaluation method

2.1. Elastic Wave Method

Impact elastic wave method is an inspection method to evaluate a target by applying input wave to the target and observing the acoustic waves that is reflected at a defect or a boundary. This method uses the elastic waves that have relatively low frequency component that is emitted by using a hammer or a steel ball as an excitation. The use of the lower frequency component is suitable for the investigation of concrete structures, since the concrete structures are the composite materials and low frequency component is less, affected by the internal structure. Furthermore, the attenuation of lower frequency component is relatively small and propagates in longer distance as consequence.

Figure 1 shows a schematic diagram of the impact elastic wave method. Elastic waves are emitted by striking the concrete surface with an iron ball, and the emitted elastic waves are received near the excitation point. Focusing on the dominant frequency of the frequency spectrum obtained by frequency analysis of the response waveforms of multiple reflections in the depth direction, the method is widely used mainly to detect the thickness and internal defects of plate structures [1]. Figure 2 shows an example of a frequency spectrum diagram obtained using the FFT (Fast Fourier Transform). The vertical axis is the spectral intensity and the horizontal axis is the frequency. From the dominant frequency F_{max} at this time, the thickness D_p can be obtained from the relationship with the elastic wave velocity V_p, as shown in the following equation:

$$D_p = \frac{V_p}{2F_{max}} \tag{1}$$

However, it should be noted that the length and the width of the plate are sufficiently large in comparison with the thickness of the plate according to the Japanese Society Non-Destructive Inspection standard "NDIS2426-2: Non-destructive examining of concrete-elastic wave method-Part 2: Impact elastic wave method" [3]. This is because, the reflection path of the waves is complicated if the shape pf the is not like a plate, the assumption is not fulfilled and the dominant frequency corresponding to the depth direction cannot be observed.

For this reason, conventional evaluation principles are not applicable to structures of rectangular cross sections such as columns and beams. In addition, in the evaluation of internal defects focusing on the dominant frequency, the change of the dominant frequency with the size and the position of the defect has not been concretely clarified. Although it is possible to evaluate the planar position and size of the defect, it is not possible to evaluate the position and size of the defect in the thickness direction.

IOP Conf. Series: Materials Science and Engineering 1138 (2021) 012025 doi:10.1088/1757-899X/1138/1/012025

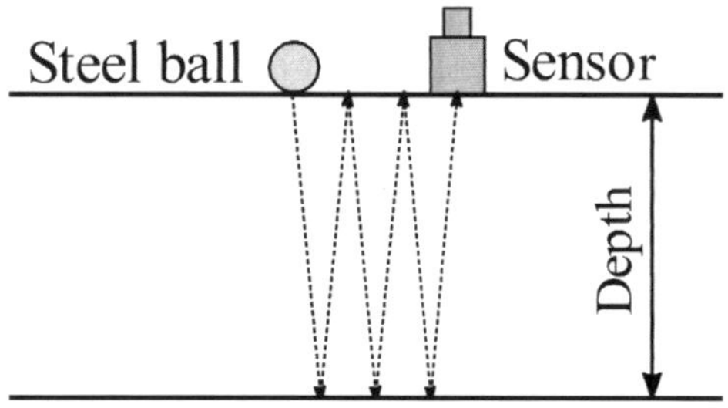

Figure 1. Measurement schematic.

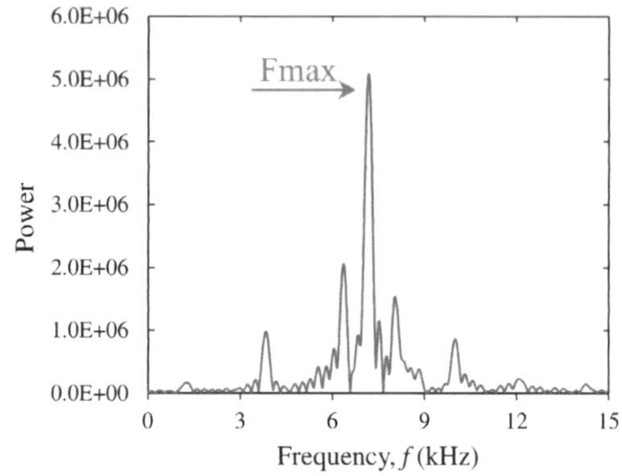

Figure 2. Result FFT.

2.2. Response Waveform Analysis in Time Domain

Figure 3 shows a schematic diagram of the measurements in the impact elastic wave method using response waveform in time domain. The index value of the soundness evaluation is the difference value that is computed by using response waveforms in time domain. This difference value shows the amount of change in the response waveform in time domain. Figure 3 (a) and (b) illustrate approximated wave paths in a model without void and with a void respectively. The difference of the model is only the existence of the void, and the change of the response wave form caused by the void is shown in figure 3 (c) as an example. The difference value (D) is proposed as equation (2), and the position and size of internal defects are evaluated on the basis of the difference value.

$$D = \frac{1}{N}\sum_{i=1}^{N}(f_i - f_i')^2 \tag{2}$$

where f_i is the amplitude value of the response waveform in the time domain without void, f'_i is the amplitude value of the response waveform in the time domain with void, and N is the number of data measured in the sampling interval values.

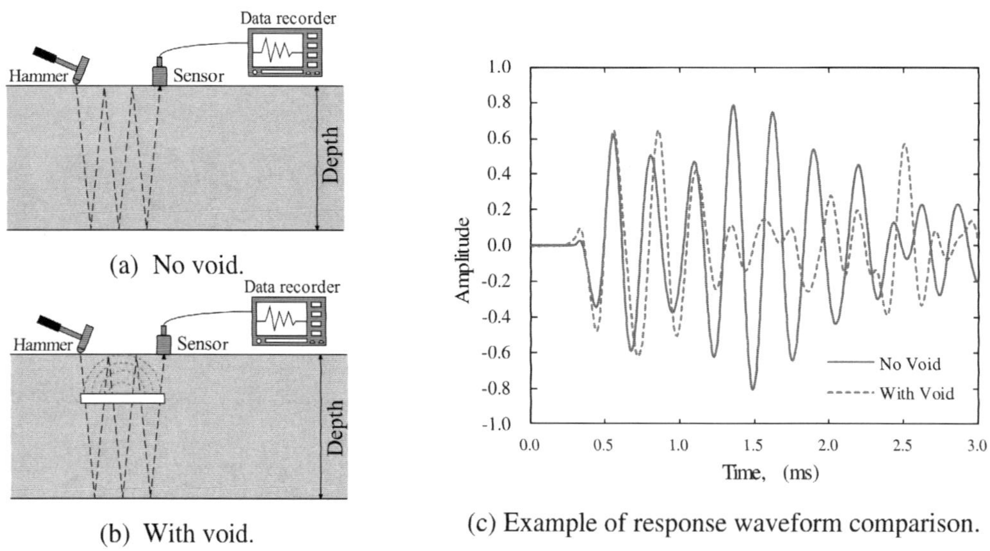

(c) Example of response waveform comparison.

Figure 3. Concept of response Waveform Analysis in Time Domain.

2.3. *Advantage of Response Waveform Analysis in Time Domain*

This soundness evaluation method based on the waveform analysis in time domain can be applied to the structures of any shape, and it is expected that the method is applied for soundness evaluation of the existing structures. In particular, this method is advantage if the structure is constructed by using products since the response waveform before using can be measured and it can be used for the other structurers that are consecrated by using the same produces. In addition, it has been confirmed from previous studies that it is possible to evaluate the size and position of internal defects by providing multiple measurement points [5-7].

On the other hand, although there are difficulties to apply this method for the evaluation of the existing structures if the response waveform is not measured before using, However, it is expected that this method can be applied to the evaluation of the existing structures if the response waveform in the soundness condition can be obtained in numerical simulation. Furthermore, the generation of intact response waveforms by the simulation eliminates the need to measure them in advance, and the effect of internal defects on the response waveforms can be understood in advance by the numerical simulation.

3. Problem in generating soundness response waveforms by numerical simulation

3.1. *Setting condition of input frequency in numerical simulation*

For the generation of response waveforms by numerical simulation, it is important to set the appropriate initial condition of the structure and determine the measurement conditions. Among these parameters, the related the analysis model and the measurement point are easily determined. On the other hand, are should be care full determined if the not observable directly. The input frequency in this measurement method is calculated from the contact time between the steel balls used for input and the concrete, and the theoretical expression for this contact time is proposed by Sansalone et al. [1].

3.2. *Input frequency by steel ball contact time*

Assuming that the collision phenomenon between steel ball and concrete is theoretically developed based on Hertz contact theory, the contact time T_c can be obtained by equation (3) [4].

$$T_c = 4.53\left\{\frac{(\delta_s + \delta_c)m_s}{\sqrt{R_s v_0}}\right\}^{\frac{2}{5}} \tag{3}$$

$$\delta_s = \frac{1 - \nu_s{}^2}{E_s} \tag{4}$$

$$\delta_c = \frac{1 - \nu_c{}^2}{E_c} \tag{5}$$

where m_s is the mass of the steel ball, R_s is the radius of the steel ball, v_0 is the impact velocity of the steel ball, ν_s is the Poisson's ratio of the steel ball, E_s is the modulus of elasticity of the steel ball, ν_c is the Poisson's ratio of the concrete, and E_c is the modulus of elasticity of the concrete. In equation (3)

$$m_s = \rho_s V_s \tag{6}$$

$$v_0 = \sqrt{2gh} \tag{7}$$

if we organize the equation as:

$$T_c = 5.97R_s\frac{\{(\delta_s + \delta_c)\rho_s\}^{\frac{2}{5}}}{h^{0.1}} \tag{8}$$

it is organized as in equation (8). ρ_s is the density of the steel ball, V_s is the volume of the steel ball, g is the gravitational acceleration, and h is the falling height of the steel ball. In equation (8), if we substitute general physical properties for each material constant, convert the steel ball radius Rs to steel ball diameter d_s (=$2R_s$), and assume that the effect of drop height can be neglected, we can obtain equation (9), which is the relationship between contact time and steel ball diameter proposed by Sansalone et al. [1].

$$T_C = 0.0043d_s \tag{9}$$

And the maximum frequency (Fmax) of the input elastic wave is gives as the inverse of the contact time as shown in the following equation.

$$F_{max} = 1/T_c \tag{10}$$

3.3. The authenticity of the theoretical value of contact time

However, since concrete is not perfectly elastic, the theoretical and actual contact times generally do not coincide. For this reason, Kubo et al. conducted a verification experiment on the difference from the theoretical value. In this experiment, a sheet sensor was attached to the impact surface of several concrete specimens with different material strengths, and the contact time of steel balls of various diameters was verified experimentally.

According to the results, the difference between the theoretical and experimental values is reported to be at most twice as large. This difference is not negligible for the generation of response waveforms by numerical simulations and should be carefully examined.

In this study, the influence of the input frequency on the response waveforms is investigated by calculating the response waveforms in are numerical models with various input frequencies and comparing the calculated response waveforms.

4. Overview of numerical simulations

4.1. Experimental model and parameters of numerical simulation

The FDTD (Finite-Difference Time-Domain) method, which is the numerical simulation method used in this paper, is one of the finite difference methods. This method was proposed as a numerical simulation method to formulate Maxwell's equations by the difference method and analyze them in the time domain. Similarly, the wave equation for propagation of elastic waves in a solid material is also formulated by the FDTD method [8].

Figure 4 shows the simulation model used in this study, and table 1 shows the model size and measurement position. This model is an off-the-shelf concrete plate that is standardized in Japan. In this study, measurements were also made under the same conditions as in the numerical simulation in order to compare the calculated and experimental waveforms. The parameters of the numerical simulation were set as shown in table 2 from the results of the experimental values.

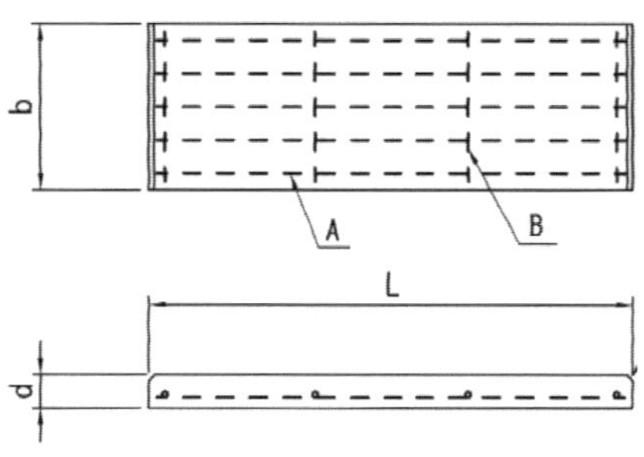

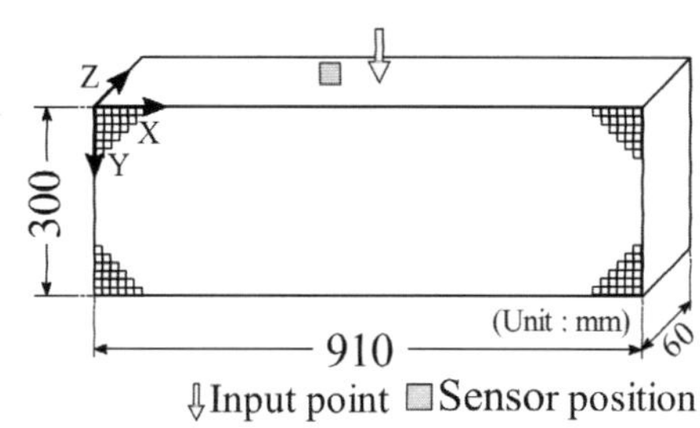

Figure 4. Simulation model.

Table 1. Model size and measurement conditions.

Length (mm)	L=X	910	Coordinates	X	Y	Z
Height (mm)	b=Y	300	Input point (mm)	455	0	30
Width (mm)	d=Z	60	Output point (mm)	400	0	30
Steel round bar (mm)	A and B	6	Grid size (mm)	10	10	10

Table 2. Parameters of numerical simulation.

Primary wave velocity: V_p	4300 m/s	Attenuation coefficient	0.03
Density: ρ	2000 kg/m^3	Sampling frequency	2.00 MHz
Lamé's constants: λ	9.24 Gpa	Grid size	10.0 mm
Lamé's constants: μ	13.9 Gpa	Calculation data: n	1000

4.2. Calculation conditions of numerical simulations

The input signal in the numerical simulation is a sinusoidal signal of one wavelength of the set input frequency. The calculated response waveform is subjected to LPF for smoothing. In this study, the Cut-off frequency was set to 80 kHz and LPF processing was performed. In addition, elastic wave method uses the artificial excitation, variation in input energy occurs. Therefore, in the experimental measurement waveform, the amplitude of the response waveform is normalized by the absolute value of the maximum amplitude. For this reason, it is normalized in the same way in this numerical simulation.

The simulation was performed under two conditions. In Case 01, the influence of changing the input frequency on the generation of the response waveform was studied. In Case 02, the experimental values measured with a steel ball diameter of 16 mm were used as a reference, and the generated of response waveforms that approximate the experimental result are compared with the reference wave. The input frequency F shown in table 4 is the theoretical input frequency by equation (3).

Table 3. Calculation condition of Case 01.

Input frequency						
(kHz)	5.0	10.0	20.0	40.0	60.0	80.0

Table 4. Calculation condition of Case 02.

Input frequency of steel ball							
ds: 16.0 mm, *F*: 14.5 kHz							
Variations of input frequency in numerical simulation							
(kHz)	5.0	7.8	10.0	12.5	14.5	16.5	20.0

5. Results and discussions

5.1. Results of Case 01

The response waveforms calculated under the conditions of Case 01 are shown in figure 5. The horizontal axis of the figure is time and the vertical axis is amplitude. It can be visually understood that there is little difference in the shape of the calculated response waveform if the input frequency is changed from 40 kHz to 80 kHz. However, the input frequency is changed from 5 kHz to 20 kHz, the change of the shape of the calculated response waveform are relatively large. Figure 6 shows the difference values between the response ware from of 80 kHz of input frequency and the other response waveforms. The number of data used for the calculation is 1000. The figure shows that as the input frequency becomes lower, the difference value also increases. This shows that the calculated response waveform changes with the input frequency, and the amount of change is smaller at higher input frequencies and larger at lower frequencies in the numerical simulation. This is due to the fact that the wavelength changes with the input frequency. Since the input frequency used in the measurement of the impact elastic wave method is relatively low, the change of the response waveform calculated by the numerical simulation would be large. These results suggest that the input frequency greatly influences the results of the response waveform.

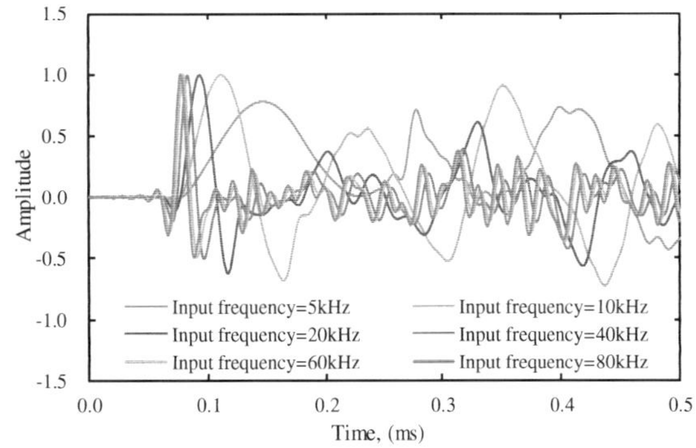

Figure 5. Case 01: Response waveforms.

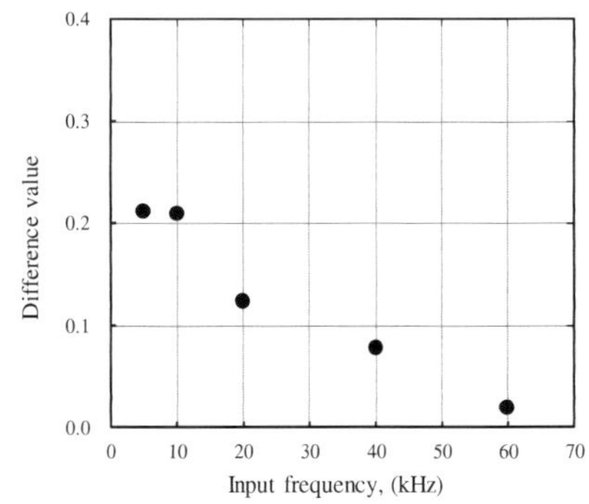

Figure 6. Difference value of the response waveform of Case 01 with based on 80 kHz waveform.

5.2. Results of Case 02

The response waveform calculated for Case 02 is shown in figure 7. The red dashed line in the figure is the experimental value measured using a steel ball with a diameter of 16 mm. figure 8 shows the results of the difference values calculated based on the experimental waveforms. The theoretical value of the frequency input from the steel ball used is 14.5 kHz. From the figure, the difference value calculated with respect to the experimental value decreases as the input frequency decreases from 20 kHz to 10 kHz, and then increases, with the value reaching its minimum at 10.0 kHz. This result suggests that a lower frequency than the theoretical value is also input in the results of numerical simulation.

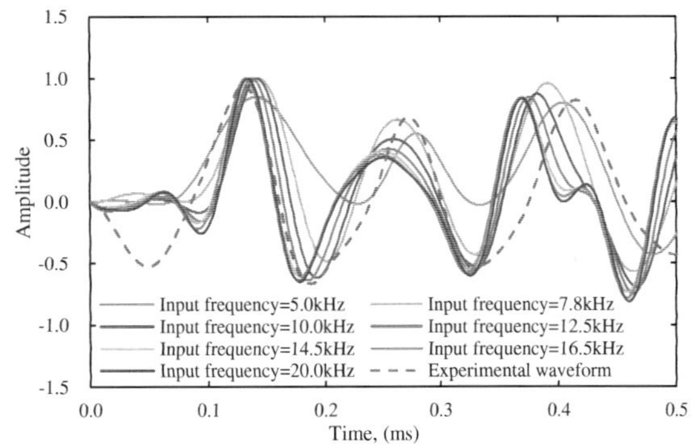

Figure 7. Case 02: Response waveforms.

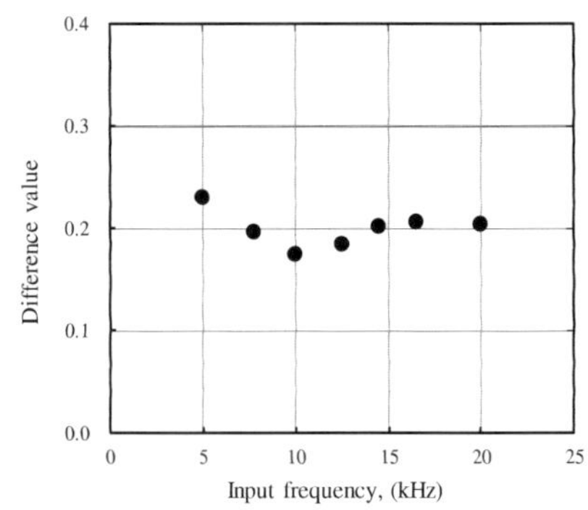

Figure 8. Difference value of the response waveform of Case 02 with based on experimental waveform.

6. Conclusion

In this paper, the influence of the input frequency on the generation of the response waveform under soundness conditions is investigated by numerical simulation a concrete plate model. The results are summarized as follows:

(a) the shape of the calculated response waveform changes with the change of the input frequency, and the change is large at the lower input frequency;

(b) as a result of comparing the experimental waveform and the numerical simulation waveform, the difference value was minimized when the frequency was calculated at a frequency lower than the theoretical value. As a result, the difference between the theoretical value and the experimental value was seen in this study as well, and it was confirmed that it is necessary to pay attention to the input frequency setting even in the numerical simulation.

From the above results, it can be predicted that the numerical simulation model of this study requires the input frequency to be set lower than the theoretical value in order to calculate the response waveform of the experimental value, and by setting appropriately, it is possible to generate the response waveform during soundness by numerical simulation.

References

[1] Mary J S and William B S 1997 IMPACT-ECHO: Nondestructive Evaluation of Concrete and Masonry *New York: Bullbrier Press* Ithaca

[2] Kubo G, Uchida S, Iwano S, Yamashita K and Sumitani K 2017 Experimental Study on Frequency of Elastic Wave Input to Concrete by Steel Ball *Proceedings of the Concrete Structure Scenarios, JSMS* **17** pp 515-520 (*in Japanese*)

[3] Japanese Society Non-Destructive Inspection Standard 2014 *NDIS2426-2: Non-destructive examining of concrete-elastic wave method-Part 2: Impact elastic wave method* (Tokyo: Japanese Society Non-Destructive Inspection)

[4] Goldsmith W 1960 *Impact: The Theory and Physical Behavior of Colliding Solids* (London: Edward Arnold)

[5] Ikebata K, Kobayashi Y, Oda K and Nakamura K 2019 Experimental Study on Detection of Voids in Concrete Slab by Using Responses in Time Domain *Inspection Annual Meeting of the Japanese Society for Non-Destructive* pp 93-96 (*in Japanease*)

[6] Ikebata K, Kobayashi Y, Oda K and Nakamura K October 2020 Detection of Defects in Rectangular Cross-Section Structure by Using Responses in Time Domain on the Basis of Numerical Investigations *Proceedings of the Concrete Structure Scenarios, JSMS* **20** (*in Japanese*)

[7] Ikebata K, Kobayashi Y, Oda K and Nakamura K 2020 Detection of Defects in Concrete Slab by Using Responses Time Domain on the Basis of Numerical Investigations *Inspection Annual Meeting of the Japanese Society for Non-Destructive* (*in Japanese*)

[8] Fujioka T, Nagata Y and Abe M 2017 Experimental and Numerical Analysis for Non-Destructive Inspection of a Void in a PC Sheath by using an Impact Elastic Wave Method *Journal of Japanese Society for Non-Destructive Inspection* **66** pp 443-450 (*in Japanese*)

IOP Conf. Series: Materials Science and Engineering 1138 (2021) 012026 doi:10.1088/1757-899X/1138/1/012026

The study on stability of composite panel structure under axial compression load

J Jiang[1,2*], C Chen[1], X Zhang[2] and Z Yang[2]

[1] Shenyang University of Technology, Institute of Mechanical Engineering, Laboratory of Mechanical Reliability and Dynamics Research Center, Shenyang, China;
[2] Shenyang City University, Institute of Intelligent Engineering, Mechanical Design and Automation, Shenyang, China.

* sherojjj@163.com

Abstract. The axial compressive load test and finite element simulation are made to study the mechanical properties of aircraft stiffened composite panel. Firstly, the numerical simulation model of composite stiffened panel structure is constructed, the accuracy and feasibility of the simulation model are verified by calculating the value of MAC (Modal Assurance Criterion) between the computational free modal and experimental free modal. Then, the nonlinear post-buckling numerical simulation is carried out to extract the buckling load, the post-buckling load path and the failure load. Then, the axial compression test of the panel structure is carried out by the testing machine, and the load-displacement curve and the load-strain curve of the loading position are obtained. Finally, the correctness of the analysis method is proved by comparing the numerical simulation value with the experimental analysis value. The experimental research method of the mechanical properties of the composite stiffened panel under axial load is explored, which can be used as a reference for the study of the bearing capacity of the stiffened panel.

1. Introduction

Composite material is a new high-performance material, which developed to meet the needs of aviation, aerospace and other high-tech fields. Compared with metal materials in the traditional sense, it has the advantages of light weight, high specific strength and high rigidity [1, 2]. At the same time, it can be layered according to the needs of practical engineering products and the characteristics of bearing loads, and various curvatures and shapes can be stacked to give full play to the mechanical properties of composite materials. It has been widely used in aviation, aerospace, automobile and other industrial structures [3, 4].

However, when the composite stiffened panel structure is subjected to axial compression load, buckling instability often occurs and fails [5, 6]. At the same time, the post-buckling behavior is nonlinear, the analysis is complicated and the workload is heavy, so the initial buckling load of the structure has always been taken as the design allowable load in actual engineering design. However, as we all know, the composite stiffened panel structure will not be destroyed immediately after buckling, but will enter the post-buckling stage, at which time the structure still has great bearing capacity [7]. Therefore, it is very necessary to study the post-buckling behavior of composite stiffened panel structure, including finite element analysis method and experimental analysis method, so as to fully explore and utilize the post-buckling bearing capacity of composite stiffened panel structure, not only effectively reduce the structural weight, but also improve the structural design, which has important economic and practical significance for modern engineering structural design [8].

Since advanced composite materials were put into aerospace applications, there have been three achievements worth mentioning. The first is an eight-seater commercial aircraft Learjet 2100 made of carbon fiber composite materials in the United States and successfully tested. The second is the Columbia space shuttle made of a large number of advanced composite materials. This space shuttle uses carbon fiber. /Epoxy resin makes the main cargo door of 18.2 m in length and 4.6m in width, and various pressure vessels are made of Kevlar/epoxy resin. Resin, metal and ceramic matrix composite materials were used on this space shuttle, which represents the most cutting-edge technological achievements in modern times; the third part is the use of advanced composite materials as the main bearing structure to manufacture a large Boeing-767 capable of carrying 80 people Passenger aircraft not only reduces the weight, but also improves the various flight performance of the aircraft [9].

After nearly 40 years of development, composite materials for aircraft have developed from the original non-load-bearing component to the application of secondary load-bearing and main load-bearing components, which can achieve a significant effect of reducing mass (20% – 30%) [10]. At present, it has entered a mature application period, and there is no doubt that it can improve the technical and tactical level, reliability, durability and maintainability of the aircraft. Its design, manufacturing and use experience has become richer. So far, the composite materials used in fighter jets account for about 30% of the total materials used, and the new generation fighter jets will reach 40%; the amount of composite materials used in helicopters and small aircraft will reach (70% - 80%), and even full-composite aircraft will appear. 70% of the fuselage of the Comanche helicopter is made of composite materials, but it is still planned to reduce the weight of the aircraft by another 15% by reducing the mass of the lower front part of the fuselage and expanding the composite materials into the accessories and bearings [11]. In order to reduce the quality of the "Apache", composite materials will be used instead of the metal body.

2. Finite element modeling and analysis

2.1. Organization introduction

As shown in figure 1, a series of composite stiffened panel structure test pieces are manufactured, the overall dimensions of which are all 400 mm x 500 mm, and three T-shaped stringers are uniformly arranged on the flat panel skin, in which the skin and stringers are manually laid and then assembled together, and the autoclave co-curing molding process is adopted [12].

The allowance of 30mm shall be designed at both ends of all stringers, and the allowance of 3mm shall be designed for the web of T-stringer, so as to facilitate milling after co-curing and eliminate the boundary effect of test pieces.

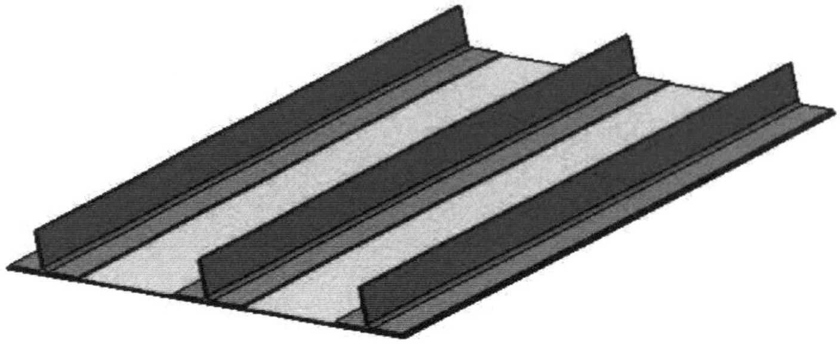

Figure 1. Schematic diagram of composite stiffened panel structure.

The skin and stringer are made of CMS-CP-306 material and epoxy prepreg toughened by medium-sized high-strength carbon fiber, which is a unidirectional tape material. The thickness of each layer of unidirectional tape is 0.13 mm and Poisson's ratio is 0.31. The mechanical properties of the material are shown in table 1. See table 2 for the sequence of structural paving.

Table 1. Material parameters of carbon fiber unidirectional tape compression (MPa).

Compression modulus E_{11}	Compression modulus E_{22}	Shear modulus G_{12}	Compression strength E_{xc}	Compression strength E_{yc}	Shear strength G_{xy}
135000	10000	6280	1100	230	136

Table 2. Structure layup sequence.

Region	Laying sequence
Web	[45/-45/0/-45/90/45/0] s
Cap	[45/-45/0/-45/90/45/0] s
Skin	[45/-45/0/90/45/-45/0] s

2.2. Verification of simulation model

The two-dimensional shell element is used to model the composite stiffened panel structure. Because the thin shell can calculate the "missing" film internal force of the thin plate when buckling and post-buckling analysis is carried out under compressive load, the composite stiffened panel model constructed by the two-dimensional shell element has higher authenticity and feasibility [13].

Under compression load, the ideal failure mode of composite clamped panel structure must be from the middle of stiffened panel, and all stringers and skin are broken together, so the part between stringer and panel is treated by common node in modeling to improve calculation efficiency. For the composite laminate structure, as long as it does not fail under load (that is, the laminate is not buckled), the strain value of each layer in the laminate is the same, but the stress of each layer is not the same [14]. Therefore, the allowable strain is used as the design limit value when designing composite structures.

The rationality and feasibility of the simulation model can be verified by comparing the finite element strain data and the test strain data and calculating the SAC value of the two.

The rationality and feasibility of the model are verified by comparing the values of finite element free mode and experimental free mode. The hammering method used in the experiment is a typical single-point excitation method. Since the excitation device is not directly connected to the structure being tested, it will not add additional mass to the structure, so it will not affect the structure Dynamic characteristics. Fix the test piece on the testing machine, select several test points and one excitation point according to the structure, use the hammer to excite, and the data acquisition system picks up the vibration signal of the test point [15].

According to the modal assurance criterion [16]:

$$MAC_{ti,fj} = \frac{\left(\varphi_{ti}^{T} \times \varphi_{fj}\right)^2}{\left(\varphi_{ti}^{T} \times \varphi_{ti}\right)\left(\varphi_{fj}^{T} \times \varphi_{fj}\right)} \quad (1)$$

Where, φ_{ti} is the i modal vector of the experimental analysis, φ_{fj} is the j modal vector of the finite element analysis, $MAC_{ti,fj}$ is the linear correlation between the i modal vector of the experimental analysis and the j modal vector of the finite element analysis.

It is generally considered that when it is $MAC_{ti,fj} \geq 0.7$, it means that the two vectors have good consistency, the constructed finite element simulation numerical model can better reflect the real situation of the structure [17].

Figure 2. Loading and testing strain value of test piece.

As shown in figure 2, gauges are used to extract the data of stiffened panel structure. According to the comparison results of the two modes, it can be seen that the MAC values of the experimental and finite element modes are all greater than 0.7, indicating that the constructed finite element simulation model has a high approximation to the real situation, and the calculation and analysis model Reasonability and feasibility, that is, it can reflect the actual load status, and lay the foundation for further research.

2.3. Analysis of simulation model

In the finite element analysis, one end of the structure is clamped and restrained, and the axial compressive displacement load of 3mm is applied to the other end of the structure. Based on the finite element analysis software Patran/Nastran, the failure and failure of structural materials are determined by using Hoffman material progressive damage criterion.

There are five kinds of commonly used failure criteria for composite materials [18], which are maximum stress criterion, maximum strain criterion, Tsai-Hill criterion, Tsai-Wu tensor criterion and Hoffman criterion. Among them, the first two kinds of criteria think that once the strength limit of material stress and strain is reached alone, the material will be destroyed immediately, without considering the interaction between material and geometry, so the accuracy of calculation results is relatively low; Tsai-Hill criterion considers the combination form to judge the damage of materials, but it is more suitable for laminates with the same tensile strength and compressive strength in the main direction of materials; Based on Tsai-Hill criterion, Tsai-Wu tensor criterion considers the different tensile and compressive strengths of composite materials, but it can not give the damage evolution process of laminates, and then it can not be determined whether the damage of laminates comes from matrix failure or fiber fracture. Hoffman criterion takes into account the difference of delamination criterion and tensile and compressive strength in the main direction of composite materials. Compared with the test results, its analysis results are closer to the test values, which can accurately determine the failure load of materials. Its material failure criterion is as follows:

$$f(\sigma)=\frac{\sigma_1^2-\sigma_1\sigma_2}{\sigma_{1t}^s\sigma_{1c}^s}+\frac{\sigma_2^2}{\sigma_{2t}^s\sigma_{2c}^s}-\frac{\sigma_{1c}^s-\sigma_{1t}^s}{\sigma_{1t}^s\sigma_{1c}^s}\sigma_1+\frac{\sigma_{2c}^s-\sigma_{2t}^s}{\sigma_{2t}^s\sigma_{2c}^s}\sigma_2+\frac{\tau_{12}^2}{\left(\tau_{12}^s\right)^2} \tag{2}$$

Extract the reaction force at the application point of displacement load and make its load-displacement curve, as shown in figure 3:

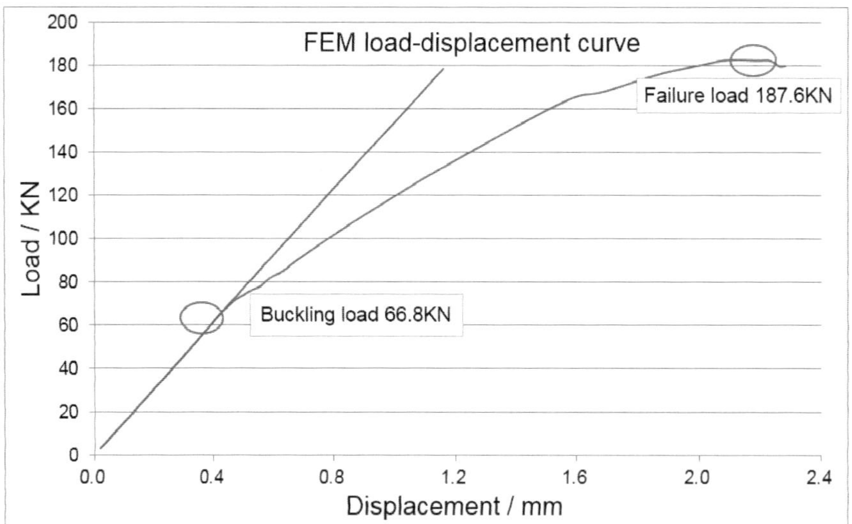

Figure 3. Finite element analysis-load-displacement curve.

It can be seen from figure 3 that the load and displacement of the structure change linearly in the range of 0 ~ 66.8 kN; When the load is in the range of 66.8 kN ~ 187.6 kN, the structure enters the post-buckling state, and the structure still has great bearing capacity at this time. Among them, the failure load value is about 3 times the buckling load value, which is also consistent with the engineering experience value.

Tracing back the post-buckling path of the composite stiffened panel structure, it can be seen that in the whole compression process, the composite skin structure first enters the local buckling state, and then with the compression load exceeding the buckling load and increasing, the structure enters the post-buckling state until it is destroyed, and the failure form is also ideal, which is from the middle of the stiffened panel, all stringers and skin are broken and destroyed together [19, 20] .

3. Experimental analysis

3.1. Experiment product preparation

All the experiment product were laid by hand and co-cured, and the parts manufacturing tooling was designed. The triangular gap of stiffened panel structure must be filled with twisted strips, and the filling amount of twisted strips directly affects the molding quality of products. According to the experiment, the most reasonable dosage of twister sliver is the calculated value of gap theory, which is multiplied by the correction factor of 1.15. As shown in figure 4, the twister strip is made of 0° carbon fiber prepreg, which is flattened and compacted by a stringer die, then covered with a cover plate and compacted in vacuum. Then, after vacuum compaction for 24 hours, the test piece is sent into hot-pressing irrigation for co-curing, and then taken out of the can and demoulded, as shown in figure 5.

Figure 4. Compaction of Twins and Vacuum Compaction of experiment product.

Figure 5. Curing and molding of experiment product.

Check the thickness of the demoulded experiment product first to ensure that the thickness of the skin and stringer is uniform, and the tolerance is controlled within 5%. If it exceeds the test piece, it will be discarded. Then the test piece is milled to remove the allowance of 30mm at both ends of the stringer

IOP Conf. Series: Materials Science and Engineering 1138 (2021) 012026 doi:10.1088/1757-899X/1138/1/012026

and 3mm at the web of the T-shaped stringer, so as to eliminate the influence of boundary effect of the test piece on the test results.

3.2. Design of test fixture

According to the form of the Test Machine, the test fixture is as shown in figure 6. The Test Fixture is made of precipitation hardening Martensitic stainless steel plate with 15-5PH and its allowable compressive yield stress is 1034 Mpa, it is manufactured by machining.

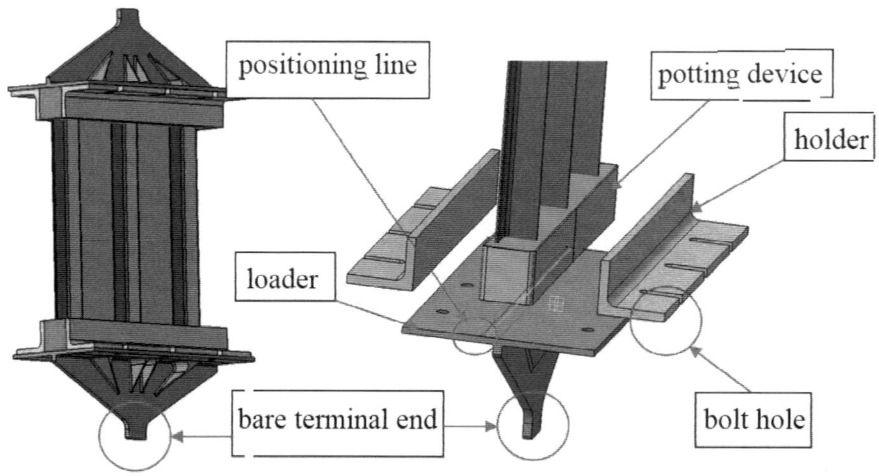

Figure 6. Schematic Diagram of test fixture.

The clamping end is designed according to the size of the testing machine, which is 30 x 50 x 10 mm. The loader is used to load the experiment product, the clamp is used to fix the experiment product on the loader, the clamp and the loader parts are fixed by 6 bolts, and the loader parts are round holes, on the gripper is a long round hole, different configuration of the wall plate shape center position, so the position of the gripper relative to the loader is also different, the long round hole can make the gripper move in a certain range. In order to ensure the loading at the centroid position of the specimen in the test, the locating line is carved on the loader according to the centroid position of the wall plates of different configurations, which is convenient for the location of the specimen.

3.3. Design of test fixture

Before attaching the strain gauge, first draw a line and draw the edge of the strain gauge according to the size of the strain gauge to ensure that the center of the strain gauge is in the position of the test design. The strain gauge is pasted along the radial direction of the test piece, and the wire is thrown to the bottom of the test piece when the test piece is loaded. Prevent the strain gauge from being damaged; before sticking, wipe the position of the patch with a wiper moistened with acetone to ensure that the patch area is clean and tidy and free of dust. If the bond between the patch and the test piece is not good, use 5000 series sandpaper to gently rub, wipe with acetone and then paste it. At the same time, try to ensure the verticality of the strain gauge; when attaching the strain gauge, take out the strain gauge and hold the strain gauge Lead wire (be careful not to touch the strain gauge), put a drop of glue on the back of the strain gauge, and shake off the excess glue with medium strength. Easy to fail; too much force will result in less glue, there will be bubbles between the strain gauge and the aluminum plate, which will make the strain gauge ineffective); put the strain gauge on the drawing line, find the verticality, and squeeze the excess glue from left to right. Make the strain gauge and the test piece perfect bonding [21].

According to the result of finite element analysis of the experiment product, the deformation form of the experiment product is predicted, the dangerous section and position are found out, combining with practical experience, the key inspection position is finally determined, in order to work out the plan of the experiment product, the gage is numbered in combination with the channel sequence of the gage, as shown in figure 7.

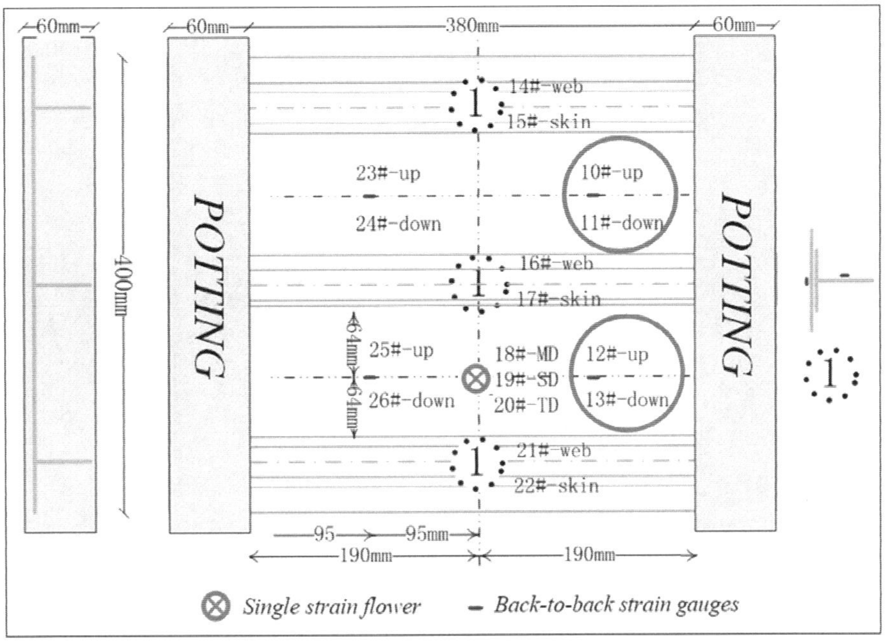

Figure 7. The map of the positions of resistive strain gauge.

It should be noted that it takes a long time to paste the strain gauges, and it is necessary to debug whether each strain gauge is working properly after the pasting, especially the static test piece, which requires a lot of strain gauges, too many channels, dense wires, and difficulty in patching It is very large, so each time a strain gauge is pasted, the wire must be fixed to prevent winding, and at the same time, be careful of the patch series, as shown in figure 8.

Figure 8. Location and theoretical number of the experiment product.

IOP Conf. Series: Materials Science and Engineering 1138 (2021) 012026 doi:10.1088/1757-899X/1138/1/012026

Using Model BX120-3AA, resistance 120, non-welding strain gauge, pasting at room temperature, the strain gauge must be firmly pasted on the tested point to ensure that the strain gauge and the tested object produce deformation together, as shown in figure 9.

When connecting the strain gauges, the wires of the strain gauges should be connected to the two binding posts AB according to the corresponding numbers (note that the numbers must be unified for subsequent test data processing), and the compensation plates should be connected at the same time. Because the resistance strain gauge is very sensitive to temperature changes. When the ambient temperature changes, because the linear expansion coefficient of the strain gauge is different from the linear expansion coefficient of the test piece, and the resistance value of the sensitive grid changes with temperature, the measured strain will include the effect of temperature change and cannot reflect actual strain of the test piece, therefore, must try to eliminate the influence of temperature change in the measurement.

Figure 9. Strain gauge paste map.

3.4. *Axial compression test*

First, connect the strain gauge. In this process, it is necessary to ensure that the theoretical number of the strain gauge is consistent with the number of the strain gauge to ensure the correctness of the test data processing.

At the same time, a compensation piece should be connected to the test piece before the test, because the resistance strain gauge used in the test is sensitive to the change of ambient temperature and humidity, the adverse effects of environmental changes on the test shall be eliminated by means of compensating plates of the same material as the test piece [22, 23].

Then, the specimen is fixed on the loading machine and preloaded with 50% buckling load, which is 33.4 kN axial compression load, to eliminate the initial torque and bending moment of the specimen. After debugging, load at the rate of 100 N/s, and destroy it.

IOP Conf. Series: Materials Science and Engineering 1138 (2021) 012026 doi:10.1088/1757-899X/1138/1/012026

Figure 10. Figure of structural failure under axial compression.

When the load is 64.2 kN, local buckling of the skin takes place, and when the load is 177.1 kN, the specimen is completely destroyed, and the final failure form is as shown in figure 10, which fits the profile [24, 25].

4. Comparison of analysis results

According to the test loading data, the load-displacement curve is drawn, as shown in figure 11. The buckling load is 64.2 kN and the failure load is 177.1 kN.

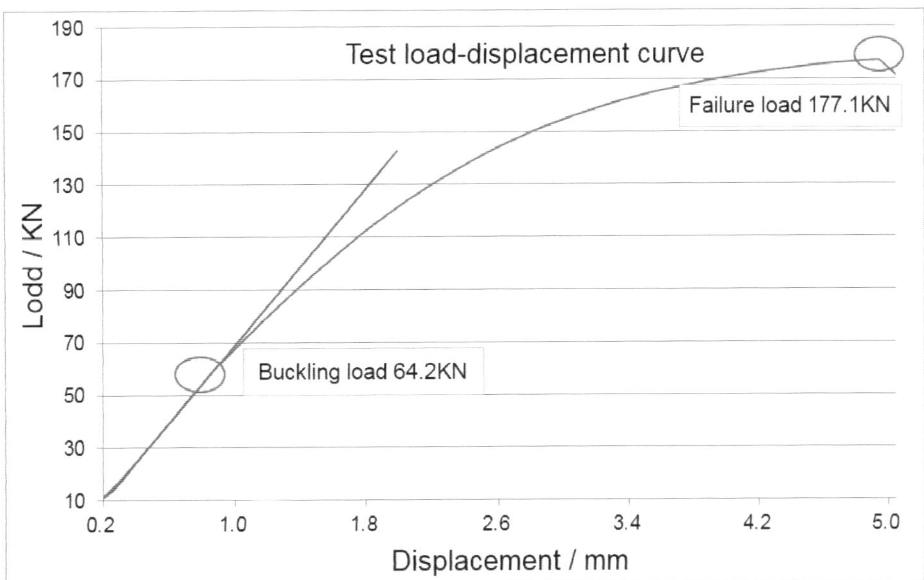

Figure 11. Test analysis-load-displacement curve.

It can be clearly seen from table 3 that the error of buckling load and failure load calculated by finite element analysis is very small, which is in good agreement with the test value, that is, the finite element analysis method can truly reflect the stability of composite stiffened panel structure.

Table 3. Results of finite element analysis and test analysis.

	Post Buckling Analysis	Test Analysis	Error
Buckling load (kN)	66.8	64.2	4.0%
Failure load (kN)	187.6	177.1	5.9%

5. Conclusion

The general procedure, points for attention and relevant test standards for the study of the stability of composite stiffened panel structures under axial compression are summarized by using finite element method and test analysis method, in particular, the making, preparation, testing, mounting, loading test and data processing of the test pieces have an important role in guiding the study of related issues.

Comparing the results of finite element analysis and test analysis, the feasibility of Patran/Nastran finite element analysis software and correlation analysis method is proved. At the same time, the failure load of the two analysis methods is about 3 times of the buckling load, which shows that the composite stiffened panel structure still has a great potential. Considering the post-buckling load-carrying capacity in the design of composite structures will give full play to its light weight, high specific strength and high stiffness, which is of great practical significance to promote the use of composite structures.

Acknowledgments

The work was supported by the National Natural Science Foundation of China (Grant Nos. 51675350), and thanks to AVIC SAC Commercial Aircraft Co. Ltd for their supports in the manufacture of specimens and NDT inspection.

References

[1] Anandan A, Dhaliwal G, Huo Z, Chandrashekhara K, Apetre N and Iyyer N 2018 *Applied Composite Materials* **25** 1155
[2] Maciej T, Ryszard B and Michal K 2017 *Composites Part B: Engineering* **109** 238
[3] Konstantinos N and Nicholas G 2012 *Applied Composite Materials* **19** 219
[4] Yang Z, Zhang J, Xie Y, Zhang B, Sun B and Guo H 2017 *Applied Composite Materials* **24**, 1447
[5] Jia L 2017 *Aeronautical Manufacturing Technology* **15** 102
[6] Chen W 2017 *Large aircraft* **10** 14
[7] Jeevan K, Ramesh B and Ratnakar P 2014 *Applied Composite Materials* **21** 259
[8] Qiao J and Brian H 2015 *Composites Part B* **69** 13
[9] Chen N and Guedes S 2007 *Reinf Plast Compos* **26** 1021
[10] Fernass D and Mircea C 2010 *International Journal of Structural Stability and Dynamics* **10** 905
[11] Tang J 2013 *Spacecraft Environment Engineering* **30** 352
[12] Huo S, Wang F and Wang P 2010 *Chinese Journal of Applied Mechanics* **27** 423
[13] Wang F, Cui D and Xiong Q 2013 *Journal of Beijing University of Aeronautics and Astronautics* **39** 494
[14] Xun W and Jonathan D 2018 *Composites Part A* **111** 62
[15] Fuller J and Wisnom M 2015 *Composites Science and Technology* **112** 8

IOP Conf. Series: Materials Science and Engineering 1138 (2021) 012026 doi:10.1088/1757-899X/1138/1/012026

[16] Liu C, Yue Z and Geng X 2015 *Journal of Experimental Mechanics* **30** 757
[17] Shen H 2002 *Post buckling Behavior of Plates and Shells* (Shanghai: Shanghai Scientific & Technical Publishers)
[18] Shen G and Hu G 2006 *Mechanics of Composite Materials* (Beijing: Tsinghua University Press),
[19] Orifici A, Thomson R and Degenhardt R 2008 *Composite Structure* **82** 217
[20] Wang H, Chen H and Lei A 2018 *Acta Materiae Compositae Sinica* **35** 2014
[21] Kong B, Ye Q and Chen P 2010 *Acta Materiae Composite Sinica* **27** 150
[22] Bisagni C and Davila C G 2014 *Composite Structure* **108** 493
[23] Orifici A C, Thomson R and Degenhardt R 2009 *Journal of Composite Materials* **43** 3239
[24] Li L, Li S and Chang F 2016 *J. of Nanjing University of Aeronautics & Astronautics* **48** 563
[25] Tifkitsis K, Mesogitis T, Struzziero G and Skordos A 2018 *Composites Part A* **112** 383

IOP Conf. Series: Materials Science and Engineering 1138 (2021) 012027 doi:10.1088/1757-899X/1138/1/012027

Spalling of high strength concrete in fire

O Lalu[1], **R Darmon**[2*] **and T Lennon**[1]

[1] Building Research Establishment, Watford, United Kingdom
[2] Technical University of Cluj-Napoca, Romania

* ruxandra.darmon@ccm.utcluj.ro

Abstract. This paper reviews a series of experimental tests programmes undertaken at BRE in the UK on the performance in fire of high strength concrete with a particular focus on mix design to assess the effects of spalling. The experiments include a series of fire tests on reinforced and unreinforced concrete columns and tunnel linings. The paper will provide information on the parameters that lead to explosive spalling of tunnel linings in a fire situation and on the measures available to mitigate this hazard.

1. Introduction
Concrete structures generally perform well during and after a fire and the majority of fire-damaged concrete structures have been repaired and reused [1, 2]. Although spalling of concrete structures following real fires has been observed, it does not necessarily have implications for the stability of the structure [3, 4]. There is a need to establish the circumstances under which spalling would have serious consequences. The use of high strength concrete in buildings is an important innovation that can reduce the size of structural elements compared to those made from normal strength concrete. This can provide for more efficient construction such as maintaining a constant column size throughout the building by using high grade concrete for the lower storey columns carrying the highest loads. One restriction to its widespread use however is concern over its performance in fire, in particular its increased susceptibility to spalling. A consideration of the literature and experience gained over the course of many years suggest that spalling is more likely in the event of high moisture content, high concrete strength, rapid increase in temperature or high levels of restraint (or applied load) or any combination of these factors.

The purpose of this paper is to bring together information derived from a number of experimental programmes, to identify the significant parameters that influence spalling and to highlight areas where further research is required. The information and expertise derived from experimental programmes into the behaviour of high strength concrete in fire has been used to undertake a number of high-profile commercial testing programmes to assess and evaluate the performance of tunnel lining segments in the event of a fire.

2. Mechanisms of spalling
Given certain conditions virtually all types of concrete will spall in a fire situation. The most significant parameters contributing to the spalling of concrete structures are heating rate, moisture content, permeability, strength, restraint to thermal expansion, aggregate type, curing regime, geometry and nature and extent of reinforcement.

It is generally recognized that there are three main categories of spalling; aggregate, corner and explosive. Of these explosive spalling is the most significant in terms of its impact on structural performance. For normal strength concrete and for high strength concrete covered by European Class 1 and 2 with silica fume contents less than 6% by weight of cement explosive spalling is assumed to be principally a function of the moisture content and the environmental conditions. EN 1992-1-2 [5] states that where the moisture content is less than 3% by weight and where members are designed to exposure class X0 (concrete inside buildings with very low air humidity) and XC1 (concrete inside buildings with low air humidity) then explosive spalling does not need to be considered explicitly. The current revision of the fire part of the Eurocode for the design of concrete structures will revisit these assumptions and update the rules based on the current state of knowledge in relation to spalling and the material behaviour of high strength concrete.

The difference between high strength concrete and normal concrete lies principally in the water to cement ratio used in the mix. For high strength concrete, lower water to cement ratios are used, the required workability of the fresh concrete being provided by superplasticisers. The effect of lowering the water to cement ratio is a reduction in the permeability of the hardened concrete, which, in most structural contexts, increases both strength and durability. Unfortunately, the reduction in permeability has been found to be detrimental to performance in fire. This is because pore pressure is produced in concrete at high temperatures. Unless there is an escape route for the steam, internal pressures are generated that, in conjunction with other stresses can exceed the tensile strength of the concrete. The overall result of all the factors relating to the increase in concrete grade is an increased susceptibility to spalling. The danger of spalling of concrete in fire is not only the loss of section but also the possibility of early yield of the steel reinforcement as it becomes directly exposed to high temperatures.

3. Fire performance of high strength concrete columns

Initial research on the performance of high strength concrete columns in fire aimed to investigate the effect of the use of monofilament polypropylene (p/p) fibres to improve performance. At this time (1997-1998) the use of p/p fibres for fire performance was an innovative application of an existing technology generally used to prevent early age cracking of floor slabs.

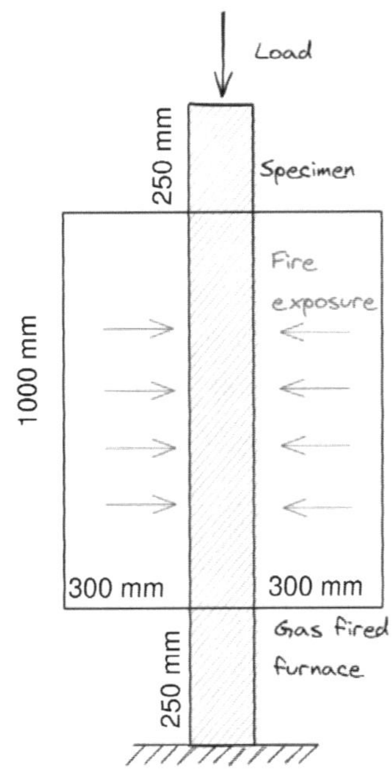

Figure 1. Experimental set-up for fire testing of concrete columns.

Two different experimental programmes were undertaken on both unreinforced and reinforced columns [6]. The fire test facility is shown in figure 1. The columns were placed inside a gas fired furnace with the fuel input controlled to follow the standard fire curve 4 for a period of 45 minutes. The columns were subject to an axial compressive load kept constant for the duration of the fire exposure.

The initial programme of tests was undertaken on unreinforced columns to provide a worst-case scenario in terms of spalling. The columns measured 1500 mm high x 200mm x 200mm. The concrete mixes were C85 and C105 (with micro silica). The coarse aggregate was crushed limestone and the moisture content at the time of test was 3.5%. Half the columns included 3 kg/m3 of p/p fibres (12mm long x < 0.1 mm thick). Half the columns were tested whilst under 40% of the calculated ultimate load and the other half tested under 60% of the calculated ultimate load.

The results are illustrated in figure 2. Only four specimens (of 30) remained intact at the end of the test and of these three included p/p fibres. None of the specimens under 60% load survived the test regardless of the inclusion of p/p fibres. Figure 3 is a comparison of the one specimen without fibres that survived the test and the corresponding sample including fibres. The results show two things very clearly. Firstly p/p fibres can reduce, delay or eliminate spalling in high strength concrete and secondly the predominate factor in terms of fire performance is the load ratio or degree of utilization of the columns.

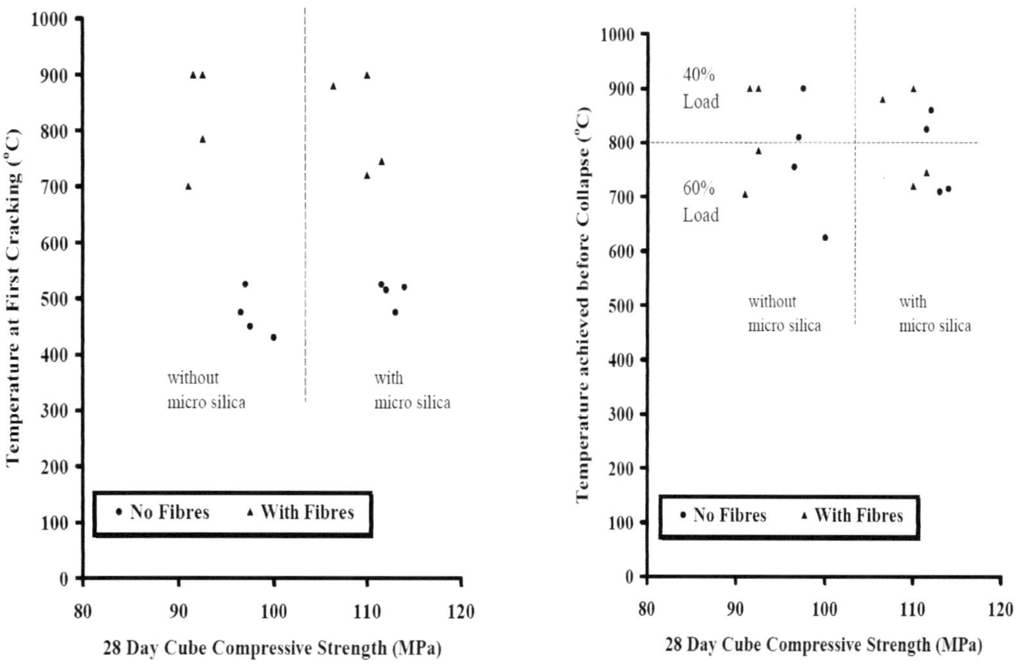

Figure 2. Temperature at first cracking and collapse temperature for unreinforced columns.

The second programme of tests were on reinforced columns of the same overall size as the unreinforced specimens. The reinforcement details are illustrated in figure 4. The range of concrete grades was extended to cover C45 and C65 grades as well as the C85 and C105 grades used previously. Two different fibre dosages were used, 3 kg/m³ and 6 kg/m³. the moisture content was 3% and the test procedure was the same as that used previously. In all cases the applied axial load was one third of the calculated ultimate capacity.

No collapses occurred for any of the reinforced columns although the degree of spalling varied according to the mix design. Figure 5 is a comparison of three of the C85 columns containing from left to right no fibres, 3 kg/m³ p/p fibres and 6 kg/m³ fibres.

IOP Conf. Series: Materials Science and Engineering 1138 (2021) 012027 doi:10.1088/1757-899X/1138/1/012027

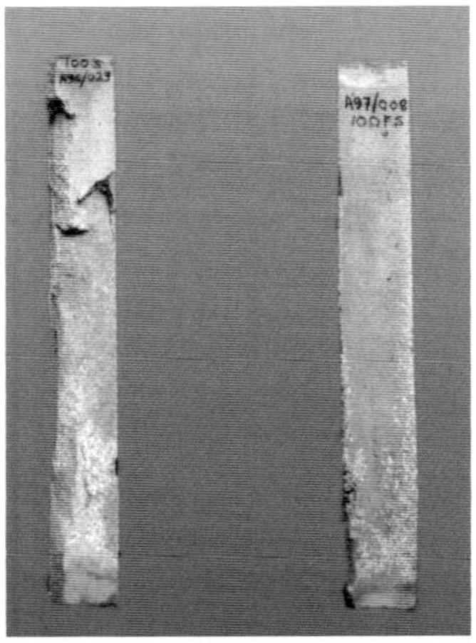

Figure 3. Comparison between high strength unreinforced columns with (right) and without (left) fibres.

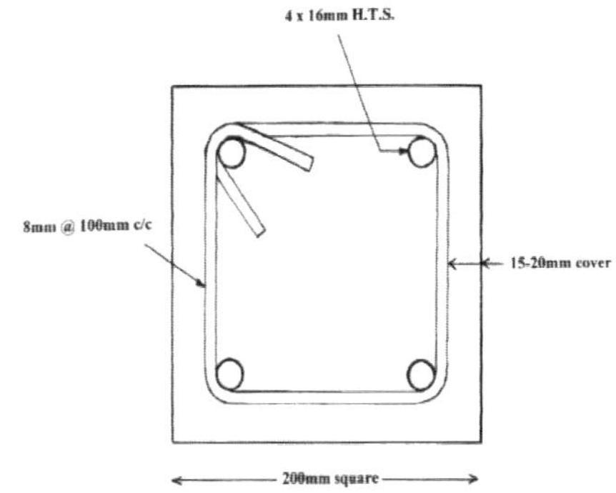

Figure 4. Reinforced concrete columns – section, with high tensile strength reinforcement

Figure 5. Comparison of C85 test specimens after test (no fibres on left, 3 kg/m³ p/p fibres in middle and 6 kg/m³ p/p fibres on right).

Table 1. Residual compressive strengths of the fire tested columns.

Concrete Type (F = 3 kg/m³ p/p fibres, FF = 6 kg/m³ p/p fibres)	Residual Strength (percentage of calculated ultimate failure load prior to testing)
C45	70%
C45F	75%
C45FF	75%
C65	60%
C65F	65%
C65FF	65%
C85	55%
C85F	70%
C85FF	70%
C105	55%
C105F	60%
C105FF	65%

IOP Conf. Series: Materials Science and Engineering 1138 (2021) 012027 doi:10.1088/1757-899X/1138/1/012027

4. Fire performance of high strength concrete columns

A large-scale fire test was undertaken on the ground floor of the European Concrete Building Project (ECBP) at BRE's large scale test facility at Cardington in 2001 [7]. The flat slab was supported by a number of high strength (C85) concrete columns containing 2.7 kg/m³ monofilament p/p fibres to reduce the tendency for explosive spalling. The test area included one completely exposed internal column and eight partially exposed columns at the boundaries of the compartment.

All of the columns survived the fire test without any significant spalling suggesting that the p/p fibres had a significant impact particularly in the light of spalling to the soffit of the flat slab (figure 6).

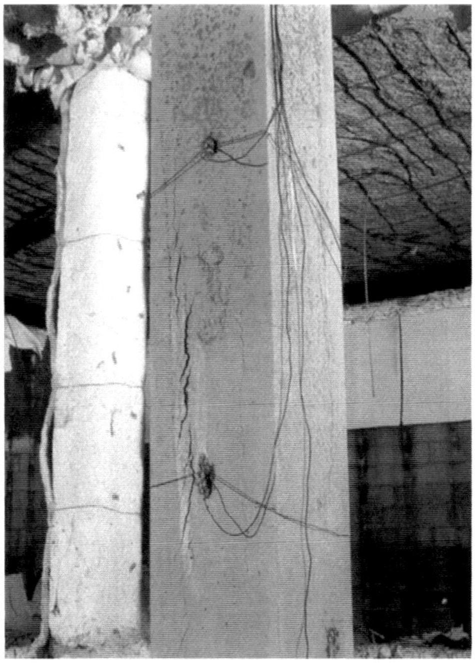

Figure 6. Longitudinal cracks in internal column from Cardington concrete building fire test.

5. Fire performance of tunnel linings and explosive spalling

Explosive spalling is a very complex phenomenon which poses a great threat to structural performance of some reinforced concrete structures in fire [8]. The most significant in relation to explosive spalling (as opposed to spalling in general) is the rate of temperature rise, the presence of restraint against thermal expansion and the permeability (related to porosity, density and strength) of the concrete. Tunnels often involve the use of high strength concrete. The nature of the construction involves significant restraint to thermal expansion and the nature of the potential fire load (petro-chemical fuel tankers) leads to a rapid rise in temperature in the event of a fire.

Given the potential consequences of failure to a tunnel lining, the nature of tunnel fires and the form of construction used in tunnels it is essential to ensure that the proposed design solution is capable of resisting a hydrocarbon fire exposure while under load.

There is currently no standardised approach for the testing and approval of tunnel lining segments in terms of fire performance. Some attempts have been made to produce standardised procedures while still allowing some flexibility in the fire exposure curve to be adopted. However, the specification does not require the specimens to be tested under load. For major infrastructure projects the requirements are often tailored to the specific circumstances of the project.

5.1 Tunnel fire testing

BRE have been involved in fire tests in a number of high-profile infrastructure projects to demonstrate compliance with client requirements for both road and rail tunnels. Figure 7 is a comparison of

IOP Conf. Series: Materials Science and Engineering 1138 (2021) 012027 doi:10.1088/1757-899X/1138/1/012027

specimens with and without p/p fibres tested under load and subject to a hydrocarbon fire exposure for a high-speed rail tunnel. Figure 8 is a comparison between (from left to right) panels subject to a hydrocarbon fire exposure while under load and containing 1 kg Polyvinyl alcohol (PVA) fibres, no fibres and 1 kg p/p fibres. The expertise derived from projects such as this provided input to the specification for fire testing for sprayed concrete linings for the Crossrail project.

Figure 7. Comparison of test specimens with (right) and without (left) p/p fibres.

Figure 8. Comparison of specimens following hydrocarbon fire test (1 kg/m³ PVA fibres on left, no fibres in middle and 1 kg/m³, p/p fibres on right).

Two different specifications were provided for fire testing of specimens for the Crossrail project. One for precast linings and one for sprayed concrete linings (SCL). Both test methods specify exposure to the Eureka time-temperature curve illustrated in figure 9. The specification for SCL defined the parameters for both large panel tests for compliance and smaller scale tests on unloaded cylinders to establish the potential suitability of trial mixes. The specification for testing of large panels for SCL consists of loading panels with nominal dimensions of 1.5 m high by 0.75 m wide by 0.3m thick to the prescribed load level and subjecting one face of the panel to a hydrocarbon fire exposure corresponding to the Eureka fire curve. The performance criteria were based on depth of spalling, temperature rise within the slab and an evaluation of the impact of temperature rise on the integrity of the waterproof membrane and the residual strength of the heated concrete.

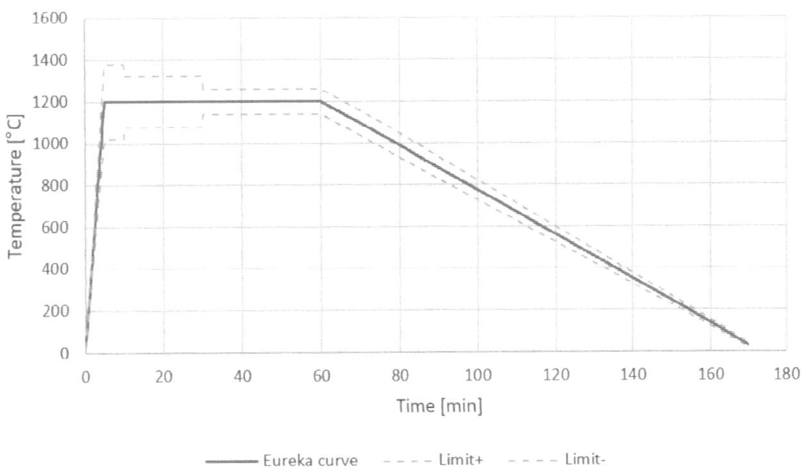

Figure 9. Eureka fire exposure curve in relation with time.

Thermocouples are installed in the specified locations on site (figure 10) at the time the panels are cast. Prior to testing the samples are stored and cured under controlled conditions (40°C and 60% RH) (figure 11) until the test date (nominally 28 days). The specification for precast linings used for Crossrail was completely different and required facilities not currently available within the UK.

Figure 10. Thermocouple installation on site.

Figure 11. Curing samples under controlled conditions (40°C and 60% RH).

The next major infrastructure project is the HS2 high speed rail link. A specification for fire testing of concrete used for underground tunnels has been prepared based largely on the specification used for Crossrail.

6. Conclusions

This paper has reviewed a number of experimental programmes undertaken at BRE to investigate the spalling of concrete structures with a particular focus on high strength concrete. The expertise derived was used to provide a means of test and assessment for a number of high-profile infrastructure projects where evidence was required of concrete performance when subject to a hydrocarbon fire exposure.

There is a need for a standardised approach to the assessment of the performance of tunnel linings in fire to be developed. However, any standardised test and assessment procedure should allow for the flexibility required for specific projects in terms of fire exposure and load level to be applied.

The Crossrail SCL specification and the current HS2 specification could provide a useful starting point to develop a standardised procedure that considers the most significant aspects of structural behaviour in fire without imposing an unreasonable financial burden on contractors.

References

[1] Tovey A K and Crook R N 1986 Experience of fires in concrete structures, in *ACI Symposium on Evaluation and repair of fire damage to concrete, San Francisco, 16-21 March 1986*, American Concrete Institute, Detroit, SP-092

[2] Iffat S and Bose B 2016 A Review of Concrete Structures in Fire, World Academy of Science, Engineering and Technology *International Journal of Structural and Construction Engineering* **10** 2, pp. 123-128

[3] Jansson R 2013 Fire spalling of concrete – A historical overview, Concrete Spalling due to Fire Exposure *Proceedings of the 3rd International Workshop, MATEC Web of Conferences* **6**

[4] Technical Note on Fire Damaged Reinforced Concrete–Investigation, Assessment and Repair 102, April, 2011

[5] British Standards Institution, BS EN 1992-1-2: 2004, Eurocode 2: Design of concrete structures – Part 1-2: General rules – Structural fire design, BSI, London

[6] Clayton N and Lennon T 2000 Effect of polypropylene fibres on performance in fire of high grade concrete, *BR 395* Published by Construction Research Communications (CRC)

[7] Lennon T, Bailey C and Clayton N 2002 The Performance of High Grade Concrete Columns in Fire, *Proceedings of the 6th International Symposium on Utilization of High Strength/High performance Concrete,* **1** pp. 341-354
[8] Deeny S, Stratford T, Dhakal R, Moss P, Buchanan A 2009 Spalling of Concrete - Implications for Structural Performance in Fire *Proceedings of the International Conference Applications of Structural Fire Engineering* pp. 202-207

Advanced technology of municipal solid waste conversion for a circular economy

M L Lupu[1,4*] , D N Isopescu[1], S G Maxineasa[1], I-R Baciu[1], I Antonescu[2] and I Cucos[2]

[1] "Gheorghe Asachi" Technical University of Iasi, Department of Civil and Industrial Engineering, Romania
[2] "Gheorghe Asachi" Technical University of Iasi, Department of Geodesy and Environmental Engineering, Romania.

* mariuslucianlupu@gmail.com

Abstract. One of the most important environmental problems of the big urban areas is represented by the large volume of waste resulted from different human activities. In the last decades, different technologies have been developed to resolve this significant issue, and the most representative is the conversion of municipal solid waste (MSW) in different types of by-products. The first part of this paper presents the current state of plasma conversion technology that can be used for processing waste, and also, its potential in the construction sector which could be taken advantage of the by-products (syngas and slag) resulted from the plasma gasification of municipal solid waste (MSW), to enhance energy-efficiency in this domain. In the second part of the paper, there are analyzed different technologies for managing and processing municipal solid waste (MSW), which can be used nowadays, and the parameters, related to the costs, the period of use, the energy consumption and the environmental impact. All these aspects are compared with those of plasma conversion technology. The plasma conversion system can be commercialized successfully, and its costs can be reduced by generating income in the form of by-products and added-value in producing performant products in a circular economy.

1. Introduction

In the last decades, all over the world, it was observed that one of the most important environmental problems of the big urban areas is represented by the large volume of waste that is resulted from different human activities, domestic and industrial. Eurostat shows that in 2016 only in Romania were generated 9012 kg per inhabitant of waste and in the European Union on average 4968 kg per inhabitant of waste [1]. According to the World Bank Group statistics, more than 2000 million tones of waste were generated worldwide in 2018, and it is expected to increase to more than 3.4 billion tons by 2050, as shown in figure 1 [2].

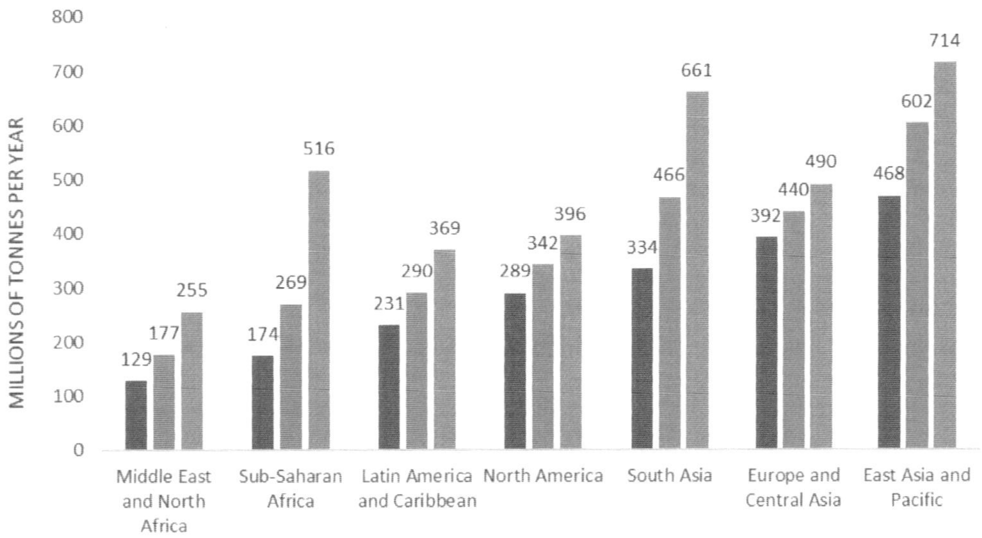

Figure 1. Waste generated worldwide estimation [2].

To reduce this alarming amount of municipal solid waste that goes to landfills, the available solutions are re-use, recycle and usage of waste to energy technologies [3].

Recycling is an alternative to conventional waste management. It is the process of transforming waste into new materials and objects that can help to lower greenhouse gas emissions [4].

Waste to energy concept is the process of converting waste into energy in the shape of electricity, heat and/or synthetic fuels [5].

2. The principal thermochemical technologies for treating municipal solid waste

2.1. Incineration and co-processing

Incineration is a combustion process that transforms waste into ash, exhaust gases and heat, reducing the initial volume to almost 5%. Incineration sterilizes the dangerous wastes, generating at the same time thermal energy that can be in the form of heat (hot water or steam), electricity or both [6].

In modern waste management incineration has the role of treating the waste that can't be recycled with the following advantages [6, 7]:

- all municipal solid waste, as well as some hazardous and industrial waste, can be disposed of, unsorted;
- municipal solid waste volume is reduced to 5%, and is mainly composed of ash which can be recycled as a filler in road construction;
- well-known process, installed worldwide, with high availability and stable operating conditions;
- if cogeneration of heat and electricity is used, or only heat results in a high-efficiency energy recovery up to 85%;
- ash and other waste materials resulted are inert;
- by incinerating waste, the release of methane is prevented [8].

Disadvantages:

- Incinerators demand substantial financial investments and a long time to recover the initial investment, generating a long-term lock-in. Incinerator lifetimes normally range from 25–30 years. Substantial financial investments.

- A large quantity of toxic fly ash and exhaust gas, which must be disposed of by storage at a suitable warehouse (approximately 2-5% of the weight of the input waste) [8].
- NO_x and other gases and small particles.
- Bottom Ash with medium toxicity with concerns of releasing heavy metals to the environment [9].
- Very often the local communities are hostile with the idea of having waste incinerating plants near them, this is called Not in My Back-Yard phenomenon. A study made in Massachusetts, US, showed that the property close to the incinerator plant devaluated with 10% [10].

In the European Union, approximately 29% of the municipal solid waste generated was incinerated in 2017, compared to Sweden, Denmark, and Finland that incinerated more than 50% of their municipal solid waste [11].

To ensure that incineration plants don't release in the air major CO_2 and other Greenhouse Gas Emissions the European Union has created standards for incineration emissions showed in table 1, table 2, table 3 and table 4, and also for discharging wastewater that was used in emission cleaning showed in table 5 [6, 12].

Table 1. European Union limits for hazardous substance emission from waste-to-energy incinerators burning solid fuels [6, 12].

Pollutant	Values in mg/Nm^3 according to Plant Output		
	50 - 100 MWth	100 - 300 MWth	>300 MWth
SO_2	850 or rate of desulfurization > 90%	850 to 200 (linear decrease for 100 to 300 MWth) or rate of desulfurization > 92%	200 or the rate of desulfurization >95%
NO_x	400	300	200
Dust	50	30	30

Table 2. European Union limits for hazardous substance emission from waste-to-energy incinerators burning biomass [6, 12].

Pollutant	Values in mg/Nm^3 according to Plant Output		
	50 - 100 MWth	100 - 300 MWth	>300 MWth
SO_2	200	200	200
NO_x	350	300	300
Dust	50	30	30

Table 3. European Union limits for hazardous substance emission from waste-to-energy incinerators burning liquid fuels [6, 12].

Pollutant	Values in mg/Nm^3 according to Plant Output		
	50 - 100 MWth	100 - 300 MWth	>300 MWth
SO_2	850	850 to 200 (linear decrease from 100 to 300 MWth)	200
NO_x	400	300	200
Dust	50	30	30

Table 4. European Union limits for hazardous substance emission from waste-to-energy incinerators burning solid, liquid or biomass fuels [6, 12].

Pollutant	Limit (mg/Nm^3 except dioxins/furans in ng/Nm^3)
Cd+Tl	0.05
Mercury	0.05
Sb+As+Pb+Cr+Co+Cu+Mn+Ni+V	0.05
Dioxins and Furans	0.1

Table 5. European Union emission limit values for discharges of wastewater from the cleaning of exhaust gases [6, 12].

Polluting substances	Emission limit values expressed in mass concentrations for unfiltered samples	
	95 %	100%
Total suspended solids as defined by Directive 91/271/EEC	30 mg/l	45 mg/l
Mercury and its compounds expressed as mercury (Hg)	0.03 mg/l	
Cadmium and its compounds expressed as cadmium (Cd)	0.05 mg/l	
Thallium and its compounds expressed as thallium (Tl)	0.05 mg/l	
Arsenic and its compounds expressed as arsenic (As)	0.15 mg/l	
Lead and its compounds expressed as lead (Pb)	0.2 mg/l	
Chromium and its compounds expressed as chromium (Cr)	0.5 mg/l	
Copper and its compounds expressed as copper (Cu)	0.5 mg/l	
Nickel and its compounds expressed as a nickel (Ni)	0.5 mg/l	
Zinc and its compounds expressed as zinc (Zn)	1.5 mg/l	
Dioxins and furans, defined as the sum of the individual dioxins and furans	0.3 mg/l	

2.2. Pyrolysis

Pyrolysis is a thermal method of pre-treatment, which can be applied to convert organic waste into a medium calorific gas, into liquid and a carbonized fraction aiming at the separation or binding of chemical compounds to reduce emissions and leachate [13, 14]. The word pyrolysis comes from the Greek language, pyro means “fire” and lysis means “separating” [15]. Usually, after pyrolysis results volatile products and solid residue high in carbon named char. Extreme exposure to pyrolysis leaves mainly carbon as residue, this process is called carbonization [16].

Advantages:
- better retention of heavy metals in carbonized residues [17];
- a smaller quantity of emission gases than in incineration [17];
- hydrochloric acid can be retained in or distilled from the solid residue [17, 18];
- all resulted residues are inert [18].

Disadvantages:
- waste must be sorted and shredded before entering the pyrolysis reactor to prevent the blocking the feeding and transport flow [17];
- resulted oils and tars could have dangerous and carcinogenic elements [17];
- relative high cost [17];
- fuel supply is required at least during start-up [17].

2.3. Plasma gasification
Gasification is a method of heat treatment, which can be applied to convert organic waste into a medium calorific gas, recyclable products, and residues. Gasification is normally followed by combustion of the gases produced, in a furnace and internal combustion engines or simple gas turbines after a proper purification of the gas produced [14, 19].

The principal difference between gasification and pyrolysis is that by gasification the fixed carbon is also gasified [13, 19]. For commercial use, the available gasification technologies are counter-current fixed bed, co-current fixed bed, fluidized bed, entrained flow, plasma, and free radical [20, 21].

Plasma gasification is a thermal process that uses plasma to transform organic compounds into synthesis gas (syngas) which mainly contains H_2, CO, and CH_4, that can be used for direct combustion or raw material [22].

Plasma, also known as the fourth state of matter, is formed by a strong electric current that passes through a gas who is ionized and converts organic compounds into syngas, with slag as a by-product [22].

Plasma is formed by the interaction between the electric current and the gas which is ionized and causes the temperature of the gas to increase significantly to more than 5500°C, almost as hot as the sun's surface [23].

Plasma torches can be supplied with various gases, air, oxygen, nitrogen, argon, and others. This flexibility allows the plasma unit to be adapted for various applications [24].

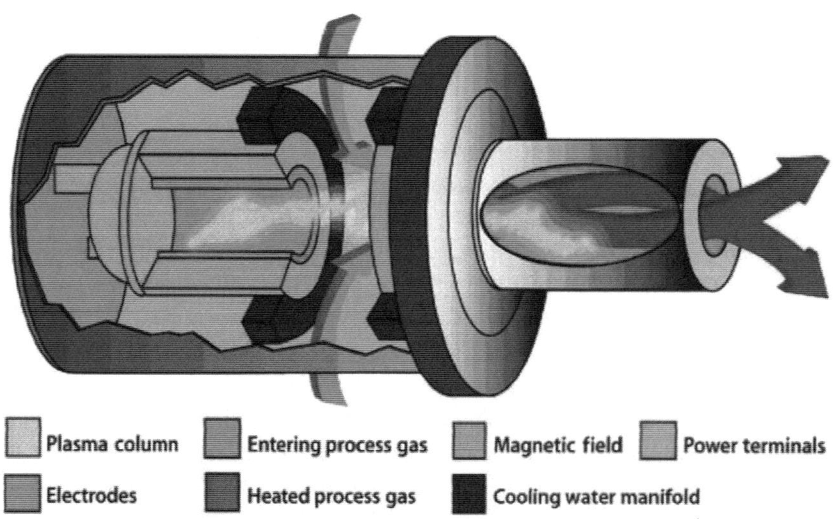

Figure 2. Plasma torch schematic [24].

IOP Conf. Series: Materials Science and Engineering 1138 (2021) 012028 doi:10.1088/1757-899X/1138/1/012028

Plasma gasification technology is a highly efficient process of heating the waste that is introduced in the plasma reactor at an extremely high temperature until they dissociate resulting in primarily chemical elements, the equipment can operate with minimal maintenance in a different industrial environment.

The plasma torches offer a high level of flexibility for plasma reactors, as it allows temperature control, regardless of the flow of waste or oxygen in this process [25].

The main benefits of the plasma gasification are:

- high reliability – more than 500,000 hours of continuous operation in commercial use [26];
- available in a wide range of powers input, from 80 to 2,400 kw [26];
- input power can be adjusted to meet the process requirements;
- long electrode life [26];
- auto stabilization and non-transferred arc (figure 3) [27];
- process gases can be air, oxygen, nitrogen, etc [28];
- long-life electrode [29];
- the syngas with very low in NOx, SOx, dioxins, and furans [30];
- high reliability - plasma torches have no moving parts, consumables are quickly replaced by maintenance personnel

Plasma gasification is clean and efficient of converting different waste into energy in a responsible way towards the environment [30 - 32].

Plasma gasification process breaks the molecular structure of any carbon-contained materials - such as solid municipal waste, tires, hazardous waste, biomass, river sediment, coal, and petroleum coke and converts them into syngas, which can be used to generate energy, liquid fuels or other sustainable energy sources [30, 31].

Gasification takes place in a low oxygen environment so that raw materials are vaporized, not incinerated [30, 32].

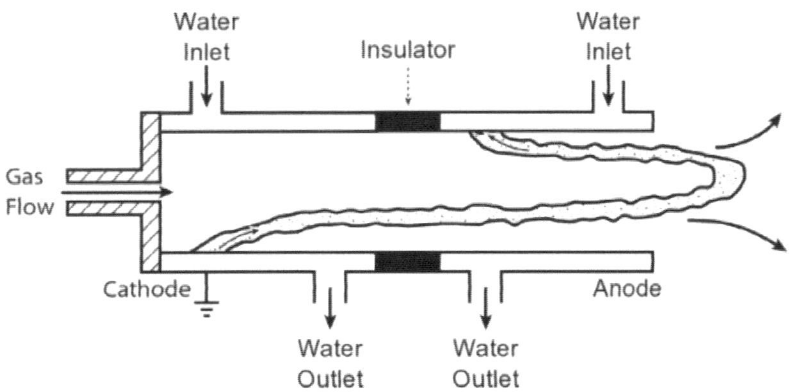

Figure 3. Schematic of a non-transferred arc discharge [27].

Due to the high operating temperatures in the plasma gasification reactor, we can observe from an environmental point of view that:

- no ash that requires treatment or disposal in a landfill occurs [33];
- metals and non-combustible inorganic materials are melted and captured in a slag medium, which can be used as a construction aggregate [31];
- the plasma gasification plant has very low emissions of NOx, SOx, dioxins, and furans (figure 4) [30, 31];
- in the CO_2 plasma gasification process can be captured and used for various industrial purposes [34].

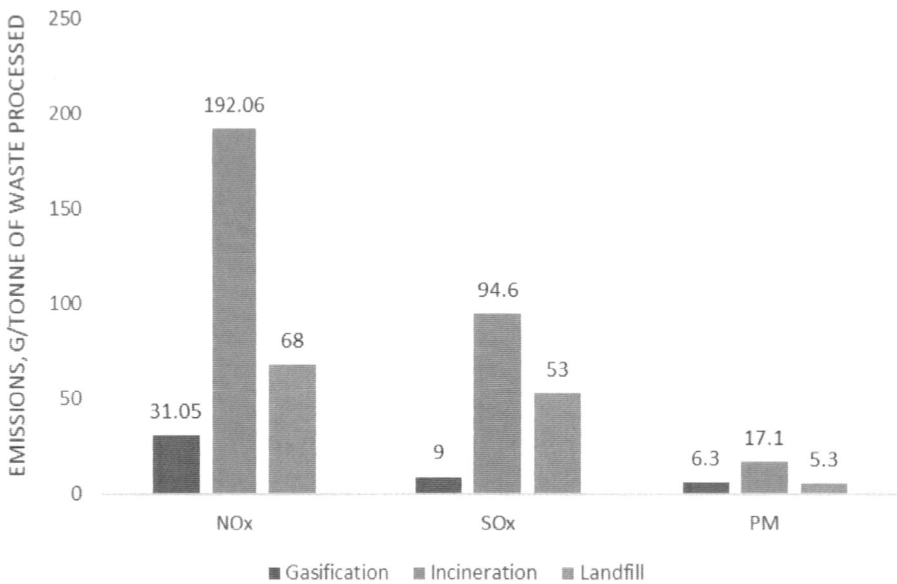

Figure 4. Emission comparison between different processes [31].

3. Circular economy

It is known that the current economy consumes a significant amount of raw materials and non-renewable energy. Now, when humanity's negative influence on the environment is intensified, the implementation of an efficient circular economy is a highly debated topic internationally.

The circular economy focuses on eliminating waste by creating a closed-loop system (figure 5) that involves reusing, refurbishing, recycling, reducing the consumption of valuable resources and the production of waste, pollution and greenhouse gases such as CO_2, NO_x, SO_x, CH_4, and others [35-37].

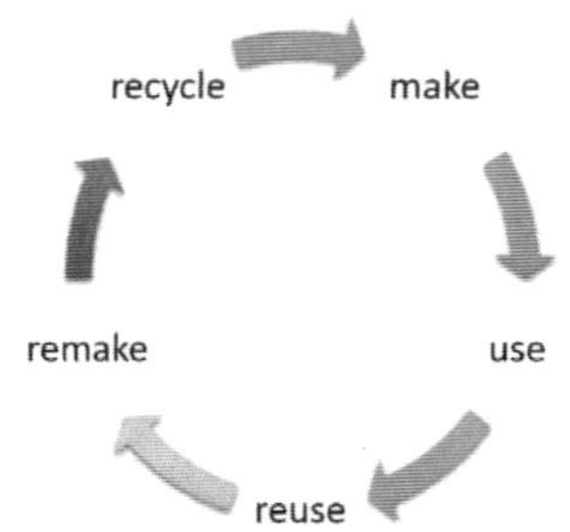

Figure 5. Circular economy (closed-loop system) [37].

To achieve a circular economy, it is necessary to recycle materials from waste as much as possible to have this closed-loop. For non-recyclable, hazardous and medical waste, an important role goes to waste-to-energy technologies, such as plasma gasification, that converts waste into energy and other inert by-products that can be utilised in construction or other fields, avoiding this way landfill disposal and reducing greenhouse gas emissions [38, 39].

This closed-loop demands primarily reusing, remaking and recycling. Another important aspect is that waste disposal should be eliminated as much as possible, and if not, it must be properly managed to ensure safety for human health and the environment.

In this context, the applications of life cycle assessment (LCA) of waste management plays an important role, and these can appear as buying bespoke LCA services; paying for staff training in this direction. These actions can be viable for the economy of large companies. The strategy is designed and can be implemented taking into account the basis of the systematic application or it can be implemented ad-hoc when it is necessary [39].

While reusing, remaking and recycling have an important role, they will never solve all the problems with waste. Numerous packaging wastes simply cannot be recycled without degradation, such as multilayer or metallized plastics, and laminates incorporating foils and fibre. Even with traditional methods of recycling, it is needed a process to recover materials that cannot accomplish their functions and finally become waste [37].

Usually, the main outlet for wastes is into landfill sites. Other choices exist, incineration and pyrolysis, but of all achievable options, exclusively plasma gasification can accomplish circular economy sustainability. Incineration produces a large quantity of ash, which is usually toxic, along with complex exhaust that must be treated to accomplish emissions regulations [7].

As the industry progressively sources of plastic from renewable materials, emissions become significantly close to zero net carbon. Also, fossil fuel usage is balanced by producing clean energy with plasma gasification. Energy efficiency is achieved when syngas is used to power combined cycle turbines. The combined cycle means that a gas-fired turbine is used to produce electricity using syngas, and from this process occurs high-temperature witch is then used to produce steam to engaged a second turbine witch produces more electricity [32].

The availability of plasma gasification technologies will reduce the need for fuel imports, reducing waste collection, and transform waste into clean energy. Savings in fuel imports will provide means to enhance waste collection to magnify the clean energy production from municipal solid waste using plasma gasification [27].

From an economic point of view, how the regulatory framework and its enforcement are applied can be the principal hindrance. On the other hand, the implementation of local economic tools has important premises as good management skills, innovation, and last but not least, environmental awareness. The instruments at the local level can take the form of tax modulation or taxes, waste pricing, refunds, and product levies. By using these applications, there are following the environmental indicators, like stimulated good behaviour, and sharing by using the voluntary instrument [39].

4. Conclusions

As we realize that reusing, remaking and recycling are insufficient to achieve our sustainability aspirations, plasma gasification stands out as the most promising municipal solid waste treatment solution.

Considering all the applications of plasma conversion for waste to value processing, the potential environmental benefits could be enormous. Public and/or private companies could build waste treatment units having substantial financial benefits and also improving the actual waste collection and treating infrastructure.

Plasma conversion systems are the future in waste management because they have a considerable low impact on the environment, having the lowest emissions from all waste treatment technologies available at the moment. Another important aspect of this technology is that it can treat medical and hazardous waste, resulting in syngas, metals and inert slag.

Even though there are some conflicts between different scientists and governmental environmental strategy makers, plasma gasification has all the advantages for a circular economy.

Acknowledgements

This paper was elaborated with the support of the "Eco-innovative Products and Technologies for Energy Efficiency in Constructions – EFECON" research grant, project ID P_40_295/105524, Program co-financed by the European Regional Development Fund through Operational Program Competitiveness 2014-2020.

5. References

[1] Eurostat 2019 *Waste statistics*

[2] Kaza S, Yao L, Bhada-Tata P and Van Woerden F 2018 *Urban Development* **1** p 28
[3] Grosso M 2016 *Waste Manage Res* **34** pp 977-979
[4] Das T K 2020 *Industrial Environmental Management: Engineering, Science, and Policy* (New Jersey: John Wiley & Sons) 266
[5] US Department of Energy 2019 *Waste-to-Energy from Municipal Solid Wastes* 1-36
[6] Knox A 2005 *An Overview of Incineration and EFW Technology as Applied to the Management of Municipal Solid Waste* (Ontario: ONEIA Energy Subcommittee)
[7] Chandler A J, Eighmy T T, Hjelmar O, Kosson D S, Sawell S E, Vehlow J, van der Sloot H A and Hartlén J 1997 *Municipal Solid Waste Incinerator Residues* vol **67** (Amsterdam: Elsevier Science)
[8] Themelis N J 2003 An overview of the global waste-to-energy industry, *Waste Manage* 40-47
[9] Abdullah M H, Rashid A S A, Anuar U H M, Marto A and Abuelgasim R 2019 *IOP Conf. Ser.: Mater. Sci. Eng.*
[10] Hsu S L 1999 *Brownfields and Property Values* (Washington DC: Environmental Law Institute)
[11] Eurostat 2019 *Municipal waste statistics*
[12] Directive 2000/76/EC 2000 *Waste incineration*
[13] CJ DJ 2019 Alternative tehnice privind gestiunea deşeurilor municipal (Dolj: Plan Judeţean de Gestionare a Deşeurilor)
[14] Gumisiriza R, Hawumba J F, Okure M and Hensel O 2017 *Biotechnol Biofuels* **10** pp 1-29
[15] Oasmaa A, Qureshi MS, Pihkola H, Deviatkin I, Mannila J, Tenhunen A, Minkkinen H, Pohjakallio M and Laine-Ylijoki J 2020 *J Anal Appl Pyrol*
[16] Zhou H 2017 *Combustible Solid Waste Thermochemical Conversion* (Beijing: Springer Theses)
[17] Crowe M, Nolan K, Collins C, Carty G, Donlon B, Kristoffersen M, Brøgger M, Carlsbæk M, Hummelshøj R M and Thomsen C D 2020 *Biodegradable municipal waste management in Europe* (Copenhagen: European Environment Agency)
[18] Chen D, Gao D, Capareda S C, Huang S and Wang Y 2019 *Bioresource Technol* **277** pp 37-45
[19] Lettner F, Timmerer H and Haselbacher P 2007 *Biomass gasification – State of the art description* (Graz: European Commission)
[20] NNFCC project 09/008 2009 *Review of Technologies for Gasification of Biomass and Wastes* (Lausanne: E4tech)
[21] Beychok M R 1975 *Process and environmental technology for producing SNG and liquid fuels (Oregon: U.S. EPA)*
[22] Ibrahimoglu B and Yilmazoglu M Z 2020 *Int J Hydrogen Energ* **45** pp 3532-3548
[23] Munir M T, Mardon I, Al-Zuhair S, Shawabkeh A and Saqib N U 2019 *Renew. Sust. Energ. Rev.* 116
[24] NwachinemeluChibuzor A F 2013 *Int J Eng Res Technol* **2**
[25] Huang H and Tang L 2007 *Energ Convers Manage* **48** pp 1331-1337
[26] Young G C 2010 *Municipal Solid Waste to Energy Conversion Processes: Economic, Technical, and Renewable Comparisons* (New Jersey: John Wiley & Sons)
[27] Fridman A 2008 *Plasma chemistry* (New York: Cambridge University Press)
[28] Petrie E M 2013 *Adhesive bonding of textiles: principles, types of adhesive and methods of use* (Woodhead Publishing Series in Textiles) pp 225-274
[29] Golish V I, Karpenko E I, Luk'yashchenko V G, Messerle V E, Ustimenko A B, Ushanov V Z 2009 *High Energy Chem* **43** pp 318–323
[30] Anyaegbunam F N C 2013 Proc. Natl. Acad. Sci. 6
[31] Lupu M L, Isopescu D N, Cucos I, Antonescu I, Maxineasa S G and Baciu I-R 2020 *Researches on energy conversion of municipal waste by plasma decomposition for energy-efficiency in civil engineering* IOP Conf. Ser.: Mater. Sci. Eng. 789
[32] Smith A D and Gill G 2011 *Toward Zero Carbon: The Chicago Central Area Decarbonization Plan* (Mulgrave: Images Publishing)
[33] Hlina M, Hrabovsky M, Kavka T and Konrad M 2013 *Waste Manage*

[34] Ma Q 1998 *Greenhouse Gases: Refining the Role of Carbon Dioxide* (New York: NASA GISS)
[35] Geissdoerfera M, Savaget P, Bocken N M P and Hultink E J 2017 *J CLEAN PROD* **143** pp 757-768
[36] Weetman C 2016 *A Circular Economy Handbook for Business and Supply Chains: Repair, Remake, Redesign, Rethink* (London: Kogan Page Publishers)
[37] European Parliament 2017 *Towards a circular economy – Waste management in the EU*
[38] EU Science Hub 2019 *Waste and Recycling*
[39] Dri M, Canfora P, Antonopoulos I S and Gaudillat P 2018 *Best Environmental Management Practice for the Waste Management Sector* (Seville: European Commission JRC)

IOP Conf. Series: Materials Science and Engineering 1138 (2021) 012029 doi:10.1088/1757-899X/1138/1/012029

Assessment of seismic performance of RC frame structure with masonry infill panels partially uncoupled from the surrounding frame members

H A Mociran[1*] and N Cobîrzan[1]

[1] Technical University of Cluj-Napoca, Faculty of Civil Engineering, Romania

* horatiu.mociran@mecon.utcluj.ro

Abstract. The present study numerically analyses the influence of physical and mechanical properties of the materials used for masonry infill panels upon the seismic response of low rise reinforced concrete frame structure. A four-storey residential building (benchmark), located in Timisoara, a moderate seismic area, has been designed according to the current Romanian design codes. Then, infill panels considered uncoupled from the frame members along their lateral and upper edges, have been added to the structure. Infill panels have been investigated in four variants, by using 2 types of masonry units (hollowed ceramic blocks and AAC), alternatively built with mortars specific for regular bed joints (G) and thin bed joints (T), for which a comparative modal response spectrum analysis has been carried out. The seismic behaviour of buildings in all these cases has been analysed in terms of peak inter-storey drifts under two limit states and of the out-of-plane strength of infill panels. The results show that the peak inter-storey drifts reductions compared to the benchmark structure are up to 59% at SLS, and 61% at ULS. Regarding the out-of-plane behaviour of the infill panels it was noted that safety verification is fulfilled only for one of the abovementioned cases. As a final remark, the results obtained in this study highlight the importance of physical and mechanical properties of the masonry infill panels in the assessments of seismic response of frame structure buildings.

1. Introduction

The infill walls for reinforced concrete frame structures are widely used in seismic areas all over the world [1, 2], due to their good thermal properties, low density and cost. Their role is to separate interior and outer space and enhance at the same time the energetic performance of the building.

Field observations after recent earthquakes have shown numerous damages to infill wall panels. The costs of repairs were high and sometimes exceeded the cost of structural members strengthening [3], [4], and their failure can threaten the occupants' life [5].

The masonry infill undertakes the own weight and could be seismically designed in two different ways: first, interacting with the surrounding frame members in order to improve their strength and lateral stiffness [6] and secondly, with partial interaction with the frame members, by providing gaps in between on one or more edges [7], with low increase of building stiffness.

The technical literature provides numerous studies on the seismic behaviour of frame structures interacting with infill panels, concerning the development of various simplified computational techniques (diagonal strut) [8], appraisal of the influence of panels upon the seismic response, investigation of possible types of unfavourable collapse mechanisms (short columns, soft storey mechanisms) [7], the influence of openings dimensions and positions, the influence of vertical and horizontal irregularities, damage assessment etc. [9]. Instead, it has been investigated the behaviour of infill panels isolated from the main structure by means of various materials and construction details, when subjected to earthquake excitations, in fewer experimental and numerical studies [5, 10, 11].

The paper aims to investigate the influence of materials properties used for masonry infill panels, restrained by the surrounding frame members only at the bottom edge, upon the seismic response of low rise reinforced concrete frame structure.

2. Materials and Methods

The reinforced concrete frame structure of a building placed in Timisoara, was designed with medium ductility (DCM), in compliance with the provisions of code P100/1-2013 [12] and standard SR EN 1992-1 [13]. First, the bare structure (benchmark) has been designed according to the usual design practice with uniformly distributed load on beams (figure 1a), instead of modelling the infill panels.

In plane, the structure has two bays of 5.7 m along Y direction and five bays range between 3.2 m and 4.6 m along X direction, the storey height being 3 m.

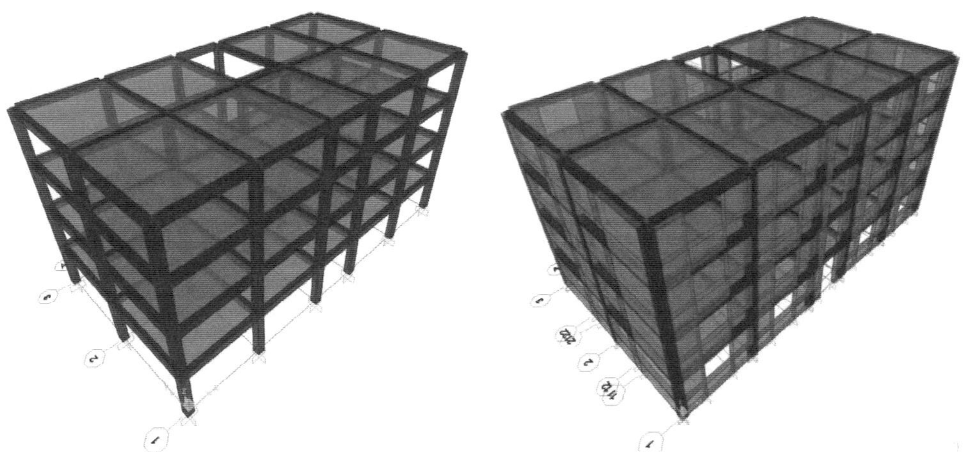

(a) (b)

Figure 1. Reinforced concrete frame structure: benchmark (a), with infill panels (b).

The seismic action has been considered by modal response spectrum analysis. The main features of the design response spectrum are the following: seismic ground acceleration ($a_g = 0.20 \cdot g$); corner period (Tc = 0.7 s); behaviour factor (q = 4.725). The columns have square cross section (40x40 cm), and the beams have rectangular cross section (30x40 cm) in both, longitudinal and transversal direction. Their dimensions have been established considering stiffness, ductility and fire requirements. The concrete class is C20/25 and steel grade S500.

In the second version, infill panels have been added to benchmark structure (BS) with 45 mm wide gap at the frame-infill interface, established from inter-storey drifts and tolerance consideration, in order to avoid the interaction (figure 1b).

In Romania, various materials are used for masonry infill panels, their choice and wall thickness depending on seismic and climatic zones.

Two types of masonry units and mortars have been used for masonry infill walls, as follows:

- hollowed ceramic blocks (with normalized compressive strength f_b = 10 N/mm^2, density ρ=850 kg/m^3) and mortar class M5 in regular bed joints (compressive strength f_m = 5 N/mm^2) (case 1) and respectively in thin bed joints (case 2).

- infill panels made of AAC, with compressive strength $f_b = 3$ N/mm^2, $\rho = 500$ kg/m^3 and mortar class M5 in regular bed joints (case 3), respectively with $f_m = 5$ N/mmm^2 and mortar in thin normal bed joints (case 4).

The mechanical properties of masonry have been calculated according to CR6-2013[14], resulting the following:

- design compressive strengths (f_d): 1.92 N/mm^2 (case 1); 2.79 N/mm^2 (case 2); 1.07 N/mm^2 (case 3); 0.83 N/mm^2 (case 4);
- design flexural strength of masonry, with the plane of failure parallel to the bed joints (f_{xd1}): 0.126 N/mm^2 (case 1); 0.079 N/mm^2 (case 2); 0.055 N/mm^2 (case 3); 0.053 N/mm^2 (case 4);
- short term secant modulus of elasticity (E_z), of 3653 N/mm^2 (case 1); 5300 N/mm^2 (case 2); 1628 N/mm^2 (case 3); 1259 N/mm^2 (case 4).

The structures have been analysed by using Finite Element Method (FEM) with SAP 2000 Educational software [15]. Shell elements have been used for the modelling of infill panels.

3. Results and discussions

The safety verifications of infill panels partially decoupled from the surrounding frame members=are in terms of inter-storey drifts at two limit states (Serviceability Limit State - SLS and Ultimate Limit State -ULS) and their out-of-plane strength.

3.1. Inter-storey drifts

Figure 2 illustrates the lateral displacements along Y direction for the benchmark (figure 2a), respectively with infill panels - case study 2 (figure 2b).

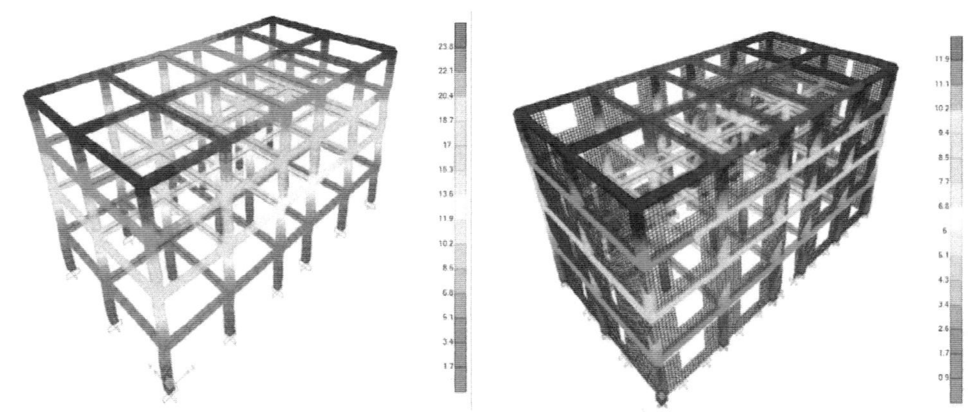

(a) (b)

Figure 2. Lateral displacements along Y direction: benchmark (a), with infill– case study 2 (b).

In each of the analysed cases the maximum lateral displacements of all storeys along the X and Y directions have been extracted. The inter-storey drifts in SLS and ULS have been computed with the previous values. Considering that the masonry ductility is lower than that of reinforced concrete, the inter-storey drifts of the infill have been corrected with the ratio between the behaviour factors of the structural and non-structural members.

In figures 3-4, the inter-storey drift profiles for benchmark and the four cases are displayed.

In comparison with the reference structure, the addition of infill panels has reduced the peak inter-storey drifts at SLS by: 36.47% - 49.55% along X direction and 38.92% - 56.84% along Y direction for case 1; 43.15% - 58.78% along X direction and 43.94% - 65.06% along Y direction for case 2; 34.4% -

51.04% along X direction and 42.12% -59.58% along Y direction for case 3; 38.4% - 51.93% along X direction and 42.12% - 58.29% along Y direction for case 4.

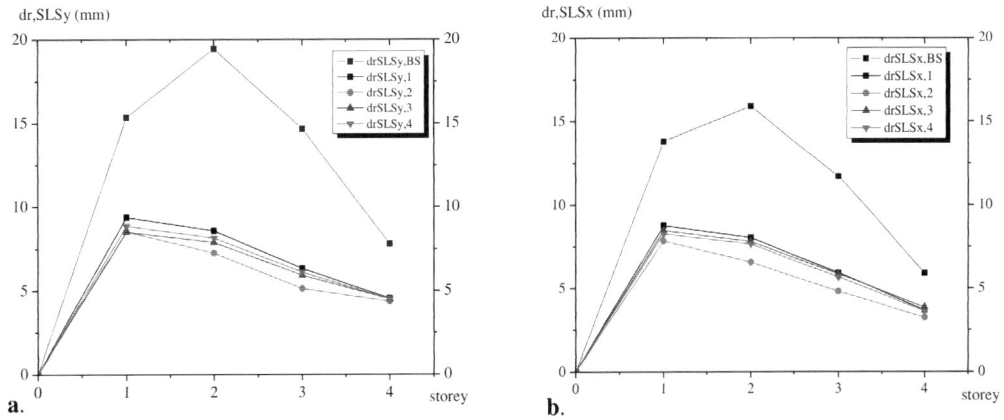

(a) (b)

Figure 3. Inter-storey drift of benchmark and structures with infill panels at SLS: along Y (a) and X directions (b).

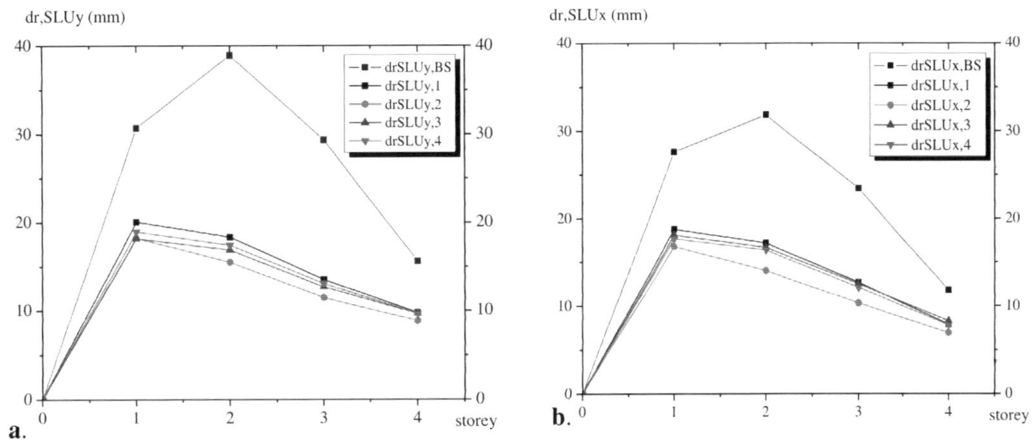

(a) (b)

Figure 4. Inter-storey drift of benchmark and structures with infill panels at ULS: along Y (a) and X directions (b).

For ULS, slightly lower decrease has been obtained: 32.03% - 46.02% along X direction and 34.65 -53.82% along Y direction for case 1; 39.17% - 55.90% along X direction and 40.57% - 60.89% along Y direction; 29.81% - 47.61% along X direction and 38.07% - 56.75% along Y direction; 34.09% - 48.56% along X direction and 38.07% - 55.37% along Y direction.

The amount of the previously mentioned reductions is due to different elastic moduli (E_z) and specific weights of the materials.

From the above figures it can be seen that the reductions obtained for the cases 3 and 4 are similar, explained by the fact that the physical and mechanical characteristics of the materials used for infill walls are not very different, E_z being slightly higher in case 3, while the specific weight is slightly higher in case 4.

Analysing the variation of reductions along the structure height, it can be observed that in all cases and for both limit states, the maximum values for decrease have been recorded at the intermediate floors

(2nd and 3rd) and the smallest one at the ground floor and at the top floor. It is worth mentioning that in the case of reference structure, the maximum displacement appears on the 2nd floor, while for the infill frame structures, the maximum displacement is reached at the first floor. The values of the previously mentioned reductions result because of the different elastic moduli (E_z) and specific weights of the materials.

By analysing the obtained results, it can be seen that inter storey drifts along Y direction are higher than those along X direction, both for SLS and ULS. It should also be mention that inter-storey drifts of infill panels and columns are different, but lower than the width of the gap, so the interaction between them is avoided.

3.2. Out-of-plane verification of infill panels

The out-of-plane strength of the infill panel is a key issue for the seismic performance of buildings. Due to the forces of inertia generated by ground motions, infill panels could be expelled from the structure in the out-of-plane direction, which pose a threat to human lives because of the falling of rubble and can lead to substantial economic losses, as demonstrated during recent earthquakes.

Figure 5 shows the values of the normal stresses perpendicular to the bed joints.

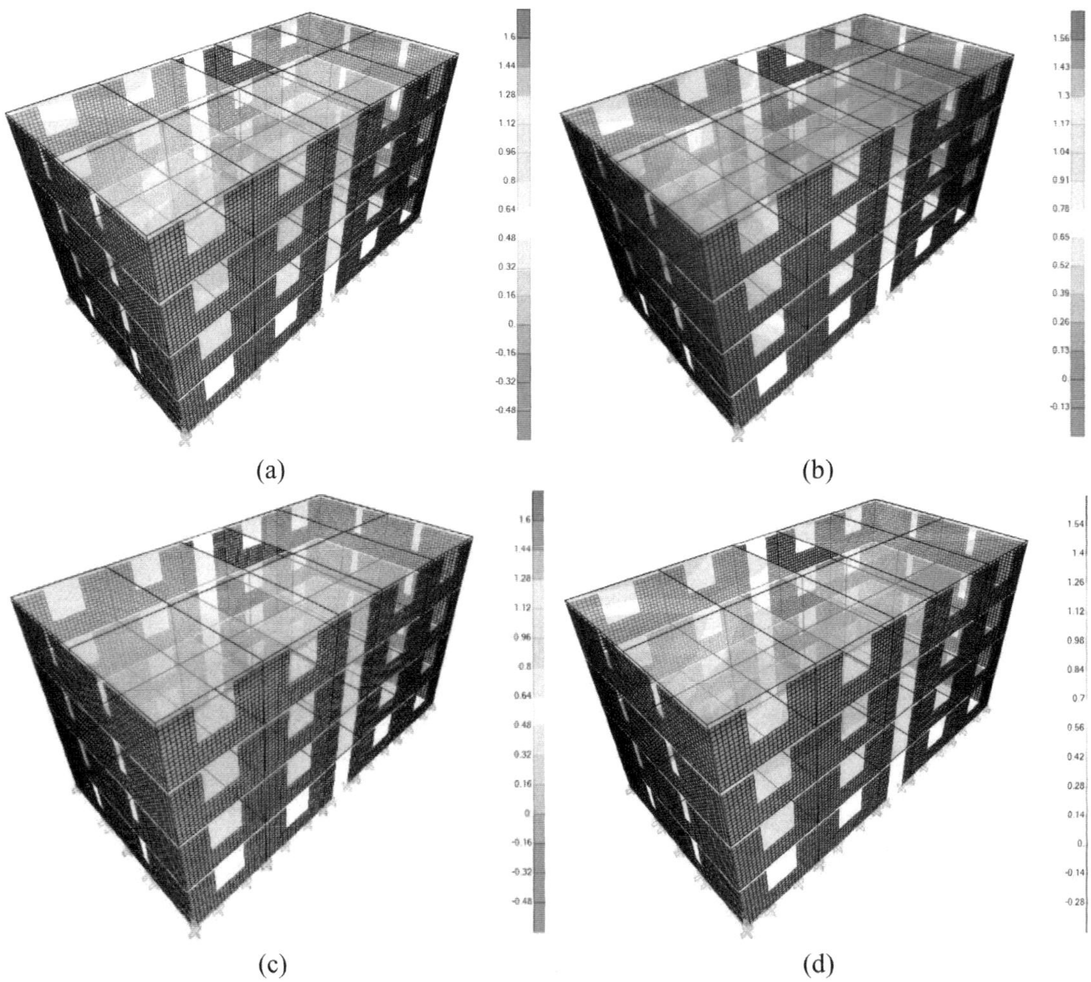

Figure 5. Normal stresses perpendicular to the bed joints: case 1 (a); case 2 (b); case 3 (c); case 4 (d).

The maximum values of the normal stresses perpendicular to bed joints at each floor of the 4 analysed cases are summarized in Table 1.

Table 1. Maximum values of the normal stresses [N/mm^2] perpendicular to bed joints at each floor of the 4 cases.

Floor	Case 1	Case 2	Case 3	Case 4
1	0.042	0.028	0.008	0.018
2	0.051	0.044	0.013	0.023
3	0.087	0.072	0.037	0.051
4	0.104	0.091	0.053	0.089

The out-of-plane strength of infill panel is verified by comparing the maximum normal stresses with design flexural strength of masonry, with the plane of failure parallel to the bed joints (f_{xd1}). In all cases the maximum normal stresses appear at the base of the 4th floor; the highest values being recorded along Y directions.

The verifications in cases 1 and 3 are fulfilled on both directions. In case 2, the verification is not fulfilled at the base of the 4th floor, on a length of about 1m starting from corners, along Y direction. In case 4 the verifications are not satisfied on the base of the 4th floor, on the entire length along Y directions and on a length of about 80 cm starting from corners along X direction.

4. Conclusions

Nowadays, on the market there are a wide range of building materials with various physical and mechanical properties that can be used for infill in case of frame structures. Their seismic performance depends on the perimeter restraining conditions. In Romania and in other Eastern European countries the infill panels are usually restrained at the base and free on the other three sides. The gaps between the panels and the surrounding members of the frames are filled with various insulation materials (foams, polystyrene etc.) thus, their interactions during seismic events are prevented, therefore the major issue is the out-of-plane verification of these panels.

In seismic design practice the structure is modelled as a bare frame, with uniformly distributed load on perimeter beams, hence the stiffness of infill panels is neglected, and their out-of-plane verifications are omitted. The disadvantages of this simplified design method could be avoided by modelling the infill panels. In this case, the frame-infill interaction could be verified, as well as the inter-storey drifts reductions and the contribution of infill panels to the lateral stiffness of the building. Also, out-of-plane infill failures could be highlighted, making possible to strengthen the weak zones during the building's execution. The results obtained in the present research have shown that seismic response of buildings is influenced by the materials properties (units and mortars) and masonry type (with regular or thin bed joints).

References

[1] Dautaj A, Kadiri Q and Kabashi Q 2018 Experimental study on the contribution of masonry infill in the behaviour of RC frame under seismic loading *Eng. Struct.* **165** 27-37
[2] Furtado A, Rodrigues H, Arede A, Varum H, Grubisic M and Sipos T K 2018 Prediction of the earthquake response of a three-storey infilled RC structure *Eng. Struct.* **171** 214-235
[3] Villaverde R, 1997 Seismic design of secondary structures: state of the art *J. Eng. Struct.* **123** (8) 1011-19
[4] Filiatrault A and Sullivan T 2014 Performance-based seismic design of non-structural building components: The next frontier of earthquake engineering *Earthq. Eng. Eng. Vib.* **13**1 7-46
[5] Tsantilis A V and Triantafillou 2018 Innovative seismic isolation of masonry infills using cellular materials at the interface with the surrounding RC frames *Eng. Struct.* **155** 279-297
[6] Di Trapani F, Macaluso G, Cavaleri L and Papia M 2015 Masonry infills and RC frames

interaction: literature overview and state of the art of macromodeling approach *Eur. J. Environ. Civ. Eng.* **19** 1059-1095
[7] Uva G, Porco F and Fiore A 2012 Appraisal of masonry infill walls effect in the seismic response of RC framed buildings: A case study *Eng. Struct.* **34** 514-526
[8] Fiore A, Spagnoletti G and Greco R 2016 On the prediction of brittle collapse mechanisms due to the infill-frame interaction in RC buildings under pushover analysis *Eng. Struct.* **121** 147-159
[9] De Risi M T, Gaudio C D, Ricci P and Verderame G M 2018 In-plane behaviour and damage assessment of masonry infills with hollow clay bricks in RC frames *Eng. Struct.* **168** 257-275
[10] Peng Q, Zhou X and Yang C 2018 Influence of connection and constructional details of masonry-infilled RC frames under cyclic loading *Soil Dyn. Earthq. Eng.* **108** 96-110
[11] Marinkovic M and Butenweg C 2019 Innovative decoupling system for the seismic protection of masonry infill walls in reinforced concrete frames *Eng. Struct.* **197**
[12] P100-1/2013 Seismic design code-Part 1: Design rules for buildings (in Romanian), Bucharest
[13] SR EN 1992-1:2004 -Design of concrete structures-Part 1: General rules and rules for buildings. National annex (in Romanian), Bucharest
[14] CR6 - 2013 -Design code of masonry structures (in Romanian), Bucharest
[15] SAP2000 educational version

IOP Conf. Series: Materials Science and Engineering 1138 (2021) 012030 doi:10.1088/1757-899X/1138/1/012030

Pushover analysis of R.C. framed structures with infill panels made of masonry having various properties

H A Mociran[1*] and N Cobîrzan[1]

[1] Technical University of Cluj-Napoca, Faculty of Civil Engineering, Romania

* horatiu.mociran@mecon.utcluj.ro

Abstract. The seismic performances of R.C. frame structures were determined in terms of inter-storey drifts at two limit states, internal forces and moments (axial forces and bending moments), and the in-plane behaviour of infills. A nonlinear static analysis (pushover) was carried out in order to assess the seismic performance of buildings, to identify the expected plastic mechanisms and to plot the capacity curves. In pushover analysis, the vertical distribution of the monotonically increasing horizontal loads, was performed according to the first mode pattern. The nonlinear behaviour of structural members for pushover analysis was modelled by using the lumped plasticity, from where it derives that the potential plastic hinges were situated at the beams ends (which yield due to the bending moment), at the top and bottom of the columns at any storey (which yield based on the interaction of axial force and bending moment) and in the struts (which yield due to the axial force).

1. Introduction

Traditionally infill walls made of masonry, widely used on a large scale in the residential reinforced concrete framed structures [1], had a low performance at seismic actions [2, 3], in the last decades. In general, in the design practice, the infills are considered non-structural elements, thus they don't influence the seismic response of buildings and are neglected in the structural analyses. As a function of the execution infills panels surrounding - frame connection, the predicted behaviour of the building structure can be altered from the one considered in the design phase [4], leading to different failure mechanism patterns in wall plane (sliding shear, diagonal cracking or corner crushing) and out-of-plane [5]. Significant damages to infill wall panels, caused by even the moderate earthquakes, yielded to large costs of repairing and strengthening [2, 3, 5], having as consequences the intensification of researches in this area.

In order to take into account, the favourable contribution of infills (energy dissipation capacity, increases of lateral stiffness and consequently diminishing the inter-storey drifts) to seismic response of reinforced concrete frames structures, many structural designers consider the infill-frame interactions with the surrounding frames on all edges of the infilled panels.

Regarding the modelling of infilled panels there are two main possibilities: micro-models and macro-models. In micro-models, the infilled panels are discretized in numerous finite elements (FE) [6], aiming a more accurate response of elements, at a high computational cost.

In contrast, in macro-modelling, the infills are considered as equivalent diagonal struts. Historically, first model being a diagonal strut, connected with hinges in the beam-columns joints, with the disadvantage of transferring the loads from the beam to the strut. Different simplified proposed models range from single to multiple diagonal pin-jointed struts, with various cross section width, disposed

IOP Conf. Series: Materials Science and Engineering 1138 (2021) 012030 doi:10.1088/1757-899X/1138/1/012030

concentric or eccentric [7-12] with respect to the beam-column joint. Single equivalent strut-model assures a good balance between simplicity and accuracy in numerical analyses [4].

The seismic performance of structure is significantly influenced by the thickness and mechanical properties of the materials (units and mortars) used for infills. Considering the dynamics of occurrence of new masonry units with high thermal properties, with the aim to increase the energy performance of buildings, it is necessary to carefully investigate the seismic response of framed buildings with infilled panels made of such materials, placed in different seismic areas.

In that respect, the paper aims are to examine the influence of the mechanical properties of the materials used for masonry infills panels of reinforced concrete frame structure, upon the building seismic performance.

2. Materials and numerical simulations

A plane reinforced concrete frame, extracted from a typical 3D frame structure, located in Piatra Neamț, designed in accordance with regulation in force (code P100/1-2013 [13] and standard SR EN 1992-1 [14]) with medium ductility (DCM), was considered in this analysis. The frame has 5 floors and two bays of 5,0 m, the story height being 3,0 m (figure 1a).

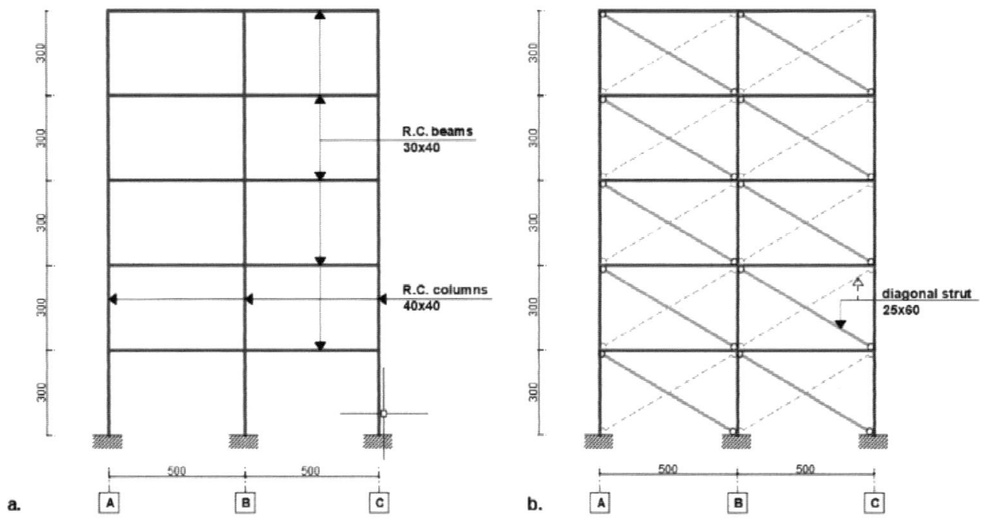

(a) (b)

Figure 1. Reinforced concrete frame: benchmark (a), with infill panels (b).

The 2D frame was analysed in three case studies (table 1).

Table 1. Mechanical and physical characteristics of materials.

Case Study	Frame structure	Mechanical compressive strength [N/mm^2]			Density [kg/m^3]
		Masonry units		Mortar [f_m]	
		∥ to the bed joints [f_{bh}]	⊥ to the bed joints [f_b]		
Benchmark	Bare structure (benchmark)				
1st	With infills	2.5	10	5	850
2nd	With infills	2	10	10	850

The main characteristics of the design response spectrum used in the seismic analysis are: ground acceleration (a_g = 0.25g); corner period (T_c = 0.7 s); behaviour factor (q = 4.725).

In the benchmark reinforced concrete frame (1st case study), the effect of the infill panels was modelled as uniformly distributed loads on beams. The cross sections of all columns and beams are 40x40cm, respectively 30x40cm, determined from stiffness, ductility and fire considerations. The concrete class is C20/25 and steel grade is S500. The reinforcement of the frame members was computed in the benchmark structure, resulting for the longitudinal reinforcement of the columns 8 Φ16.

For the 1st and 2nd case studies, the dimensions of the cross-section members and their reinforcement was taken from the benchmark structure. The infill panels have been considered in full contact with the surrounding frames and have been modelled in accordance to seismic code P100-1/2013, through diagonal compression strut. The dimensions of diagonal strut were 25x60 cm. The values of the elastic moduli of the diagonal struts for the 1st and 2nd case studies have been 3382 N/mm^2 and 2667 N/mm^2, respectively.

The numerical analysis has been performed by using Finite Element Method (FEM) with SAP 2000 Educational [15].

3. Results and discussions

The seismic performances of reinforced concrete frame structures, are determined in terms of inter-storey drifts at two limit states (Serviceability Limit State (SLS) and Ultimate Limit State (ULS)), internal forces and moments (axial forces and bending moments), and the in-plane behaviour of infills. Also, a pushover analysis was carried out in order to identify the distribution of damages and to plot capacity curves.

3.1. Inter-storey drifts

The inter-storey drifts over the height of all analysed frames are displayed in figure 2.

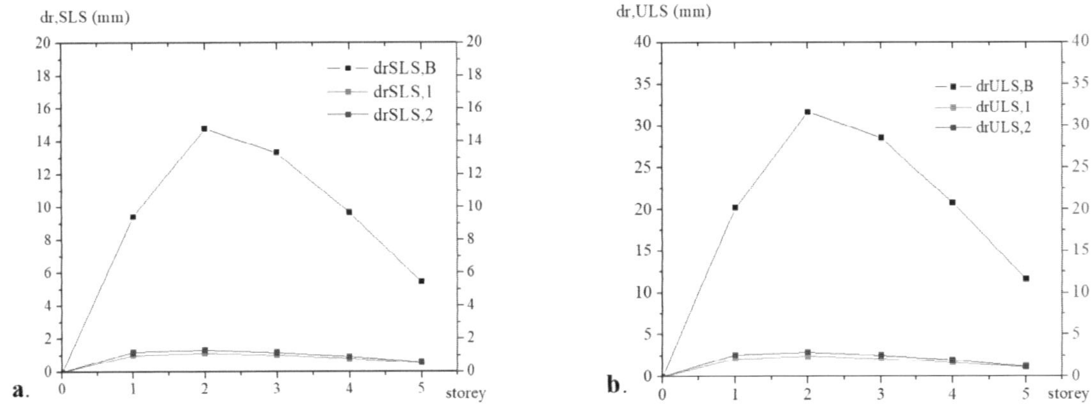

(a) (b)

Figure 2. Inter-storey of all frame structures at: SLS (a) and ULS (b).

It should be mentioned that the inter-storey drifts at SLS, in the 1st analysed case study, are closer to allowable drifts (d_{ra},SLS = 0.005h) given by seismic design code, the stiffness conditions being fulfilled. By adding the infill panels, the inter-storey drifts at SLS (d_r,SLS) and ULS (d_r,ULS), were diminished by up to 89% - 93% in case study 1 and up to 87% - 91% in case study 2, compared with benchmark. As can be observed in figure 2, the reductions are approximately uniform at all stories.

3.2. Internal forces and moments

The axial forces (N) and bending moments (M), for central (cc) and marginal columns (mc) are shown in figures 3-4.

Considering the infill panels effect, the values of axial forces for marginal columns (N_{mc}) were increased at the first 4 floors up to 10% and diminished at the roof level up to 5%, compared with the bare frames. In contrast with the marginal columns, the axial forces of the central columns (N_{cc}), records decrease at all levels up to 24% (figure 3).

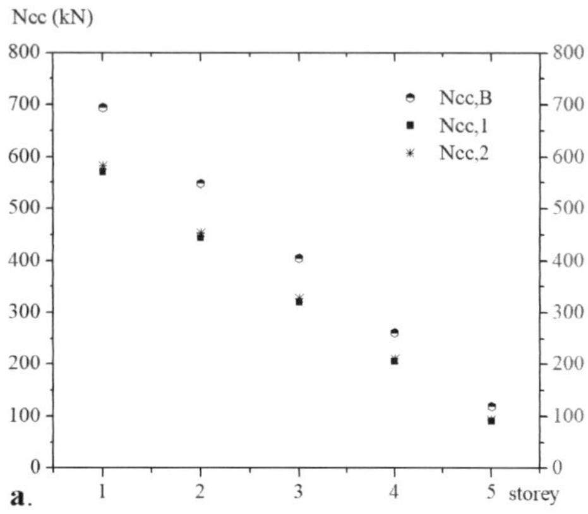

(a)

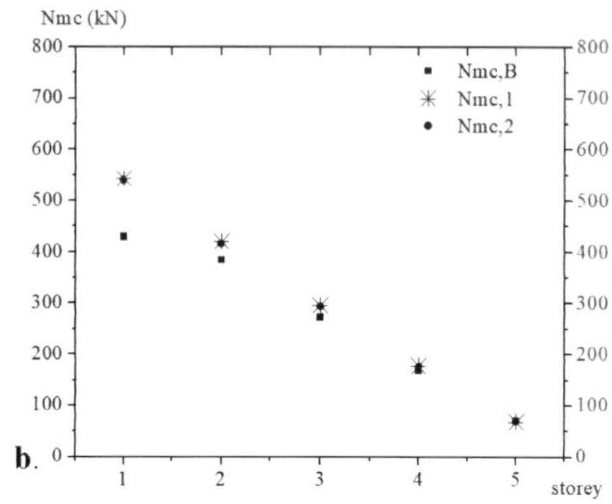

(b)

Figure 3. Axial forces for: central, N_{cc} (a); marginal columns, N_{mc} (b).

Regarding the values of bending moments for marginal (M_{mc}) and central columns (M_{cc}) were diminished up to 80%, respectively up to 97% in comparison with the benchmark (figure 4).

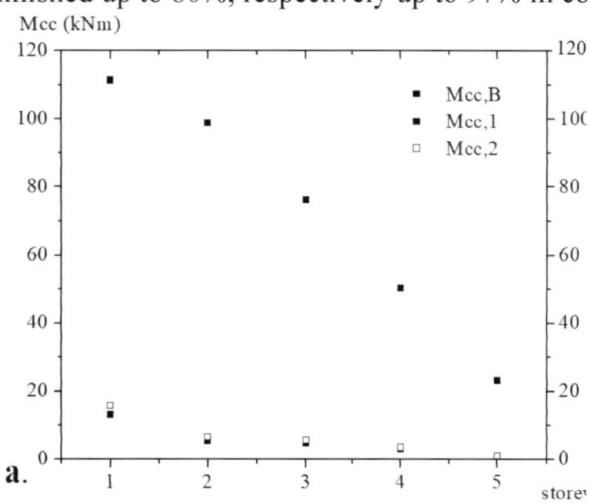

(a)

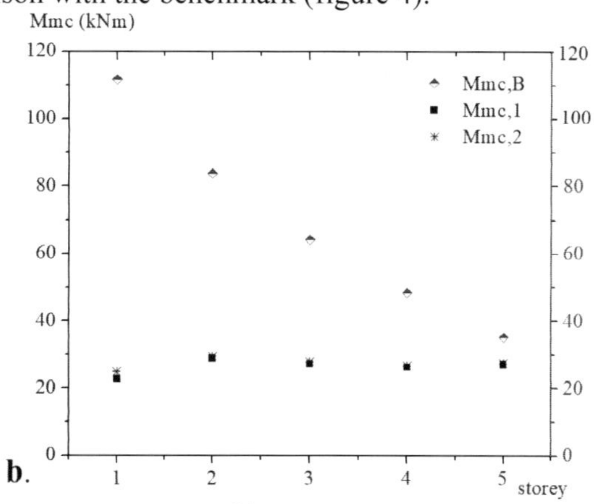

(b)

Figure 4. Bending moments for: central, M_{cc} (a); marginal columns, M_{mc} (b).

By analysing the previous results, it should be noted that by adding the infills, the behaviour of the structure is changed from a moment resisting frame, subjected mainly to bending moment, into a truss type structure. This yields to important reductions of the values of bending moments in all columns and negligible changes of axial forces.

3.3. In-plane performance of infills

The in-plane performance of infills, was analysed by investigating the three possible failure patterns: sliding shear, diagonal cracking, and corners crushing (figure 5b).

The axial forces in diagonal struts (N_d) versus the design values of resistance of infill panels (FR_{d1}, FR_{d2}, FRd_{31}, FRd_{32}), for frames analysed in case studies 1.2, are presented in figure 5.

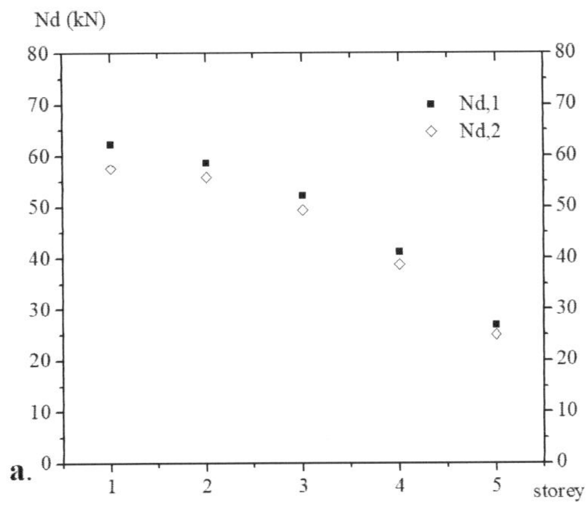

(a)

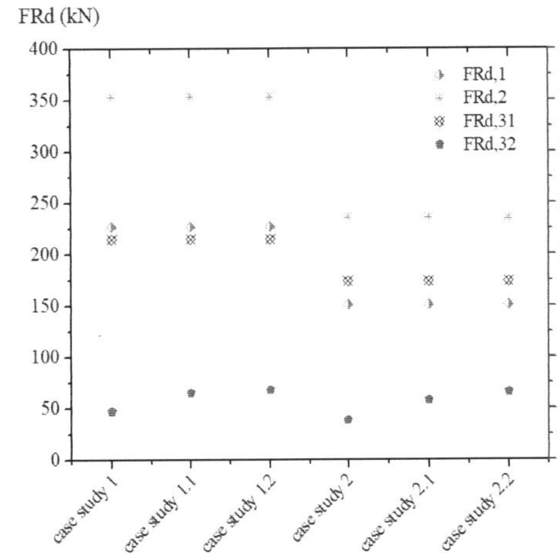

(b)

Figure 5. Axial force in the diagonal strut (a) and design values of resistance of infills (b).

In both cases, at the lower level, the axial forces in diagonal strut exceed the design values of resistance of infills. In order to avoid the corners crushing failure of infills three other solutions have been proposed: masonry made of clay units from group 2S, with superior compressive strength, $f_{bh} = 4N/mm^2$ (case study 1.1); masonry made of units from group 2S, with superior compressive strength, $f_{bh} = 3.5\ N/mm^2$ (case study 2.1) respectively, units from group 2 instead of 2S with $f_{bh} = 4N/mm^2$ (case study 2.2). The elastic moduli of the diagonal struts of the case studies 1.1, 2.1 and 2.2 have been determined as follows: 3647 N/mm^2, 2917 N/mm^2 and 2963 N/mm^2, respectively.
By considering infills panels made of materials with improved mechanical properties, the axial forces induced in struts by seismic actions increase, in case of study 1.1 up to 2.88% with respect to case study 1, and in case studies 2.1 and 2.2 up to 3.56% respectively 4.20% in comparison with the case study 2 (figure 5a). It is worth to mention that the new obtained values are lower than the design values of resistance of infills (figure 5b).

3.4. Pushover analysis
Also, a nonlinear static analysis (pushover) was carried out in order to assess the seismic performance of the building, to identify the expected plastic mechanism and to plot the capacity curves.

In pushover analysis, the vertical distribution of the monotonically increasing horizontal loads, was according to the first mode pattern. The nonlinear behaviour of structural members for pushover analysis was modelled by using the lumped plasticity. The potential plastic hinges are situated at the beams ends (which yield due to the bending moment), at the top and bottom of the columns at any storey (which yield based on the interaction of axial force and bending moment) and the strut (which yield due to axial force). Also, the geometrical nonlinearity (second order effect) was considered.

In figure 6 is displayed the plastic hinge configuration at the collapse mechanism of the structures.

The seismic performance of the building was investigated at SLS and ULS, considering the allowable interstorey-drifts corresponding to the both limit states, prescribed by the seismic code, the pushover curves and the sequence of plastic hinge appearance.

Regarding the SLS performance, it was noticed in all cases that damages to infill panels extended from the base to upper floors. As expected, the worst behaviour was recorded in case study 2 where the design value of resistance of infill panel was the lowest, while the best performance was attained in the proposed solutions (case studies 1.1, 2.1 respectively 2.2), where the infills were destroyed entirely in the first three floors and partially at the fourth level.

Concerning the ULS, the behaviour of frame members was similar for all cases, with plastic hinges at beams ends in the first three levels and at the base of the columns. The worst seismic performance of the infill was achieved in case study 2, where all the panels were damaged.

In contrast, the best seismic response was noted in case study 2.2, with the maximum design value of resistance of infill panel, where only the panels for the first fourth floors were damaged.

Other pushover analyses carried out by several authors [16], had revealed that infilled damages are first meet at the lower level of the buildings.

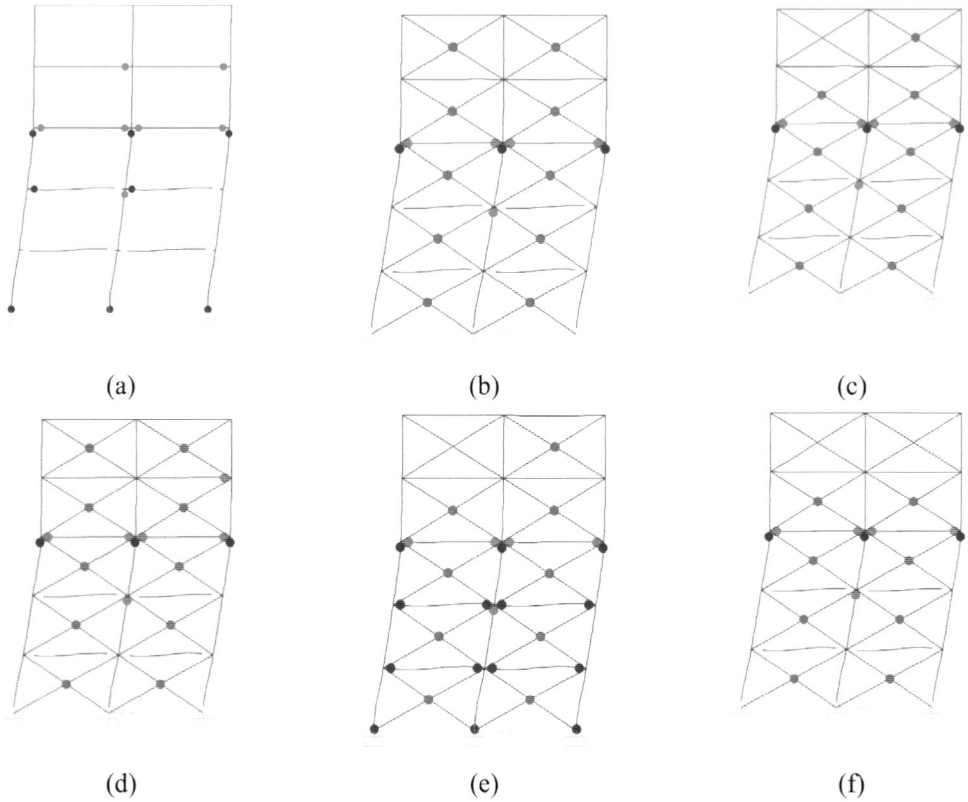

Figure 6. Plastic hinge configuration at collapse mechanism: benchmark (a); case study 1 (b); case study 1.1 (c); case study 2 (d); case study 2.1 (e); case study 2.2 (f).

4. Conclusions

In this paper the influence of the material properties used for masonry infill panels of R.C. frames upon the overall seismic performance of the buildings has been investigated by means of modal seismic response spectrum and pushover analyses.

The following conclusion can be drawn:

- the inter-storey drift at SLS and ULS were diminished approximatively uniform at all stories up to 93% in both case studies compared with benchmarks;
- the values of bending moments in all columns were reduced up to 97% in case studies 1 and 2 while the values of axial forces were changed insignificantly in comparison with the benchmark. Thus, the behaviour of the structure has been changed from a moment resistance frame into a truss type structure;
- the mechanical properties of masonry units used for infilled panels influence significantly the in-plane performance of infills;
- in all cases the seismic performance of infills was related to design values of their resistance.

The results obtained in these numerical analyses highlight that seismic performance of the buildings with R.C frame structure is depending on mechanical properties of masonry used for infill panels.

IOP Conf. Series: Materials Science and Engineering 1138 (2021) 012030 doi:10.1088/1757-899X/1138/1/012030

References

[1] Dautaj A, Kadiri Q and Kabashi Q 2018 Experimental study on the contribution of masonry infill in the behaviour of RC frame under seismic loading *Eng. Struct.* **165** 27-37
[2] Villaverde R, 1997 Seismic design of secondary structures: state of the art *J. Eng. Struct.***123** (8) 1011-19
[3] Filiatrault A and Sullivan T 2014 Performance-based seismic design of non-structural building components: The next frontier of earthquake engineering *Earthq. Eng. Eng. Vib.* **13** 17-46
[4] De Risi M T, Gaudio C D, Ricci P and Verderame G M 2018 In-plane behaviour and damage assessment of masonry infills with hollow clay bricks in RC frames *Eng. Struct.* **168** 257-275
[5] Furtado A, Rodrigues H, Arede A, Varum H, Grubisic M and Sipos T K 2018 Prediction of the earthquake response of a three-storey infilled RC structure *Eng. Struct.* **171** 214-235
[6] Peng Q, Zhou X and Yang C 2018 Influence of connection and constructional details of masonry-infilled RC frames under cyclic loading *Soil Dyn. Earthq. Eng.* **108** 96-110
[7] Di Trapani F, Giordano L, Mancini G and Malavisi M 2020 Modelling failure analysis of RC frame structures with masonry infills under sudden column losses *Proc. Struct. Integr.* **26** 393-401
[8] Holmes M 1961 Steel frames with brickwork and concrete infilling *Proc. of the ICE* **19** 473-478
[9] Smith B S 1967 Methods for predicting the lateral stiffness and strength of multi-storey infilled frames *Build. Sci.* **2** 247-257
[10] Mainstone R J 1974 *Supplementary note on the stiffness and strength of infilled frames* (Garston: Building Research Station) CP 13/74
[11] Pennelis A J and Kappos A J 2019 *Earthquake resistant concrete structures* (Boca Raton: CRC Press Press)
[12] Paulay T and Priestley M J N 1992 *Seismic design of reinforced concrete and masonry buildings* (New York: John Wiley & Sons, Inc.)
[13] P100-1/2013 Seismic design code-Part 1: Design rules for buildings (in Romanian), Bucharest
[14] SR EN 1992-1:2004 -Design of concrete structures-Part 1: General rules and rules for buildings. National annex (in Romanian), Bucharest
[15] SAP2000 educational version
[16] Perrone D, Leone M and Aiello M A 2017 Non-linear behaviour of masonry infilled RC frames: Influence of masonry mechanical properties *Eng. Struct.* **150** 875-891

IOP Conf. Series: Materials Science and Engineering 1138 (2021) 012031 doi:10.1088/1757-899X/1138/1/012031

Calculation of disc friction power in pumps at operation on viscoelastic fluids

V Morozov[1*] **and A Morozov**[1]

[1] Southwest State University, Kursk, Russia

* natalybredikhin@yandex.ru

Abstract. The flow of viscoelastic fluid in the gap between the impeller disc and the pump wall for laminar flow mode is considered. Waste water sludge (WWS) was used as a working fluid. Its flow is described by viscoplastic fluid model. The results of theoretical and experimental studies of disk friction at WWS are presented and formulas are proposed for calculation of disk friction power between wheel disks and pump body.

1. Introduction

There are inevitable and high energy consumption for friction of impellers disks against liquid at operation of centrifugal pumps on any liquid [1 - 3]. This is especially for viscous and viscous plastic liquids. Disk losses can be up to 30% of energy consumption in a total pump energy balance [4, 5].

There are dependencies to determine the disc friction coefficient, confirmed by numerous experimental studies for Newtonian viscous liquids. There are formulas for determining the coefficient of disk friction not taking into account the second rheological factor - limit shear stress. These formulas are used for viscoelastic fluid [5, 6].

2. Calculation of disk friction power between the wheel disk and the pump body

Let us consider the flow of viscoplastic fluid in the gap between the impeller disc and the pump wall for laminar flow mode.

As it is known, disc losses depend on the flow mode, i.e. on the Reynolds number in the gap, gap width, the amount of leakage through the gap, the surface area contacting the liquid and the amount of roughness [7].

In our case, the flow state will be determined by rheological properties, i.e. depends on two similarity criteria: Reynolds number and Saint-Venant number.

The effect of disc roughness in laminar mode does not depend on roughness. The effect of overflows has not been studied in our case because there was no fluid overflows as a result of the use in the installation of Patent Decision No. 139427 "Shaft Seal" [7].

The effect on the disc friction coefficient of rheological properties of viscoplastic fluid and gap width between the wheel disc and the pump body was thus studied.

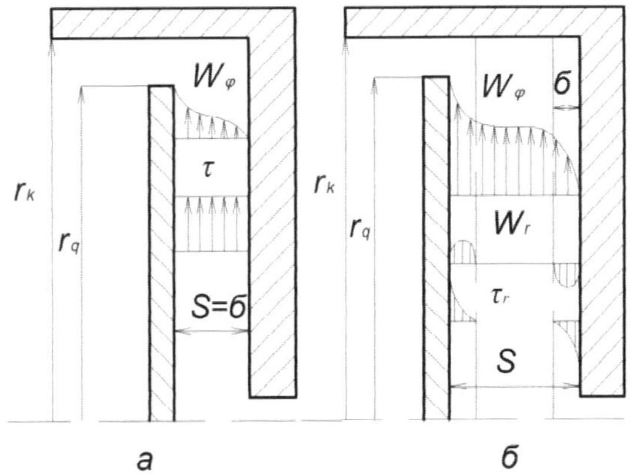

Figure 1. Flow diagram between disc and the body where W_r is the radial velocity projection; W_φ is the circular (conditionally shown in the same plane) velocity projection; δ is the boundary layer thickness; S is the gap width; τ is the shear stress; τ_m is the yield strength; r_q is the disc's radius;

Figure 1 shows the flow diagram in the gap between the pump body and the impeller disc. Figure 1 a shows velocity distribution in the narrow gap between the disc and the body. In this case we have Cuett's laminar flow with linear velocity profile. Distribution of circumferential velocities is calculated by the following formula:

$$w_r = \omega r = \frac{S}{\mu}(\tau - \tau_m) \tag{1}$$

in case of $S \ll \delta$, where μ is viscosity and ω is the angular velocity.

In this case, tangential stress at the distance from the axis is calculated by the following relation:

$$\tau_\varphi = \tau_m + \frac{u\mu}{\delta} \tag{2}$$

where u is the circumferential velocity.

As the power of disc friction is calculated by following formula:

$$N = \frac{1}{4}\rho\omega^3 {r_q}^5 c \tag{3}$$

where ρ is the density, r_q is the disk's radius and c is disk friction coefficient.

So c disk friction coefficient is:

$$c = \frac{4N}{\rho\omega^3 {r_q}^5} \tag{4}$$

Capacity of the disc friction of one side of pump disc is calculated by following formula:

$$N = 2\pi \int_0^r \tau_\varphi\, r_q w_r d_r = \frac{\pi}{2}\omega^2 {r_q}^4 (1 + \frac{4\tau_m \delta}{3u\mu})\frac{\mu}{\delta} \tag{5}$$

So the disc friction coefficient is calculated as:

$$C = \frac{2\cdot\pi\cdot\omega^2\cdot {r_q}^4\cdot\left(1+\frac{4}{3}\cdot\frac{\tau_m\cdot\delta}{u\cdot\mu}\right)\cdot\frac{\mu}{\delta}}{\rho\cdot\omega^3\cdot {r_q}^5} = \frac{2\cdot\pi\cdot\left(1+\frac{4}{3}\cdot\frac{\tau_m\cdot\delta}{u\cdot\mu}\right)\cdot\mu}{\rho\cdot\omega\cdot r_q\cdot\delta} = \\ = \frac{2\cdot\pi\cdot r_q}{R_e\cdot\delta}\left(1+\frac{4}{3}Se\right) = \frac{2\cdot\pi\cdot r_q}{R_e^*\cdot\delta} \tag{6}$$

Where $R_e = \rho\cdot u\cdot r / \mu$ is the Reynolds number and $S_e = \tau_m\cdot\delta / u\cdot\mu$ is the Saint-Venant number.

There are separate boundary layers on the disc and the pump body with sufficiently large gaps between the pump body and the disc wheel [6, 7].

There will also be the movement of a fluid in separate boundary layers as in the first case. The movement is on the disc from the center to disc's periphery and on the wall of the body is to the center.

We will consider laminar flow of viscoplastic fluid at sufficient gaps exceeding thickness of boundary layer as circulating with interlayer between boundary layers. This interlayer rotates as a rigid body at an angular velocity less than disc's angular velocity. This is for viscous fluids.

Figure 1 b shows flow diagram at sufficient clearance and velocity distribution diagram. We integrate the known equations of the boundary layer of Newtonian liquid with substitution of Shvedova-Bingam formula into them. It is done to derive the formula of disc friction coefficient:

$$\begin{cases} \rho\int_0^\delta W_r\cdot\frac{dW_r}{dr}\cdot dz + \frac{\rho}{r}\int_0^\delta W_\varphi^2\cdot dz = \int_0^\delta \frac{\partial P}{\partial z}\cdot dz - \tau_m \\ \rho\int_0^\delta W_r\cdot\frac{\partial W_\varphi}{\partial r}\cdot dz + \rho\int_0^\delta W_z\cdot W_\varphi\cdot dz = \tau_\varphi \end{cases} \tag{7}$$

As a result of integration, we determine tangential stresses on the disc. Pressure gradient is calculated from circumferential velosity of interlayer rotation.

Speed profiles on disc and fixed wall of the body are calculated taking into account the fact that thickness of boundary layer does not depend on disc's radius and maximum radial speed depends linearly on radius.

Frictional power of the disc against viscoplastic fluid will be:

$$N = 4\cdot\pi\cdot\int_0^r \tau'_\varphi\cdot r_q\cdot\omega_r\cdot dr^2 = \frac{2\cdot\pi\cdot\mu\cdot\beta\cdot r_q\cdot(\varepsilon-1)}{\delta} + \frac{4}{3}\cdot\pi\cdot\tau_m\cdot r_q\cdot dr_q \tag{8}$$

So disk friction coefficient:

$$c = \frac{4\cdot\pi\cdot(\varepsilon-1)\cdot(3\cdot\xi+7)^3}{\sqrt{15}\cdot 20\cdot {}^1\!/_4\cdot(\varepsilon-1)^{1/4}\cdot\xi^{3/2}\cdot\sqrt{Re}}\cdot\sqrt{1+\frac{2}{3}\cdot Se\cdot\frac{\xi}{\xi-1}} \tag{9}$$

where ξ is angular rate of interlayer; $\xi = \omega / 2$ is used for practical use in calculations.

In conclusion

$$c = \frac{2{,}82}{\sqrt{Re}}\cdot\sqrt{1+\frac{4}{3}\cdot Se} = \frac{2{,}82}{\sqrt{Re^*}} \tag{10}$$

In this case analysis of obtained formulas (6) and (10) shows that disc friction coefficient depends on both the actual Reynolds number and the Saint-Venant number in the presence of both closed and separate boundary layers. Reynolds number increasing reduces disc friction coefficient and Saint-Venant parameter increasing increases it.

where
$$Re^* = \frac{Re}{1+\frac{4}{3}Se} \tag{11}$$

As it can be seen from formulae (6) and (10) both equivalent Reynolds numbers coincide. Thus it is possible to use formulae to determine the power of disc friction for Newtonian fluids substituting generalized Reynolds number and to determine disc friction coefficient.
It should be noted that equations (6) and (10) are also true for Newtonian fluid when Saint-Venant parameter is zero. Thus the relation of disc friction coefficient is universal for both classes of fluids.

3. Results of pilot studies

The installation is a pump body in which a disc mounted on the shaft can rotate [7 - 9]. The aim of experimental studies is to determine disk friction coefficient to confirm calculated relation.

Figure 2 shows the results of experimental studies of disc friction, where generalized Reynolds numbers are on axis X and values of disc friction coefficient are on axis Y. Studies are carried out at the installation where waste water sludge (WWS) is used as viscoplastic fluid.

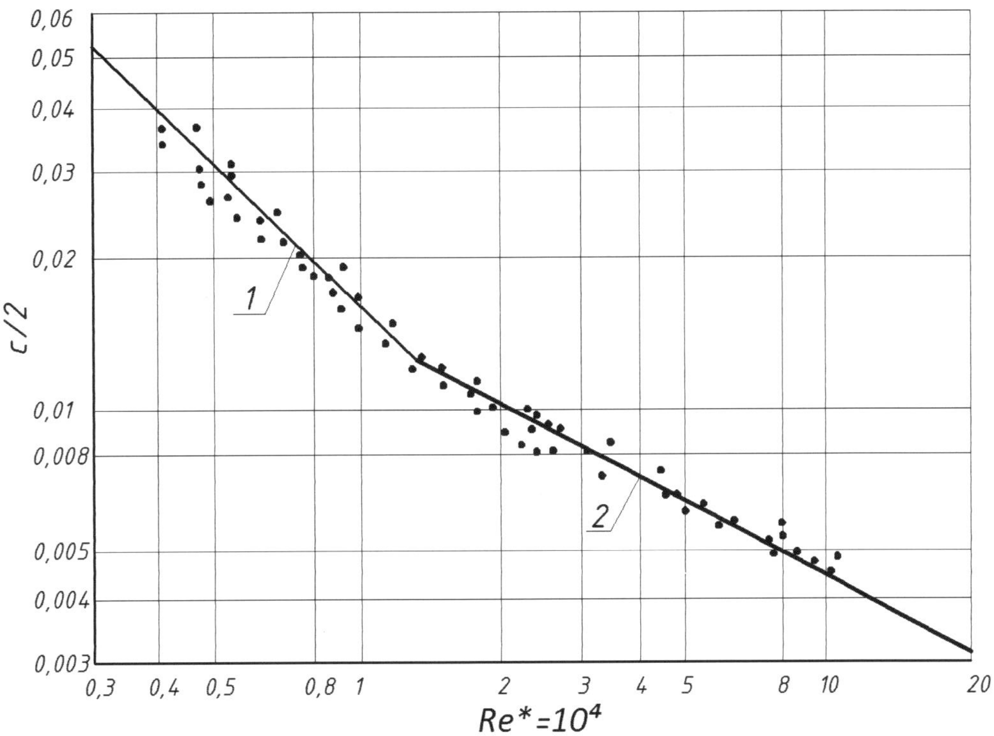

Figure 2. Comparison of experimental data and design values of disk friction coefficient for WWS.– $\begin{cases} 1 - \frac{C}{2} = \frac{\pi r_q}{Re^* S} \\ 2 - \frac{C}{2} = \frac{1.41}{\sqrt{Re^*}} \end{cases}$, where 1 - closed interface, 2 - not closed interface according to experimental results.

Test data match up the calculation data according to equation (9) for Reynolds numbers less than $2 \cdot 10^4$. This fact confirms the theory of individual boundary layers closing. In this case the distribution of circumferential velocities is the same as to the distribution of velocities at Couette flow, as there is Couette laminar flow with a linear velocity profile at $R_e \leq 10^5$.

There are two separate boundary layers - on the disc and the pump body at $R_e \geq 2 \cdot 10^4$ around disc wheel at sufficiently large axial clearance.

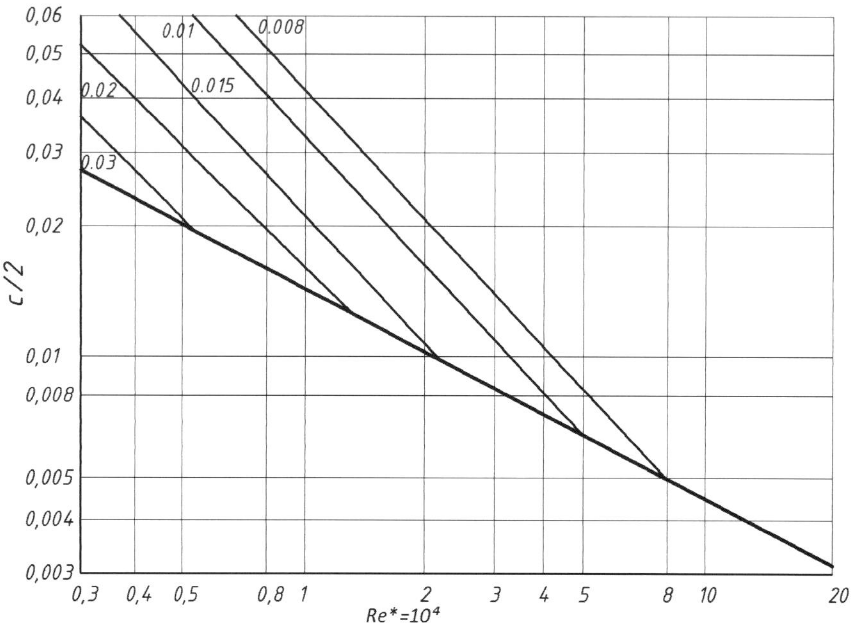

Figure 3. Dependence of disc friction coefficient on generalized Reynolds number in the gap between disc wheel and body wall.

There is a misrepresentation of experimental data when $R_e > 10^5$. An increase in a gap leads to an increase in disc friction coefficient which depends on gap width between the disc and the body. Figure 3 shows a graph of c dependence on generalized R_e^* for closed boundary layers depending on S gap width between the disc and the pump body and R_e. This graph is done to ease the calculation of disc friction coefficient.

4. Conclusions

The relations of disc friction coefficient of viscoplastic fluid for centrifugal pumps are theoretically obtained for the case of separate boundary layers and for the case of closed boundary layers for laminar flow mode. Experimental studies of disc friction in the gap between the impeller and the pump body confirm obtained theoretical dependences for the disk friction coefficient. Generalized similarity criteria are proposed for viscoplastic fluid flow in the gap between the pump impeller and its body. These criteria allow using known dependencies of disc friction coefficient for viscous fluid.

References

[1] Morozov A 2009. Calculation of the Sewage Pumping Stations parameter *Industrial and Civil Construction Engineering* **5** pp. 42-43

[2] Morozova E 2015 Design and Construction Improvement of Sewage Pumping Stations Reliability as a Factor of Improving Cities Ecological Safety *Designing and Construction: a collection of abstracts of scientific and practical conference reports* Kursk, pp. 39-40

[3] Scherbakov V 2015 Prediction of Performance Parameters of Sewage Sludge Centrifugal Pumps *X Scientific and Technical Conference. Collection of reports* Moscow ASV Publishing House, pp. 61-65

[4] Sherstyuk A 1998 Calculation of the Centrifugal Pumps Performance Parameters for viscoplastic fluids transportation *Power engineering* Minsk **1** pp. 123-125

[5] Morozov V, Morozov A, Morozova E 2017 Some Aspects of Calculation Of Joint Operation Of Sewage Sludge Pumps And Pipeline Networks *Journal of Applied Engineering Science* **15** 3 pp. 272-275

[6] Morozov A 2018 Hydraulics of water supply and water disposal systems South West State University Kursk 148 p.

[7] Morozov V 2016 The work of centrifugal pumps on viscoplastic liquids, South West State University Kursk 167 p.

[8] Polivanova T, Polivanova S, Kobelev N, Akul'shin A and Kobelev V 2015 Innovative Environment Friendly Refurbishment Technologies for Water Supply and Sewage Systems of Sugar Factories *Applied Mechanics and Materials* T. 725-726. C. 1332-1337

[9] Akulshin A, Kobelev N, Polivanova T, Kretova V and Kobelev V 2015 Designing Filtraring Equipment Forwater Wells of The Municipal Water Intakes in The City of Kursk *Applied Mechanics and Materials* T. 725-726. C. 1344-1349.

Static and dynamic plate loading tests of stabilized soil samples used for riverbank consolidation

A-Cs Nagy[1*], N M Ilieş[1], A-P Cîrcu[1], V-C Ciubotaru[2] and B M Crăciunescu[2]

[1] Technical University of Cluj-Napoca, Faculty of Civil Engineering, Romania
[2] Technical University of Petroşani, Faculty of Mines, Romania

[*] andor.nagy@dst.utcluj.ro

Abstract. Building of an experimental slope protection system on Căpuş river, near Cluj presented the opportunity of in situ testing of stabilized soils. Different hydraulic binder was used on each side of the river, so this testing series provided valuable data about deformability properties of the encountered soil types. There are some deformability parameters used in Romania which were "imported" at the specific request of foreign investors. The strain modulus, E_v, is a parameter expressing the deformation characteristics of a soil and is calculated taking values from the load-settlement curve obtained from the first and second loading cycle, on a static plate load test. In earthworks, the dynamic plate load test using the Light Falling Weight Deflectometer may be used for testing load-bearing capacity as an alternative to the static plate load test which provides the strain modulus, E_v parameter. Dynamic modulus of deformation E_{vd} is a parameter for the deformability of soil under a defined, vertical impact load with the impact duration. Its value is being calculated with the maximum settlement of the load plate. None of these methods made way into official Romanian standardisation, the E_{vd} is subject of a technical agreement, which also contains the correlation of it with E_v. Conducting both static and dynamic loading tests enabled the verification of validity of relation between strain modulus and dynamic modulus.

1. Introduction

1.1. Objective of the study

This study is part of an experimental project which was carried out on "Căpuş" river from Cluj county, by building a slope protection system on both sides of the riverbank. Riverbank stabilization was made from stabilized soils, using hydraulic binders from our partners the Lafarge Holcim Group Romania. The left bank was stabilized mixing the local soil with Doroport TB25© [1], thus building a small dam, as seen in figure 1. On the right side considering the steeper slope, a protection retaining wall was more suitable from a technologic point of view. This system was built using soil-Dorosol C30© [2] admixture. Construction of the dam and slope protection was made regarding values obtained from laboratory analysis. Hydraulic binder was spread across the width and length of the structures, then a heavy mixer added optimal moisture content and created the desired admixture on both sides of the river. Compaction works were made according to Romanian regulations [3], with successive layers of equal thickness. Figure 2 presents one step of the spreading and mixing of hydraulic binder. The construction process was described in detail in other publications of this research group [4].

Stabilized soils are commonly used in road works, but as the need for riverbank consolidation is arising, this method could be a cheap and environmentally friendly solution, because it uses local soils rather than concrete or crushed stone structures, therefore having a much lower carbon footprint. Some of the river dams from Transylvania were developed for cycling tracks (in Satu Mare for example). There are also maintenance works concerning the river bed, on which construction machinery can climb on the protection system. Consequently both systems were subjected to on site tests typically used on road beds.

Figure 1. Riverbank stabilized with Doroport TB25©.

Figure 2. Technology used for soil stabilization.

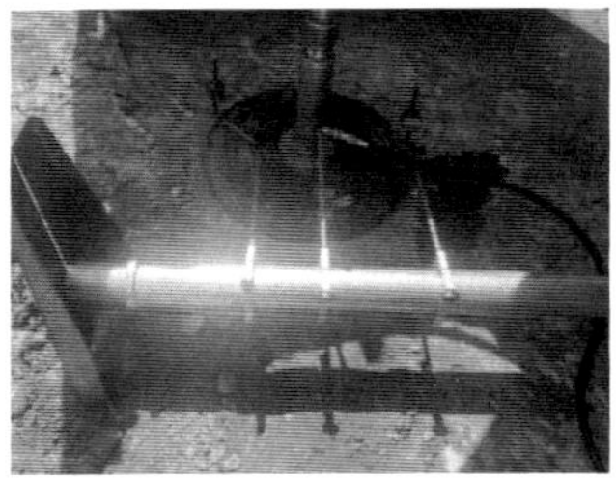

Figure 3. Static plate testing with the 300 mm Lucas plate.

1.2. Review of standards and procedures

1.2.1. State of the art. There are several studies concerning the conversion of static and dynamic load bearing capacity moduli. Tompai [5] proposed a correlation formula, and required target values. In another paper Abdulkareem [6] used Artificial Neural Network model to calculate dynamic deformation modulus, Evd and comparing with the regression statistical model. The results indicate that ANN model have the capability of predicting dynamic deformation modulus, Evd with a high degree of accuracy. Good correlations were obtained, which demonstrated that the LFWD (Light Falling Weight Deflectometer) can be reliably used to predict the modules obtained from plate load test and degree of compaction values, and hence can be used to evaluate the stiffness/strength parameters of shallow subgrade layers.

1.2.2. Static plate loading test (SPLT). This test consist in applying load in increments to a soil sample using a circular loading plate as seen in figure 3 and a loading device, then releasing the load in decrements, and the entire process being repeated. The average normal stress below the plate, σ_0, is plotted against the settlement, s, for each load increment so as to obtain a load-settlement curve.

According to DIN 18134 [7] to determine the strain modulus, E_v, the load shall be applied in not less than six stages, in approximately equal increments, until the required maximum normal stress is reached. Each increase in load (from stage to stage) shall be completed within one minute. The load shall be released in stages, to 50% and 25% of the maximum load and then to the load corresponding to the zero reading. Following that, a further (2nd) loading cycle shall be carried out, in which the load is to be increased only to the penultimate stage of the first cycle.

The strain modulus, E_v, is a parameter expressing the deformation characteristics of a soil, and is calculated taking values from the load-settlement curve obtained from the first and second loading cycle, from the gradient of the secant between points $0.3 \cdot \sigma_{0max}$ and $0.7 \cdot \sigma_{0max}$.

Calculation of the strain moduli of the first and the second loading cycle shall be based on smooth load - settlement curves. These shall be expressed by calculating the settlement, *s*, at the centre of the loading plate using the equation:

$$s = a_0 + a_1 * \sigma_0 + a_2 * \sigma_0^2 \tag{1}$$

where:

σ_0 - is the average normal stress below the plate, in MN/m^2;

a_0, a_1, a_2 - are factors, in mm/MN^2/m^4, which consider different sums of total and partial settlement, respectively sums and sum of products of applied pressure, and the multiplication of pressure by settlement;
For determining the factors, a value of *s* equal to zero shall be ignored.
The strain modulus, E_v, in MN/m^2, shall be calculated using the following equation:

$$E_v = 1.5 * r * \frac{1}{a_1 + a_2 + \sigma_{0\,max}} \quad (2)$$

where:
r - is the radius of the loading plate, in mm;
$\sigma_{0\,max}$ - is the maximum average normal stress, in MN/m^2;
As the testing procedures describe three stages, a loading stage, an unloading stage and a reloading stage we can describe two values for the strain modulus value, distinct for every loading stage: E_{v1} and E_{v2}.

1.2.3. Dynamic plate loading test (DPLT). Dynamic test procedures have been developed in various countries to enable fast testing of constructed layers of earth under load conditions approximating those imposed by road traffic. The following described dynamic plate load test differs from the static plate load test (according to DIN 18134 [7]) in that the load is generated by a damped impact [8]. The impact activates forces of inertia in the soil and in the tester, and these have an effect on the movements which have been generated. In analogy to the static plate load test, also for the dynamic plate load test, a load plate with a 300 mm diameter is used, as seen in figure 4.

The dynamic plate-load test is suitable to determine the bearing capacity and the reached compaction quality of soils in underground/substructure in earthwork and road construction.
Dynamic modulus of deformation E_{vd} is a parameter for the deformability of soil under a defined, vertical impact load with the impact duration t_{max}. Its value is being calculated with the maximum settlement s_{max} of the load plate as follows according to the formula:

$$E_{vd} = 1.5 * r * \frac{\sigma_{max}}{s_{max}} \quad (3)$$

where:
s_{max} - average value of the settlements s_{max1}, s_{max2}, s_{max3} out of 3 measuring impacts (after 3 preloading impacts);
r - is the radius of the loading plate, in mm;
σ_{max} - normal stress under the load plate (0.1 MN/m^2);

Figure 4. Dynamic plate loading test equipment.

IOP Conf. Series: Materials Science and Engineering 1138 (2021) 012032 doi:10.1088/1757-899X/1138/1/012032

1.2.4. Correlation between testing procedures. The situation of having two different types of testing procedures, based on different principles of load application, and one of them being significantly shorter and more convenient than the other led to the necessity of finding a relation between the strain modulus, E_v, and the dynamic modulus of deformation E_{vd}. A technical agreement [9] was put in practice, which contained the technical sheet of the equipment [10], defining the link between the two parameters in the following relationship:

$$E_{v2} = 600 * ln\frac{300}{300-E_{vd}} \quad (4)$$

where:
E_{v2} - strain modulus from the second loading cycle of the static plate loading test;
E_{vd} - dynamic deformation modulus from the dynamic plate loading test.

2. Materials and methods

2.1. In situ conditions

Two different soil types were encountered on the sides of the riverbank. This fact led to using two different hydraulic binders, after performing a series of geotechnical analysis. Stabilization effects on shear strength parameters (cohesion, internal friction angle), and oedometric modulus were presented in previous papers [11]. Freeze-thaw resistance and testing also made the object of a separate study [12]. Doroport TB25© was used on the left side, because the soil there was identified as sandy clay, and Dorosol C30© was mixed with the soil from the right side which had a higher content of sand, being described as a clayey sand. From an economic point of view 3% and 5% hydraulic binder content was considered.

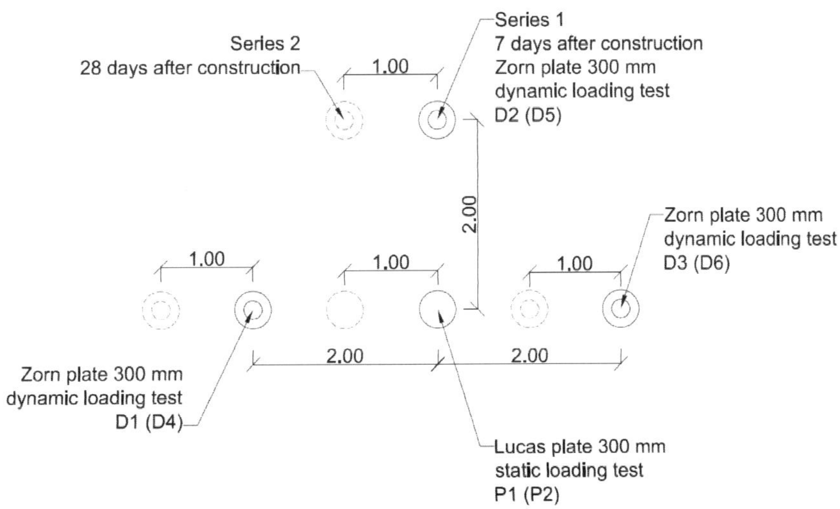

Figure 5. Testing point displacement on site.

Stabilized soil needs a curing time in order to reach its full mechanical parameters. In situ testing was scheduled at 7 and 28 days after construction. A test point for static plate loading test was selected on each side, roughly on the center of the embankments. To this test point 3 dynamic loading test points were added, at a distance of 2.00 m measured from the center of the 300 mm Lucas plate from the static loading test, as seen in figure 5. The testing points were marked with colored spray in order to find them for the second set of determinations, which took place 21 days after the first set. It was considered, that testing on the very same spot would have vitiated the results, so the whole system of points was shifted 1.00 m to the left on both sides of the embankment.

2.2. Static plate loading test (SPLT)

Performing the static plate loading test lead to the results presented in figure 6 and 7 for Point P1, located on the left side (blue points - first loading cycle, green points - unloading cycle, red points - second loading cycle):

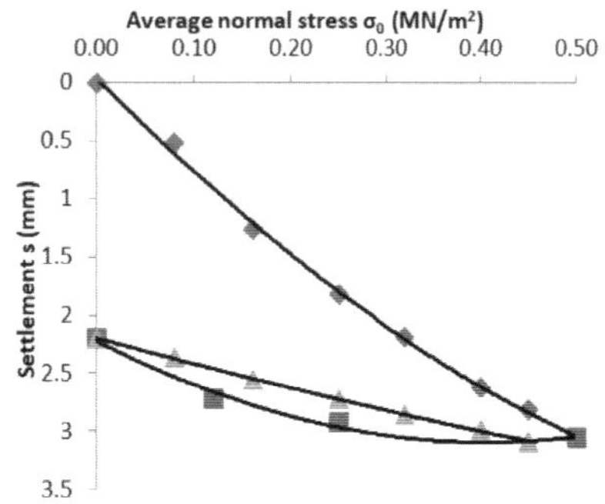

Parameter	First loading cycle	Second loading cycle
σ_{0max} (MN/m^2)	0.500	0.450
a_0 (mm)	-0.150	2.197
a1 (mm/(MN/m^2))	9.362	2.294
a2 (mm/(MN^2/m^4))	-6.039	-0.725
Plate radius (mm)	150	150
E_v (MN/m^2)	**35.5**	**116.5**
E_{v2}/E_{v1}	**3.28**	

Figure 6. SPLT test results after 7 days curing time on the left embankment, point P1.

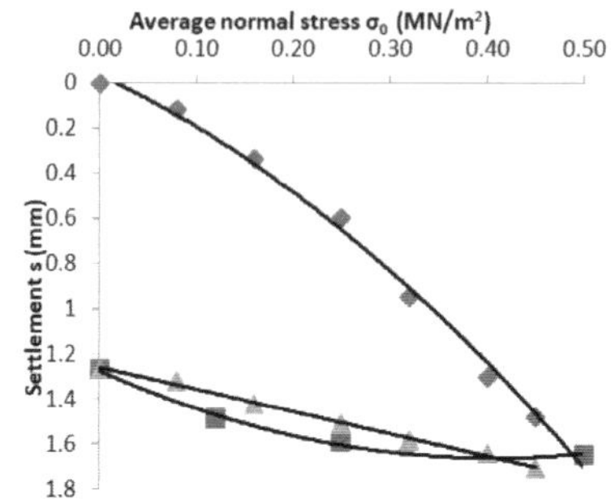

Parameter	First loading cycle	Second loading cycle
σ_{0max} (MN/m^2)	0.500	0.450
a_0 (mm)	-0.138	1.260
a1 (mm/(MN/m^2))	2.760	0.968
a2 (mm/(MN^2/m^4))	1.777	0.036
Plate radius (mm)	150	150
E_v (MN/m^2)	**61.7**	**228.2**
E_{v2}/E_{v1}	**3.70**	

Figure 7. SPLT test results after 28 days curing time on the left embankment, point P1.

The difference between strain modulus is notable on both first and second loading cycle. Testing on the right side provided slightly better results, as it seems sandy soil combined with Dorosol C30© is less deformable. Although initially the right embankment had slightly larger settlements, it seems that the stabilized sandy soil has greater reserves in terms of strain modulus E_{v2}, on the second loading cycle. Results and stress-settlement functions are presented in figures 8 and 9.

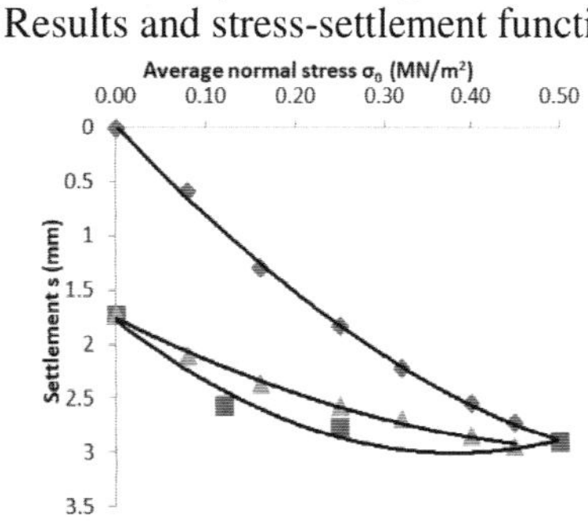

Parameter	First loading cycle	Second loading cycle
σ_{0max} (MN/m^2)	0.500	0.450
a_0 (mm)	-0.110	1.758
a1 (mm/(MN/m^2))	9.636	4.189
a2 (mm/(MN^2/m^4))	-7.332	-3.588
Plate radius (mm)	150	150
E_v (MN/m^2)	**37.7**	**93.9**
E_{v2}/E_{v1}	**2.49**	

Figure 8. SPLT test results after 7 days curing time on the right embankment, point P2.

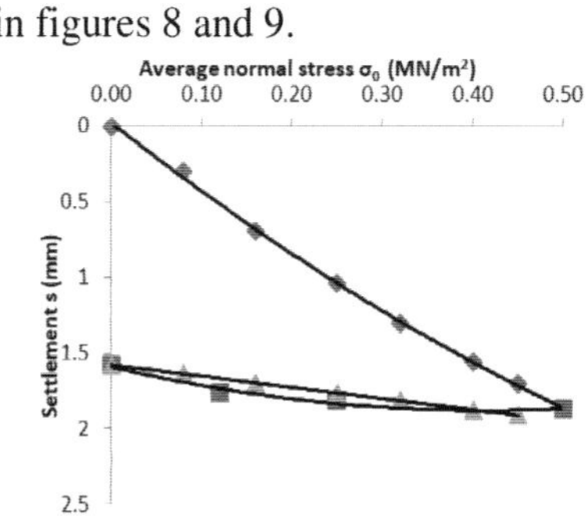

Parameter	First loading cycle	Second loading cycle
σ_{0max} (MN/m^2)	0.500	0.450
a_0 (mm)	-0.084	1.578
a1 (mm/(MN/m^2))	5.158	0.738
a2 (mm/(MN^2/m^4))	-2.586	-0.017
Plate radius (mm)	150	150
E_v (MN/m^2)	**58.2**	**308.6**
E_{v2}/E_{v1}	**5.30**	

Figure 9. SPLT test results after 28 days curing time on the right embankment, point P2.

IOP Conf. Series: Materials Science and Engineering 1138 (2021) 012032 doi:10.1088/1757-899X/1138/1/012032

2.3. *Dynamic plate loading test (DPLT)*

The results of the DPLT tests performed on the left side are shown in figures 10, 11, 12:

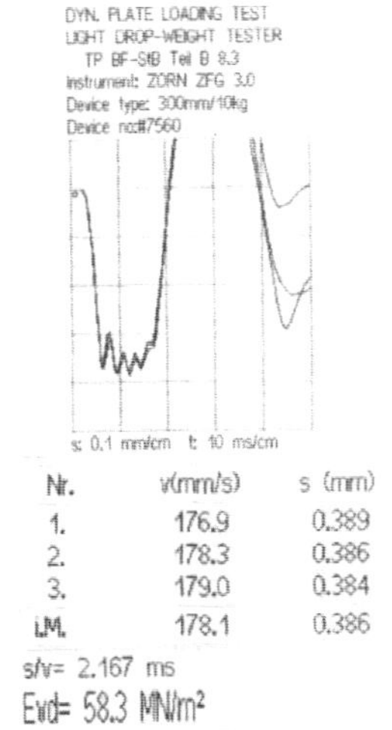

Figure 10. DPLT test results after 7 days curing time on the right embankment, point D1.

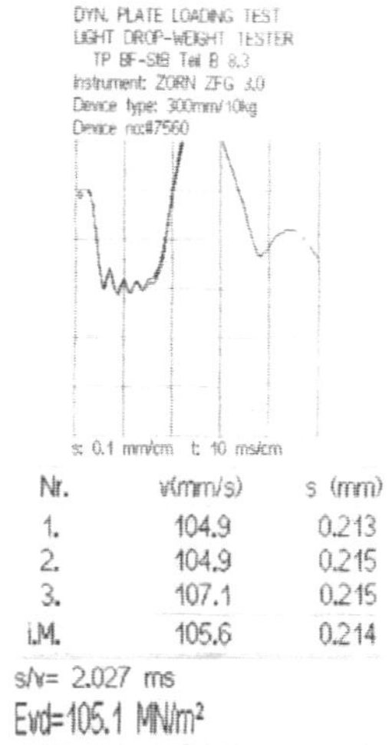

Figure 11. SPLT test results after 7 days curing time on the right embankment, point D2.

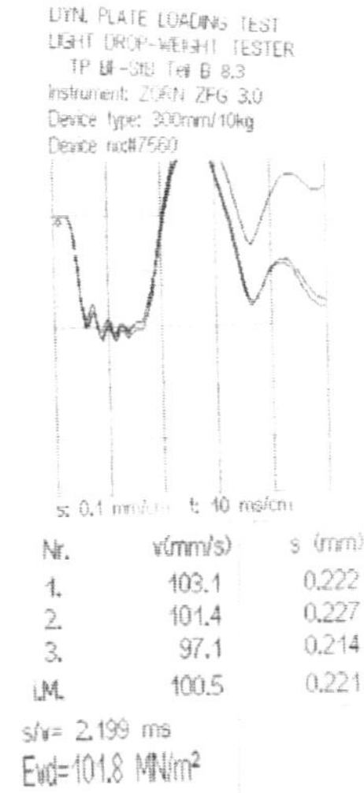

Figure 12. SPLT test results after 7 days curing time on the right embankment, point D3.

Point D1 was probably set on a zone which wasn't homogeneously mixed, this being the cause of the difference between the resulting dynamic modulus of deformation E_{vd}. Due to the execution technology which requires a chained mixer which follows a spreader device the presence of such points on top layers is inherent. The set of tests performed after 28 days from construction showed an unquestionable improvement of the dynamic modulus of deformation E_{vd}, on all three zones (considered after shifting points 1.00 m leftward). The results are presented in figures 13, 14, 15.

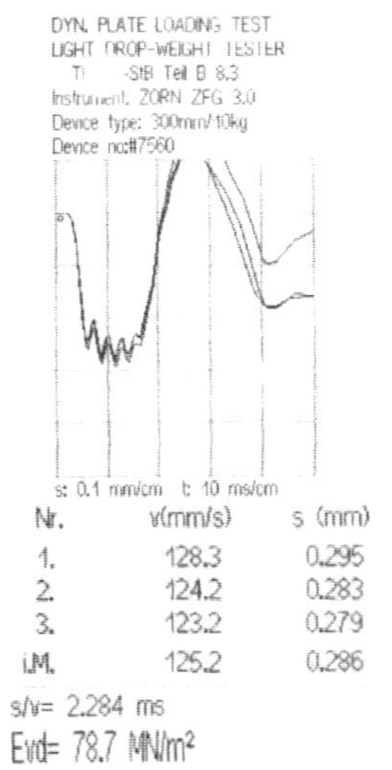

Figure 13. DPLT test results after 7 days curing time on the left embankment, point D1+1.00 m.

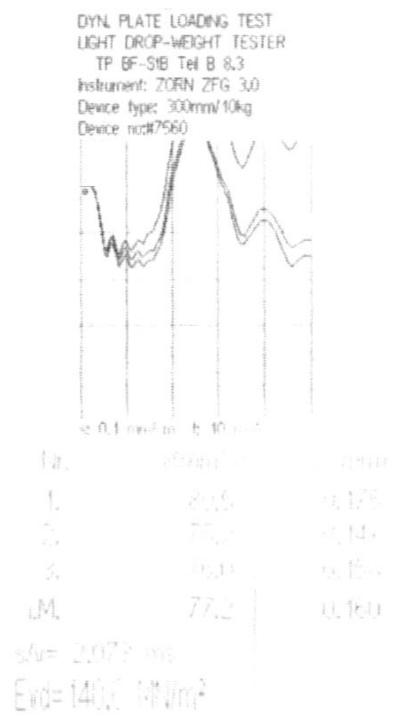

Figure 14. DPLT test results after 28 days curing time on the left embankment, point D2+1.00 m.

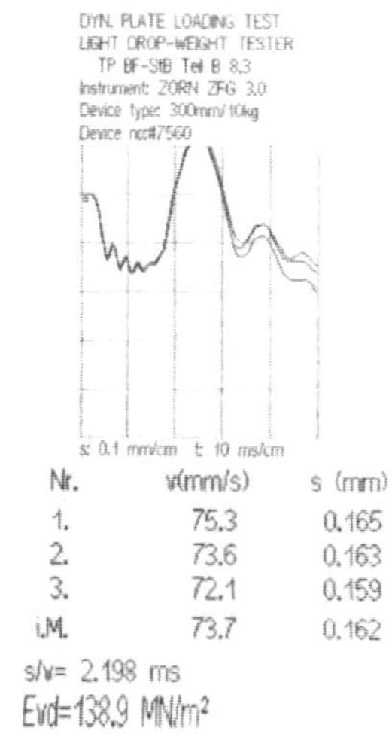

Figure 15. DPLT test results after 28 days curing time on the left embankment, point D3+1.00 m.

IOP Conf. Series: Materials Science and Engineering 1138 (2021) 012032 doi:10.1088/1757-899X/1138/1/012032

Testing procedures were identical for the right embankment, and the results are presented in figures 16, 17, 18:

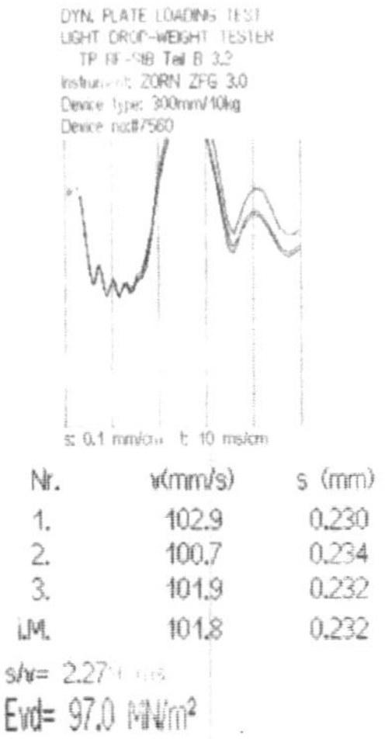

Figure 16. DPLT test results after 7 days curing time on the right embankment, point D4.

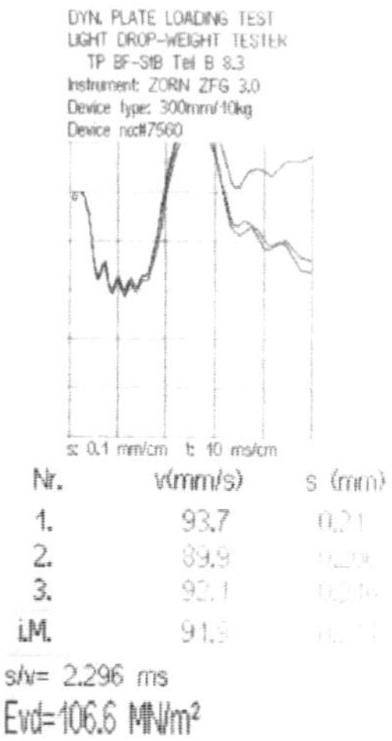

Figure 17. DPLT test results after 7 days curing time on the right embankment, point D5.

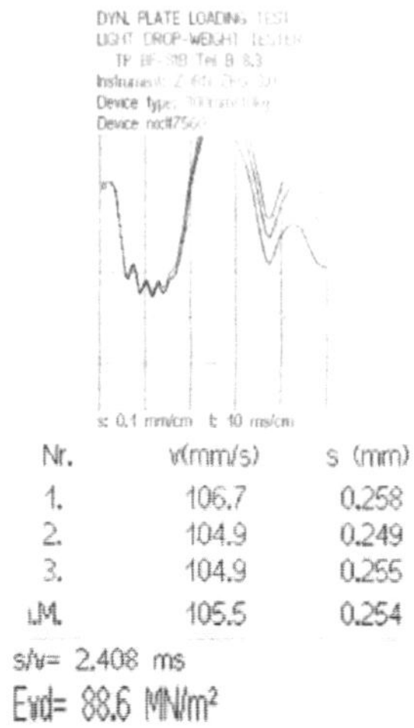

Figure 18. DPLT test results after 7 days curing time on the right embankment, point D6.

A weaker zone is also present on the right side, on the zone around point D6. The values of the set of tests performed after 28 days show the same evolution as in the case of the left embankment. The results are presented in figures 18, 19, 20.

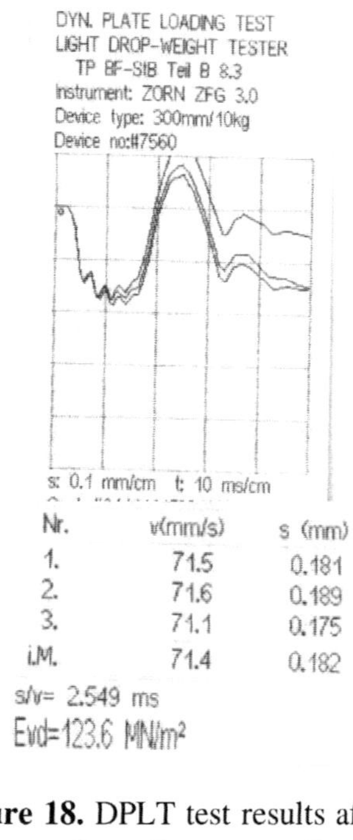

Figure 18. DPLT test results after 7 days curing time on the left embankment, point D4+1.00 m.

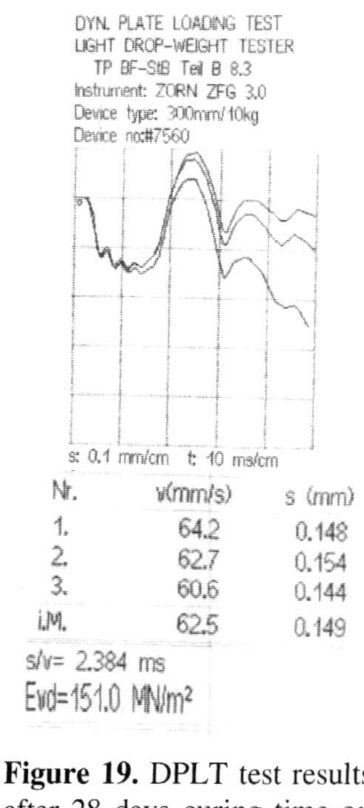

Figure 19. DPLT test results after 28 days curing time on the left embankment, point D5+1.00 m.

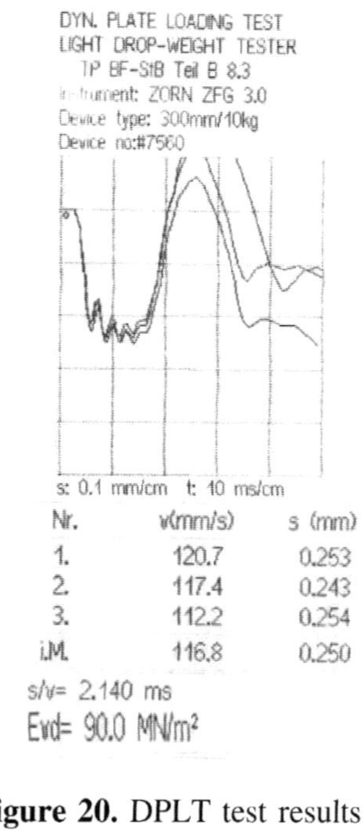

Figure 20. DPLT test results after 28 days curing time on the left embankment, point D6+1.00 m.

IOP Conf. Series: Materials Science and Engineering 1138 (2021) 012032 doi:10.1088/1757-899X/1138/1/012032

2.4. Correlation between strain modulus (E_{v2}) and dynamic modulus of deformation (E_{vd})
The results obtained in situ permit a verification of relation (4) for the test performed at 7 and 28 days after stabilization.

2.4.1. Point P1 - left embankment, after 7 days

Table 1. Measured and deducted strain modulus (E_{v2}).

	E_{vd}	E_{v2} from relation	E_{v2} measured on site
	[MN/m^2]	[MN/m^2]	[MN/m^2]
D1	58.30	129.65	
D2	105.10	258.78	116.50
D3	101.80	248.70	
Average	88.40	209.45	

2.4.2. Point P1 - left embankment, after 28 days

Table 2. Measured and deducted strain modulus (E_{v2}).

	E_{vd}	E_{v2} from relation	E_{v2} measured on site
	[MN/m^2]	[MN/m^2]	[MN/m^2]
D1+1.00	78.70	182.56	
D2+1.00	140.60	379.42	228.20
D3+1.00	138.90	373.05	
Average	119.40	304.50	

2.4.3. Point P2 - right embankment, after 7 days

Table 3. Measured and deducted strain modulus (E_{v2}).

	E_{vd}	E_{v2} from relation	E_{v2} measured on site
	[MN/m^2]	[MN/m^2]	[MN/m^2]
D4	97.00	234.35	
D5	106.60	263.41	93.90
D6	88.60	210.02	
Average	97.40	235.53	

2.4.4. Point P2 - right embankment, after 28 days

Table 4. Measured and deducted strain modulus (E_{v2}).

	E_{vd}	E_{v2} from relation	E_{v2} measured on site
	[MN/m^2]	[MN/m^2]	[MN/m^2]
D4+1.00	123.60	318.62	
D5+1.00	151.00	419.90	308.6
D6+1.00	90.00	214.00	
Average	121.53	311.62	

3. Results and discussion

Results for deducted and measured strain modulus from point 2.4 are quite different. This seems to confirm the assumption from the German standard [4], which states at point 1, Field Application and Purpose: "The application of the dynamic plate load test is not permitted for the dynamic modulus of deformation E_{vd} above 70 MN/m^2, because in this range the LDWT can not be calibrated sufficiently". This statement, however, doesn't appear in the user manual [8], nor in the technical agreement [7]. As we can see, having values consistently above 70 MN/m^2, by applying relation (4) one dangerously over-estimates the value of the strain modulus E_{v2}, from the second loading cycle. The relationship between the values of the two parameters is interesting even considering that testing points where the dynamic deformation modulus was below or slightly above the specified limit value of 70 MN/m^2. On point D1, from the left embankment, E_{vd} is 58.30 MN/m^2, resulting an E_{v2} of 129.65 MN/m^2. The measured E_{v2} in the same zone is 116.50 MN/m^2, 10% below the calculated value. Considering another point, D1+1.00 m E_{vd} is 78.70 MN/m^2, resulting an E_{v2} of 182.56 MN/m^2. The measured E_{v2} in the same zone is 228.20 MN/m^2, 25% above the calculated value. On the right embankment all results after 7 days curing time seem irrelevant, determined and measured values for strain modulus being far in between. Nonetheless the average value of the three points is very close to the measured E_{v2} strain modulus. The comparison of the two test types aimed to study the validity of the correlation, when high values of dynamic deformation modulus are obtained, through soil stabilization with hydraulic binder. As the results from point 2.4 show, the relation (4) is mostly non-applicable.

4. Conclusions

Studying the results obtained with the SPLT, two conclusions can be drawn. On one hand there is a considerable improvement between strain moduli (E_{v2}) obtained on both sides of the riverbank, between the tests conducted after 7 days from construction date, and 28 days respectively. On the left side which was stabilized with Doroport TB25© the strain modulus on the second loading cycle increased from 116.5 MN/m^2 to 228.2 MN/m^2, an addition of 49%. This tendency was even greater on the right side, where Dorosol C30© was used. The E_{v2} parameter went from 93.9 MN/m^2 to 308.6 MN/m^2, a growth of almost 70%. Therefore, both stabilized layers mobilize important deformability resistance reserves through their curing time.

This can also be seen in terms of settlement. On point P1, the average final settlement on the second loading cycle varied from 2.94 mm after 7 days to 1.91 mm after 28 days. Across the river, on point P2 this difference was 3.09 mm after 7 days, and 1.71 mm after 28 days. In both cases the settlement from the static plate loading test decreased about 30-40% on the given time interval.

Although proven as unreliable for determining the strain modulus values, the dynamic plate tests revealed high values for the E_{vd} parameter, peaking the usual values obtained in road sub-bases. A clayey or sandy soil as found in its natural state on both embankments wouldn't have a value above 40 MN/m^2, so hydraulic binder stabilization was a strong input in these structures. If this technology is chosen for riverbank consolidation, there is no problem in further using the structure for cycle tracks or maintenance works, as described in the introduction.

Acknowledgments

The authors would like to thank our colleagues Adrian Dârgău, Anca Gheorghe and Daniel Stănică from the Lafarge Holcim Group Romania for all their support during the execution phase, and for the documentation provided.

References

[1] Doroport – technical sheet
[2] Dorosol – technical sheet
[3] GE-026-97 Ghid pentru execuţia compactării în plan orizontal şi înclinat a terasamentelor
[4] Ciubotaru V C, Cîrcu A P, Crăciunescu B M, Nagy A Cs, Topan G and Dârgău A 2019 Metode de apărare împotriva inundaţiilor. Sector experimental pe râul Căpuş Revista Construcţiilor

nr. 156, March 2019, p. 58-59

[5] Tompai Z 2008 *Conversion between static and dynamic load bearing capacity moduli and introduction of dynamic target values, Periodica Polytechnica, Civil Engineering 52/2 (2008) 97–102*

[6] Abdulkareem A H 2018. *Assessment of Relationship Between Static and Dynamic Load Using Regression Analysis and Artificial Neural Network Model, Soil Testing, Soil Stability and Ground Improvement, Sustainable Civil Infrastructures, DOI 10.1007/978-3-319-61902-6_21*

[7] DIN 18134 – Deutsche Norm. Determining the deformation and strength characteristics of soil by the plate loading test.

[8] TP BF-StB Part B 8.3 – Technical testing regulations for soil and rock in road constructions. Dynamic Plate Load Testing with the Light Drop-Weight Tester

[9] Agrement Tehnic nr. 004-07/935-2007 – Echipament ZORN ZFG 02 pentru determinarea caracteristicilor mecanice "in situ"

[10] User Manual for the Light Weight Deflectometer (LWD) ZFG 3.0 according to German standards Technical Test Code for Soils and Rock Mechanics in Road Construction TP BF-StB Part B 8.3

[11] Nagy A Cs, Cîrcu A P, Ciubotaru V C, Crăciunescu B M, Topan G, Chereches T and Fărcaş V S 2019 *Applications of stabilized soil in hydraulic constructions* (Conference paper, in course of publication)

[12] Nagy A Cs, Cîrcu A P, Ciubotaru V C, Crăciunescu B M, Topan G, Moldovan D V, Moldovan L E and Paşca D M 2020 *Freeze-thaw testing of stabilized soil samples used for riverbank consolidation, IOP Conf. Ser.: Mater. Sci. Eng.* **789** 012044

Validation of AE tomography based on model tests of soil specimen with heterogeneous elastic wave velocity distribution

K Nakamura[1*], Y Kobayashi[1], K Oda[1], S Shigemura[1] and K Ikebata[1]

[1] Department of Civil Engineering, College of Science and Technology, Nihon University, Tokyo, Japan

* nakamura.katsuya@nihon-u.ac.jp

Abstract. In order to investigate the progressive failure of soil-structures, non-destructive inspections have been applied in recent. Use of Acoustic Emission (AE), the elastic wave which is generated by the occurrence of cracks, is attractive in this field to detail the progressive failure if the measurement is conducted with sufficiently small time interval. Although this technique potentially detects the cracks clearly, conventional methods of AE source localization consider that elastic wave velocity distribution is homogenous, and there are the limitations of its application due to the diffraction and the refraction of elastic wave if the heterogeneity of the elastic wave velocity distribution is not negligible. In this study, AE-Tomography (AET), which performs the identification elastic wave velocity distribution and the localization of AE sources with consideration of the diffraction and the refraction of elastic wave, is validated about the performance of the identification of the elastic wave velocity distribution and the AE sources localizations on numerical tests and model tests. Artificial AE events are generated in the soil specimen that have some parts of different relative density and validation of AET is conducted by using the arrival times of the artificial AE. According to the results of the validation, it is suggested that AET identifies elastic wave velocity distributions and AE source locations adequately on the heterogeneous elastic wave velocity distributions.

1. Introduction

The failure of soil-structures is well known as the progressive behavior of granular materials. However, the consideration of the progressive failure of soil-structures is difficult for the design, because the mechanism of soil particles slide and crush is unclear.

In the studies about the mechanism of the progressive failure, experimental and numerical approaches have been attempted. For instance, Digital image correlation (DIC) method is quoted as a technique that measures the deformation of a specimen from a series of images which are continuously taken. This technique has been applied to detect the evolution of shear bands [1, 2]. Although the results visualize the motion of the surface of the soil specimens, it is difficult to reveal the inside failures. DIC also requires sufficient framerate of cameras to detail motions of particles. To observe the process of inside failure of soil, X-ray CT, it reveals difference of densities and atomic compositions by X-ray absorptions, have been applied [3, 4]. Although, it detects and visualizes the progressive failure inside soil specimens, a larger space in comparison with the specimen for triaxial tests is required in the X-ray apparatus, and the apparatus should be installed in a secure area which is a barrier surrounded by the walls. Even though the technique that combines X-ray CT and DIC was proposed and it observed the correlation of surface and inside behavior of soil specimens [5], the difficulties on the use of X-Ray still remains.

Use of AE, the elastic wave which is emitted at the occurrence of the cracks and/or the frictions is attractive to reveal the mechanism of failures [6]. AE is measured by AE sensors located on the surface of targets. The measurement system is portable, and the special facilities to stop X-Ray is not required. Further, it is able to apply any specimens as far as the wave can be detected. Due to the above characteristic, AE tests have been applied to geomaterials and observe the failure of soil-structures by AE parameters [7, 8, 9]. However, the conventional methods of AE source localization, it detects the failure locations, assume that elastic waves are transmitted in homogenous materials, and it is limitedly applied on heterogeneous elastic wave velocity distributions due to the diffraction and the refraction of the elastic waves.

AET simultaneously identifies the elastic wave velocity distribution and AE source locations with the consideration of the diffraction and the refraction of elastic waves [10]. In this paper, the performance of AET is validated on the heterogenous elastic wave velocity distribution to develop a method to visualize the progressive failure in soil-structures. It is expected that AET reveals the correlation between local density and failure locations by identifying elastic wave velocity distributions and AE source locations. For the validation, numerical investigations respected for the heterogeneous soil-structure are conducted. In addition, the method is validated by model tests in which the model has similar elastic wave distributions with the numerical models for the numerical investigations to check the performance of AET on heterogeneous dried sands.

2. Acoustic Emission Tomography (AET)

2.1. Ray-Tracing

In order to consider the diffraction and the refraction of elastic wave, Ray-Tracing which have been applied to seismic tomography [11] is adopted in AET for AE source locations. The outline of Ray-Tracing is illustrated in figure 1. It is assumed that the elastic wave velocity distribution consists of the cells in which the elastic wave velocity is constant, and elastic the wave velocity is identified in each cell to reproduce the heterogeneous velocity distribution. Thus, the ray-paths are approximated as straight line on each cell. In the first step of AET, initial velocity distributions are set to a homogeneous distribution of the elastic wave velocity in soundness conditions in the target. In the second step, travel times from AE sensors to candidates of AE sources are computed by using Ray-Tracing technique. The travel time t_{ij} from the AE sensor i to the candidate j is defined as

$$t_{ij} = \sum_{k=1}^{n} S_k \cdot l_{ijk} \tag{1}$$

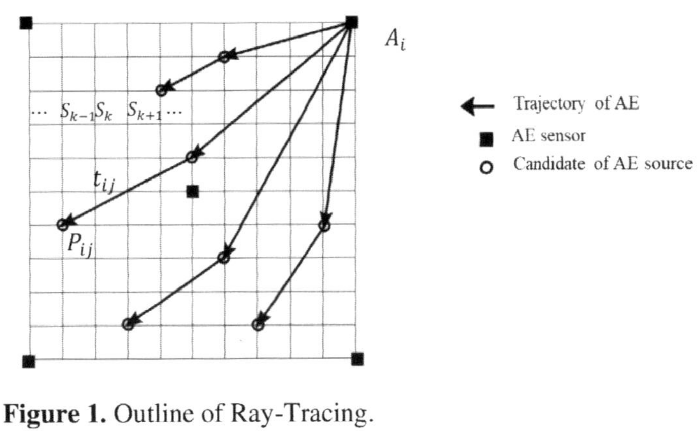

Figure 1. Outline of Ray-Tracing.

where S_k is slowness in cell k, and l_{ijk} is length of the ray-path from the AE sensor i to the candidate j in the cell k. The ray-paths are an approximated as polygonal lines which vertices are at the nodal points, and a path that gives the first travel time is sought from all of possible ray-paths on the heterogeneous velocity distribution. The possible pulse originated time is computed as subtraction of the arrival time of the first travel times. The possible pulse originated time P_{ij} is defined as

$$P_{ij} = A_i - t_{ij} \tag{2}$$

where A_i is the arrival time on the AE sensor i. The number of the estimated originated times at all candidates is same with the number of the sensors. AE source is selected as a nodal point where the variance of the originated time is minimum. The procedure is illustrated in figure 2. It should be noted that accuracies of trajectory of AE depended on the mesh in this algorithm.

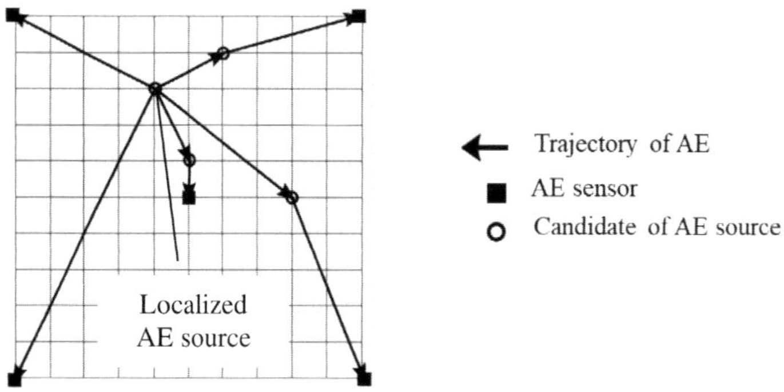

Figure 2. Localized AE sources and simulated trajectory of AE.

2.2. Identification velocity distribution

The measured travel times is obtained as the sum of the first travel time and the average of the originated times at the localized points. The velocity distribution is updated to minimize the difference of observed and computed first travel times by inverse techniques. In this study, Simultaneous Iterative Reconstruction Technique (SIRT) is adopted as the identification technique because it is simple and solving matrix equations is not required, and consequently cost is lower. According to the figure 3, the velocity distributions of the cells where ray-paths cross, are only updated. Therefore, the identification technique requires sufficient number of AE events since all of the cells should be covered by the ray-paths for updating the velocity distribution accurately. SIRT update the velocity on the cells to minimize difference of the measured and computed first travel times. In order to prevent that velocity is identified negative values, upper limitation of the velocity is applied in this study.

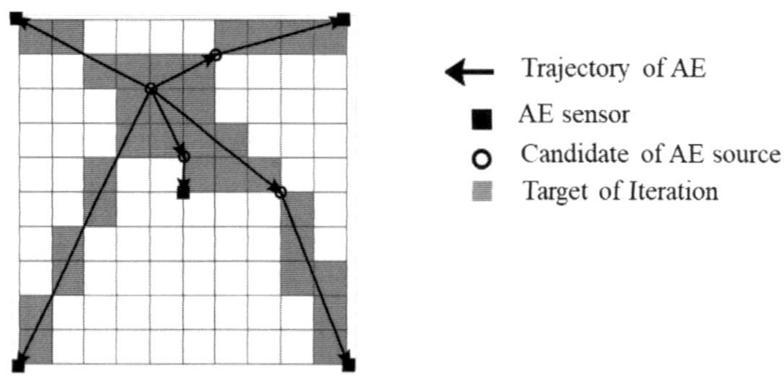

Figure 3. Targets of iteration for identification elastic wave velocity distributions.

2.3. Minimizing mesh dependency

Accuracy of AE sources localization is related with the cell size if the nodal points is associated as the candidate of the source locations. However, shrinking the cell size increases the computational cost. In this study, Lagrange interpolation is applied to interpolate AE sources localization for minimizing the mesh dependency. Interpolation function $L(u)$ is defined as

$$L(u) = \sum_{j=1}^{n} v_j \left(\prod_{\substack{i=1 \\ j \neq i}}^{n} \frac{u - u_i}{u_j - u_i} \right) \tag{3}$$

where u is coordinate of source location, and v the variance of the originated time. $L(u)$ is applied around selected AE source in horizontal and vertical direction, respectively. According to the figure 4, AE sources is localized by the vertex of $L(u)$. In order to compute the clear vertex of $L(u)$, $L(u)$ should be consisted by 3 coordinates.

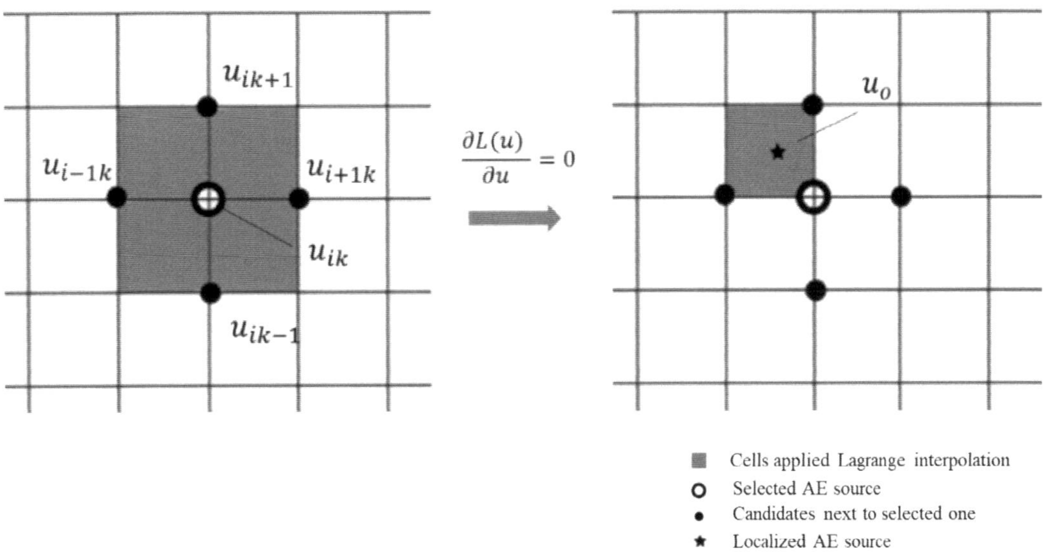

Figure 4. Outline of AE source localization in the cell.

3. Validation of AET by numerical tests

In order to validate performance of AET on heterogeneous elastic velocity distributions, a series of numerical tests is conducted.

3.1. Initial conditions

Figure 5 shows a model that is used for the numerical tests. Elastic velocity distributions on the model consist of high velocity area located in centre of the model and low velocity area on the left and the right side of the model. The elastic wave velocities in the areas are shown in table 1. The model simulates that dried sand with high relative density D_r exists in the centre. It is assumed that the artificial AE events are generated here. The low velocity is determined by referring results of preliminary experiments in which AE were measured to compute the elastic wave velocity in dried sands D_r of 88%. The preliminary experiments use Toyoura sand which is popular in experimental approaches [12, 13]. The high velocity is assumed as a velocity of Toyoura sand which is highly compacted by loads. 6 AE sensors are installed as illustrated in figure 5 around the model. The location of the sensors is related with the accuracy of the source localization because of the localization based on difference of arrival times. If the sensors are located on side of models, the iteration of AE source localization is limitedly performed because of iteration reducing the errors of the AE source localization on the specific aspects. Thus, the error of iteration is remained on AE source locations. The simulated area is divided by square meshes with a size of 1.25×10^{-2} m. Although increasing number of the cells is expected to identify detailed velocity distributions, it leads to increasing unknown quantities in SIRT and difficulty of obtaining accurate solutions, and further, calculation cost rises. In this study, 64 cells are adapted in consideration of the computation time. Initial conditions of the numerical tests are shown in table 2. In Case 1, AE source localization is conducted on the original distribution to validate the performance of AE source localization in which Lagrange interpolation is applied. In Case 2, the identification of AE source locations and elastic wave velocity distribution is performed. In order to validate the performance of Lagrange interpolation on the source localization, the interpolate function is applied in (a) and not applied in (b). Initial velocity distribution is homogenous and the velocity is the same with the upper limit of the velocity. The limit is set to 200 m/s which is the highest velocity in the original distribution. AE events is located randomly in the model. The arrival times are computed by performing Ray-trace from the generated source locations to the sensor locations on the original distribution.

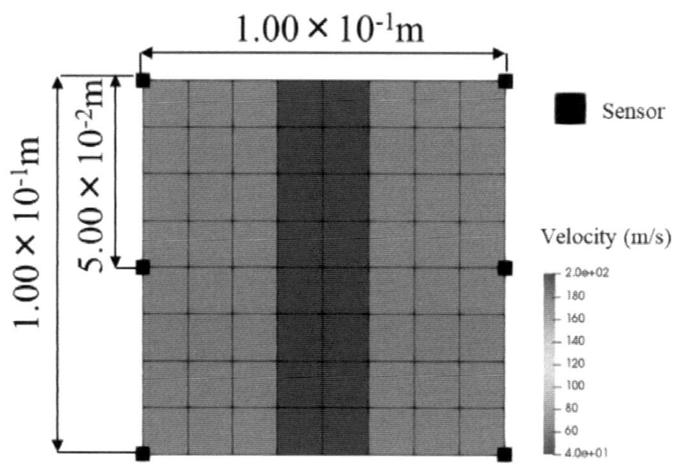

Figure 5. Original distribution.

Table 1. The elastic wave velocities in the areas.

High velocity(m/s)	Low velocity(m/s)
200.0	160.0

Table 2. Initial conditions.

Case	One side of the square (m)	Initial velocity distribution(m/s)	Number of cells	Number of events	Upper limit of velocity(m/s)	Number of sensors
Case 1	1.25×10^{-2}	-	64	1000	-	6
Case 2(a, b)	1.25×10^{-2}	200.0	64	1000	200.0	6

3.2. Results of numerical tests

According to the figure 6, it is confirmed that the source locations are identified accurately. It is noteworthy that the source locations are identified inside of the cells and this is achieved by using the interpolation function with lower computational cost. These results suggest that AE source location are accurately without installing the other candidate points, if identified elastic wave velocity distributions is identical to the real distribution. According to the table 3, the average of the errors in Case 2(a) is smaller than (b). Furthermore, results of Case 2(a) shows that the localized sources are closer to the exact sources than the results of Case 2(b) according to figure 7. The results of case 2(b) shows the source locations are identified at the nodal points and it illustrates the source locations depend on the mesh. In results of identified elastic wave velocity distributions in Case 2, the high velocity area is identified more clear at the centre of the model in (a) in comparison with (b). This suggested that interpolation function contributes to raise the accuracy of AET results and minimize the influence of the mesh dependency. The elastic wave velocity is identified lower than the original velocity in the side of mode of (a). This is caused by the difficulty of localization in the vicinity of AE sensors. The localization is conducted with updating the velocity distributions. Hence, if the source around the sensor is localized, the cells in which wave pass is few and the influence of the path is stronger related to the other path. In order to prevent above problem, the localized source around the sensor would be excluded in the identification of the elastic wave velocity distribution.

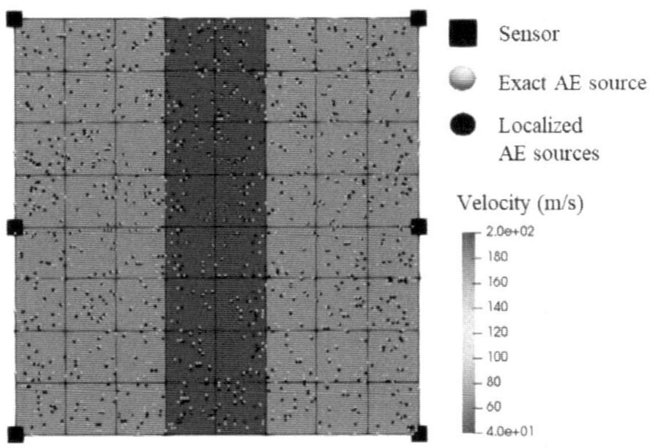

Figure 6. Result of Case 1.

Table 3. Error of AE source localization.

Case	Average of errors (m)
Case 1	9.22×10^{-4}
Case 2(a)	8.73×10^{-3}
Case 2(b)	1.14×10^{-2}

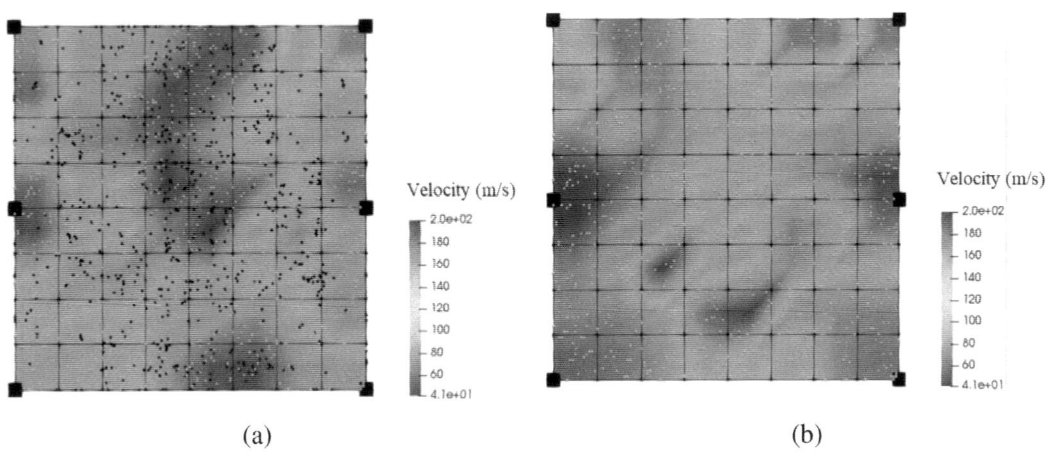

Figure 7. Result of Case 2.

IOP Conf. Series: Materials Science and Engineering 1138 (2021) 012033 doi:10.1088/1757-899X/1138/1/012033

4. Validation of AET by the model test

4.1. Testing material and apparatus

AET is validated with model tests on dried sands. The soil specimen is made by air pluviation method and consists of dried Toyoura sands which D_r is 88%. The size of the soil specimens is 4.90×10^{-1} m in length, 3.40×10^{-1} m in width, and 3.00×10^{-1} m in height. Figure 8 illustrates 6ch of the AE sensors that are located at 1.00×10^{-1} m, 1.50×10^{-1} m, and 2.00×10^{-1} m in depth. The location of the AE sensors is determined to make a sufficient margin from soil tank frame to avoid observing the waves that propagates the soil tank frame. Operating frequency of the AE sensors is 150-400 kHz. Measured AE is amplified 60 dB by the preamp and sampling frequency is 1 MHz. Therefore, arrival times include errors of minimum 10^{-6} s.

4.2. The method of generating AE and detecting arrival time

AE is generated artificially by penetrating a pile which is 5.0×10^{-3} m in length, 5.0×10^{-3} m in width, and 1.0 m in height as illustrated in figure 9. The amount penetration is 1.80×10^{-1} m by hammering and centre of the specimen is compacted as consequence. The elastic velocity distribution of the specimen is supposed to be same as the original distribution in the previous section. Measured AE is shown in figure 10 as an example, and it is separated to a noise part and a signal part, respectively. Arrival time is determined as the start of the signal part, and it is detected by visual confirmation.

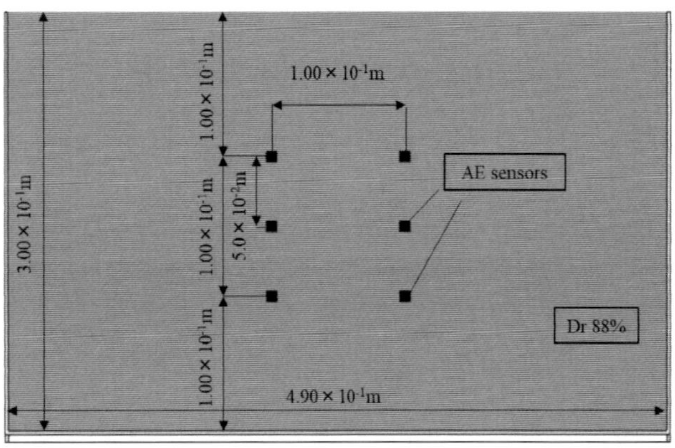

Figure 8. Dimension of the soil tank and AE sensors' locations (the sensors are located centre of width 3.4×10^{-1} m).

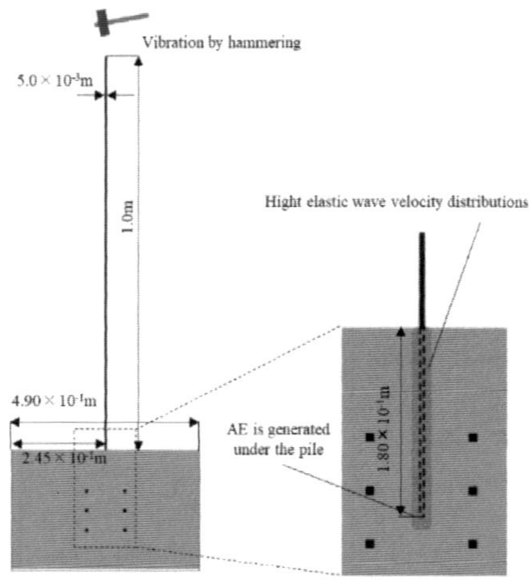

Figure 9. The method of generating AE occurred in heterogenous elastic wave velocity distributions.

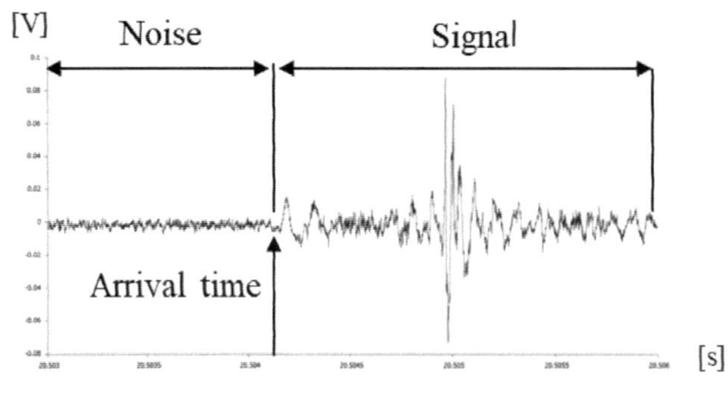

Figure 10. Example of measured AE.

4.3. Results of the model tests and AET with measured AE

According to results of detection of arrival times by visual confirmation, 4 AE events are chosen from all AE for the validation. It should be noted that the noise part and the signal part is clearly separated in the events. Other Initial conditions are the same as Case 2 in the previous section. The localized AE sources and the identified elastic wave velocity distributions are shown in figure 11 and table 4 shows the identified depth and originated time of AE events, respectively. According to the results, AE sources is localized deeper with the pulse originated time. It implies that the progress of the pile penetration is adequately detected. Further, the identified distribution estimates high velocity distributions in centre of the specimen, and it may suggest that the compression of soil structures at the top of the pile is detected. However, the identified high velocity distribution is unclear than the result of Case 2(a) in figure 7. This would be caused because the number of events is smaller than Case 2(a) and the ray-paths did not cover all of cells. Thus, it is expected that move events would be necessary to rise accuracy of the identified velocity distribution.

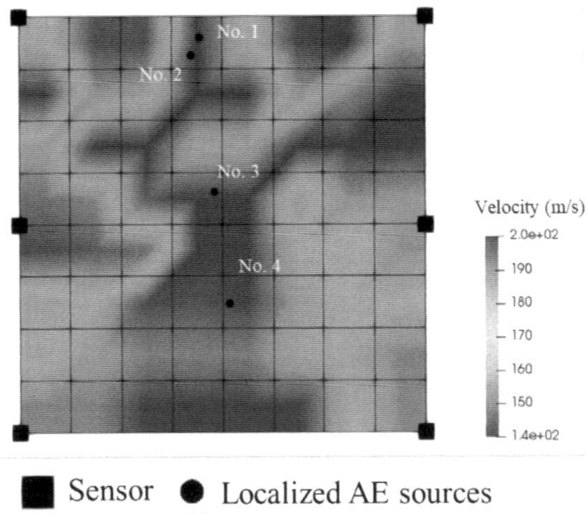

Figure 11. Results of AET with measured AE.

Table 4. Pulse originated time and depth of AE sources in the specimen.

Events	Pulse originated time (s)	Depth (m)
No. 1	20.503508	1.05×10^{-1}
No. 2	21.980975	1.09×10^{-1}
No. 3	23.263172	1.42×10^{-1}
No. 4	24.731889	1.69×10^{-1}

5. Conclusions

In this paper, AET is validated by performing the numerical investigation and the model tests on the heterogenous elastic velocity distribution to develop a method for visualizing the progressive failure in the soil-structures. The conclusions that are obtained in this study are written below.

- According to results of Case 1, it suggests that if the identified elastic wave velocity distributions is identical to the real distributions, AE source locations are accurately without installing the other candidate points.
- According to results of Case 2, it is revealed that the interpolation function rises the accuracy of AET results and lead minimizing mesh dependency.
- If the source is localized around the sensor, the cells in which the waves crossed is few and updated of the velocity is focused on these cells.
- According to results of AET for the model test, AET successfully shows the progress of the penetration of the pile and the compression of the sand around the tip of the pile. These results suggest that the progressive failure and the change of local density in soil-structures would be able to visualize by the identified elastic wave velocity distributions and AE sources.
- Updating the clear elastic wave velocity distributions as Case 2(a) is required enough events more than 4 events.

References

[1] Rechenmacher A, Abedi S and Chupin O 2010 Evolution of force chains in shear bands in sands *Ge´otechnique* **60** 5 pp 343–351 DOI: 10.1680/geot.2010.60.5.343

[2] Bhandari A R, Powrie W, and Harkness R M 2012 A Digital Image-Based Deformation Measurement System for Triaxial Tests *Geotechnical Testing Journal* **35** 2 Paper ID GTJ103821

[3] Hall S A, Bornert M, Desrues J, Pannier Y, Lenoir N, Viggiani G and Suelle P Be´ 2010 Discrete and continuum analysis of localised deformation in sand using X-ray CT and volumetric digital image correlation *Ge´otechnique* **60** 5 pp 315–322

[4] Ando` E, Viggiani G, Hall S A and Desrues J 2013 Experimental micro-mechanics of granular media studied by x-ray tomography: recent results and challenges *Ge´otechnique Letters* **3** 3 pp 142–146 http://dx.doi.org/10.1680/geolett.13.00036

[5] Takanoa D, Lenoir N, Otanic J and Hall S A 2015 Localised deformation in a wide-grained sand under triaxial compression revealed by X-ray tomography and digital image correlation *Soils and Foundations* **55** 4 pp 906-915

[6] Leia X, Masuda K, Nishizawa O, Jouniaux L, Liu L, Ma W, Satoh T and Kusunose K 2004 Detailed analysis of acoustic emission activity during catastrophic fracture of faults in rock *Journal of Structural Geology* **26** pp 247–258

[7] Mao W, Aoyama S, Goto S and Towhata I 2015 Acoustic emission characteristics of subsoil subjected to vertical pile loading in sand Journal of Applied Geophysics *Journal of Applied Geophysics* **119** pp 119–127

[8] Mao W, Goto S and Towhata I 2020 A study on particle breakage behavior during pile penetration process using acoustic emission source location *Geoscience Frontiers* **11** pp 413-427

[9] Wenli L, Liu A, Mao W and Koseki J 2020 Acoustic emission behavior of granular soils with various ground conditions in drained triaxial compression tests *Soils and Foundations*. **60** pp 929–943

[10] Kobayashi Y and Shiotani T 2016 Computerized AE Tomography: Innovative AE and NDT Techniques for On-Site Measurement of Concrete and Masonry Structures *State-of-the-Art Report of the RILEM Technical Committee 239-MCM* Springer pp 47-68

[11] Sassa K, Ashida Y, Kozawa T and Yamada M 1989 Improvement in the accuracy of seismic tomography by use of an effective ray-tracing algorithm *MMIJ/IMM Joint Symposium Volume of Papers*. Kyoto. pp 129-136

[12] Chiaroa G, Kiyota T and Koseki J 2013 Strain localization characteristics of loose saturated Toyoura sand in undrained cyclic torsional shear tests with initial static shear *Soils and Foundations* **53** 1 pp 23-34

[13] Shogaki T 2017 Mechanism of sample disturbance caused by tube penetration: Model tests on Toyoura sand *Soils and Foundations* **57** 4 pp 527-542

IOP Conf. Series: Materials Science and Engineering 1138 (2021) 012034 doi:10.1088/1757-899X/1138/1/012034

Study on determination of an object material using the microwave doppler sensor

K Oda[1*], K Nakamura[1], K Ikebata[1] and Y Kobayashi[1]

[1] Department of Civil Engineering, College of Science and Technology, Nihon University, Tokyo, Japan

* oda.kenichi@nihon-u.ac.jp

Abstract. In this study, to consider a method for detecting the sediment portion of avalanches containing sediment using microwave Doppler sensors, we confirmed whether the signal processing obtained by the sensors can differentiate the objects. Since microwaves in this frequency band are not easily affected by rain and snowfall, these are being developed for use in dashboard camera and other applications. However, it is necessary to understand the attenuation of the microwave for the snow in order to apply this technology to detect the sediment in the avalanche. It is also necessary to know the rate of reflection of microwaves from the snow and sand. Therefore, it is necessary to verify whether the reflection of the microwave from the sediment is correctly measured or not. In this report, we discuss the results of the verification to see if the distance between the sensor and the object, which is known in advance, can be detected correctly.

1. Introduction

Avalanches involving sediment and snow are often treated in the fields of geotechnical engineering and geology that deal with slope disaster prevention as one of the sediment collapses with snow accumulation during the snowmelt season in Japan. In addition, it has been reported that the reach distance of collapsed sediment exceeds that of normal sediment due to the inclusion of snow [1]. On the other hand, there have been few reports focusing on avalanches containing sediment, and the phenomenon of avalanches reaching longer distances than usual has recently begun to attract attention in the field of glaciology, so the mechanical interpretation of the effect of sediment on avalanches has not progressed. A typical example is the avalanche that occurred in Tokamachi and Tsunan towns in Niigata Prefecture and Sakae town in Nagano Prefecture on March 12, 2011, which was the first reported case of heavy snowfall over 2 m due to the earthquake in northern Nagano Prefecture. After this case, there is a growing need to consider the simultaneous occurrence of landslides and avalanches [2]. In order to elucidate the motion mechanism of avalanches containing sediment, it is necessary to make experimental observations on how the shape of the avalanche changes during flow and how sediment is distributed in the avalanche during flow.

On the other hand, there is an example of analyzing the shape of an avalanche by measuring the avalanche's path and area of influence from monitoring images [3]. For the observation of the inside of avalanches, model slope experiments with artificial simulated avalanches in laboratory tests have been conducted to understand the internal structure of avalanches during flow from image data [4], and structures equipped with a number of monitoring instruments have been installed in advance on slopes where avalanches are empirically predicted to occur [5]. However, there are still no sufficient reports for avalanches containing sediment. In this study, we have been developing an avalanche flow prediction method and have been conducting flow observation experiments of artificial avalanches using a small-scale model slope for the purpose of constructing an analytical model [6]. In this experiment, we are observing the basic flow of avalanches, especially changes in flow velocity and shape, to collect data that can be used to verify the accuracy of avalanche replication analysis, and to verify the effects of avalanche and snowpack incorporation during flow. In addition, experiments have been conducted in which sand and gravel, which are assumed to be sediment, are simultaneously flowed with snow to see how they affect the flow velocity. However, the process by which sediment contributes to avalanche motion is still unclear, and there are still issues such as how sediment is distributed and under what conditions it is fluidized over long distances in a downstream avalanche.

In this study, we focused on the microwave Doppler sensor, which has been used in the field of non-destructive inspection (e.g., [7] and [8]), in order to understand how much energy of the sediment is contributed to the avalanche flow by mixing the sediment with the snow. The purpose of this study is to observe how the distribution of sediment and snow inside the avalanche changes during the flow, and to differentiate the snow and sediment inside the avalanche by comparing the frequency characteristics of the electromagnetic waves obtained by the microwave Doppler sensor with the results observed visually. In this paper, as a preliminary step to differentiate between sediment and snow using microwave Doppler sensors, we discuss the results of organizing the frequency characteristics of sand and water as objects for measurement at the laboratory experimental.

2. Experimental system

2.1. Microwave Doppler sensor

In this study, the Microwave Doppler Sensor that can emit microwaves in the 24 GHz band was used (NJR-4262J: New Japan Radio). Figure 1 shows the Microwave Doppler Sensor used in this study. This sensor can detect the position of an object by using the fact that the frequencies of the transmitted and reflected waves are shifted by the Doppler effect when the radio wave is reflected from the object (e.g., [7] and [8]).

The sensor irradiates microwaves in the 24 GHz band to the measurement object and receives the reflected microwaves. The theoretical equation for the transmitted wave T_s and the received wave R_s is given as follows:

$$T_s(t) = A_s \cos(\omega_s t) \tag{1}$$

$$R_s(t) = A_r \cos([\omega_s + \omega_d]t + \varphi) \tag{2}$$

where A_s and A_R are the respective amplitudes, ω_s is the angular velocity of the transmitted wave, ω_d is the Doppler angular velocity, t is the time, and φ is the phase depending on the distance to the object. The microwave Doppler sensor used in this study is equipped with a function to output the frequency fluctuation components of mixing the transmitted and received waves as I and Q signals. The mixed signal is expressed by the following theoretical equation.

$$\begin{aligned} T_s(t)R_s(t) &= A_s A_r \cos(\omega_s t)\cos([\omega_s + \omega_d]t + \varphi) \\ &= \frac{A_s A_r}{2}\{\cos(\omega_d t + \varphi) + \cos([2\omega_s + \omega_d]t + \varphi)\}. \end{aligned} \tag{3}$$

The theoretical equation of the output I and Q signals after mixing is expressed by the following equation, in which only the term including the low frequency Doppler angular velocity is left out of equation (3), and the phase difference between the Q output and the I output is given by π/2.

$$I(t) = \frac{A_s A_r}{2}\cos(\omega_d t + \varphi) \tag{4}$$

$$Q(t) = \frac{A_s A_r}{2}\cos(\omega_d t + \varphi - \frac{\pi}{2}) \tag{5}$$

In this case, the I-Q signal, which is the phase of the transmitted wave and the received wave, can be obtained from the sensor. This signal is detected when the target object is approached, the I output is $+\pi/2$ greater than the Q output, and when the object is separated, the I output is detected $-\pi/2$ greater than the Q output. This effect can be used to monitor the movement detection of an object. If we define the phase θ of the Lissajous waveform synthesized from the I and Q output signals as shown in figure 2, the Doppler angular velocity ω_d is the time derivative of the phase θ. In general, the displacement of the value of phase θ is compared when the state of motion of an object is grasped using a microwave Doppler sensor, but it is said that it is possible to grasp the state of an object in more detail by tracking the change of Doppler angular velocity ω_d [9].

In this study, we examine whether the Doppler angular velocity ω_d in the I-Q signal obtained by the microwave Doppler sensor can be used to differentiate between objects. Mitani proposed the following approximate formula for the Doppler angular velocity to improve the efficiency of the calculation [9].

$$\omega_d \approx \frac{(I - I_{offset}) \times \Delta Q - (Q - Q_{offset}) \times \Delta I}{(I - I_{offset})^2 + (Q - Q_{offset})^2} \tag{6}$$

where ΔI is the time variation of I output, ΔQ is the time variation of Q output, and I_{offset} and Q_{offset} are the central values of I-Q Lissajous. In order to analyse the behaviour of I-Q Lissajous, we need the offset coordinates of I-Q, which is the center of the circular path. In order to analyse the behaviour of I-Q Lissajous, the offset coordinate of I-Q, which is the center of the circular path, must be calculated in accordance with the time variation of I-Q output, because this I-Q offset coordinate fluctuates with the movement of the object. In this study, we use the fact that the behaviours of the I-Q Lissajous is a circular path to estimate the I-Q offset at three points using the I-Q value of $\pi/2$ and π periods ahead of the I-Q value based on the I-Q value at the time to be determined. The formula for estimating the I-Q offset is as follows.

$$\begin{pmatrix} I_{offset} \\ Q_{offset} \end{pmatrix} = \frac{1}{2(\alpha\delta - \beta\gamma)} \times \begin{pmatrix} \delta & -\beta \\ -\gamma & \alpha \end{pmatrix} \begin{pmatrix} |\boldsymbol{X}_1|^2 - |\boldsymbol{X}_2|^2 \\ |\boldsymbol{X}_2|^2 - |\boldsymbol{X}_3|^2 \end{pmatrix} \tag{7}$$

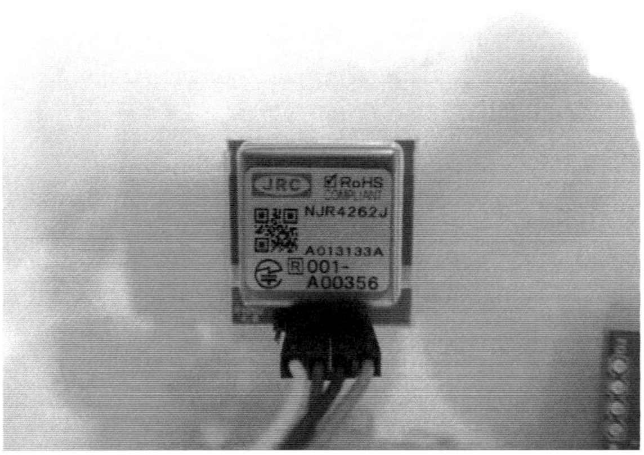

Figure 1. Microwave Doppler sensor.

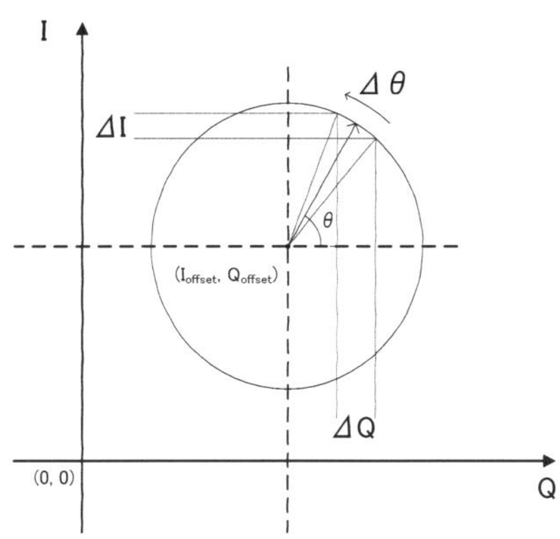

Figure 2. I-Q Lissajous waveform.

where

$$\begin{pmatrix} \delta & -\beta \\ -\gamma & \alpha \end{pmatrix} = \begin{pmatrix} x_1 - x_2 & y_1 - y_2 \\ x_2 - x_3 & y_2 - y_3 \end{pmatrix} \tag{8}$$

$$|\boldsymbol{X}_i|^2 = x_i^2 + y_i^2 \tag{9}$$

x is the value of I, and y is the value of Q.

2.2. Measurement System

In this study, we used Arduino uno to realize a simple measurement system. Arduino uno is one of the general-purpose microcontroller boards. In addition, the microwave Doppler sensor used in this study is an analog output type. Therefore, we used an AD converter to get the digital data. An overall diagram of the measurement system is shown in figure 3. The sampling rate of the measurement data was set to 50 Hz.

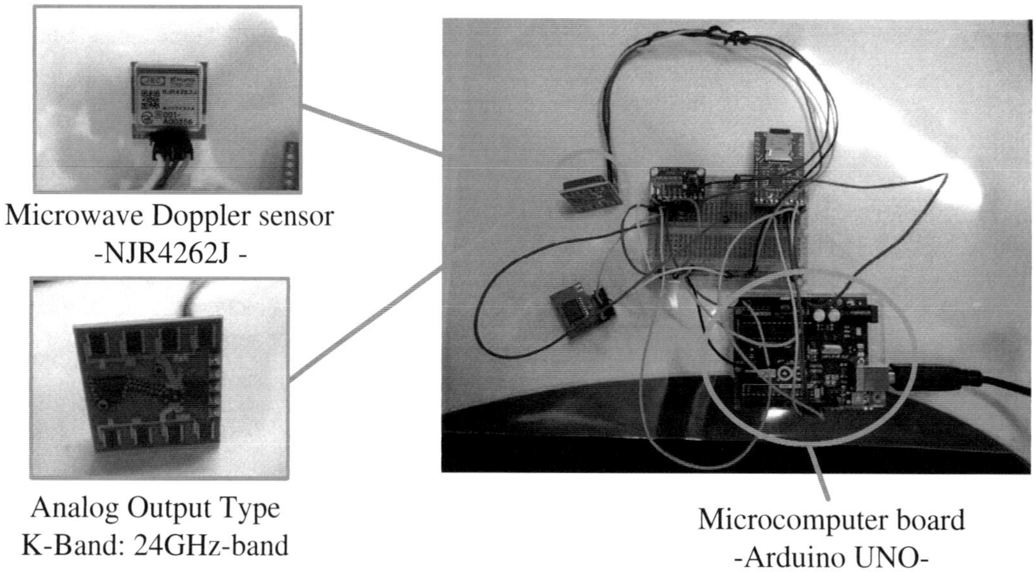

Figure 3. Measurement system.

3. Experimental conditions

In this experiment, sand and water were placed in a 500 ml glass beaker, and the beaker and the object were moved by means of the elevating machine that operated at a constant speed. The beaker was filled with four cases: (1) empty, (2) only water, (3) only sand and (4) water with sand. Table 1 shows experimental condition of this study. These objects were irradiated with microwaves from the top of the beaker to detect the signals. Figure 4 shows an overall view of the experimental system. The displacement speed of the elevating machine is 0.01 m/s. The object used in the experiments is shown in figure 5. The object filled with sand in the beaker was prepared in two cases in amount of 150 ml and 300 ml of sand. The distance between the sensor and the top plate of the elevating machine is 16 cm, and the distance from the position where the content of the beaker is 500 ml to the sensor is 7 cm, and the distance from the position where the content of the beaker is 300 ml to the sensor is 10 cm. Each object was measured by repeating move up and down 3 times.

Table 1. Experimental conditions.

	Case1	Case2	Case3	Case4	Case5	Case6
Sample	None	Sand 150 ml	Sand 300 ml	Water	Sand 150 ml with water	Sand 300 ml with water
H	15.0	15.0	15.0	15.0	15.0	15.0
D_1	14.5	12.0	10.0	14.5	12.0	10.0
D_2	-	-	-	7.00	7.00	7.00

Unit: cm

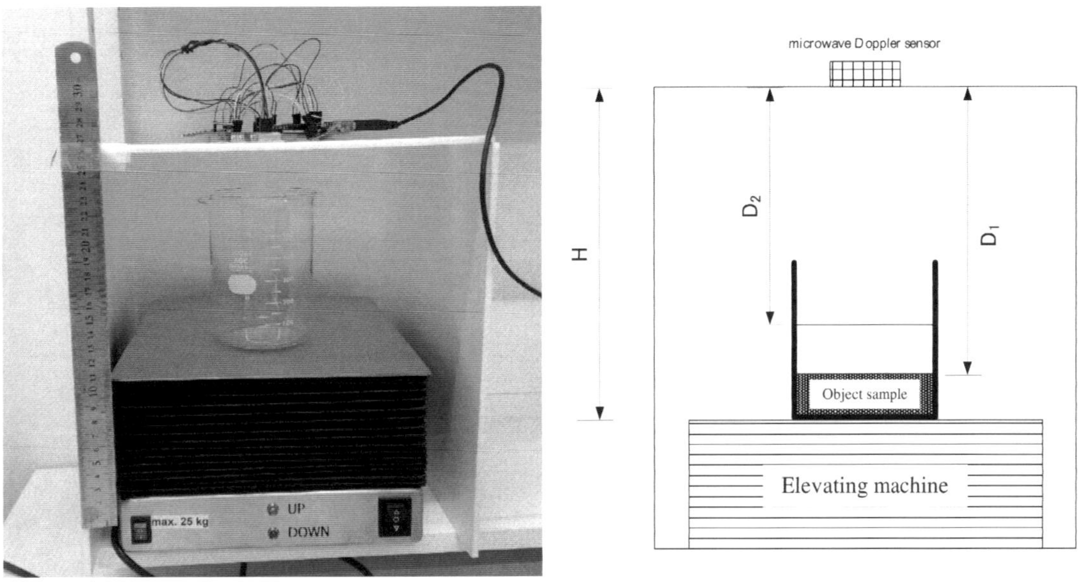

Figure 4. Overall view of the experimental system.

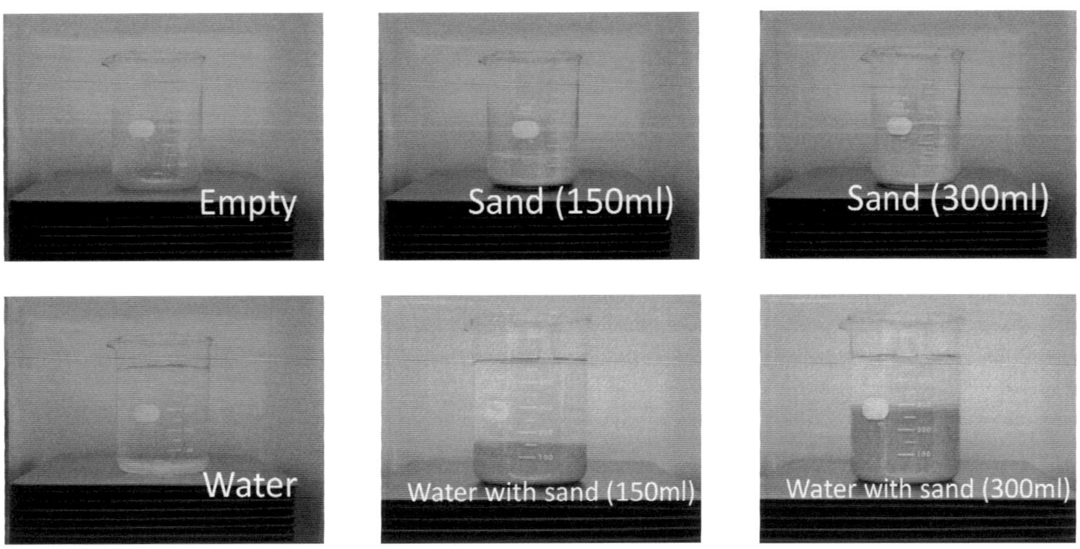

Figure 5. The object used in the experiments.

4. Experimental results and discussion

4.1. Time variation of I and Q signals

Figure 6 and figure 7 show the output results of the I and Q signals and the resultant value of the I and Q signals obtained in the experiment. In figures, red shows the I signal, blue shows the Q signal, and green shows the resultant value of the I signal and the Q signal. The resultant value of the IQ signal is defined by the following equation,

$$V_{IQ} = \sqrt{V_I^2 + V_Q^2}. \tag{10}$$

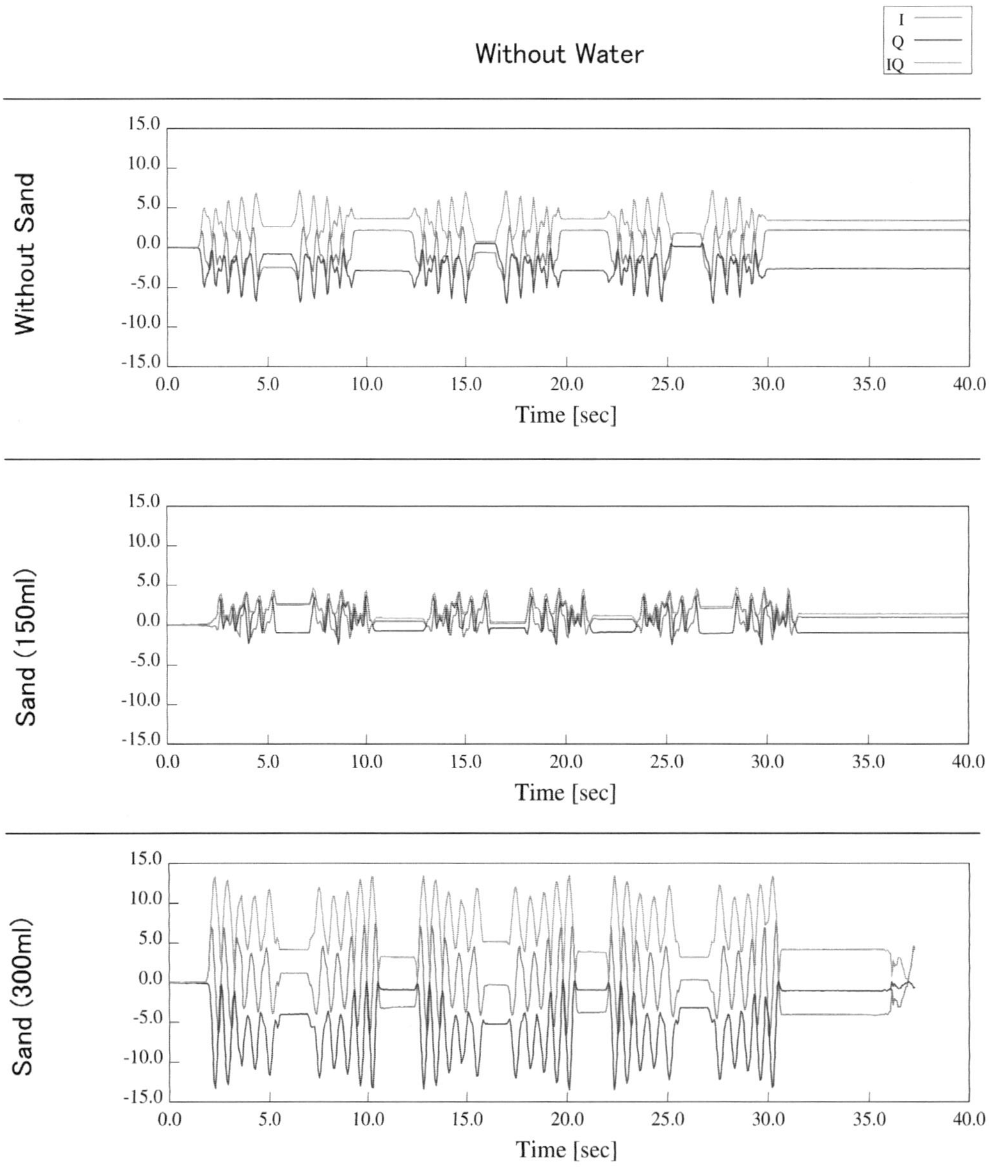

Figure 6. I-Q signal without water.

It is known that the amplitude of the signal obtained from the microwave Doppler sensor varies with the presence of the object. In the result of the measurement, we found three points where the amplitude oscillated greatly. In addition, each point was symmetrical and the time of the large amplitude oscillation was about 3 seconds. In this experiment, the raising and lowering of the elevating machine was done in 3 seconds each in all the experiments. It can be seen that the three points where the waveform changed greatly in all the measurements responded to the movement of the elevating machine. In other words, we can see that this measurement system is able to detect the movement of the object.

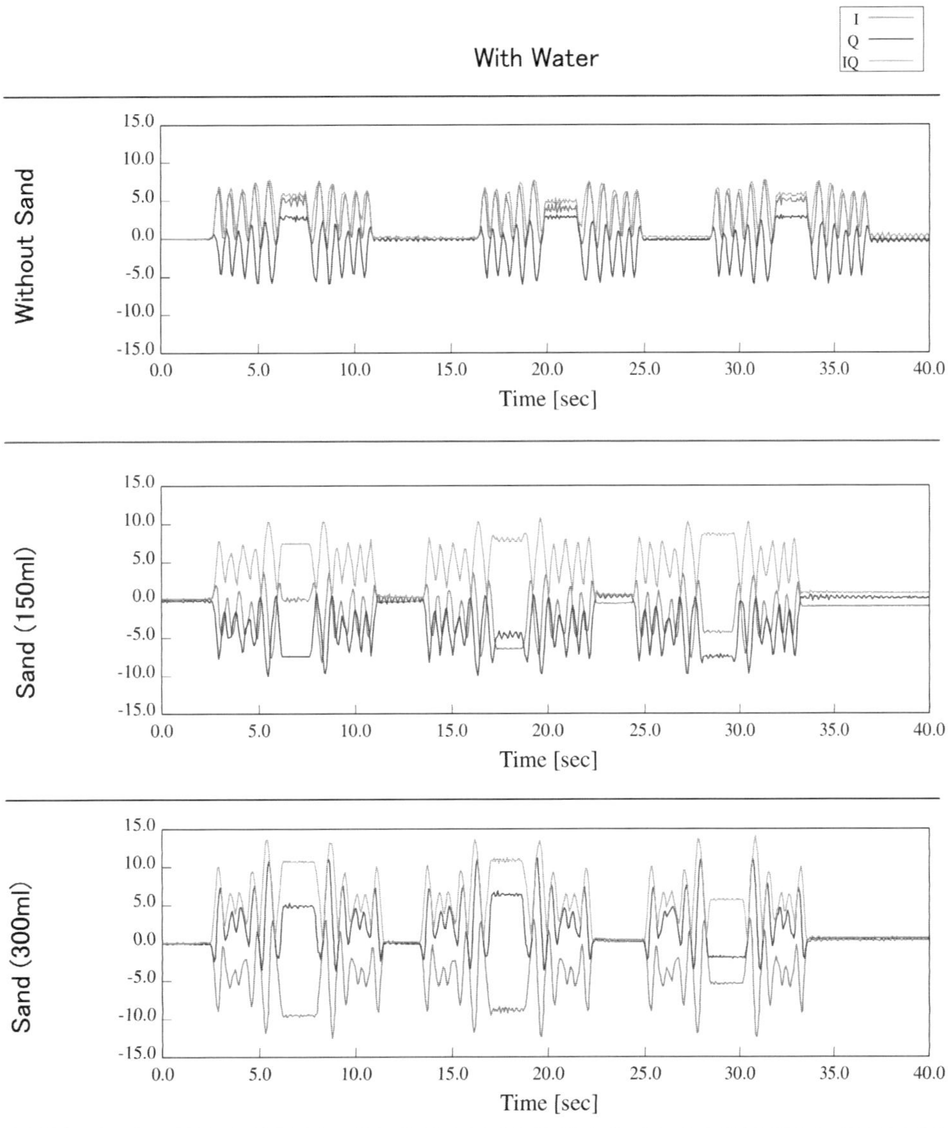

Figure 7. I-Q signal with water.

In the next section, we analyse the results of the measurement with the object containing sand only in the beaker. It can be seen that there is a difference in the amplitude of the waveform obtained by the movement of the elevating machine between the case with and without the sand. The difference in amplitude is smaller when the sand is as little as 150ml than when the sand is empty, and larger when the sand is 300ml. Accordingly, it can be seen that the difference in amplitude is dependent on the type

IOP Conf. Series: Materials Science and Engineering 1138 (2021) 012034 doi:10.1088/1757-899X/1138/1/012034

of moving object and also on the distance to the sensor. However, the voltage obtained from the first step-up cycle to the second and third cycle was different from the initial value.

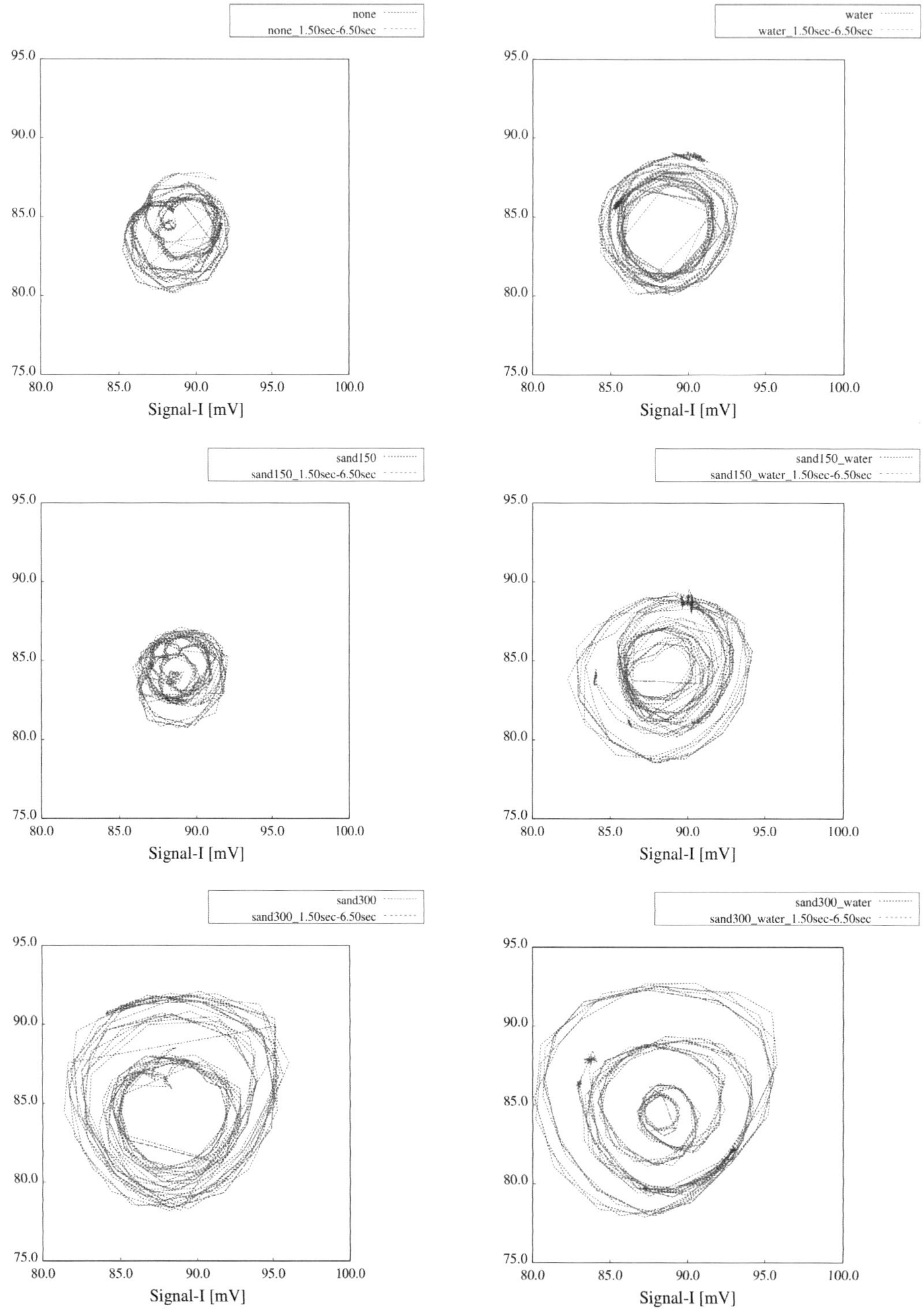

Figure 8. The results of the I-Q Lissajous waveforms.

Then, the object filled with water up to 500 ml scale is analyzed. The first thing that can be confirmed from these results is that the voltages obtained by all the I and Q signals returned around the initial values after three cycles. This indicates that the signals obtained tend to be stabilized when water is included in the object. In addition, the amplitude that responds to the movement of the elevating machine is larger in the case of water only than in the case of an empty object. This indicates that the sensor used in this experiment is affected by water. On the other hand, in the case of the object containing sand and water, the difference in amplitude of the waveform obtained is different as in the case of the object containing sand only.

4.2. Comparison of Doppler angular velocities calculated from I-Q Lissajous waveforms

Figure 8 shows the results of the I-Q Lissajous waveforms obtained in each case. The broken line in blue shows the trajectory for all the times when the ascent and descent were repeated three times, and the broken line in red shows only the first ascent part. In Cases 1 to 3 without water, the shape of the Lissajous waveform becomes larger as the amount of sand in the object increases.

On the other hand, in Case 4 to Case 6 where water is added, as the amount of sand increases, the shape of the Lissajous waveform increases as in the case without water, indicating that the waveform starts differently from the case without water. These mean that the I-Q Lissajous obtained by changing the object are different, and it is confirmed that the results are largely related to the differentiation of the object. In the case with water, the circular path fluctuation is less than the case without water, which raises the possibility that the water absorbs the noise contained in the I-Q signal.

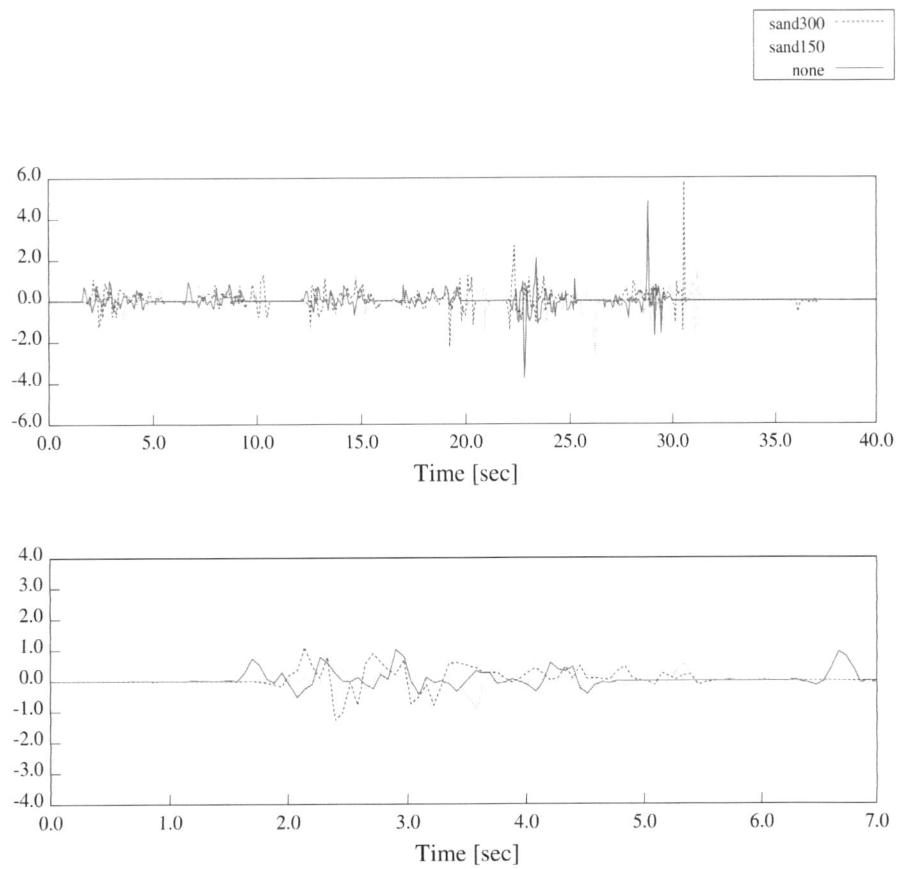

Figure 9. The time variation of the Doppler angular velocity obtained from Case 1 to Case 3.

Next, figure 9 shows the time variation of the Doppler angular velocity obtained from Case 1 to Case 3 where no water is added, and figure 10 shows the time variation of the Doppler angular velocity obtained from Case 4 to Case 6 where water is added. Note that the graphs shown below each figure represent an enlarged view of the first moving part where the elevating machine is raised. The results without water show not much difference in the variation of the Doppler angular velocity, but the results with water show that the Doppler angular velocity fluctuates with a fine period in proportion to the amount of sand. These results suggest that it may be possible to differentiate the objects by comparing the results with the changes in the I-Q Lissajous waveform.

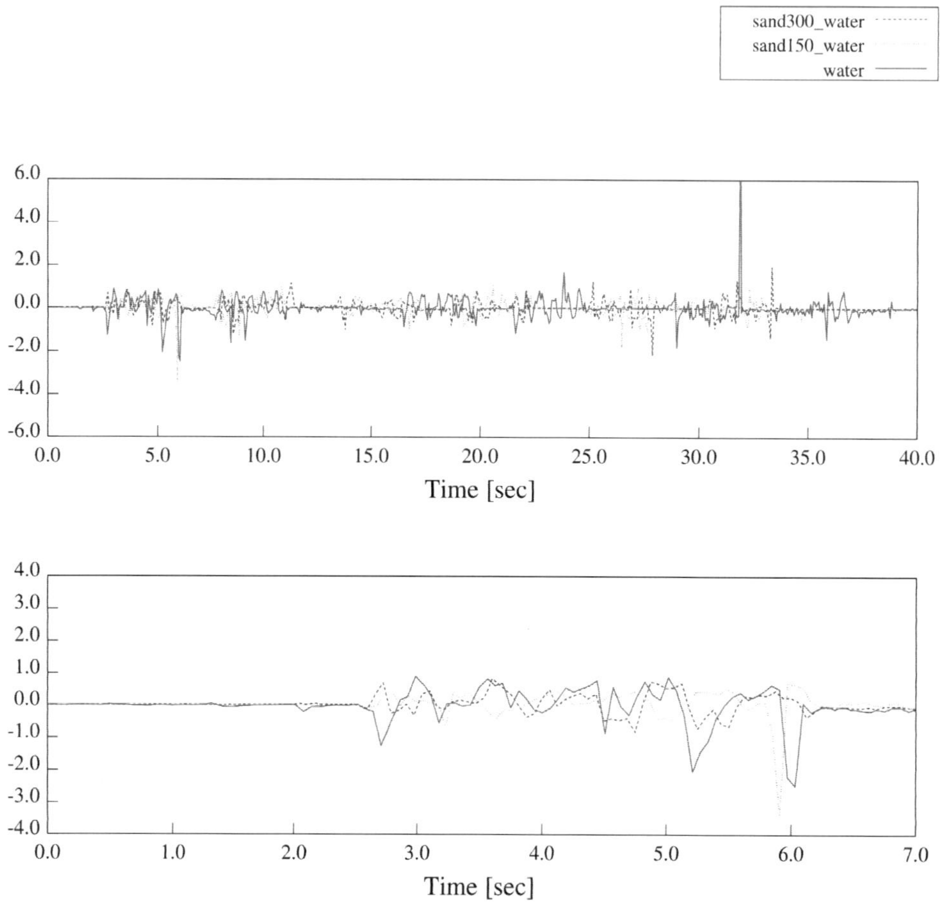

Figure 10. The time variation of the Doppler angular velocity obtained from Case 4 to Case 6.

5. Conclusion

In order to understand whether the microwave Doppler sensor is effective as a method to confirm the structure inside the avalanche including sediment in a non-contact way, we verified the basic analysis method to differentiate the material of the object from the detection signal of the moving object obtained by the microwave Doppler sensor. In addition, a simple material of water and sand were used for the basic verification of whether the microwave Doppler sensor can differentiate the object. Furthermore, we used an inexpensive and simple microcomputer board as a system to measure the signals obtained from the microwave Doppler sensor, assuming the measurement at the field level.

In the measurement of object detection, the object was moved by an elevating machine that can maintain a constant speed moving environment, and the movement of the object was detected by the microwave Doppler sensor. The object was prepared as a specimen in a beaker with water or sand. In this measurement, the I-Q signal output by the microwave Doppler sensor was obtained even in an environment where the microcomputer board was used as the measurement system, and it was confirmed that the movement detection of the object was approximately correct.

We also analysed whether the material of the object can be differentiated from the I-Q signal obtained from the microwave Doppler sensor. In this analysis, the I-Q Lissajous waveform and the Doppler angular velocity were calculated from the obtained I-Q signals, and it was confirmed that the magnitude of the I-Q Lissajous waveform and the period of the Doppler angular velocity changed depending on the condition of the object. In particular, it was found that the waveform and the magnitude of the period were different when with and without water in the object. This suggests that the method of comparing I-Q Lissajous waveform and Doppler angular velocity may be effective as a method to differentiate the material of the object.

On the other hand, in order to understand the internal structure of avalanches, which is the purpose of this study, it is necessary to confirm the detection of snow and ice objects. In the future, we will investigate the possibility of understanding the internal structure of avalanches by confirming the signals obtained when snow, ice and sand are used as objects.

References

[1] Sekiguchi T, Sato H and Akiyama K 2005 Geomorphological Features of Avalanche Furrows in Heavy Snow Region in Japan *Bulletin of the Geographical Survey Institute* **52** pp 37-47
[2] Kamiishi I, Motoyoshi H and Ishizaka M 2012 Avalanches Induced by North Nagano Prefecture Earthquake on 12 March 2011 *Natural Disaster Research Report of the National Research Institute for Earth Science and Disaster Resilience* **47** (Japanese)
[3] van Herwijnen A, Heck M and Schweizer J 2016 Forecasting snow avalanches using avalanche activity data obtained through seismic monitoring *Cold Regions Science and Technology* **132** pp 68-80
[4] Oda K, Nakamura K, Kobayashi Y and Suzumura J 2020 Inverse Simulation for Extracting the Flow Characteristics of Artificial Snow Avalanches Based on Computation Fluid Dynamics *Geosciences* **10**(6) 221
[5] Lacroix P, Grasso J. R, Roulle J, Giraud G, Goetz D, Morin S and Helmstetter A 2012 Monitoring of snow avalanches using a seismic array: Location, speed estimation, and relationships to meteorological variables *Journal of Geophysical Research: Earth Surface* **117** F1
[6] Oda K, Moriguchi S, Kamiishi I, Yashima A, Sawada K and Sato A 2011 Simulation of snow avalanche model test using computational fluid dynamics *Annals of Glaciology* **52** 58 pp 57-64
[7] Izumi S, Okano T, Matsunaga D, Kwaguchi H and Yoshimoto M 2019 Non-Contact Instantaneous Heart Rate Extraction System Using 24-GHz Microwave Doppler Sensor *IEICE Transactions on Communications* **E102.B** 6 pp 1088-1096
[8] Nguyen Thi Phuoc Van, Liqiong Tang, Veysel Demir, Syed Faraz Hasan, Nguyen Duc Minh and Subhas Mukhopadhyay 2019 Review-Microwave Radar Sensing Systems for Search and Rescue Purposes *Sensors (Basel)* 10.3390/s19132879
[9] Mitani S 2019 Development of the in-vehicle pulse sensor *OMRON TECHNICS* **50** 006

IOP Conf. Series: Materials Science and Engineering 1138 (2021) 012035 doi:10.1088/1757-899X/1138/1/012035

Using energy of compressed natural gas as an environmentally safe source of heat energy of the heating system of the industrial premises

E Pakhomova[1], N Grigorova[1], P Monastyrev[2] and N Semicheva[1]*

[1] Southwest State University, Kursk, Russia
[2] Tambov State Technical University, Tambov, Russia

* nsemicheva@yandex.ru

Abstract. The housing and utilities system is extremely energy-intensive and currently impedes the implementation of energy efficiency policies. In this regard, the development of science-based technology for environmentally friendly resource-saving consumption of compressed natural gas, as a source of thermal energy, is becoming relevant. The article describes the technical solution for using the energy of compressed natural gas as a source of thermal energy, which consists in using a vortex heat exchanger in the heating system of a gas control station, which allows you to generate thermal energy due to technologically unused energy, which is the pressure drop between high and medium gas pipelines as well as medium and low pressure natural gas. This technical solution is aimed at reducing energy costs and ensuring environmental safety when using compressed natural gas as a source of thermal energy. To determine the possibility of practical use of this technical solution in the production room for installing gas control equipment, an analytical expression of the temperature of the wet gas mixture after the process of condensate-evaporative heat and mass transfer in a vortex heat exchanger has been obtained, which allows us to predict the parameters of the coolant in the heating system of the production room.

1. Introduction

Currently, much attention is paid to environmental safety and energy saving in all areas of the fuel and energy complex related to the generation, transportation and consumption of energy resources, including the reduction of the energy intensity of heating systems, especially small industrial buildings [1 - 5].

The production of thermal energy for communal needs is invariably associated with a negative impact on the environment. This problem is especially relevant in the face of a tense environmental situation in most large cities of Russia. The introduction of autonomous heat supply can improve the current environmental situation and increase the efficiency of thermal energy production up to 85 - 97%. The total yield of combustion products during autonomous heat supply is significantly less and safer than emissions from existing heat and power plants (TPP).

IOP Conf. Series: Materials Science and Engineering 1138 (2021) 012035 doi:10.1088/1757-899X/1138/1/012035

The most widespread in the case of autonomous heat supply are capacitive water heaters of the AOGV series, which provide hot water of a given temperature and in sufficient volume for room heating systems. This quality has become the basis for the predominant use of AOGV for heating an industrial building, in particular a gas control unit (GCU), however, this heat exchanger requires fuel consumption for burning gas when heating water in a heating system.

A feature of the operation of the GCU room is the regulation of the pressure of the gas coming from the pipeline to consumers. At the same time, pressure regulators operate at a sufficiently high (from 3.5 and more times) differential pressure of the input and output pressure values with unclaimed repayment of excess energy. Due to the significant energy potential of natural gas, which consists in the need to reduce its pressure before the consumer by 0.35 MPa or more, the article proposes one of the technical solutions to the problem of reducing energy consumption for heating a gas distribution building.

The proposed technical solution consists in the use of a vortex heat exchanger in the heating system of a gas control station, which allows you to generate thermal energy due to technologically unused energy, which is the pressure drop between the high and medium, as well as medium and low pressure natural gas pipelines.

2. Materials and methods

An analysis of the known works [6, 7] shows that the heat transfer coefficient or the degree of heat transfer intensity of a heating and heated medium in a heat exchanger depends largely on the largest of the thermal resistances of the heat exchange surface. Of particular urgency is the task of intensifying heat transfer when using gaseous working media, which are characterized by a reduced intensity of metabolic processes and high energy costs to overcome hydraulic resistance during gas movement.

In addition, the decrease in the intensity of heat exchange of the heating surface with the heated room air [8, 9] is affected by the corrosion of heat exchange surfaces, which refers to the processes of chemical and electrochemical corrosion and leads to damage to their heating surface.

The most common and widely used in heating systems of industrial buildings is a tubular heat exchanger-air heater. It is easy to use and operate, but has a large mass and a large volume. In tubular heat exchangers made of steel tubes, significant low-temperature corrosion is observed. When exposed to a cold coolant flow on a metal surface, a sharp decrease in the intensity of heat transfer occurs. This is due to the transition of the "cold" flow to the "hot" one, which reduces the heat exchange between swirling gas and the air of the heated room [10].

Another common structural type of heating apparatus-air heaters is represented by a plate heat exchanger. It consists of steel sheets 1.5–2.0 mm thick. Such a heat exchanger is difficult to manufacture due to the formation of a laminar layer. When the temperature gradient is observed warpage of the sheets, which reduces the reliability of the air heater.

An effective way to protect low-temperature heating surfaces from corrosion is to make them from corrosion-resistant materials, namely low-alloy steel with copper additive 10xndp, oh23n28m3d3t, nickel alloy om70m27f.

However, the use of these materials significantly increases the cost of heat exchange equipment. This necessitates the use of new, cheaper, corrosion-resistant materials. For example: the use of non-metallic materials and coatings, such as organosilicon varnishes and enamels based on furan, epoxy and other resins, as well as the use of polymeric materials. However, these materials have low heat resistance.

In this regard, an original design of an air heater in the form of a vortex tube was developed, which provides the conditions for intensive and reliable heat transfer under the corrosive effect of a heat carrier [11, 12].

The determination of the possibility of the technological use of this constructive expansion in the production room for the placement of the gas control equipment leads to the need to formulate the problem of mathematical modeling of the heat exchanger with maintaining the initial temperatures. For example: at the inlet = 100°C and at the outlet = 70°C, which is the closest with the standard in heating systems.

In a vortex tube, the thermodynamic separation of the moving gas proceeds from the well-known theoretical assumption that the energy exchange process at the exit of the swirl is complete, the result of which is the adiabatic distribution of temperature gradients along the radius of the heat exchanger [13].

Currently, the resource-saving policy pays special attention to the development and use of both energy-saving technologies and equipment that ensures the implementation of processes with minimization of costs, including heat, for the heating system of industrial premises [14].

The ratio between the heat transfer rate and the hydraulic resistance of the apparatus near the wall of the heat exchanger enclosure is more favorable in that case, the more significant is the difference in the temperature distribution and the velocity of the coolant particles [15].

When using a vortex tube heating system as a heat exchanger, one of the effective solutions to intensify heat transfer is to make swirlers in the form of blades.

We study the nature of the movement at the place of rotation of the swirl blade at the entrance to the vortex tube of the elementary volume of the coolant in the form of compressed natural gas moving at a speed $\omega_{v.g}$, revolutions per second, whose mass is equal:

$$dm_{v.g} = \rho d_f dz, \tag{1}$$

where d_f – cross-sectional area of elementary volume, m^2;
dz – direction of movement at the moment, m;
ρ – heat carrier density, kg / m^3.

The elementary centrifugal force acting on this object is:

$$\frac{\delta d_f d_z \omega_{v.g}^2}{z} \tag{2}$$

where z - is the radius of rotation of the spinning blade, m

This force is balanced by the pressure difference on the faces of the volume in question:

$$dp = \left(\frac{dp}{dz}\right) d_f dz \tag{3}$$

Based on the d'Alembert principle, we design all the forces in the Z direction in the direction of the moving coolant, write:

$$d_f dz\omega/z - dp/dzd_f dz = 0,$$

where does it come from:

$$\frac{\rho\omega_{v.g}^2}{z} = \frac{dp}{dz} \tag{4}$$

A change in static pressure based on the Bernoulli equation can occur only as a result of a change in the speeding pressure head [16]:

$$p + \frac{\rho\omega_{v.g}^2}{2} = \text{const} \tag{5}$$

Differentiating (5), we get:

$$\frac{dp}{dz} = -\frac{\rho\omega_{v.g} d\omega_{v.g}}{dz} \tag{6}$$

Substituting (6) in (4), we obtain:

$$-\frac{\rho\omega_{v.g}d\omega_{v.g}}{dz} = \frac{\rho\omega^2}{z} \cdots \frac{\omega_{v.g}}{z} + \frac{d\omega_{v.g}}{dz} = 0 \quad (7)$$

Integrating (7), we arrive at the result:

$$z\omega_{v.g} = \text{const} \quad (8)$$

Accept c_p=const, since the process of heat and mass transfer occurs in a particular element under study, moving under the action of centrifugal forces, on stain of a fluid at a constant pressure for these conditions:

$$M_{CM}h_{CM} = m_{v.g}h_{v.g} + m_{oh}h_{is}, \quad (9)$$

$$h_{CM} = g_1h_{v.g} + g_2h_{is}, \quad (10)$$

where g_1, g_2 – mass fractions of wet gas at the exit of the tapering nozzle and in the boundary layer of the swirl blade during the evaporation process, %:

$$h_{CM} = C_{p_{zh}}T_{CM};\ h_{v.g} = C_{p_{v.g}}T_{v.g};\ h_{is} = C_{p_{is}}T_{is},$$
$$c_{p_{zh}}T_{CM} = g_1c_{p_{v.g}}T_{v.g} + g_2c_{p_{is}}T_{is}. \quad (11)$$

Then, at c_p = const, we have the temperature of the wet gas mixture after the process of condensate-evaporative heat and mass transfer:

$$T_{CM} = \frac{g_1c_{p_{v.g}}T_{v.g} + g_2c_{p_{is}}T_{is}}{g_1c_{p_{v.g}} + g_2c_{p_{is}}} \quad (12)$$

For the temperature of thermodynamic separation of the mixture of "cold" and "hot" flows of wet gas, the obtained analytical expression of the temperature of the mixture of wet gas after the process of condensate-evaporative heat and mass transfer in a vortex heat exchanger allows predicting the parameters of the heat carrier in the heating system of the industrial building [17]. The resulting equation allows you to get the parameters of the heat carrier of the heating system of the production room closest to standardized. The experimental examination of the analytical dependencies was carried out in laboratory and industrial conditions on the installation simulating the heating element of the heating system of an industrial building, in particular a gas control station (GCU).

The developed mathematical model for the process of heat transfer in a vortex heat exchanger from a swirling flow of natural gas to the air of a heated room became the basis for creating the concept of a vortex heat exchange element [18].

3. Results

The operation of the gas control station is carried out under conditions when the pressure regulators operate at a sufficiently high (3 - 5 times or more) differential pressure input and output values in gas pipelines with unclaimed discharge of excess energy [19] before supplying the consumer. The use of this energy of a moving gas stream is possible when using an excess pressure of a vortex tube as a partial absorber.

The operation of the gas control station is carried out under conditions when the pressure regulators operate at a sufficiently high (3 - 5 times or more) differential pressure input and output values in gas pipelines with unclaimed discharge of excess energy [19] before supplying the consumer. The use of this energy of a moving gas stream is possible when using an excess pressure of a vortex tube as a partial absorber.

The vortex heat exchange element operates as follows (figure 1). Depending on the place of extraction and production conditions, the compressed natural gas used in the vortex method of heat transfer (without preliminary drying) is always saturated to a certain concentration with moisture vapor,

IOP Conf. Series: Materials Science and Engineering 1138 (2021) 012035 doi:10.1088/1757-899X/1138/1/012035

both atmospheric and technological, therefore, as shown by the analysis of known vortex heat exchangers, in a cold stream after thermodynamic separation of natural gas, condensation of moisture vapor with the release of heat of condensation is observed.

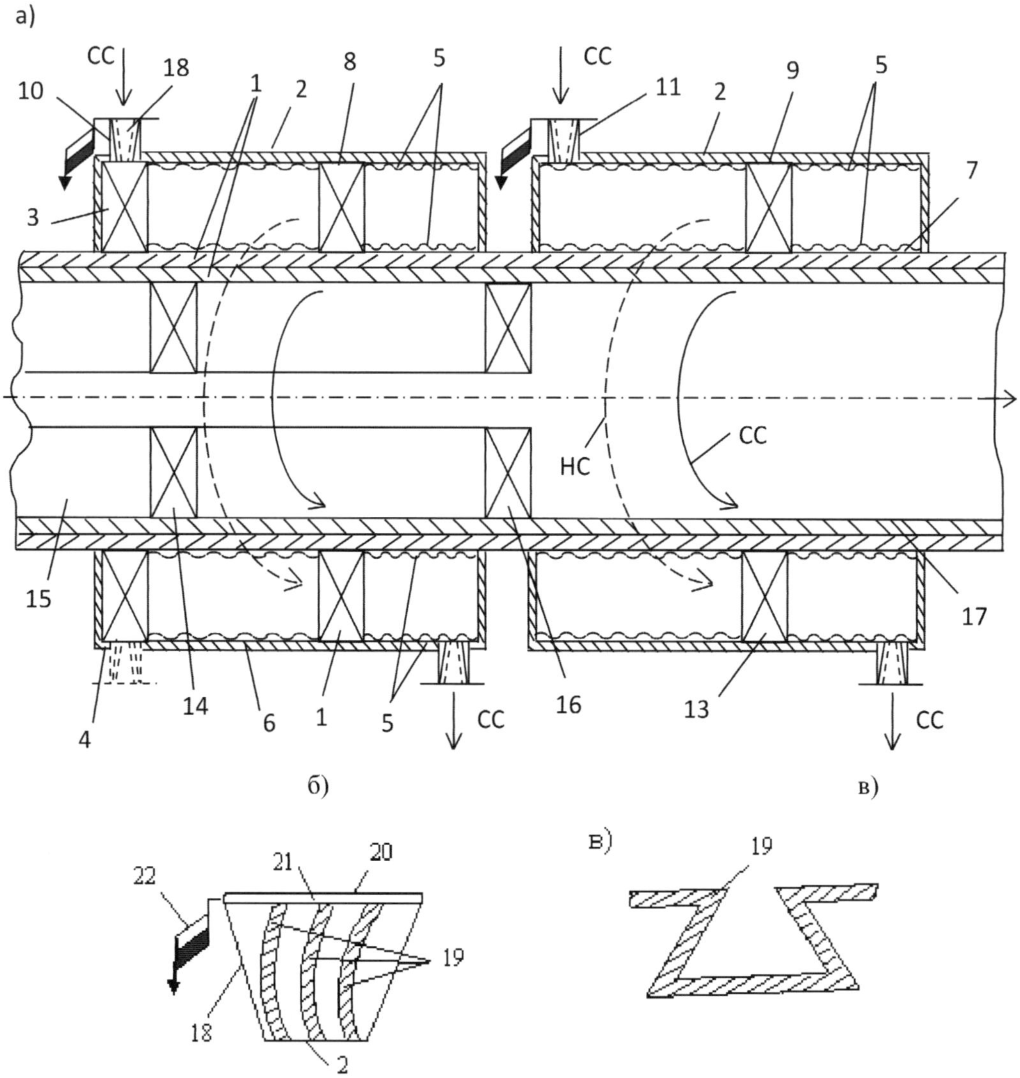

Figure 1. Vortex heat exchange element: a) - a schematic diagram of a vortex heat exchange element; b) - the inner surface of the cold coolant supply pipe with a round groove connected to both the contaminant removal device and longitudinally placed curved grooves; c) - a profile of a helical groove in the form of a “dovetail”; 1 - inner pipe; 2 - pipe of larger diameter; 3 - swirl; 4 - input section; 5 - peripheral zone of the flow of cold coolant (CC); 6 - the inner surface of the pipe 2 of larger diameter; 7 - the outer surface of the inner pipe 1; 8, 9 - pipe sections 2, equipped with cold coolant supply pipes; 10, 11 - nozzles supply cold coolant; 12, 13, 14 - swirlers; 15 - input section; 16- swirl; 17 - the inner surface of the inner pipe 1; 18 - narrowing nozzle; 19 - curved groove; 20 - inlet; 21 - groove; 22 is a device for removing contaminants.

The heat of condensation obtained in conjunction with the heat of the "hot layer" of an additionally stratified cold stream acts on the cylindrical wall of the inner pipe 1. As a result, the thermal ratio of the counter-directed heat flows of the "hot" and "cold" heat carriers transferred by the thermal conductivity through the bimetallic fence in the form of a cylindrical the walls of the inner pipe 1. This virtually eliminates the possibility of efficient use of the vortex method of heat transfer, because the condensation process, especially drip, has a random nature of the release of condensation heat over the heat exchange surface.

To eliminate this phenomenon, i.e., to remove contaminants from the cold coolant (CC) in the form of condensed moisture vapor, the CC is sent to the supply pipes 10 and 11, made in the form of tapering nozzles 18, where it increases its speed and moves along curved grooves 19 and twists. As a result, the particles of contaminants (droplet-like moisture) fill the cavities of the curved grooves 19 and under the action of centrifugal forces move, enlarging, towards the groove 21 located at the inlet 20 of the nozzles 10 and 11, from where they accumulate into the pollution removal device 22 for discharge into environment manually or automatically.

The resulting natural gas or compressed air as a result of separation of droplet-like moisture at the outlet of the swirl 14 is stratified into layers: "hot" peripheral and "cold" axial. The heat is transferred by convection from the "hot" layer of the HC to the inner surface 17 of the inner pipe 1 and then, through the thermal conductivity, the thickness of the material of the inner pipe 1 is heated.

At the same time, CC passing through swirlers 3, 12, and 13 also stratifies into a "hot" peripheral layer located in zone 5 and a "cold" axial layer, while the "hot" layer contacts the outer surface 7 of the inner pipe 1, giving it its heat convection and, further, thermal conductivity.

The flows of hot coolant (HT) and cold coolant (HT) are twisted and mixed in the axial direction, while simultaneously performing a rotational movement. Due to the intense heat exchange between the rotational flow CC in the pipe 2 and the outer surface 7 of the inner pipe 1, an even greater heating of the peripheral layer of CC in zone 5. This results in the formation of CC with an inhomogeneous density field, which leads to the continuous replacement of less heavy particles of CC heavier and this process continues until the rotational motion of the flow attenuates.

From the peripheral "hot" layer of HC, thermal energy is transferred to the inner surface 17 of the pipe 1 with convection. Further, the thermal conductivity of the bimetal material with an increased value of the coefficient of thermal conductivity and has a higher temperature gradient than the heat transferred from the peripheral flow of CC to the outer surface 7 of the inner pipe by the thermal conductivity of the material of the bimetal with a lower value of the coefficient of thermal conductivity.

The contact area of the counter-directed heat flows in this case is shifted towards the outer surface 7 of the inner tube 1 and is about 20% of the distance from the outer surface 7 (see figure 1). As a result, heat losses caused by the movement of heat energy along the thickness of the inner tube 1 are significantly reduced, which allows to significantly increase the efficiency of using the method of heat transfer in recuperative heat exchangers, for example, with the location of swirlers inside the cavity of both the pipe 2 with a large diameter, and inside the inner tube 1. The novelty of the developed technical solutions applicable for the constructive calculation of environmentally safe and resource-saving equipment for technological systems for transportation and consumption of natural gas is protected by a number of patents of the Russian Federation for inventions and utility models [20, 21].

4. Conclusions

The developed technical solution allows you to: save gas as fuel in the heating system of the gas control point, eliminating the need to install GCU in which water heating in the heating system is carried out by gas combustion; provide more comfortable working conditions for the pressure regulator due to a decrease in the differential controlled pressure of gas entering the consumer, since there is a significant decrease in the output pressure due to the operation of the vortex tube as a heat exchanger in the heating

IOP Conf. Series: Materials Science and Engineering 1138 (2021) 012035 doi:10.1088/1757-899X/1138/1/012035

system of the production building; increase environmental safety in the production of thermal energy, since there is no need to burn natural gas for heating the production premises. Production of heat energy to maintain a microclimate in the production room is carried out by a regulated pressure drop of natural gas.

References

[1] Dubrakova K, Monastyrev P. Klychnikov R and Yezerskiy V 2019 Optimizatsiya teplovoy modernizatsii gruppy zdaniy s ispolzovaniyem imitatsionnogo modelirovaniya *Zhurnal prikladnykh tekhnicheskikh nauk*. Sbornik 17. Vypusk 2. S. 192-197

[2] Mishchenko E, Monastyrev P Evdokimtsev O 2018 Povysheniye kachestva podgotovki spetsialistov dlya energoeffektivnogo stroitelstva. IOP Conf. Ser.: Mater. Sci. Eng. 463 032046 https://doi.org/10.1088/1757-899X/463/3/032046/

[3] Erofeyev A, Yartsev V and Monastyrev P 2017 Dekorativno-zashchitnyye plity dlya fasadnoy otdelki zdaniy Izvestiya vysshikh uchebnykh zavedeniy. Tekhnologiya tekstilnoy promyshlennosti **1 (367)**

[4] Gusev B, Ezerskiy V and Monastyrev P 2005 Teploprovodnost mineralovatnykh plit v usloviyakh ekspluatatsionnykh vozdeystviy *Promyshlennoye i grazhdanskoye stroitelstvo* **1**

[5] Gusev B, Ezerskiy V and Monastyrev P 2004 Izmeneniye lineynykh razmerov mineralovatnykh plit v usloviyakh ekspluatatsionnykh vozdeystviy *Promyshlennoye i grazhdanskoye stroitelstvo* **8**

[6] Kovalenko L and Glushkov A 1986 Teploobmenniki s intensifikatsiyey teplootdachi *Energoizdat*

[7] Pisyumennyy E 1990 Sposoby sovershenstvovaniya teploobmennikov iz poperechnorebristykh trub *Promyshlennaya teplotekhnika* **36**

[8] Kutateladze O 1979 Osnovy teorii teploobmena *Atomizdat*.

[9] Voronin G and Dubrovskiy M 1973 Effektivnyye teploobmenniki *Mashinostroyeniye*

[10] Pozin G 1980 Printsipy razrabotki priblizhennoy modeli teplovozdushnykh protsessov v ventiliruyemykh pomeshcheniyakh *Stroitelstvo i arkhitektura* **11**

[11] Isachenko V, Osipova A and Sukhomel M 1980 Teploperedacha *Vysshaya shkola*

[12] Sudarev A and Kuznetsov L 1968 Teploperedacha zakruchennoy strui vozdukha pri dvizhenii po vnutrenney poverkhnosti tsilindra *Energomashinostroyeniye* **1**

[13] Merkulov A 2002 Vikhrevoy effekt i ego primeneniye v tekhnike *Samara* .385 s

[14] Kobelev N 2012 Osobennosti teploobmena vozdushnykh prosloyek iz paneley s vnutrennimi krivolineynymi kanavkami *Izvestiya Yugo-zapadnogo gosudarstvennogo universiteta* **5** Kursk

[15] Koryakin E 2003 Promyshlennoye gazovoye oborudovaniye *Gazovik*.

[16] Kobelev N 2013 Innovatsionnyye resheniya po povysheniyu effektivnosti sistem gazosnabzheniya i klimatologii Monografiya Yugo-Zap. gos. un-t. Kursk

[17] Kobelev N, Krygina A and Kobelev V 2011 Energosberegayushchiye konstruktivnyye elementy naruzhnykh ograzhdeniy *Izvestiya Yugo-Zapadnogo gosudarstvennogo universiteta* **2**

[18] Klassifikatsiya istochnikov vybrosov zagryaznyayushchikh veshchestv v atmosfernyy vozdukh 2010 Sankt-Peterburg

[19] Bondarenko V and Kolchunov V 2013 Kontseptsiya i napravleniya razvitiya teorii konstruktivnoy bezopasnosti zdaniy i sooruzheniy pri silovykh i sredovykh vozdeystviyakh *Promyshlennoye i grazhdanskoye stroitelstvo* **2** Moskva

[20] Kobelev N and Grigorova N.P 2017 Pat. 2615878 Rossiyskaya Federatsiya: MPK F 28 D 7/10 Vikhrevoy teploobmennyy element; zayavitel i patentoobladatel Kursk. gos. tekhn. un-t. №2016110870; zayavl. 04.07.2016; opubl. 11.04.2017. Byul. № 11

[21] Kobelev N and Grigorova N.P 2017 Pat. 2622340 Rossiyskaya Federatsiya: MPK F 28 D 7/10. Vikhrevoy teploobmennyy element zayavitel i patentoobladatel Kursk. gos. tekhn. un-t. №2016128870953; zayavl. 15.07.2016; opubl. 14.06.2017. Byul. № 17

IOP Conf. Series: Materials Science and Engineering 1138 (2021) 012036 doi:10.1088/1757-899X/1138/1/012036

Restoration and preservation of the authentic traditional housing model. Case study - the peasant house in the Danube Delta area

B D Pericleanu[1], M Pericleanu[1*] and A M Grămescu[2]

[1] Faculty of Civil Engineering, Ovidius University of Constanta, Romania
[2] Institute of Doctoral Studies, Ovidius University of Constanta, Romania

* dragoi.mihaela@gmail.com

Abstract. The paper develops a new concept, relatively unique in approaching the concept of rehabilitation of the peasant house at the beginning of the XXI century, namely the traditional architecture of the Danube Delta as a system of integrated research, preservation of cultural and historical values. As it is known, the Vernacular Heritage Charter, ratified in 1999 in Mexico, draws attention to the fact that vernacular architecture around the world is valuable and requires protection, positive attitude in preserving, preserving and transmitting technologies, materials, architecture combined with authentic elements can even unique. This valuable built fund is increasingly subject to the pressures generated by urbanization, globalization (a phenomenon that affects local specificity) and lack of information or interest. The vernacular architecture of the Delta area was executed empirically, within the community using local materials, techniques and traditional crafts is what the Austrian American architect Bernard Rudofsky called it, in the title of his 1964 book, "*Architecture without architects*". The heritage reflected in the built environment of the villages in the Delta belongs to a type of vernacular architecture, and, of course, shares the general characteristics of the genre: a close communion with the natural landscape and the society in which it was "*born*". If we talk about contemporary ideology in architecture, in the context of very present discourses on sustainable development and sustainability (in absolutely all areas). The paper presents two approaches; high-tech, which supports the use of modern materials and advanced technology in construction and the low-tech approach, which aims at methods and techniques with low impact on the environment. The authors detail the concept of restoration corresponding to traditional materials and technologies.

1. Introduction

When we talk about Dobrogea, we are referring to 3 main areas – North Dobrogea, Central Dobrogea and South Dobrogea [1-4]. In this studied area we highlighted the construction techniques and methods, the materials specific to vernacular architecture that has withstood over the years of modernity. Popular architecture is highlighted in [1] by professor Andreşoiu Bruno as *"... more fragile, less versatile and less glorious"*, but no less significant, authentic and even unique.

In the history of Dobrogea, several periods can be distinguished regarding the evolution of the Dobrogean settlements, including the houses characterized by the plan structure and the materials used. We can refer to the following periods [5]: before 1878, period characterized by huts and houses with 1-2 rooms built of "*paiantă*", wicker twigs, "*ceamur*" and covered with straw, reeds, awnings predominate. Rarely 3 room dwellings of adobe and tile roofs; the period between 1878 – 1950

IOP Conf. Series: Materials Science and Engineering 1138 (2021) 012036 doi:10.1088/1757-899X/1138/1/012036

characterized by the houses with 3 rooms predominate and at the same time the disappearance of the huts as living houses is noticed; the period after 1950 – 2000 characterized by three-room houses also predominate, but there is the introduction of new construction materials alongside the traditional ones, a tendency to increase the living space with new rooms and well-defined functions and decorative elements on the façade and the current period characterized by houses that are built on a load-bearing brick structure, BCA and / or with a structure in reinforced concrete frames, to the detriment of traditional materials that were handy and more environmentally friendly.

From the point of view of traditional architecture, the household of the house in the Danube Delta is composed of the house itself and the economic constructions arranged either in alignment with the house or behind the house, but in this situation, we find in the continuation of the house. The porch house plays a key role in the architecture of the Danube Delta as in figure 1. The studied building is characteristic of the vernacular architecture not subjected to the pressures generated by urbanization, globalization (phenomenon that affects the local specificity) and the lack of information or interest. It is made empirically within the community and uses natural materials, local resources, techniques and traditional crafts (or, as the American architect of Austrian origin Bernard Rudofsky called it, in the title of his 1964 book, "*Architecture without Architects: a brief introduction to architecture without pedigree*").

The paper presents two approaches; high-tech, which supports the use of modern materials and advanced technology in construction and the low-tech approach, which aims at methods and techniques with low impact on the environment. The authors detail the concept of restoration corresponding to traditional materials and technologies.

Figure 1. Example of specific houses of the studied area.

As in most rural areas, the heritage reflected in the built environment of the villages in the Delta belongs to a type of vernacular architecture, and, of course, shares the general characteristics of the genre: a close communion with the natural landscape and society in which it was "*born*". The house is located close to the road, perpendicular to it, with the short side on the street, which has a narrow porch that is partially found also in the facade of the yard [5].

2. Typical peasant houses in Dobrogea

The study highlights the use of local materials: wood, twigs, reeds, rushes, earth. These are some of the materials used by popular architecture along with pantiles [1], tiles, bricks masonry, etc. During the 20th century, other materials were brought, such as the stone from the Măcinului mountains and the Casimcei plateau. In this case study, aspects related to physical condition, aspects relevant to the evaluation of the strength and safety of a construction with such a structure, as well as the establishment of restoration solutions were investigated.

In the imagines below (figures 2 and 3) you can see the degradation survey for the entire house.

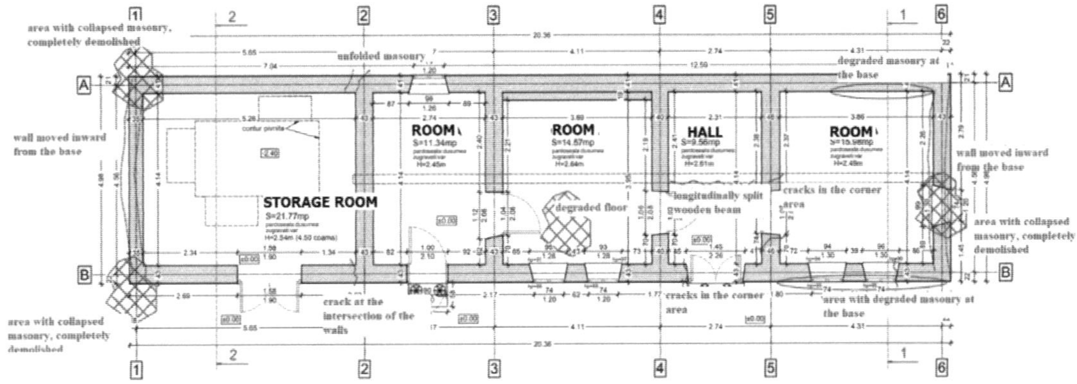

Figure 2. Survey degradation - horizontal plane.

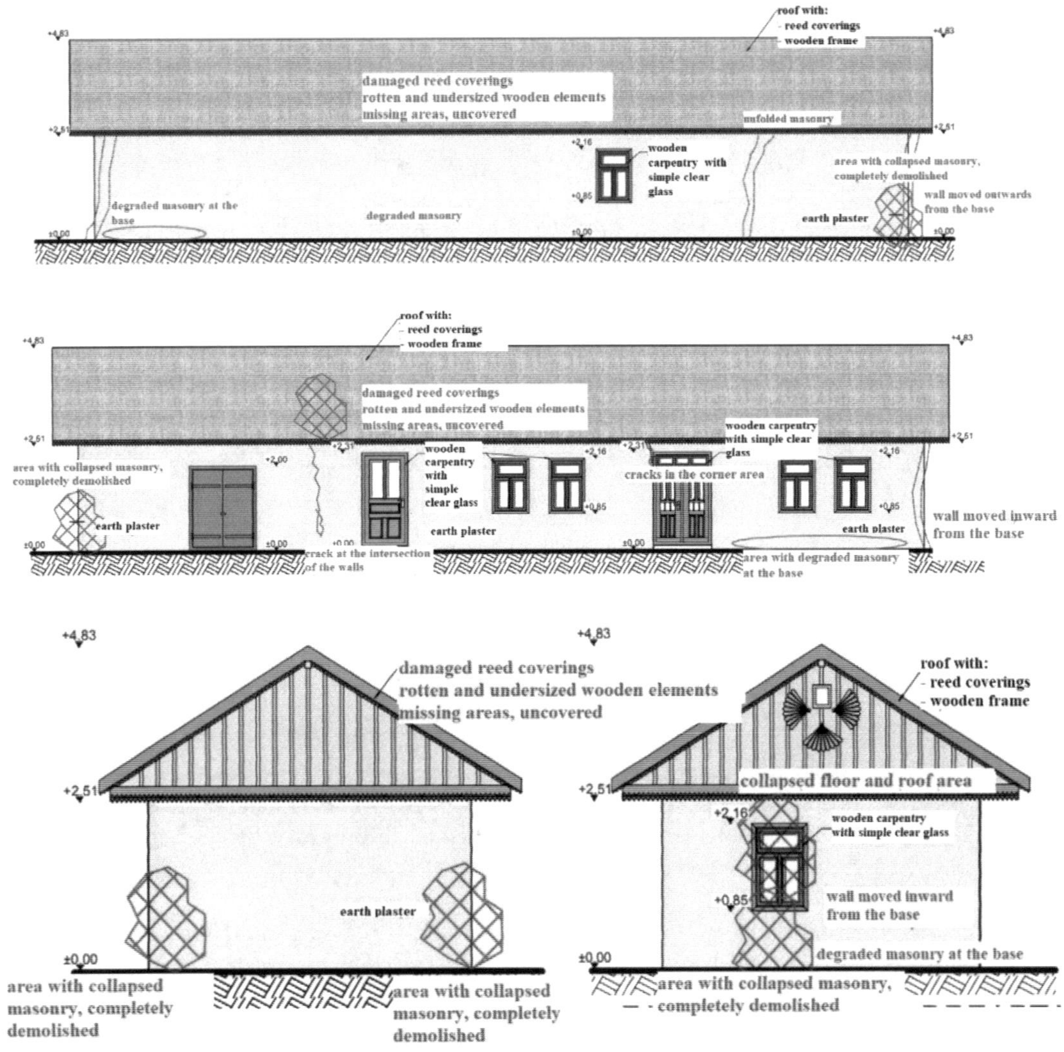

Figure 3. Survey degradation – house facades.

IOP Conf. Series: Materials Science and Engineering 1138 (2021) 012036 doi:10.1088/1757-899X/1138/1/012036

Most of the buildings do not have a foundation like in fact the one studied, where the structural walls are placed on a layer of sand. At other times, the houses were placed on a stone and sand platform over which was made a bed of beaten earth with a thickness of 1 - 1.5 m in which the pillars of the resistance structure were embedded, so that the building was located at a higher level than the ground natural. The function of the house is shaped by the existence of the porch, the living room, the dormitory room but also the guest room. The shape of the building is rectangular, with the placement of wagon-style rooms, usually with two entrances. Sometimes through successive extensions of the house they could get the L shape. The construction technique was made of forks and reeds glued to the ground. The forks were made of oak, which sewed together. Above ground they were tied with horizontal ties. Between the forks he placed on the outside a reed placed vertically and tied with wire on the inside of the walls, after which leashes were beaten and on the inside of the walls to fix the reed. The finish consists of clay plaster, later after their drying other finishes are applied.

Another technology for making houses in this area was "*paianta*" [6]. The "*paianta*" as a constructive solution consists in building a wooden structure - slats and balls with a diameter of 10 - 15 cm, after which it is filled inside with "*ceamur*" (clay mixed with straw). The adobe appears as an alternative to the "*paianta*", making clay blocks with straw with dimensions of 35 - 40 cm long, 20 - 25 cm wide and 12 - 15 cm high. For adobe, clay must be cleaned of impurities. Thus, the "*ceamur*" [6] is poured into the formwork and left to dry for at least 30 days to make the adobe. The finishing of these constructions was done with clay and lime in which the "*sineala*" was added. Technologically, 3-4 rows of adobe sawing were made, after which a minimum of 1 day was left to dry the joints [7]. Perimetrically, the chips were placed on the side of the wall to achieve a greater wall thickness. At the studied building, two rows of adobe were found, placed so that the thickness of the wall is about 80 cm perimeter. The information on the specific construction details of the building highlights the following aspects:

- the building has identity cuts, a replica of compositions that combine Turkish tradition ("*ceamur*") with Russian elements;
- the roof is made of reeds;
- the facade is painted in bright colors with blue or green of Slavic connotation with whitewashed exterior walls;
- the decorations on wooden pediments / eardrums are of Slavic origin;
- the presence of "*lijanca*" is of Slavic origin;
- composition of the structural material: clay with straw and organic material is of Turkish tradition that has gone from the structure of lumps on wooden skeleton or twig structure to blocks made of clay, Slavic procedure - the adobe that is not tied with wire similar to the area southern Europe, but is relatively free but in a technological weaving taken from the Asian craft.

A characteristic element of the local Lipovan house is the presence of a completely original heating system, "*lijanca*". It is an earthen bed built next to the stove and which is heated due to an ingenious installation.

The volume, although simple in composition, is personalized by the space of the porch, delimited by thin painted wooden pillars, which surround the house on one or two sides. Instinctively, the craftsmen found unstudied proportions, but which satisfy the eye. Through the intermediate space of the porch, the transition from the outside to the inside is made, in a subtle and skillful way, this becoming an element of connection between the outside and the inside.

A reed cover is placed over the frame [8], which has two layers of reed: a long layer arranged as an asterial over which the second layer of medium-sized (colored) reed is applied. This layer has a special technology, it is uniform by beating from the bottom to the ridge. This layer will have a thickness of about 25 - 30 cm made by adding successive bundles. The roof ends at the ridge with a ridge protection arrangement also made of reed. The service life of such a roof system is about 20 years unlike pantiles or tiles that have more life service [1]. Regarding the construction elements the roof specific to the Delta is the roof with reed coverings, made in specific techniques (figure 4): German beaten method and Russian beaten method [8].

IOP Conf. Series: Materials Science and Engineering 1138 (2021) 012036 doi:10.1088/1757-899X/1138/1/012036

Figure 4. Specific techniques for reed coverings.

The architecture of the house attracts through the decoration of the house made of perforated boards. The most decorated parts are the pediments. The constructive system specific in the area for this type of houses is characterized horizontally by bi and tri cellular houses and vertically by one level or two / less common.

The models are specific to villages with Lipovan population and refer to Slavic myths as in figure 5, although today the inhabitants no longer make any connection with them. The sun's rays meet, the siren's tail.

In Romanian houses, the pediments are decorated with geometric, floral, zoomorphic figures.

Figure 5. Specific lipovan decoration for house fronton.

The floor above the ground floor has a longitudinal beam on which the ridge beam unloads. On the rafters and the longitudinal central beam, unload transversal beams on which slats are beaten and a layer of reed about 40 cm thick is arranged as in figure 6.

"*Ceamur*" is placed over the reed and this whole system forms the garret/attic. Inside, the ceiling is plastered and whitewashed so that the median longitudinal beam remains visible, which in most cases is made of poplar wood and less often of fir.

Figure 6. Specific type of floors above the ground floor.

Another construction element: the foundation. Most of the buildings do not have a foundation, like the one studied in fact, where the structural walls are placed on a layer of sand. At other times, the houses were placed on a stone and sand platform over which was made a bed of beaten earth with a thickness of 1 - 1.5 m in which the pillars of the resistance structure were embedded, so that the building was located at a higher level than the natural ground.

3. State of degradation and implementing restoration measures

Considering the presented aspects, the case study is the analysis of the technical condition of a construction for restoration with reference to: the structural conception of the building, the degree of degradation, but also the request made by the beneficiary to refurbish the building and introduce it in the tourist circuit, corroborated with the necessary rehabilitation and modernization works, intervention measures were proposed that aimed at reducing the seismic risk, ensuring the resistance and stability requirements characteristic of the function.

The establishment of construction vulnerabilities was analyzed together with:

- required level of performance;
- degree of impairment of the function;
- the real technological possibilities of the location but also of the building and the economic possibilities of the beneficiary.

In this case study, issues related to physical condition, issues relevant to the assessment of the strength and safety of a construction with such a structure, as well as the establishment of restoration solutions were investigated.

During the investigations on site made we could observe and analyze that:

- the heel wall facing north was affected by weather and the presence of stored snow and may not always be removed in time, rain and the lack of gluing / plastering at the base performed continuously annually - led to the weakening of the wall on the lower area with loss effect of its own stability as in figure 7, the yielding with the appearance of the thrusts at the transversal walls that cracked at the upper part, deforms in the area of the doorway, the yielding of the lower window truss in axis 6;
- vertical plane deformations of the axis wall A but slightly and B motivated by the high load of the snow-driven roof but also by the loss of the quality of the reed layer by impregnation with water, yielding, mold, etc.
- the degradation of the roofing led to the degradation of the wooden eaves, but also of the floor beams over the ground floor, which show obvious signs of exceeding the load-bearing capacity, especially the secondary ones;

- degradation of the carpentry due to the lack of ensuring its protection through painting;
- degradation at the base of the plasters with serious damage to the base of the structural walls, washing of the soil both in axis A but also in axis B. In axis A there were even displacements fact for which the restoration of this part considers the application of the anastylosis procedure;
- degradation of the turbocharged wall between axis 1 and axis 2 by local yielding and expulsion of the adobe due to the weight of the roof in the wet state;
- floor degradation and the presence of a strong musty smell.

Figure 7. State of degradation of the walls.

Figure 8. State of degradation of the North-facing wall.

North-facing wall - has deteriorated sharply especially at the base as shown in figure 8.
The main data regarding the geometry of the structure from the local material took into account:

- the plan positioning of the structural walls and their dimensions, of the solid and hollow spatulas;
- vertical and horizontal deformability;
- positioning and dimensions in the plane and elevation of the gaps (doors, windows) and
- of low-thickness wall areas (niches);

- positioning in plan and elevation of the structural elements that generate thrusts by local yielding of the structural elements that led to deformability as well as of the elements that can partially take over the thrusts;
- positioning in the plan and dimensions of the main wooden floor elements;
- the weight of the elements structurally corroborated with the lack of foundations.

4. Conclusions

The recommended intervention strategy was a minimal variant that was based on improving the mechanical characteristics of the building, on dismantling and restoring degraded areas, deformed without changing the structural and functional composition of the building, while maintaining its authenticity, rarity and possibly even uniqueness.

The restoration was proposed to be done with original materials and technologies. The authors propose that when this is technically possible, in compliance with the essential requirements under Article 5 of Law 10, to try to restore with traditional materials and technologies but also using new materials compatible with the existing structure but also environmentally friendly. The degraded areas of masonry were dismantled and rebuilt by making masonry according to traditional recipes, by weaving blocks, treatments with waterproofing solutions for elements exposed to the weather, restoring the reed covering on the affected parts, treating the wooden elements of the frame and floor, treating carpentry and wood flooring.

We believe that we can discuss specific and traditional housing in the Dobrogea area as long as we consider the structural system, local materials used, assimilated architecture, including house volumes, materials implementation technologies and tools used specific to the area and fingerprints left by ethnic groups that are found or have existed on the territory of Dobrogea.

The paper develops a restoration concept among vernacular architecture restorers of peasant houses in the Danube Delta, a relatively new and unique concept in approaching the concept of peasant house rehabilitation at the beginning of the XXI century, namely the traditional architecture of the Danube Delta as a system of integrated research, of preserving cultural-historical values, of Romanian traditions with Slavic and Asian influences.

The paper presents a non-traditional intervention that can help to develop a restoration concept among vernacular architecture restorers of peasant houses in the Danube Delta, with principles to take into consideration as a system of integrated research of preserving traditional cultural-historical values.

References

[1] Andreşoiu B, Ioan A, Şerbescu A, Dobrogeanu P and Pârâu S 2012 Pantiles. Traditional houses from Dobrogea, *Igloo* p 199
[2] Order of Architects of Romania 2016 Architectural guide for framing in the local specifics of the rural environment. Danube Delta Area p 66 http://oar.archi/despre-oar/ghidurile-de-arhitectura-pentru-incadrarea-in-specificul-local-din-mediul-rural
[3] Order of Architects of Romania 2016 Architectural guide for framing in the local specifics of the rural environment. South Dobrogea area p 71 http://oar.archi/despre-oar/ghidurile-de-arhitectura-pentru-incadrarea-in-specificul-local-din-mediul-rural
[4] Order of Architects of Romania 2016 Architecture guide for framing in the local specifics of the rural environment. Central Dobrogea Area and Măcin Mountains p 70, http://oar.archi/ despre-oar/ghidurile-de-arhitectura-pentru-incadrarea-in-specificul-local- mediul-rural
[5] Popoiu P 2001 Habitat anthropology in Dobrogea: man - nature – culture Oscar Print Publishing House
[6] Pericleanu BD and Pericleanu M 2019 Contributions to the analysis of building's durability made of local materials *Ovidius University Annals, Construction series* **21** p 121-126
[7] Minke G 2006 *Building with Earth* Springer Publishing House.
[8] Titov I and Chiselev A 2015 Traditional crafts in the Danube Delta. Research report

IOP Conf. Series: Materials Science and Engineering 1138 (2021) 012037 doi:10.1088/1757-899X/1138/1/012037

Evaluation of the quality of the road infrastructure with the photogrammetry technique

V Popa[1], G Năstase[1*], G Dragomir[1], A Brezeanu[1] and A Şerban[2]

[1] Faculty of Civil Engineering, Transilvania University of Brasov, Romania
[2] Faculty of Mechanical Engineering and Mechatronics, Thermotechnics, engines, thermal and refrigeration equipment Department, Politehnica University of Bucharest, Romania

* gabrielnastase@unitbv.ro

Abstract. Solutions that are integrating different modern techniques related to a new or an existing road, using non-destructive test brings several advantages which the content of the article highlights. Comparing the current method used in the road technique AND 540 and the aerial scanning techniques, the main element that stands out is the speed of accomplishing an ample work which consists in the 3D processing much faster and more accurately. The article presents a short current stage of these applications in the field of civil engineering. Creating 3D surfaces that it compares with what it was initially and also processing registered the deformations and displacements. A case study of a road sector is also presented to highlight everything that this technique brings.

1. Introduction

Specialists in the field are continually looking for an effective analytical tool to assist in analysing pavement structures, taking into consideration the in-service condition of the road. This paper addresses the problem of using photogrammetric systems as a support for topographic and geodetic works in the field of roads constructions.

The modernization, rehabilitation and construction of roads and works of art have required a significant increase in the last period, the need comes from the fact that the infrastructure is aging the need for investment, in this context the use of new technology brings increased speed. Photogrammetry as accurate non-contact measuring method provides powerful means for solving different tasks in road surface reconstruction and analysis. The range of dimensions concerned in road surface analysis can have great variation from tenths of millimeter to hundreds of meters and more [1]. The data provided in this way significantly shortens the time but also improves the process of collecting information from the field, taking into account the accuracy of the process as seen in figure 1.

Through this paper, the main goal is to present an efficient, working alternative to facilitate the work in terms of identifying and assessing the state of degradation of bituminous pavement for roads, based on the AND 540 Romanian standard, as a guide for identification and classification types of degradation and using elements in the field of photogrammetry for taking over and analysing field characteristics.

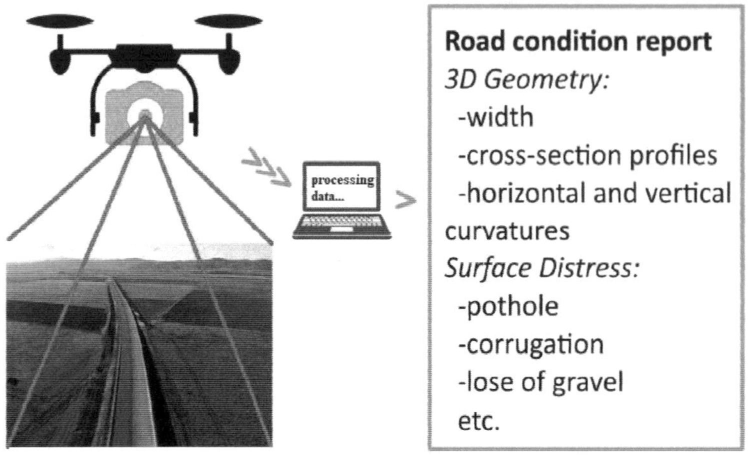

Figure 1. Work process.

2. Advantages and disadvantages

Both, the classical method of identifying and evaluating the state of road degradation, and the proposed method, have advantages but also disadvantages. Compared to the classic method, regulated by AND 540 [2], the proposed method has as its first advantage the efficiency given by the reduced working time but also the quality of determining the results. A very large amount of data can be obtained in a relatively short time compared to the classic method, without going through the entire target area. In addition, the chances of errors of the operator interpretation are much reduced because the entire area is analysed, and by the classical method, can be identified degradations that are overlooked from the negligence of the operator. Moreover, the situation in the field from the moment of taking over the data, remains at the disposal of the operator for possible ambiguities. Then, another very important advantage is the accuracy of the determinations. The evaluation of the degradations appeared at the level of the bituminous coating is done digitally on the 3d model using specialized software, which eliminates the reading errors on the roulette or the approximation errors, determining precisely the dimensions of the degradations. As a main disadvantage is the fact that working on this method is not provided in any norm, determining the user to create his own working method using his knowledge in the field but also the AND 540 norm. Then, another disadvantage is the high cost of the equipment used but also of the software.

3.Methodology for determining the quality of road infrastructure

The problems of the road structure design require complex approaches that needs a higher quantity and quality of input data in four major categories: traffic; material characterization and properties; environmental influences; pavement response and hazard models [3-7]. The problem of maintaining the viability is complex, on the one hand due to the economic side of maintaining and monitoring the state, the mechanisms for prioritizing the works, the policies that changes periodically [4].

The evaluation of the road pavement degradation condition is currently done in Romania, according to the AND 540 norm, visually by going through the analyzed surface, followed by the processing of the data collected from the field and their analysis in order to determine the technical condition of the road. After obtaining these data, the technical expertise is drawn up, based on which the amount of works necessary to carry out the improvements of the technical condition of the road is determined.

The proposed method involves the identification of the area that has to be analyzed, planning the flight as seen in figure 2a, materializing and determining the control points, performing the flight, downloading and processing the images and analyzing the data obtained after processing.

Flight planning is done using Mission Planner software. The following elements from the land will be taken into account: the inclination of the land, elements that are found in the vicinity - vegetation, constructions, lighting poles, etc. "The flights followed a conventional photogrammetric pattern with vertical photographs and image overlaps of the order of 70 - 80% to guarantee a proper vertical accuracy and enough redundancy" [8]. The take-off of the aircraft is always done from the outside of the targeted perimeter, from a flat ground, free of bushes, trees, constructions or other elements that can make it difficult to take off and later land. It is recommended that the flight has to be made around noon when the sun's rays form at right angles to the ground surface and the elements in the ground, avoiding shadows. The height from which the flight is made is done depending on the accuracy and density of the desired point cloud. The higher the flight altitude is, the lower the accuracy of the details are. In this case, the flight was made from 7 m above the ground and the obtained accuracy was 1.5mm / pixel. The area on which the flight is made, is composed of the road sector to be analyzed and areas in its immediate vicinity, so the road sector will be more easily framed in the area. The flight direction is chosen so that the route is as short as possible.

a b

Figure 2. (a) flight planning; (b) data interpretation using Agisoft Metashape software.

In order to the existing points in the digital terrain model and the points in the field to coincide, a number of control points/benchmarks is needed to connect them. The establishment of their location is done by analysing the targeted area. The control points will be uniformly located on the entire targeted surface at optimal distances. "The benchmark system should be sufficient to cover overlaps of set-ups but not more than 200 m apart. The benchmark system must be check levelled to a 2 mm/km closure [9]. In this case, the checkpoints were marked with black carioca or metal nails and highlighted with orange spray. The determination of the coordinates of these points is done with the help of the GNSS RTK system by making successive measurements for 60 seconds in point and calculating the average position.

The images are downloaded using a memory stick or SD card. Each image will contain the following instructions: the name of the image, the time when it was taken and the approximate position. This information is needed when processing the images and aligning them. The image processing is done using the Agisoft Metashape program as seen in figure 2b.

Agisoft Metashape is a program that allows you to obtain point clouds, the DEM model of the terrain or orthophotograms based on aerial photographs. Uploading photos and the first step is to align the photos. This is done using the Align photos command in the Workflow menu. At this stage, the photos are aligned with each other, depending on the approximate position and the common elements in each photo. A sparse point cloud is also formed (tie points), consisting of the identified common elements in each photo. In this case, the sparse point cloud is about 265,683 points. Markers / control points are used for a better reconstruction of the model. More precisely, the operator identifies the field markers in the photos and makes the connection between them by loading in the program their measured coordinates.

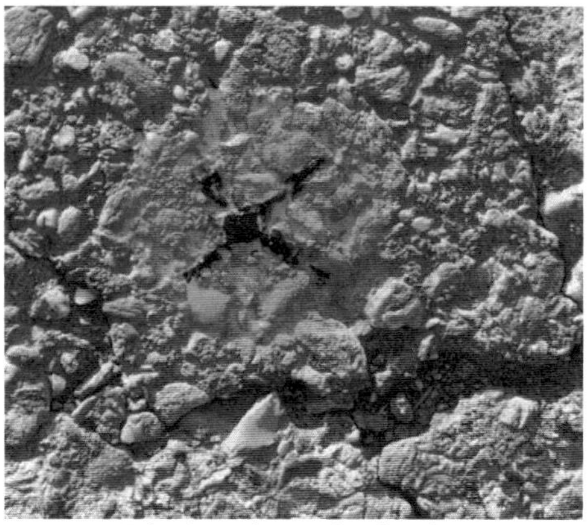

Black carioca checkpoint

Metal nail checkpoint

Figure 3. Control points / check points on the field.

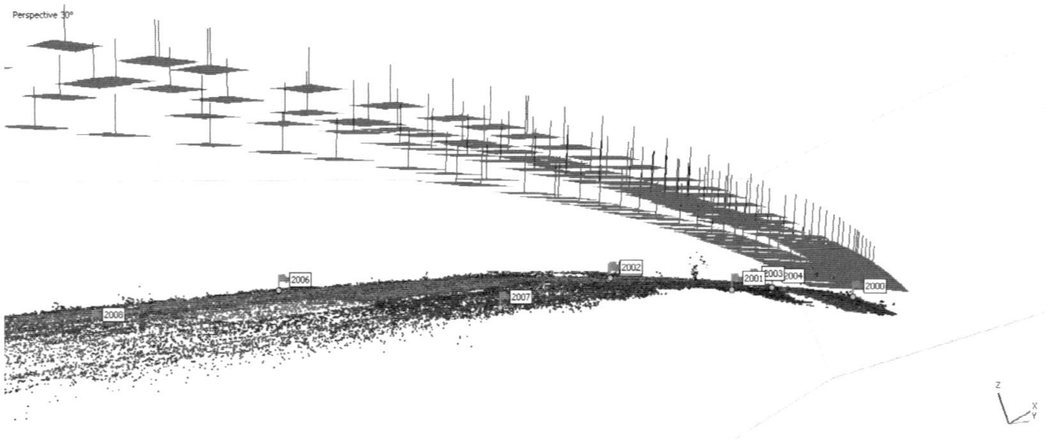

Figure 4. Photo alignment, sparse point cloud and markers.

After entering the checkpoints, the photos are realigned. After performing this process, the generated point cloud, as well as the following procedures are in the WGS84 projection system given by the coordinates of the control points.

The next process is Build dense cloud. Three-dimensional point clouds are the main result of both terrestrial and airborne laser scans [10]. The points in the dense cloud are calculated by rectifying the pairs of successive images thus according to the orientation parameters and then an algorithm is used that matches each pixel in the image pairs. In this case, the resulting point cloud is composed of 654 296 860 points. The points in the dense cloud are so detailed that they faithfully reproduce the shape of the element in the field.

From the dense cloud can be generated "Mesh" - obtained from the creation of networks of polygons formed by the points in the dense cloud. It reproduces, better than dense cloud, the shape of the details in the field, representing a three-dimensional model of the terrain as seen in figure 5.

To better highlight the texture, "Tilled model" and "Textured model" are made. The difference between all these models lies in the details.

The last stage of image processing consists in making the orthomosaic, on which, later, the defects will be identified. The resulting orthomosaic has the size of 13928 x 72346 and an accuracy of 1.5mm / pixel.

4. Results interpretation

On the orthomosaic and the other models resulting from the image processing, the next step is the interpretation of the results which consists in the identification and registration of the defects appeared at the level of the road infrastructure. Defects are recorded according to: shape, surface, length, depth, exact position, etc., according to the indications in AND540. Each defect is registered in a specific layer, depending on the type. The characteristics of each type of degradation are recorded in an attribute table corresponding to each layer.

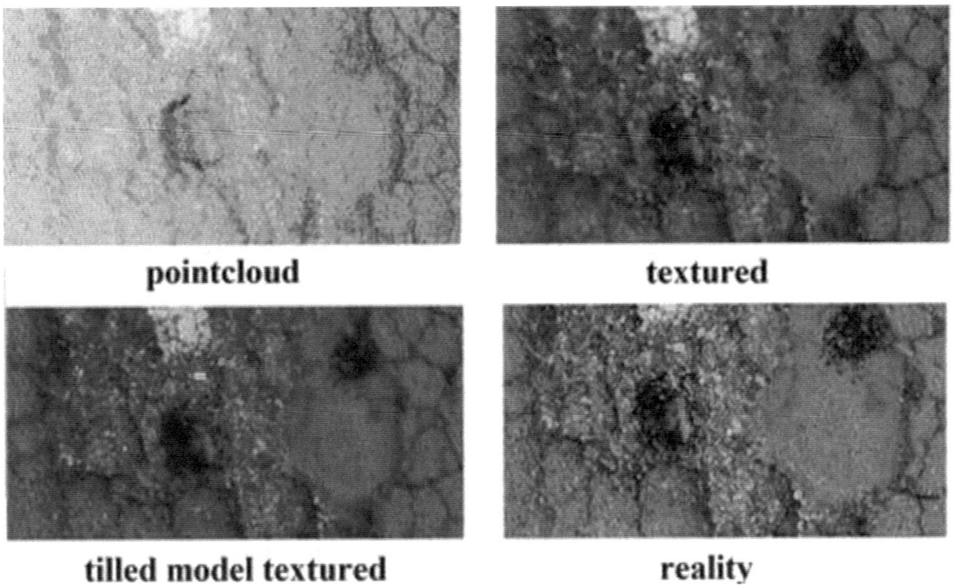

Figure 5. Dense point cloud and models.

Following the analysis, the following defects were identified: spider web tiles, slab tiles, slits and longitudinal cracks, seals and sealed surfaces, potholes, grooves, cracks and transverse cracks, cracks and longitudinal cracks due to the reflections of concrete slab joints, cracks at the edges, peeling, polished surfaces, pinched surfaces.

From the category of structural type degradations, the most common types of degradations were: tile tiles, longitudinal cracks, seals and sealed surfaces and pinches.

The pinches, according to AND540 "are the defects of bituminous layer, with a diameter of less than 750 mm, located in heavily eroded, tiled, cracked or compacted areas". These are evaluated according to the maximum depth and diameter.

The orthomosaic and the other models obtained based on which the field data is interpreted are loaded in the Qgis software. Based on them, cross sections are extracted in the desired areas to determine the depths as seen in figure 6.

If, by the method provided in AND540 "The surface of the potholes is determined by framing it in a rectangle, appreciating the size of the sides a and b, in meters", the determination of the surface of the potholes in the presented method is much more precise: the exact shape of the pothole is given by an irregular polygon, identical to its real shape whose surface is calculated automatically, thus obtaining a result much closer to reality than by the classical method, especially when it comes to pinches with irregular shapes as seen in figure 7.

IOP Conf. Series: Materials Science and Engineering 1138 (2021) 012037 doi:10.1088/1757-899X/1138/1/012037

Figure 6. Determining the depth of the pinches - cross section.

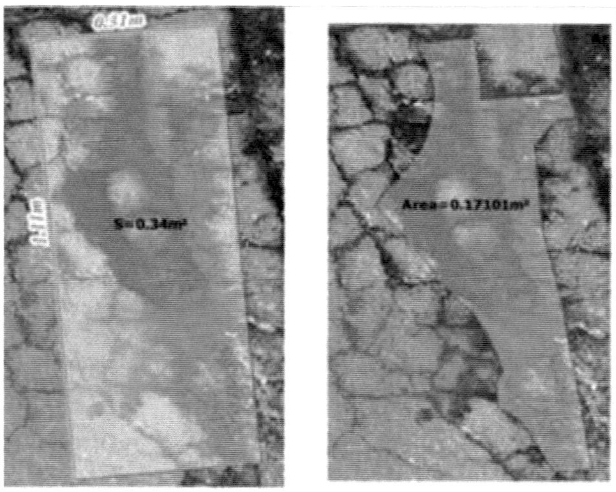

Figure 7. Determining the area of the pinches with both classical (AND540) and presented method.

At the level of the analyzed area, a number of 84 potholes with surfaces varied between 0.002 and 0.32 m² was identified. Each pinches type defect was registered in a specific layer, and for each one the coordinates, the surface is known and the depth can be determined, and depending on these details the severity level is established. After identifying all the pinches, a map of the potholes with their position and shape is obtained as seen in figure 8.

Similarly, the seals and sealed surfaces are identified. These were found in very large numbers, 62, on the analyzed area, with areas ranging between 0.0375 and 81,713 m² totaling a total area of 288,12573 m². In the same way, the alligator cracking's, the longitudinal cracks, the edge cracking, the depression, the rutting's, with pinches and beeches are also identified.

Identification and registration of the types of degradations, for which the evaluation is made by measuring the length: longitudinal cracks and fissures, transverse cracks and fissures longitudinal cracks and fissures due to the reflection of cement concrete slab joints, and cracks it is made, by the method presented, on the orthomosaic. The defects are identified on it, and the shape

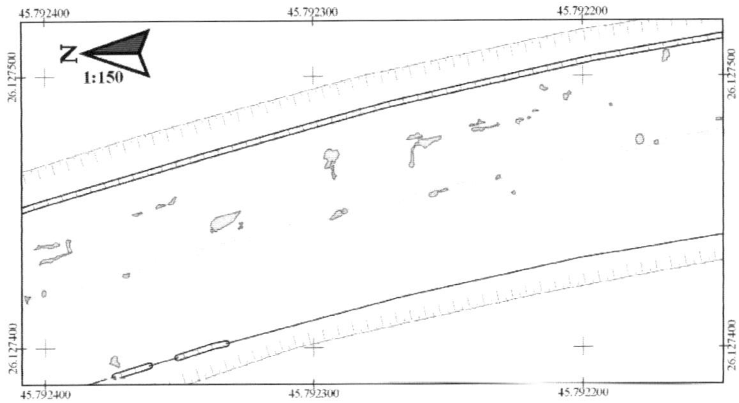

Figure 8. The map of the pinches.

is registered in a line type layer. The determination of the severity level is done by measuring, on the orthomosaic, the opening of the degradation in question. The value of the crack length is calculated automatically for each identified element and is registered in the attribute table together with the other details. Finally, the crack map is obtained, and in the present situation, the longitudinal and transverse cracks were evaluated in a single layer as seen in figure 9.

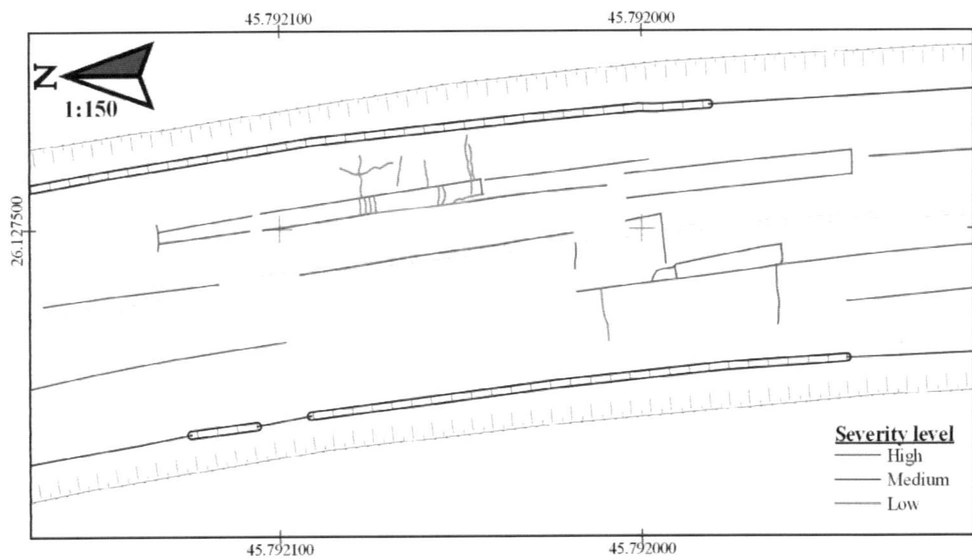

Figure 9. The map of longitudinal and transverse cracks.

Similar to cracks, the identification and registration of other defects is performed, which are evaluated according to the length.

In addition to working with the orthomosaic, to clarify the details of identifying types of degradation and severity, you can use the other models obtained for image processing: point cloud, tilled textured model, which can be traversed and viewed from any perspective and angle (figure 10).

IOP Conf. Series: Materials Science and Engineering 1138 (2021) 012037 doi:10.1088/1757-899X/1138/1/012037

Figure 10. Edge fissures and cracks.

Another important piece, but also useful for the analysis of a road sector in terms of defects but which is also useful later on in the design of rehabilitation works, is the level cube plan. Following the image processing process to obtain DEM- Digital elevation model. From this model, level curves can be generated at intervals of millimeters, centimeters or tens of centimeters as seen in figure 11. Level curves help a lot in identifying and determining depths for potholes or cracks, cracks, veils and unevenness.

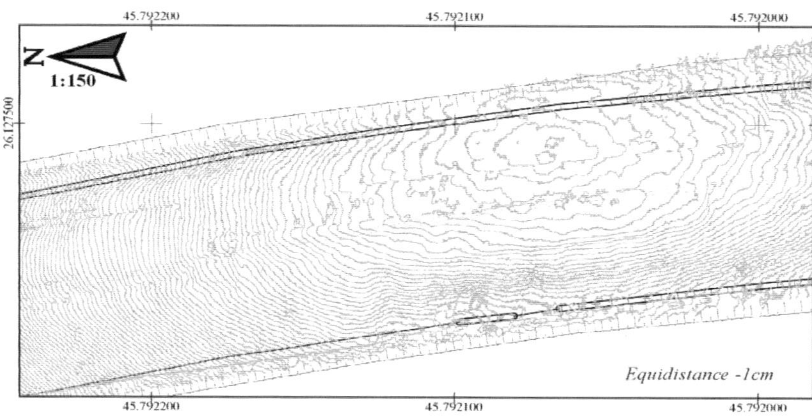

Figure 11. Level curves – detail.

5. Conclusions

The technique of measurements in the field of engineering topography, has evolved so much that in the case of elevations made to modernize roads, their use bringing important benefits both technically and economically. The time to perform the measurements is considerably reduced, this benefit is highlighted by the final costs of the work performed. The number of points acquired in a short time is very large, this helps to interpret very correctly the scanned objects even small ones. The software for processing the measurements resulting from the scan consists of several methods of compensation and registration, this fact allows the user to choose the appropriate method of processing depending on the characteristics of the study being performed. The reports with the results can be analyzed, compared and evaluated for the realization of statistics. The creation of sections, longitudinal and transversal profiles is done automatically, by imposing the sectioning interval. Sections and profiles, as a final product, can be obtained in the point cloud system (national system) or in a local coordinate system of each section for real-time relative comparisons.

The errors and mistakes due to the human staff are reduced in a very high percentage, the results being more objective and the situation in the field from the moment of making the measurements can be reviewed as many times as necessary without going on the site.

References

[1] Knyaz V A and Chibunichev A G 2016 *Photogrammetric techniques for road surface analysis* (Prague, Czech Republic)
[2] AND 540/2015 *Assessment of the state of degradation of bituminous pavement for roads with flexible and mixed pavements*
[3] Pleşcan E-L and Pleşcan C 2014 *The International Scientific Conference CIBv 2014* Implementation of mechanistic empirical pavement design guide Me-Pdg in Romania,
[5] Puşlău E and Pleşcan C 210 *Specific pavement condition indicators for rigid and flexible pavements*, "Computational Civil Engineering 2010" Iaşi, New computationalconcepts în civil engineering, (Iaşi Romania)
[6] Andrei R, Puşlău E and Pleşcan C 2010 *Evaluation of performance of flexible pavement by using specific pavement condition indicators*, Nat.cong roads and bridges (Brasov, Romania) pp. 74-83
[7] Pleşcan E-L and Pleşcan C 2015 *Roads Winter Maintenance Management System (WMMS)*, The International Scientific Conference CIBv 2015
[8] Kraus K 2007 *Photogrammetry: Geometry from Images and Laser Scans*; Walter de Gruyter (Berlin, Germany)
[9] Altus Strydom, Global Geomatics 2011 *Road rehabilitation and scanning technology*
[10] Siwiec J 2018 Comparison of airborne laser scanning of low and high above ground level for selected infrastructure objects, *Journal of Applied Engineering Sciences*

IOP Conf. Series: Materials Science and Engineering 1138 (2021) 012038 doi:10.1088/1757-899X/1138/1/012038

Reactive power compensation in the railway electrical traction system, using synchronous machines controlled by SCADA

A F Popov[1*] and M Fratu[2]

[1] Department of Automatics and Technology of Information, Transilvania University of Braşov, Romania
[2] Department of Building Services, Transilvania University of Braşov, Romania

* alexandru.popov@unitbv.ro

Abstract. This article brings forward a solution for compensating the reactive power of the railway electrical traction system. Using the V-characteristics of the synchronous machine, reactive power can be produced or absorbed from the power supply network, keeping the power factor at a value as close to 1 as possible. Given that the load consumed by the railway electrical traction system is variable, it is necessary to be able to promptly correct the power factor. The SCADA system will change the synchronous machine's operating characteristics relative as a function of the angle between the voltage and the current that are set at a given time by the consumers (electrical locomotives) passing through the railway section. By compensating the reactive power, the amount of active power supplied to the electrical network of the railway traction system is increased, without the need to modify the cross-section of the electrical conductors.

1. Introduction

The electrical railway transportation system (ERTS) employs electrical locomotives and railcars, with a 1-hour peak power between 600 and 6600 kW. In addition to the absorption of active power from the network, the functioning of the rolling stock requires absorption of reactive power. This leads to a drop in the system's power factor.

Continuous operation of the ERTS with the power factor under the neutral value (the agreed-upon limit above which expended reactive power is not billed) negatively impacts the cost price of railway transport. It has been shown that the lack of power factor compensation has led to the rise of cost price of consumed and billed electricity by approximately 110% over 30 days, over roughly 60 km of railway [1].

Power factor compensation at power supply points of the ERTS increases the available active power that can at one time be delivered to the traction system. Additionally, electrical overloads are eliminated from the power system, which in turn increases the stability of the supply and distribution of electrical power [2].

2. Methods used for power factor compensation

2.1. The need to compensate the power factor in the ERTS

Table 1 consists of data recorded for a period of 30 days at an electric traction substation (ETS). The interpretation of the data in the table shows that the ERTS consumed, per ETS, for a period of approximately one month, as follows: invoiced active energy was 549071.04 kWh (Wa) and invoiced reactive energy was 2148073.15 kVArh (Wr). Using the billing calculation methodology applied at this moment (according to ANRE - National Authority for Reglementation of Energy - Order no. 33 of May 28 2014) for the active and the reactive energy, the following values were obtained:

- invoiced active energy was 113404.58 RON
- invoiced reactive energy was 123739.72 RON

Table 1. Energy cost for July 2017 (MPS).

State	Date	W_a [kWh]	W_r [kVArh]			Cost W_a [RON]	Cost W_r [RON]
			PF < 0.65	0.65 < PF < 0 .90	PF ≥ 0.90		
Normal day	02.07.17 24h	134310.3	9680	94294.79	0.00	2770.82	6931.42
Max. Wa	08.07.17 24h	176644.26	0	73004.56	34858.11	3644.17	4102.86
Max. Wr	25.07.17 24h	146188.08	0	65647.47	77564.02	3015.86	3689.39
Month	31.07.17 744h	5497071.04	26850.60	2121222.55	1138097.76	113404.58	123739.72

Total energy invoiced for the analyzed period amounted to 237144.3 RON. An interesting point to consider is that for 549071.04 kWh, a doubling of the cost price for each kWh of consumed active energy is observed, caused by the lack of compensation of reactive energy.

The average power factor achieved for the invoiced interval is determined based on active energy consumed from the network and the reactive energy transited through the invoiced point, using the following formula:

$$cos\varphi = \frac{W_a}{W_{ap}}, \qquad (1)$$

where $W_{ap} = \sqrt{W_a{}^2 + W_r{}^2}$.

Taking the above into consideration, the resulted average power factor for the billing interval (cosφ) is 0.85. Clearly, this value shows that the ERTS has an inductive operating regime that must be corrected by compensating the reactive power in each electric traction substation.

Given that the ERTS operates on single phase current, supplied at UN = 25 kV from 16 MVA transformers in electrical traction substations (ETSs), the power factor compensation is accomplished by using single-phase compensators dimensioned to the maximum value of the apparent power provided by the power source [3].

2.2. Static compensation using capacitor banks

The static method of power factor compensation in the ERTS, using capacitor banks, requires connecting the capacitor bank with the power source and the load, in parallel [4, 5]. The reactive power provided by the capacitor bank can be dimensioned for either total compensation, or partial compensation.

The following mathematical statements hold true:

Reactive power absorbed by the ERTS is:

$$Q = U * I * sin\varphi, \qquad (2)$$

and the reactive power provided by the capacitor bank is:

$$Q_c = U * I_c = \omega C U_f^2, \qquad (3)$$

From the equality of the two powers:

$$C = \frac{I * sin\varphi}{\omega U_f^2} \qquad (4)$$

For partial power factor compensation:

$$C = \frac{\Delta Q}{\omega U_f^2}, \qquad (5)$$

where: $\Delta Q = Q_1 - Q_2$, Q_1 = total reactive power consumed by the ERTS, and Q_1 = total reactive power consumed at a power factor cosφ < 0,9.

Given that current regulations pertaining to the consumption and billing of electrical energy do not require total compensation of the power factor, it is sufficient to perform partial compensation, up to the neutral factor of cosφ = 0.9.

Through the introduction of capacitive current in the system, the inductive current demanded by the asynchronous motors of the ERTS is compensated. This method of compensation has the following drawbacks:

- the adjustment of compensation can only be done in steps, there is no possibility of continuous adjustment;
- providing additional insulation for the capacitor banks used at medium voltage entails additional costs and certain limitations;
- increased risk of loading the system with reactive capacitive electrical energy, which leads to loading the system as in the case of not compensating inductive reactive energy.

2.3. Compensation using synchronous machines

The synchronous machine (SM) is a piece of electrical equipment that can function as a generator, motor or as a synchronous compensator that produces reactive energy. The operation mode of a SM can be changed by modifying the Θ angle between the electromotive voltage induced in the stator winding and the voltage measured at the winding's posts. The Θ angle can take values between 0 degrees when idling and ± 90 degrees for maximum load in the generator or motor modes.

Synchronous compensators (SC) are idling synchronous motors, built for the purpose of providing reactive power. They are similar to capacitor banks, but with the added advantage of allowing for continuous adjustment of the reactive power.

The reactive power of a SC can be determined using this equation:

$$Q = m \cdot U \cdot Ia \cdot sin\varphi \ (6)$$

where m is the number of phases of the SC.

The synchronous reactive power compensator can introduce power to the system, or absorb power from it, according to the excitation current applied to the excitation winding. The "V" characteristics of the SC show the functioning regimen of the machine in relation to the variation of the excitation current (figure 1).

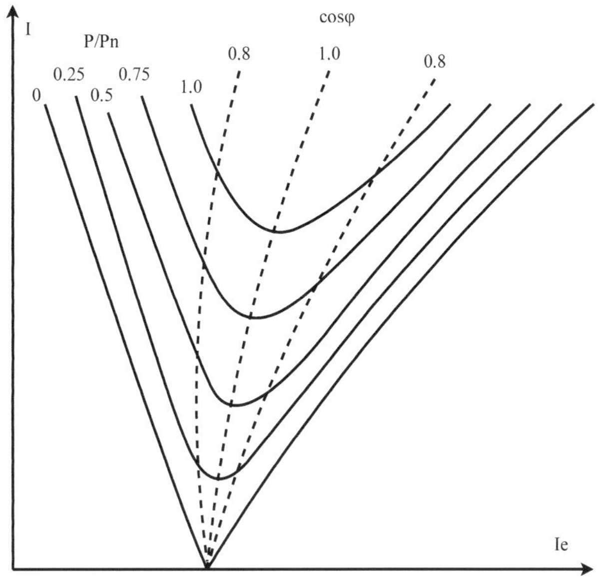

Figure 1.''V'' characteristics of the synchronous compensator.

By modifying the excitation current, the SC can be set to one of three different states:

- overexcited – the I_a current is capacitive – the SC provides reactive power;
- optimal – the I_a current is resistive – the SC doesn't absorb or provide reactive power;
- underexcited – the I_a current is inductive – the SC absorbs reactive power.

The characteristics of SC adjustment show the relationship between the load current I and the excitation current I_e when the SC functions in parallel with the power source of the ERTS.

2.4. Adjustment of the excitation current of the SC

There are two stages in the control of the excitation current.

The theoretical stage:

- First, the nominal excitation current is determined from the V characteristic and the cosφ = f (I_{ex}) characteristic of the SC, both increased to nominal operation values;
- The excitation current is determined. This current is necessary to compensate the longitudinal and transversal reaction effect of the rotor;
- The adjustment coefficients of the excitation currents are defined, depending on the current power factor and the one that is desired.

The practical stage:

- The measurement block collects the values of voltage, current, reactive and active power;
- Using the SCADA system controlling software, the collected values are analyzed. The excitation current generator is given a command, proportional to the difference between the existing power factor and the target power factor;
- The result is the modification of excitation current, increasing or decreasing the reactive power in the system.

Figure 2 shows a block diagram of the automation of the excitation current control, introduced in the SCADA system.

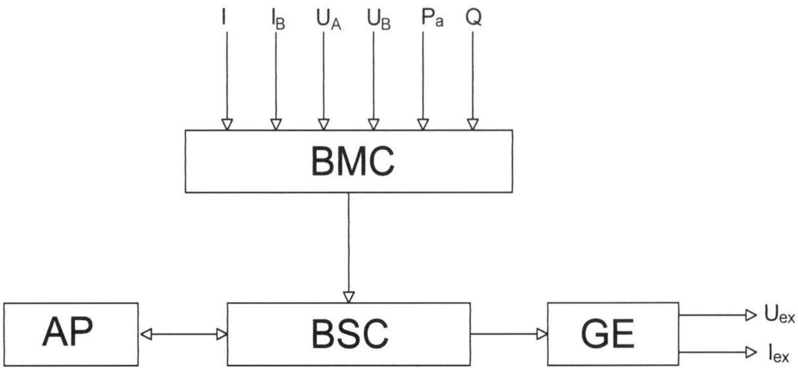

Figure 2. Diagram of SC excitation current control. BMC = common measurement block; AP = programmable logic controller; SC = supervision and control unit; GE = excitation current generator.

2.5. Advantages to the use of SC

Reactive power compensation using SCs in the ERTS has the following advantages:

- the SC allows for the continuous adjustment of reactive power provided to or absorbed from the ERTS;
- SC can be produced with a specification of $Q_{max} = 200$ MVArh, thus avoiding the limitations of using capacitor banks;
- the risk of introducing capacitive reactive energy in the ERTS through compensating inductive reactive energy is avoided.

3. Compensation of reactive power in ETSs

Electrical traction substations provide power to the ERTS using 16 MVA transformers. Typically, one transformer is functioning and the other one is kept as a backup. The functioning of ETSs at a power factor of $\cos\varphi > 0.9$ necessitates the compensation of reactive power consumed by the ERTS. This is accomplished through parallel-coupling to the STE transformer of a SC dimensioned for the apparent power provided by said transformer.

Compensating the reactive power absorbed by the ERTS permits keeping one transformer as a backup, given that it results in increased active power availability from the functioning transformer, up to the value of installed active power [6].

4. Control of the SCs utilizing the SCADA (Supervisory Control and Data Acquisition) system

The SCADA system consists of a main server and multiple Programmable Logic Controllers (PLC) and/or Remote Terminal Unit (RTU) with the purpose of monitoring and controlling the connected equipment at ETS level. The network schematic is presented in figure 4.

A PLC or RTU collects the U, I, φ, and $\cos\varphi$ and I_e values from the measurement cell of the running transformer. It then transmits the values to a microcontroller for analysis, and to a display for visual control [7].

If the microcontroller reads a $\cos\varphi < 0.9$ it will analyze the φ_0 angle between U and I. If $\varphi_0 > 0$ then the load is said to be inductive ($X_L > X_C$) and the microcontroller will increase the value of the excitation current I_{ex}, and will overexcite CS for the introduction of reactive energy in the system.

Conversely, a $\varphi 0 < 0$ means the load is capacitive ($XC > XL$) and the microcontroller will decrease the excitation current I_{ex}, and will underexcite the SC which will absorb reactive energy from the system.

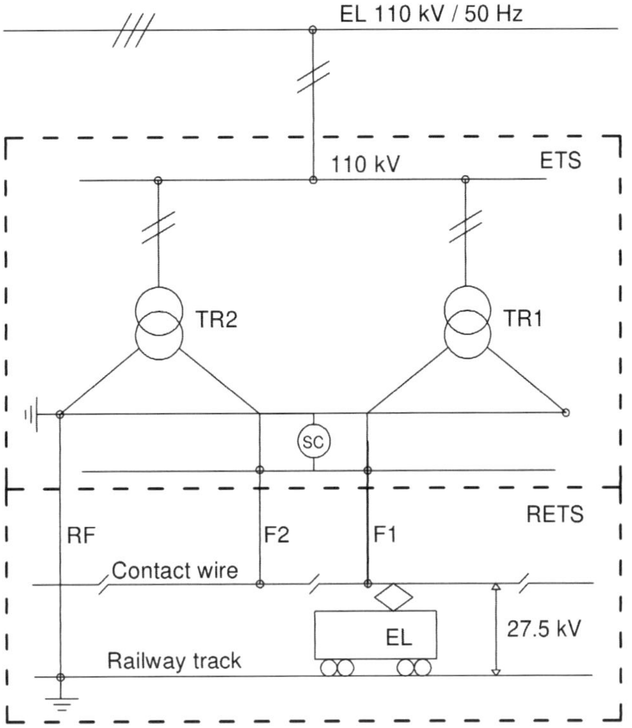

Figure 3. Single-wire schematic of an ETS containing an SC, parallel-coupled with the power transformers. EL-electrical line; ETS-electric traction substation; TR1, TR2-transformer 1,2; SC-synchronous compensator; F1, F2-Feeder 1,2; RF- Return feeders; EL-Electrical locomotive.

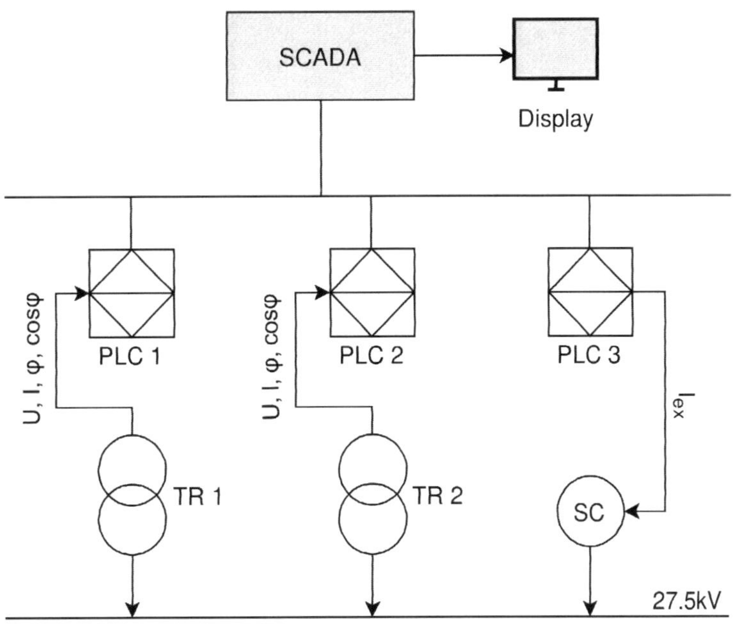

Figure 4. Schematic illustrating the use of SCADA in monitoring and controlling ETS equipment

If cosφ ≥ 0,9 and $\varphi_0 = 0$ (XL = XC) the circuit is resistive; the microcontroller will repeatedly attempt to read the parameters.

The principle of functioning of SCADA-assisted control of the SC is shown in figure 5 [7, 8].

The implementation of power factor control and SC control in the SCADA system which controls the ETSs is done in the Citect SCADA software. After programming and restarting the SCADA system, the operator can control the system, and the HMI (Human Machine Interface) software will display the real-time variation of the parameters on an LCD monitor (see figure 6) [8, 9].

5. Conclusions

Prior to power factor compensation, the ETS had a monthly electricity bill of 237144,3 RON ($55152 at the current exchange rate), of which only 113404,58 RON ($26374) represent consumed active energy.

Following the introduction of SCADA-controlled power factor compensation, the total billed value of electrical energy decreased by approximately 51% [10].

Power factor compensation decreased the load of the running transformer of the ETS. Load peaks that previously occurred at certain hourly intervals have been eliminated. Additionally, it allowed for keeping the second transformer as a backup, the power reserve of active power at the ETS has been increased, and implicitly the power supply system's stability has been improved.

Using SCADA-controlled SCs for power factor compensation permits the real-time continuous adjustment of the parameters of the electrical energy supplied to the ERTS.

Increased power flow is ensured for the involved segment of the installation, as well as the uniformity of voltage for the entire area supplied by the ETS in question. This has resulted in the supplied electricity parameters being adequate for the rolling stock of the ERTS.

The reduction of electrical energy cost for electric railway transport is a problem that concerns both the electrical energy providers and the railway traffic operators. Making the ERTS profitable through the reduction of energy cost results in the continued development and modernization of the transport system, a goal pursued by all market players.

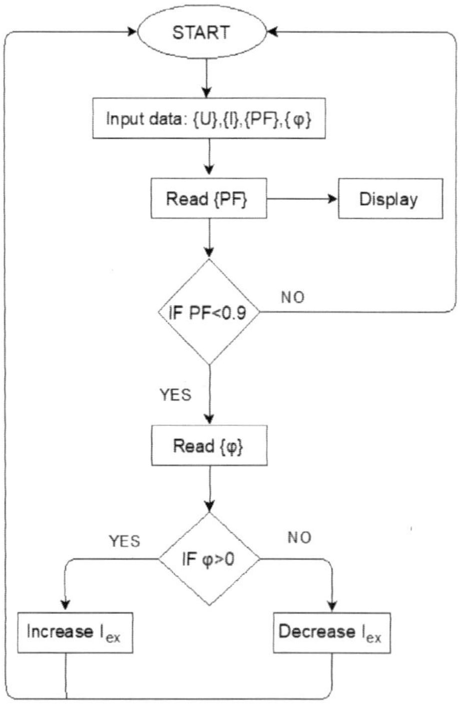

Figure 5. Logic flowchart of SCADA power factor control.

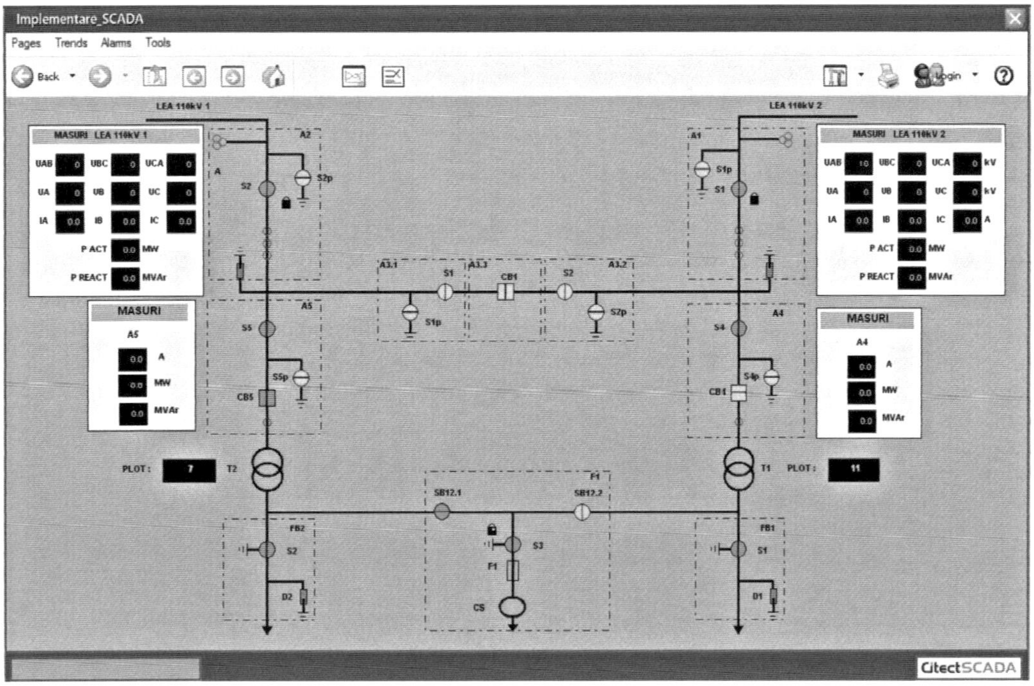

Figure 6. Single-wire schematic of an ETS introduced in the SCADA system.

References

[1] Popov A, Fratu A, Lepadat I, Helerea E and Cojanu V 2019 Monitoring the Cost of Energy for Powering the Railway Electric Traction System *Proc. 2019 8th Int. Conf. Mod. Power Syst. MPS 2019* 1–6

[2] Yousefi S, Abyaneh H A, Hosseinian S H and Zaboli A 2015 Utilization of Hybrid Filter to Eliminate Harmonics of 25 kV AC Electric Railway 8–13

[3] García O, Cobos J A, Prieto R, Alou P and Uceda J 2003 Single phase power factor correction: A survey *IEEE Trans. Power Electron.* **18** 749–55

[4] Behera S, Mohapatra S and Bisoi M 2007 *Automatic Power Factor Correction by Microcontroller 8051* (Rourkela National Institute of Technology)

[5] Ranjan R, Kumar D, Anand A and Namekar S 2017 Automatic Power Factor Correction using Microcontroller *Int. J. Res. Appl. Sci. Eng. Technol.* **4** 1666–8

[6] Chen T H and Liao R N 2016 Modelling, simulation, and verification for detailed short-circuit analysis of a 1 × 25 kV railway traction system *IET Gener. Transm. Distrib.* **10** 1124–35

[7] Yeole V, Dhurve S, Sure S, Meshram A, Meshram A and Engineering E 2017 Design of Single Phase Automatic Power Factor Controller Using Microcontroller *Int. Res. J. Eng. Technol.* **4** 1809–13

[8] Mane S R. Kolekar A, Molaj M V, Patil S V and Mestri M N, "Arduino based power factor correction 1," no. 4, pp. 4–8, 2016.

[9] Biswas R Sen and Mal S 2015 Automatic power factor improvement using microcontroller *2015 Int. Conf. Work. Comput. Commun. IEMCON 2015* 756–62

[10] Scholl M V. and Rocha C R 2016 Embedded SCADA for Small Applications *IFAC-PapersOnLine* **49** 246–53

IOP Conf. Series: Materials Science and Engineering 1138 (2021) 012039 doi:10.1088/1757-899X/1138/1/012039

Principles regarding integrated conservation of historical and archaeological sites to capitalize in the urban location

A G Simion[1*], A M Grămescu[1] and B D Pericleanu[2]

[1] Institute of Doctoral Studies, Ovidius University of Constanta, Romania
[2] Faculty of Civil Engineering, Ovidius University of Constanta, Romania

* anca_gemanaru@yahoo.com

Abstract. The promotion in Romania of the integrated conservation of the historical / archaeological sites from the urban areas represents a vector of the recognition of the importance of the archaeological patrimony, as a source of non-renewable and public development. The starting point of the article is to analyse the state of the built heritage, of the historical and natural sites in the peninsular area of Constanta in order to propose solutions for conservation / rehabilitation and their enhancement. The article will outline some principles of major importance: the need for an in-depth historical and archaeological study that precedes any intervention or enhancement, the obligation to respect with the utmost rigor the authenticity of the remains, preserving their original substance, the requirement that public accessibility not harm the quality of the sites, as the case may be, their integration in the urban structure. Case study - the wall of the Tomis fortress.

1. Introduction

The theme of the article highlights the concept of integrated patrimony starting with the definition of both terms – an element that defines the origin and age of a culture and represents what we have most precious for ourselves and for the coming generation viewed as an indivisible whole composed of inseparables elements.

To talk about integrated preservation of material heritage, in this case about above - ground or underground archaeological sites in urban sites, is to talk about all knowledge, legislation, techniques, methods, initiatives, interconnected concepts that result in the development of these sites.

The case study brings to the attention the vestiges of the ancient city of Tomis, Constanta today, the city with the highest continuity of habitation in Romania, for over 2500 years [1].

In the context of highlighting the sites classified as archaeological sites, there is the problem of developing these areas and capitalizing on the elements resulting or discovered following the archaeological research of the site during excavations, excavations for carrying out authorized works. This direction is also applied internationally, a well-known example being the Bank of Portugal in Lisbon where during the execution works was discovered underground an ancient wall of special archaeological importance which led to the restoration of the project in that area by including the new discovery wall in the turist visiting circuit underground after putting into service the preserved and restored building.

In areas with historical archaeological sites, it is desired to capitalize on the underground elements

IOP Conf. Series: Materials Science and Engineering 1138 (2021) 012039 doi:10.1088/1757-899X/1138/1/012039

(city walls, enclosures, etc.) by preserving the historical substance and inclusion in the visiting / tourist circuits, raising the issue of choosing the best technical and architectural solution respecting the legislation in force. and internationally adopted principles of restoration and conservation.

In this paper, after a brief presentation of the legislative context and familiarization with the peninsular area of Constanta, with previous in situ studies by the National Museum of History and Archeology of Constanta, the authors propose to analyze the solutions of such a site in the area on Mihail Kogalniceanu Street.

2. Generalities regarding the concepts of restoration and conservation of historical sites

Carrying a spiritual message of the past, historical and archeological monuments remain today an expression of the historical past of the peoples, the living testimony of their traditions. Humanity is becoming increasingly aware of the unity of human values, of the need to preserve and restore this common heritage, and is aware of the duty to pass them on to their descendants with all their rich authenticity.

In its short history, the notion of restoration and conservation has gone through several stages, each time consolidating its position and responsible role of society.

Since the middle of the 19th century, there is a need to establish truthful principles in the restoration action, so that the specialized literature records four representative periods according to [2]:

- the first period, characteristic of the beginning of the 19th century, also called the empirical period, in which it was mainly chosen to consolidate those monuments that had a precarious technical condition, was characterized as the period of free will in which everyone proceeded to choose;
- the second period, also called the doctrinal period, belongs to the period of the second half of the 19th century, in which two conceptions of romantic but diametrically opposed essence dominated: in a first conception the doctrinal restoration was promoted based on the idea of unity of style, the promoter being the French architect Eugen Emmanuel Viollet le Duc; the other concept promoted the idea of respect for the monument without any intervention of the nature of consolidations, but exclusively maintenance interventions;
- the 3rd period corresponds to the end of the 19th century and it is the period of historical restaurating by preserving the state in which the past left us - respecting the history of each monument;
- the 4thperiod is the contemporary period in which, along with the historical restoration, there is also the restoration with additions loyal to originality.

These periods have been shaped by international scientific events - imposed as necessary by specialists in the desire to have common procedures and methodologies. Thus were organized:

- the Congress of Italian Architects, but with international participation, held in Rome in 1883, during which the doctrine of historical restoration was formulated [3];
- the 1931 Athens International Conference, which laid the foundations of the Athens Charter;
- the Second International Congress, held in Venice in 1964, when an international Charter known as the Charter of Venice was drafted and approved, which is in fact a draft amendment to the Charter of Athens by expanding the notion of historical monument, the preservation referring not only to the monument but also to its frame [4].

Thus, according to the Venice Charter, the notion of monument includes the isolated architectural creation but also the urban or rural settlement that bears the testimony of a civilization, of a significant evolution, or of a historical event. It extends not only to the great creations but also to the modest works that have acquired a cultural significance over time. The phrase monument can refer to above-ground and underground monuments. As it is known, the archeological patrimony defines the origin and the age of a culture, histories, traditions of a people, of a state in relation to other peoples and states. At present, the archeological vestiges have exceeded the limits of a state, these being, in many situation, cultural goods of humanity.

The valorization of the archeological sites supposes the research, the safeguarding and the dissemination of the information about the discovered vestiges.

In many cases, the archaeological heritage is subject to major risks due to the influence of factors and phenomena that significantly damage its originality, value and integrity. In many cases the destruction of archaeological sites is due to human activity [5].

Romanian legislation in the field of cultural heritage protection has its roots in the interwar period, being developed in the postwar period. But most actions in this regard were taken after 1989 and are in the context of political changes in Romania that contribute to the development of a democratic society and that correspond to its European aspirations. However, the political, economic and social problems encountered by Romania in recent decades affect the most sensitive areas, including the protection of cultural heritage in general and the archaeological heritage in particular.

3. Presentation of the studied area. Historical and urban implications

Tomis is one of the richest archaeological sites in Romania (figure 1). As known, the modern city of Constanţa was built on the ancient Tomis fortress. The research conducted in the second half of the twentieth century and today has highlighted a number of cultural values of exceptional importance such as: Romanesque mosaic building, late Roman baths, network of underground galleries of Tomis, fortress walls, tower areas, basilicas with underground apses and crypts, public buildings, traces of urban housing, mosaic floors, colonnades, streets and aqueducts, commercial markets.

According to the List of Historical Monuments (LMI) 2015 - Constanţa County, only one of the streets, which is the object of the case study, respectively Mihail Kogalniceanu Street, is located in:

a. Urban site "Central commercial area" LMI code: CT-II-s-B-02811,

b. The urban site "Peninsular area Constanţa" LMI code: CT-II-a-02832.03,

c. Urban site LMI code: CT-II-s-B-02842.

In addition to the above, the buildings can be located in protected areas of some monumental constructions. The building that is the object of the case study is also in the protection area of the monument construction: Traian Bellu Law Firm, located in Constanţa, 15 M. Kogalniceanu Street, LMI code: CT-II-m-B-02815 (figure 2).

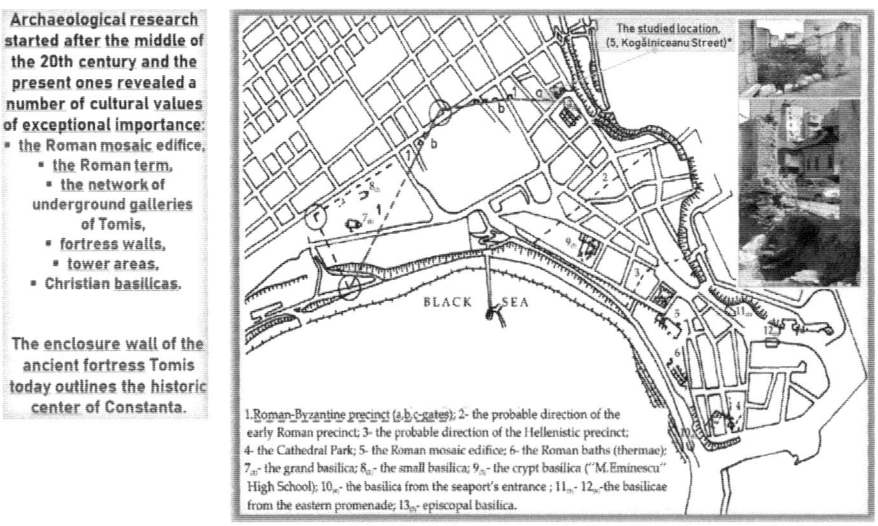

Figure 1. The plan of the ancient city with the main monuments (after Buzoianu L, Bărbulescu M, Tomis. Historical and archaeological commentary, Constanţa, Ed.Ex.Ponto, 2012).

Figure 2. Location of the case study, M.Kogălniceanu street, Constanta (photo source: internet).

4. Presentation and analysis of protected historical site. Technical implications

According to the preventive research work of the case study site carried out by the researchers from the Museum of National History and Archeology Constanta, it was discovered (figures 3, 4 and 5):

- The enclosure wall - was discovered on a portion of about 8 m in the southern corner of the researched area. The wall is strongly dismantled, but the foundations on the outside are still quite well preserved, as well as three large blocks. The mortar is whitish, different from the one used to build the tower. In the area of deep dismantling was identified the block of the foundation – construction within the period 324-328 AD [6].
- The tower - the eastern side was delimited and a limited part of the northern front - the tower is U-shaped. The foundation is made in several stages, being adapted to the situation on the ground, which requires reinforcements in the front area of the tower due to the presence a trench of appreciable size, slightly clogged at the construction stage of the tower. Thus, on the outside of the east side, up to the starting point of the arched front, the foundation of the tower is made "in the limelight" and smoothed with mortar (probably as a waterproofing solution); the operation was carried out in a large trench, based on a stone foundation tied with mortar with a thickness of about 0.70 m (well finished at the top by pouring a mortar screed from which the construction of the foundation started themselves) [6]. As the foundations of the tower were raised, the outer trench was filled with successive layers of heavily compacted earth. The tower was identified on a length of 8.30 m and a width of 5.50 m; based on the analogies with the other tower, it can be assumed that the total dimensions on the outside are 9.00 x 7.50 m. The thickness of the wall is 2.45 m;
- Masonry structures in the extramural area;
- Late Roman trenches: they were executed in the construction phase of the tower (F2) and in the next period (F3);
- Early Roman trenches: the safest frame in this period is the F4 trench, oriented in the NV-SE direction; graves M1, M3 and M5.

Another important component of the preliminary report was the precinct of the site where a new building is to be built; the new building is adjacent to other buildings on the 3 sides, buildings with a certain age that have a poor construction technology (the building in the rear area has no foundation) and at the same time a state of degradation of the structural elements arranged towards the open ground enclosure (the construction without foundation shows visible settlements generated by losses from water pipes, special aspects to be recorded on the present date, the wall of the building with similar number 7 shows significant degradation).

IOP Conf. Series: Materials Science and Engineering 1138 (2021) 012039 doi:10.1088/1757-899X/1138/1/012039

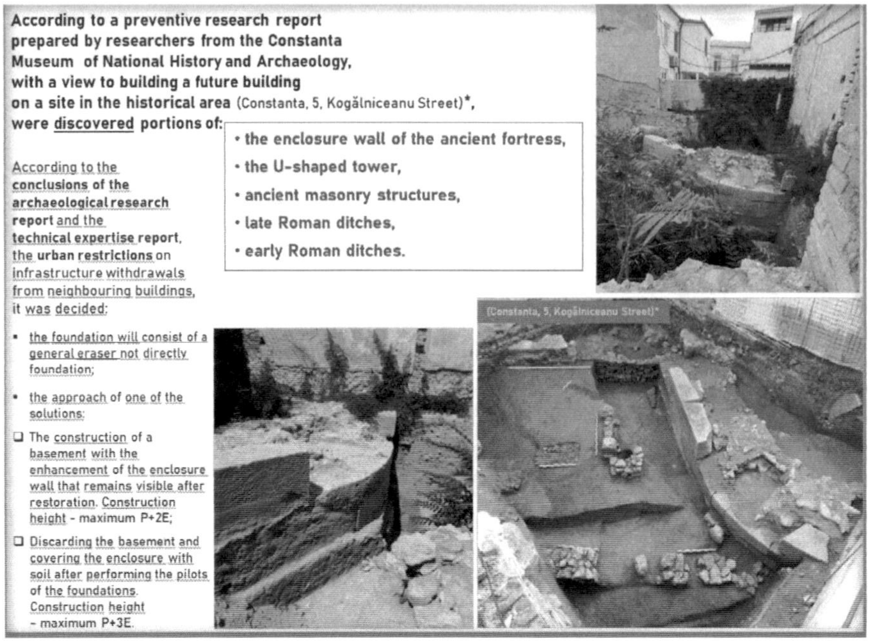

Figure 3. Highlighting historical research report and technical expertise of the studied location (photo collage).

In this context, measures were proposed and taken to protect existing buildings along the site but also in the background by removing from these existing properties the foundations of the new building proposed to be executed on site, to quantify the condition technical elements of the neighboring buildings in order to ensure the necessary findings of evidence, but also proposals for intervention works to protect the buildings identified underground, the protection of the archaeological site.

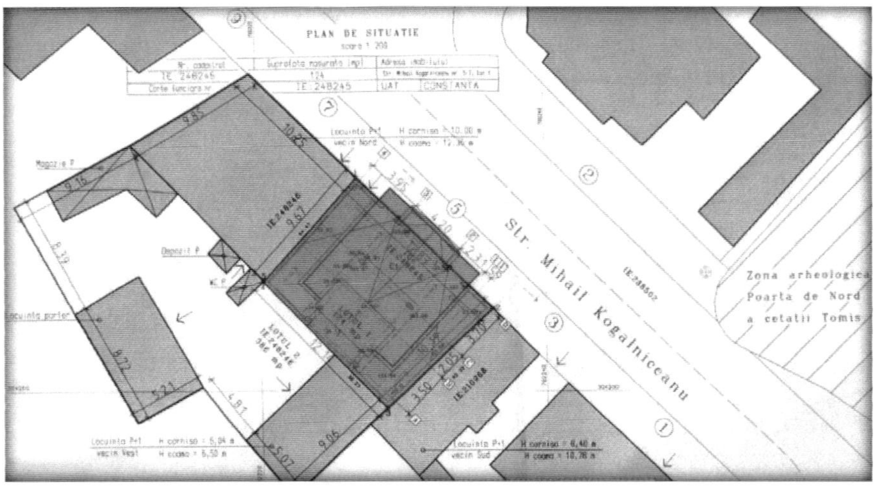

Figure 4. Situation plan studied location, M. Kogălniceanu street, Constanța.

In this context, out of a total of 140 sqm, it is found that 34 sqm are lost by distancing the future construction from the neighboring properties and 30% of the enclosure is occupied by the fortress wall. Therefore, only on an area of 70 square meters could be adopted an infrastructure solution capable of supporting GF + 3F.

These aspects do not give the possibility of direct foundation.

In this context, analyzing the available free area (excluding the archaeological constructions of historical monument to be preserved) according to the conclusions of the archaeological research report and the technical expertise report and the need to withdraw the infrastructure from neighboring buildings, it was decided to address one of the solutions [7]:

- creation of a basement with the valorization of the enclosure wall that remains visible after it is restored. In this variant it is not possible to achieve GF+3F but a building with maximumGF+2F;
- giving up the basement and covering the enclosure with earth after piles were made for the foundations of the building. In this last variant, GF+3F can be achieved.

For the realization of the infrastructure:

a) Carrying out the pilots according to a pilot plan that should take into account the protection of the existing constructions in the underground, the protection of the existing constructions perimeter of the enclosure.

b) The pilots will be joined by a reinforced concrete slab with a thickness of approx. 40 cm, with turned concrete beams arranged along the rectangular axes of the construction A, B, C and 1, 2, 3, whose height will be established in agreement with the expert AF, the structural engineer and the architect, so as to protect the walls existing archeological sites.

c) The foundation of the new GF + 2F or GF + 3F building will consist of a general slab. The construction of the slab will take into account that on a width of at least 2.60 ml from the building with number 7 to ensure the access level equal to the level of the protection sidewalk, the rest being with the level ± 0.00 by 40 cm above the level of the sidewalk so that in heavy rains there are adequate protection conditions.

In this context, the realization of pilots in diameter and number capable of supporting the new construction appears as a possible foundation solution.

The preventive research from Mihail Kogălniceanu siteallowed the identification of a new segment belonging to the late Roman enclosure of Tomis, a fragment almost similar in geometry to the butcher's tower unveiled by Vasile Pârvan, which represents the North East Gate, or the Great Gate of the Tomis fortress.

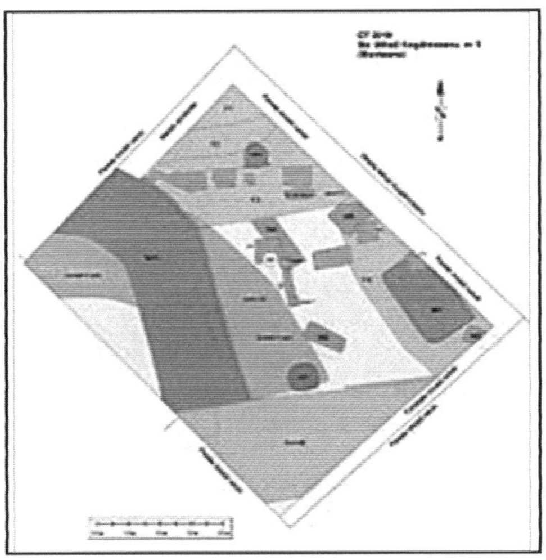

Figure 5. Image from the archaeological research report of the studied site, M.Kogălniceanu street, Constanța.

Figure 6 shows examples of conservation solutions, protection of archaeological sites in an urban site. In situations where the sites belong to the public domain or to areas adjacent to the urban areas, the archaeological site can be capitalized by making constructions with the role of protection, conservation. These structures discharge on independent foundation systems, usually pilots.

The uncovered walls are usually consolidated with stone - fillings made of quarry stone and mortar with hydraulic lime.

The mortar used for the intervention works on the existing foundations and identified by archaeological research is an M5 mortar with very low capillary absorption. Joints are made and the mortar bed is applied to the interior and exterior walls of ceramic material, brick, tuff, natural stones with compact mortar with very high hygroscopicity and breathability of pure natural hydraulic lime NHL 3.5, extra-fine natural stink and inert materials of siliceous sand and dolomitic limestone 0-1.4mm.

Other solutions for arranging and enhancing the ruins identified in the archeological sites [8]:

The researches highlighted a possible typology of interventions on archaeological sites in urban areas, a typology aimed at restitution. In the current approach restitution varies between two extreme approaches: a conservative directive promoted by John Ruskin that supports their conservation as they are at one time. Contrary to this opinion, Viollet le Duc is a supporter of complete restoration.

As shown above, a type of conservation restoration is the construction of protection structures.

Figure 6. Examples of conservation solutions, protection of archaeological sites in an urban site (photo collage, photo source - internet).

5. Conclusions

Following the analysis carried out for the site studied in the Constanța peninsular area, which was the case study of this article, it was possible to highlight the principles underlying the concept of conservation and integration of historical / archaeological sites and the concept of highlighting discoveries. archeological sites in urban areas. Following the excavations, it was found that in the site of a total of 140 sqm, 34 sqm are lost by removing the future construction from the neighboring properties and 30% of the enclosure is occupied by the fortress wall. Therefore, only on an area of 70 square meters could be realized infrastructure solutions capable of supporting the desired level of the new GF + 3F construction.

In this context, analyzing the available free area (excluding the archeological constructions historical monument to be preserved) according to the conclusions of the archaeological research report and the technical expertise report as well as the need to withdraw the infrastructure from the neighboring constructions, two possible solutions were identified. -is to create a basement with the enhancement of the enclosure wall that remains visible after it is restored (in this variant it is not possible to make GF + 3F but a building with max GF + 2F). The researches highlighted a possible typology of interventions on archaeological sites in urban areas, a typology aimed at restitution.

It is noted that the restoration and enhancement of archaeological sites aims to ensure that the intervention is minimal, reversible, justified, with the conservation of resources and respect for authenticity [9]. The reference documents are the "European Convention for the Protection of the Archaeological Heritage" and which is partly based on the "Venice Charter". Some of the elements of the archaeological heritage are components of some architectural structures that must be protected in accordance with the Venice Charter.

References

[1] Bologa A 2019 Dobrogea at 140 Years after its Union with the Romanian State - An Example of Contemporary West-Pontic Multiethnic Understanding (Ex Ponto) p 332
[2] Historical monuments - studies and restoration works http://cimec.ro.
[3] Bologa A and Grămescu AM 2020, Heritage http://mts.intechopen.com Chapter *Tropaeum Augusti (France)and Tropaeum Traiani (Romania): A Comparative Study.*
[4] Grămescu AM, Pericleanu M and Suliman S 2018, History of construction science and technology expression of the Turkish-Tatar civilization in the Dobrogea area MatrixRom p 67
[5] Simion G A 2016 Contributions regarding the reduction of the risk factors to which the constructions of the historical monument located in the active area of the Black Sea cliff are subjected. Solutions for limited security, doctoral thesis.
[6] Fălticeanu D 2016 In connection with the conservation and capitalization of historical monuments in Constanța County, Pontica **3**.
[7] Grămescu AM 2018 Technical Expertise - Inspection and assessment of construction damage - Guide for technical experts, assessors, insurers- Part I Roofs MatrixRom p 349.
[8] Grămescu AM 2020 The evolution of civil constructions in Dobrogea ExPonto *Lectures CRIFST Course organized by the Romanian* Academy p 32.
[9] Grămescu AM 2019 The history of the science and technique of constructions expression of the Turkish civilization in the Dobrogea area, presentation of the Scientific Symposium, in CRIFST Constanța, *The history and philosophy of science and technology - Romanian Academy*

IOP Conf. Series: Materials Science and Engineering 1138 (2021) 012040 doi:10.1088/1757-899X/1138/1/012040

Modelling the structure of Carol I Mosque Minaret taken into account the seismic evaluation

S Suliman[1*], A M Gramescu[1] and S Gelmambet[1]

[1] Faculty of Civil Engineering, Ovidius University of Constanta, Romania

* eusever@yahoo.com

Abstract. The paper aims to analyze the behavior of seismic actions of minaret – structural components characteristic of mosque. As it is known, the characteristics of this construction are governed by the slenderness and the distribution of the constituent materials by height. An important component is the most real quantification of the characteristics resistances as well as the impact of other loads in operation. The paper presents the case study – the minaret of the Carol I Mosque from Constanta, an emblematic historical monument of Constanta made by Eng. Gogu Constantinescu 110 years ago.

1. Introduction

The minaret is a very important component in the composition of a mosque, being defined as a high tower attached to the building, which has at the top a balcony from where, the muezzin calls the people for praying. Also, the minaret has an important role due to its height, being observed from a distance by the visitors of those lands, which thus symbolizes the presence of Muslim people.

A special worship place for the Muslim community in Dobrogea is the Carol I Mosque, which was built between 1910-1913 by the Romanian state during the leading by King Carol I, this being the reason that the mosque received his name, but is also known like the royal mosque. The architect Victor Stefanescu, one of the famous architects of the time, and the engineer Gogu Constantinescu, an exceptional inventor in the field of reinforced concrete, who used it in the construction of the dome and minaret of royal mosque, were engaged to build the project (figure 1).

It is a minaret that reaches a height of over 40m, so its behavior under the action of destructive environmental factors becomes a special concern for specialists, for this important component of the mosque that is a historical monument. So, in this paper are presented the efforts and displacements that appear in the minaret under the seismic action force, the structure being modelled in the automatic calculation program Scia Engineer.

It is necessary to know the critical point of this type of construction in order to combat the serious damages that may occur under the seismic action. We can thus refer to the minarets of other mosques in Dobrogea that have suffered very important degradations that have led to the need for significant repair works. For example, at the minaret of the Esmahan Sultan mosque from Mangalia, built in 1573 year, whose structure is made of limestone masonry, in 2008 year were carried out on this structure, which was partially demolished and rebuilt with the original materials left intact. The cause of the need for work on this minaret was the fact that the structure was tilted to one side and its resistance was severely damaged. It is also mentioning that the minaret at the beginning of the building had a height twice as high as the current one according to an image made around 1828. [1]

IOP Conf. Series: Materials Science and Engineering 1138 (2021) 012040 doi:10.1088/1757-899X/1138/1/012040

Another example can be the mosque from Cernavoda built in 1756, and built of stone masonry. Its minaret has tilted over time and it was necessary to carry out works to consolidate the minaret in critical areas. Also the mosque from Medgidia, built between 1856-1861 years, the minaret was in same situation like in Cernavoda.

Figure 1. Carol I Mosque.

2. Methodology

In this paper the following methodology was applied to obtain the results: comprehensive analysis of the references, field survey and seismic behavior analysis with Scia Engineer.

In the first stage, was studied the history of the mosque, if was some restoration or repairing process at this building. The aim was also to determine the characteristics of the materials used to build the minaret, and to identify the architectural elements.

The field study was performed to verify the information obtained from the documentation, about the dimensions and materials used for minaret building. An analysis of the degradations in the minaret structure was also performed through a photographic analysis.

In the third stage we performed the modelling of the minaret structure in the automatic calculation program Scia Engineer, whose shape was obtained as seen below (figure 2).

The outer body of the minaret was introduced into the program using the "Predefined shapes" function. The outer part is defined as a wall, its shape being like a cone trunk. In the middle of the minaret was inserted the central pillar, which like the outer wall has the shape of a cone trunk. The helical staircase is defined as concrete slabs connected in the outer wall and in the central pillar.

The following emplacement characteristics were used for the automatic calculation:

- $a_g = 0.2 \cdot g$;
- $T_c = 0.7s$.

The structure fit into class II of importance – exposure because the height of the structure exceeds 28m, and the importance factor γ is 1.2. The behavior factor q is 2, because the minaret has a medium ductility class, having the shape of an inverted pendulum type.

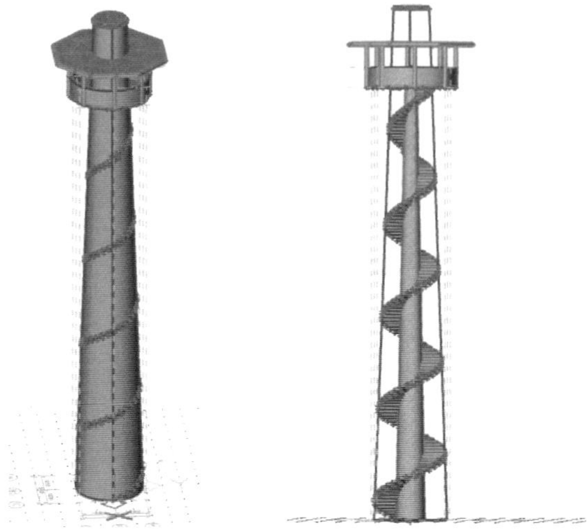

Figure 2. 3D modeled structure.

After structural modelling were introduced load cases defined in table 1, that acting on structure.

Table 1. Load Cases

Name	Description Spec	Action type Load type
LC1	Self weight	Permanent Self weight
LC2	Useful weight Standard	Variable Static
LC3	Snow Standard	Variable Static
LC4	Seism X Seismicity	Variable Dynamic
LC5	Seism Y Seismicity	Variable Dynamic

Thus, after the introduction of the loads, the load combinations were generated in order to perform the calculation for determining the efforts and displacements, in order to verify the resistance and stability conditions at SLS and ULS.

3. Results

In the first stage, was obtained information about the materials used to build the mosque, which are as follows: for the foundations of the mosque: limestone; for the walls of the mosque: brick masonry, and for the dome and the minaret: reinforced concrete. The type of concret used is C 6/8 (figure 3).

The shape of the minaret is tronconical, with a size of 3.9 m at the base, and 2 m at the top. The thickness of the minaret walls is 20 cm, and the height to the top reaches 40 m. At the top of the minaret is the balcony, which is accessed by the helical staircase inside the minaret, a staircase that consists of 140 steps. In the middle of the minaret is the central column which has the size at the base 1.5m and at the top 0.8m (figure 4). The amount of material used in the construction of the minaret can be found below in table 2.

IOP Conf. Series: Materials Science and Engineering 1138 (2021) 012040 doi:10.1088/1757-899X/1138/1/012040

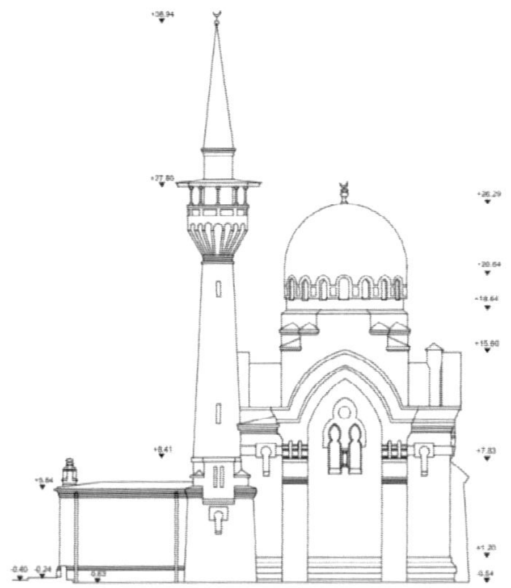

Figure 3. Plan Carol I Mosque, side view.

Table 2 Material used for minaret building.

Material	Mass [kg]	Surface [m^2]	Volume [m^3]
Concrete	249973.9	511.737	100
Total	249973.9	511.737	100

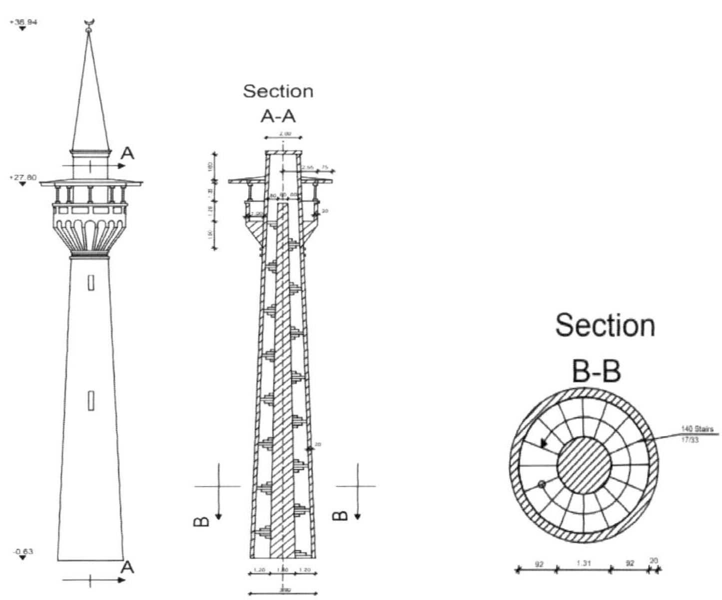

Figure 4. Minaret plan.

During the field study we noticed some degradations on the minaret inside at the finishes due to the water penetrations through the windows, and in front of the balcony door. But also, in this stage we didn't notice large cracks in the minaret structure (figure 5).

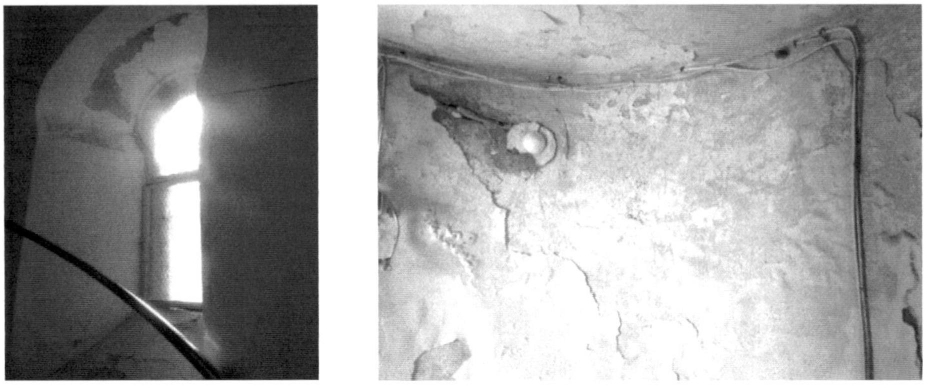

Figure 5. Degradations inside the minaret.

In Scia Engineer are obtained the 3D deformed shape of the minaret, the maximum displacement, and the values of the efforts in the walls of the minaret and in the central column, values of bending moments and axial efforts (figure 6, 7, 8)

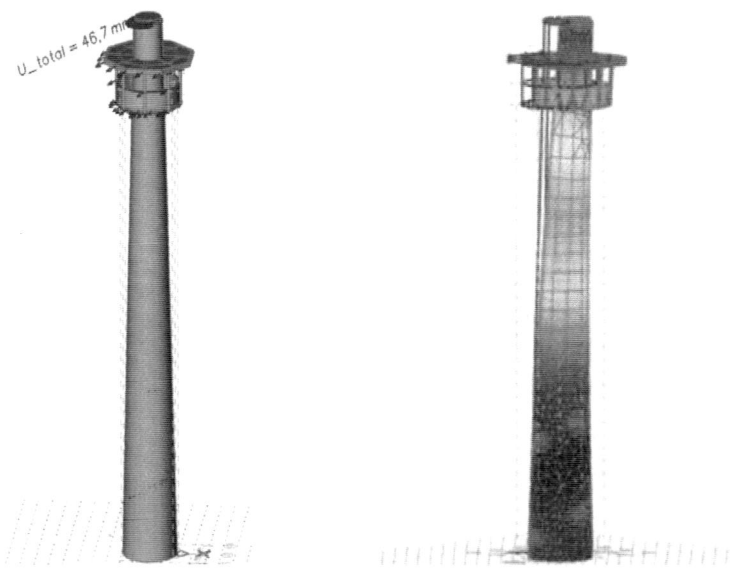

Figure 6. Deformed shape of the minaret

After obtaining the results from Scia Engineer, are checked the conditions:

Resistance conditions

$$E_d \leq R_d \quad (1)$$

where:

$$E_d = \frac{1}{q} \cdot E_e^* + E_g \quad (2)$$

E_d – total calculation effort;

E_e^* - seismic actions effort;

E_g – non-seismic actions effort;

q – behaviour factor;

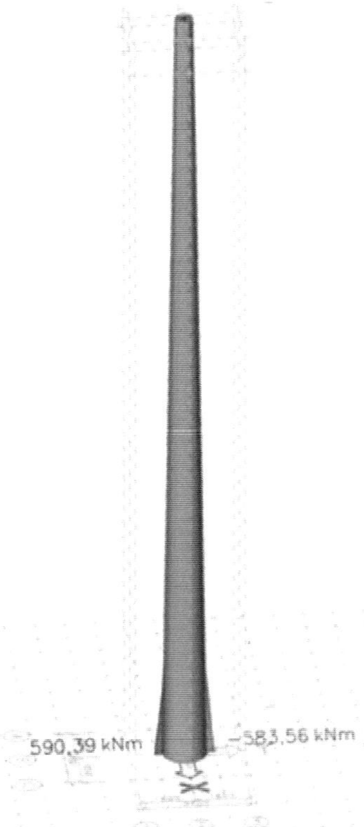

Figure 7. Maximum bending moment in central. column

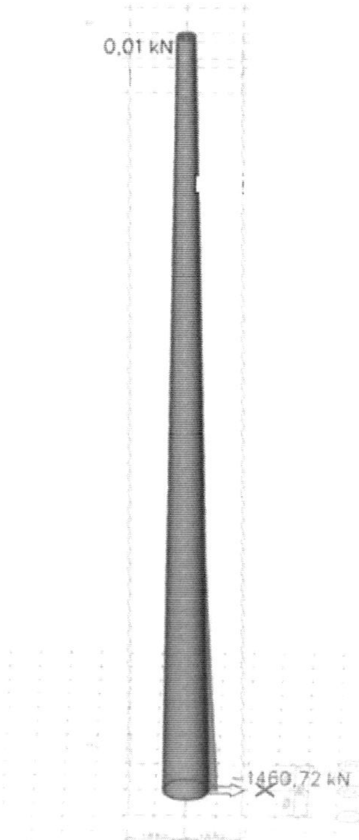

Figure 8. Axial effort in central column.

$$E_d = 1/2 \cdot 1018 + 1004{,}58 = 1513.58 \text{ kN} \tag{3}$$

$$R_d = 2816 \text{ kN } for\ \upsilon = 0.4 \tag{4}$$

$$=> E_d \leq R_d \tag{5}$$

Stability conditions:

$$SLS - d_s = \upsilon \cdot q \cdot d_e \leq d_{ra} \tag{6}$$

$$\text{SLS - } d_s = \upsilon \cdot q \cdot d_e = 0.5 \cdot 2 \cdot 46.7 = 46.7 \text{ mm} \tag{7}$$

$$d_{ra} = 0.005 \cdot \text{h} = 150 \text{ mm} \tag{8}$$

$$=> d_s \leq d_{ra} \tag{9}$$

$$\text{SLU - } d_s = c \cdot q \cdot d_e \leq d_{ra} \tag{10}$$

$$\text{SLU - } d_s = c \cdot q \cdot d_e = 0.89 \cdot 2 \cdot 46.7 = 84 \text{ mm} \tag{11}$$

$$d_{ra} = 0.025 \cdot \text{h} = 750 \text{ mm} \tag{12}$$

$$=> d_s \leq d_{ra} \tag{13}$$

4. Conclusions

Historical monuments present a very important objective for the country. This construction is very important also for Constanta being a point of attraction for tourists as well. This monument attracts a significant number of visitors, so knowing the critical points of the minaret under the seismic action is particularly important.

IOP Conf. Series: Materials Science and Engineering 1138 (2021) 012040 doi:10.1088/1757-899X/1138/1/012040

From the point of view of structural behavior under the action of seismic force the minaret shows very good stability, although it has not undergone any consolidation process, from the edification to the present day.

It is visible the need to perform repair works on the finishes, such as painting inside the minaret, achieving a proper sealing at the windows so that it cannot penetrate the humidity. The reparations process that are used to ensure the stability of this kind of structure, must not affect its originality.

The modelling of this minaret represents a first step towards the modelling of the other minarets of the mosques on the territory of Dobrogea, framed as historical monuments, so that the serious damages on these constructions can be combated.

References

[1] Gramescu A M, Pericleanu M and Suliman S 2018 *History of construction science and technology expression of Muslim civilization in the Dobrogea are,* Ed. Matrix Rom

[2] Nuredin I 2007 *Muslims from Romania,* Ed. Golden

[3] Mehmet A E, 1994, *From the history of the Dobrogea turks*, Ed. Kriterion Bucuresti

[4] Andresoiu B, Baracescu C, Maxim M, Nicolau S, Bonciocat S and Arsenie C 2012 *Mosques. Minarets in the sky of Dobrogea,* Ed. Igloo Patrimoniu

[5] Bibliography from the National Heritage Institute – Historical Monument Sheets

[6] List of historical monuments – Constanța County, Department of Culture

[7] P100-1/2013 – Seismic design code – Part 1 – *Design provisions for buildings*

[8] Mehmet A E 1981 *Turkish civilization*, Ed. Sport tourism

[9] MP025-2004 – *Methodology for risk assessment and intervention proposals necessary for the structures of historical monuments constructions within their restoration works*

[10] P100-3/2008 – Seismic design code – Part III – *Provisions for seismic assessment of existing buildings*

[11] Gramescu A M 2007 *Constructii civile*, Ed. Agir, Bucuresti

IOP Conf. Series: Materials Science and Engineering 1138 (2021) 012041 doi:10.1088/1757-899X/1138/1/012041

Reasons for the collapse of ceilings during the construction of monolithic buildings

M V Tomakov[1], A I Pykhtin[1*], V I Tomakov[1] and I A Tomakova[1]

[1] Southwest State University, Kursk, Russia

* aipykhtin@swsu.ru

Abstract. Construction works, as a rule, differ in complexity and hazards of works. Nowadays, cases of collapses of a formwork of the cast concrete section of a building under construction in the course of concreting became frequent. Accidents entail death of people and significant material losses. Purpose of the research: the purpose of the research is to study accidents occurred during installation of in-situ concrete floors of buildings and to reveal the typical causes of a formwork system collapse. The initial materials were the data on the causes of accidents at construction sites obtained on the basis of the analysis of statistical materials on injuries, as well as data on the identified violations of the requirements for safe work practices in capital construction and reconstruction of buildings and structures. A total of 148 accidents at construction sites from 1997 to 2019 were considered. The paper systematizes some main errors and violations of technologies committed in the course of in-situ concrete works. Typical causes common for all formwork collapses during concrete works for in-situ floor slabs installation are revealed. Among the objective causes of accidents and high injury rate in this type of construction works the following should be highlighted: the lack of method statements and construction plans; lack of supervision of the correct and safe execution of works; shortcomings in the organization of training and testing of labor protection knowledge; violation of technological regulations and safety rules.

1. Introduction

According to the Federal Labor and Employment Agency, 2344 people died from injuries in workplace accidents in organizations of all types of business activity in 2014, 2089 - in 2015 and 2072 - in 2016 [1, 2]. In 2017, 1722 people died due to workplace accidents in Russia. Most people died due to accidents in the construction sector - 24% of the total number of deaths at workplace in all types of enterprises, as reported by the judge of the Supreme Court of the Russian Federation [3].

Construction works, as a rule, differ in complexity and hazards of works. In this industry, a significant number of severe accidents occur annually due to the fall of the injured from a height, structure collapse, fall of objects and materials [4, 5].

Professional risks of injury of construction workers are connected with the specifics of the work, including the occurrence of emergency situations during the construction of buildings and structures accompanied with the collapse of structural elements of buildings, applied technological equipment [6].

At the present stage of construction, cast reinforced concrete is used in the construction of public facilities and residential buildings, especially multi-storey and high-rise ones. The technology of cast-in-situ construction is the most mobile technology; it makes it possible to erect buildings of different heights and shapes in the shortest possible time in cramped urban conditions, and in many cases with more favorable economic indicators [7].

Almost all the elements of buildings are constructed applying the technology of cast-in-situ construction: foundations, supporting columns and walls, floor and roof slabs. The share of facilities built using this technology in the total number of buildings under construction in Russia increases annually. This technology consists of several main stages: installation of a reinforcement cage, installation (assembly) of a formwork, concreting, removal of the formwork. For the installation of horizontal elements of in-situ concrete construction in the required specified position the system of a slab formwork is intended. The most common today is the system of slab formwork with telescopic extension legs.

However, as the practice of construction shows, in the course of the work due to the lack of Method Statement confirmed with necessary engineering calculations, violations of the technology of concrete mixture preparation, the technology of work during the construction of formwork and reinforcement, pouring, removing the formwork, can be the causes of emergency situations and collapses of structures in the process of work.

Besides, the market has a lot of "novice developers" who tend to make various kinds of "non-standard decisions" that go against safety standards, which leads to a large number of accidents.

Purposes of the research of the study were to study accidents that occurred during the installation of monolithic slabs of a reinforced concrete building and to identify typical causes of the collapse of the formwork system.

2. Materials and methods

The source materials were data on the causes of accidents at construction sites, obtained on the basis of analysis of statistical materials on injuries, as well as data on violations of the requirements for safe work practices during capital construction and reconstruction of buildings and structures.

The causes of failures of formwork systems when installing concrete slabs on site were established on the basis of statistical data from the National Association of builders (NOSTROY), the National Association of expert organizations in the field of construction (NOEX), the company "welding", other organizations, as well as from other open sources [8, 9], including the Internet. A total of 148 accidents at construction sites were reviewed from 1997 to 2019.

The results obtained were subject to statistical processing, analysis and systematization for the same type of causes of the accident. Systematization in this context is carried out for the first time.

3. Results and Discussion

3.1. Features of accidents with formwork systems of buildings and structures under construction

Some accidents at construction sites that occurred in different years and different regions of Russia are considered below.

A typical example of such situations can be the accident that occurred on August 10, 2006 in OOO "Rospromservice". During the concrete works in the addition to the structure of the shopping center "New Era" in Nizhny Novgorod, the in-situ concrete ceiling of the building under construction in the Sormov highway collapsed. The total area of the collapse was 540 sqm. During the investigation of the accident, it was found out that no Method statements (MS) for construction and installation works (for the installation of the formwork, installation of reinforcement cages, pouring concrete) had been developed. There was no substantiation of the choice of service poles (calculation of actual loads acting on the poles) and their arrangement pattern for the floor formwork. At the moment of the accident, the poles failed (broke) because they were installed improperly - either with a vertical deviation after installation above the permissible (safe) values, or with a pitch above the permissible due to the lack of the pole arrangement pattern and MS for installation work. As a result of the collapse, six construction workers suffered injuries of varying severity, two construction workers died.

IOP Conf. Series: Materials Science and Engineering 1138 (2021) 012041 doi:10.1088/1757-899X/1138/1/012041

The Technical Commission investigating the circumstances of the collapse on the 7th of November 2007 of the between floor slabs in a reconstructed building of hypermarket "Big dipper" in Novosibirsk, concluded that they had collapsed because of the buckling of the formwork shores. According to the conclusion of the Technical Commission, the cause of the collapse of the formwork while installing the floor, was the loss of stability of the shores. As a result of this there was an additional load in the place of leakage of the concrete onto the support part of the shore that led to the fall of the formwork on the slabs of the basement. The area of the collapse was more than 400 sq. m. According to the Commission, the main violations caused the accident committed by the contractor were failure to comply with the requirements of technological documentation and the lack of necessary control by the responsible officials of the execution of construction works.

Less than a year later (21.08.2008) in Novosibirsk, there was a group accident resulting in death of four workers during in-situ reinforced concrete works. When pouring concrete on the second floor of the shopping center under construction, about 400 sq. m of the floor collapsed. As it was found out by the construction expertise, the building collapsed due to the non-compliance with the technology for the formwork installation.

In Kozhevnicheskaya street in Moscow, on 03.09.2009 in the multilevel 600 sqm garage under construction one of the underground floors collapsed. Under the rubble, one worker was killed and three others were injured.

In the suburbs of Vladivostok, on 01.09.2011, during concreting works at a water pump station, a floor collapsed. As a result of the accident, one person was killed and four were injured.

In the Nevsky district of St. Petersburg, on 07.05.2012 in Krzhizhanovsky street, a formwork collapsed when concreting a floor slab of the parking lot. The consequences of the accident are shown in figure 1.

Figure 1. Collapsed formwork during the construction of a Parking lot in St. Petersburg: fragment of the consequences.

A similar situation occurred in St. Petersburg on 01.03.2018 in Krasnoselsky district in Admiral Tributs street - the formwork of the parking lot under construction collapsed. It happened while concrete was being poured into the formwork. A worker died under the collapsed structure.

In Novoshakhtinsk, the Rostov region, in December 2011, during the execution of works on concreting the floor slab of the 3rd floor collapse occurred. It was found out that during the execution of concrete works the slab of the third floor was subjected to a dynamic impact by the broken container with concrete.

A multi-storey building under construction collapsed in Moscow on April 17, 2012. During the production of filling floors with liquid concrete, the floor collapsed on an area of 2.5 thousand square meters. Six floors at once formed a "layer cake". There were workers on the floors. Among the workers there are dead and injured. The accident happened in an industrial zone on the territory of the former ZIL automobile plant. The building was to house a new office center for commercial companies.

Due to the violation of construction technology on 13.12.2012 in Taganrog, the Rostov region, a three-storey building under construction collapsed. 18 workers were trapped, five of them died. The building was built using the technology of leave-in-place expanded polystyrene formwork. The collapse occurred when builders began to erect the fourth floor, although permission was granted for the construction of a three-story apartment building with shops and an underground parking lot. According to the Town planning code, buildings with a height of up to three floors are not capital construction facilities, and their construction does not require project documentation which must be submitted for examination to the state regulatory authorities. Developers take advantage of this and go to the trick - get permits for the construction of buildings with a greater number of floors.

Collapse of a metal structure of the formwork and erected in-situ concrete floor in the addition to the shopping center "Suvorovsky" occurred on 11.06.2013 in Penza. The accident is related to the overload of the formwork during the work on pouring concrete floor of the second floor.

It was not the first case when during construction of shopping centers in Penza the situation got out of control. So, on 25.08.2007 in the center of Penza, ceilings and part of the wall of the building "Children's world" collapsed. One worker died and 13 were injured.

In the Irtysh lane in Vladivostok, on 07.03.2013 due to the violation of the technology of construction processes a cast-in-place concrete building under construction collapsed. This was caused by the removal of the supporting elements of the formwork before the concrete hardened. A few hours after the first collapse the remaining part of the structure collapsed, the lower floors failed to carry the additional load. The construction of this building was carried out without permits. A judicial decision in 2012 prohibited construction and installation works at the site. Despite everything, the work continued until the first signs of the collapse.

The next day, on 08.03.2013, there was a partial collapse of the shopping and entertainment complex under construction in Moscow in the area of Borisov ponds. The collapse of the formwork and the structure occurred during pouring of concrete on the third floor slab of the shop under construction. The total area of the collapse of the third floor was 200 sq. m. During the accident one of the bearing supports on the corner of the building collapsed. Of the six workers who were at the time of the accident at the construction site, no one was injured.

In Sochi, on 27.03.2013, during the reconstruction of the "Shopping Gallery" a floor collapsed. The accident occurred at the time of pouring in-situ concrete ceiling of the first floor. The cause of the collapse was the fact that the formwork shores failed. The consequences of the accident are shown in figure 2.

The collapse of the ceiling of the third floor and the roof fall of the building under construction occurred on 27.08.2013 in the Kirov district of Novosibirsk. Two people were trapped, both were rescued and taken to hospital. The cause of the collapse was a violation of the technology of pouring concrete floors.

In Vladivostok, in Yeniseiskaya street, on 12.08.2014 during construction and installation works at the construction site of the administrative and hotel complex "Yeniseiskaya Plaza" there was a collapse of the supporting elements of the formwork with the reinforcement of the section of the floor slab and concrete mixture.

During concreting, the load on the formwork poles exceeded the permissible load, resulting in the failure of the formwork shores, which caused the collapse of building structures. Three builders were injured. There was no Method Statement (MS) and technological maps, consequently, construction and installation works were carried out without developed and approved organizational and technological documentation. The installed formwork poles in the place of the collapse had inadmissible length of more than 7 m without spatial horizontal laterals and vertical bracing, formwork shores rested on cantilever wooden joists without considering their twisting in the case of off-center installation (relative to the axis of the joist) of the formwork poles; the poles were installed without the use of special tripods.

Figure 2. Consequences of the collapse during the construction of a monolithic ceiling of the building of the "Trade Gallery" in Sochi.

The following day, on 13.08.2014, in the same place in Vladivostok in Kalinin street, under similar circumstances during construction works, a partial collapse of the structures of the shopping center under construction occurred. Four people were injured in the accident.

In St. Petersburg, on the Okhta river embankment, on 31.01.2015, a 300-square-meter floor collapse occurred in a shopping center under construction.

In New Devyatkino, the Vsevolozhsk district of the Leningrad region, on 04.04.2015 a floor slab under construction of a three-storey parking lot collapsed. The collapse occurred during the concreting of the formwork of reinforced concrete structures.

In Rostov-on-Don, on 29.10.2015 during the construction of shopping Mall "Megamag" there was the collapse of the floor formwork during the concrete work. Three workers were taken to hospital with injuries of varying severity.

At the construction site of Sovgavanskaya CHP in Khabarovsk Krai, Sovetskaya Gavan, on 27.05.2016, the formwork collapsed. Six builders were trapped. There were no fatalities.

In Vladivostok, on 21.10.2017, the floor of the shop building under construction collapsed during pouring concrete on the 3rd floor. The regional Construction Supervision and Control body of Primorsky Krai conducted an inspection of the collapse of the floors of the building. The construction permit was issued on October 11, 2016 by Vladivostok authorities. According to the documents, the developer announced the construction of a two-storey building. As the inspection showed, the developer, contrary to the permits, began to build the third floor. During the formwork installation the in-situ concrete floor structures collapsed. The cause of the collapse was the deviation from the design.

In Kursk, on December 11, 2017, during the construction works, the roof collapsed in a factory reinforced concrete building; the area of the collapse on the premises of OOO "Kursk solod" was 400 sq. m. Workers arranged the formwork and began to fill it with concrete. At this point, the structure failed and collapsed. Four builders were injured.

In the village of Boguchany, the Krasnoyarsk region, on 18.03.2016 at the construction site of "Central district hospital" a concrete slab of the third floor of the building collapsed. Two concrete carpenters were killed.

In the Oktyabrsky district of Ivanovo on 27.12.2017, during the construction of an apartment building of the RC "Panorama", a formwork collapse occurred during the pouring of concrete, as a result of which two builders were injured.

In the Orthodox Church of the Blessed Matrona of Moscow under construction in Biscayskaya street near the "German village" in Krasnodar, on 27.04.2018, ceilings collapsed. During the in-situ concreting, the formwork failed. Rescuers pulled the injured trapped builders.

In Yakutsk, on 5.05.2018, there was a collapse of the formwork in the building of school No. 25 under construction. The area of the collapse was about 140 sq. m. The collapse of the formwork occurred in the process of concrete pouring. The installed supports failed to carry the weight of the concrete mixture. The accident happened because of the violation of the concrete work technology.

Two weeks later, on May 19, 2018, again in Yakutsk, there was a collapse of the floor slab in the "National school for 550 students"; the total area of the collapse was about 53 sq. m. During concreting, the formwork failed and collapsed, because the installed demountable jack-legs failed to carry the load.

In Kursk, on 19.01.2019 in the course of the construction works in a new school building, the formwork shores failed to carry the concrete weight. The slab collapsed avalanche-like around the perimeter. The area of the collapse was 400 square meters. The consequences of the accident are shown in figure 3.

In Yaroslavl, on 31.08.2019 in the surgery of the oncological hospital under construction there was a collapse of an intermediate ceiling. When pouring concrete, the formwork failed to carry its weight, and the structure collapsed and, in some places, it caved in and so did the reinforcement. The area of the collapse and the sagging was 350 sq. m. Among the causes there were problems with the poles that supported the formwork of the concrete structure. The poles failed due to a manufacturing defect.

There are a lot of examples of similar accidents. Of course, these kinds of accidents don't usually happen every day, but they do happen.

Figure 3. The Collapse of the floor slab during the construction of a school building in Kursk.

3.2. Systematization of identified causes of accidents

Analysis of accidents at construction sites shows their regular recurrence. As a result of this, the designers, construction participants and supervisory authorities study the causes leading to the collapses of in-situ concrete slabs of buildings and structures in order to prevent them in the future. Nowadays it is an urgent problem.

As a rule, the same mistakes are repeated in design and organizational decisions, and violations of performing work technology that could lead to similar events.

It is necessary to provide a list of such errors and violations:

- the developer has no design solutions for specific sections of slabs with a detailed study of formwork pickup fittings;
- solutions for placing formwork without the approval and confirmation by calculations of formwork suppliers;
- there is no rejection and removal of deformed or substandard molding elements from the construction site, as a result of which damaged structural elements of the molding are used for further operation;
- the presence of defects and violations during the operation of formwork locks;
- increase in the number of formwork turnover cycles;
- the wrong choice of the type of formwork systems initially, inattentive attitude to the material the formwork is made of (in particular, to steel grade and its quality, plywood, wood, components, locks);
- non-compliance (and often lack of technological maps) with the order of concreting sites and unauthorized performance of those works that are not provided for by the project of production of works;
- inadequate organization and control of various types of activities in order to ensure the necessary quality of work and the conformity of production processes with the prescribed safety requirements;
- unauthorized storage of reinforcement and construction equipment on the formwork system structures;
- a significant excess of working loads on the formwork, leading to twofold (or more) excess of the permissible vertical load on formwork poles;
- installation of the structure on an unprepared non-rigid, deformable base;
- absence of horizontal elements and diagonal connections when using the slab formwork (or low-quality elements) in the positions determined by the project of work;
- installation of slab formwork and scaffolding with deviations from the vertical and horizontal, if the final adjustment of formwork by builders was not carried out;
- the presence of sagging or deflexion of slab formwork;
- dynamic impacts on the formwork with a concrete boom when feeding concrete or a container with concrete;
- ignoring the operation instructions of formwork systems, although all professional patented products are delivered with detailed and thoroughly prescribed instructions for operation, installation and dismantling of the system.

When working with a formwork and performing concrete work the exclusion of the above and other errors can guarantee the absence of force majeure at the construction site.

It should also be noted that critical defects allowed during the installation of an in-situ concrete slab are potentially the causes that can cause a collapse of structures during the operation of buildings when an unfavorable combination of factors affect them.

3.3. Discussion of the causes

Following the global practices and trends of architecture and urban planning, there appeared buildings on the construction market, the construction of which requires the implementation of complex structural solutions. But modern and reliable formwork systems [10, 11], high-quality technology and organization of construction, as well as highly qualified construction personnel must correspond to the complexity of the design. In modern domestic construction, these requirements are not fully met. It turned out that there is a shortage of qualified personnel to carry out large volumes of work and implement new construction technologies [12 - 14]. But the market demands, and personnel can be found in other

IOP Conf. Series: Materials Science and Engineering	1138 (2021) 012041	doi:10.1088/1757-899X/1138/1/012041

countries. In 2017, 6.3 million people were employed in the construction industry in Russia, labor migrants were a significant proportion of them. Since 2019, the share of foreigners employed in construction is proposed to be limited to 80%. Previously, there were no restrictions on the engagement of foreign workers in construction.

However, the engaged labor migrants are mainly not specialists, but people who have little understanding of construction engineering. The use of labor of foreign migrants in construction creates a number of new problems for site managers. The main ones are: poor knowledge of the Russian language, low vocational and educational training, misunderstanding of instructions and orders, complication of production communication and coordination of interaction, and, as a result, low quality of work performed, violation of safety requirements. Inconsistency with the professional level of migrant workers often leads to faulty work. According to research [15], only 40% of labor migrants from Tajikistan employed in construction were familiar with the rules of safety technology, 55% were not provided with special clothing, 40% - with personal protective equipment. Thus, legal requirements for construction organizations to ensure safety are not fulfilled [16].

Execution of complex construction work by unprofessional firms without trained personnel leads to the considered consequences. Violations of construction rules and regulations during the construction of facilities of various purposes have become commonplace. Therefore, the presence of professional competencies corresponding to the envisaged work should be checked by the certification committee before admission to work [17]. Engineers and technicians should know the urgent issues of construction, including technology of in-situ concrete construction, types of formwork used in such construction and features of concrete work [10, 11].

Studying the problem of personnel in construction, the authors of the paper [18] have noted: "According to recruiting agencies, there is a large shortage of foremen, architects, design engineers, designers, and project management specialists in the construction industry today. Therefore, many companies simply "outbid" scarce specialists from competitors. Company executives complain that the majority of experienced builders working at construction sites in the Soviet era are of pre-retirement or retirement age. At the same time, the problem of young personnel remains very important. The preparedness of current graduates of construction faculties and universities is low, primarily due to the lack of motivation of students to study".

In October 2014, the IX national Congress "Modernization of Russian industry: development priorities" was held in Moscow. One of its main topics was the training of personnel necessary for the modernisation of the construction industry. As one of the speakers put it, a systematic personnel crisis has begun in the Russian construction industry. This applies not only to the training of workers and specialists with higher education, but also to the overall state of the professional level of builders, ranging from working professions to the highest level [19]. In November 2016, at the section "Housing and utilities: infrastructure and infrastructure development" of the XII National Congress "Modernization of industry in Russia. Development priorities", the speakers again drew attention to the decrease in professional competence of specialists working in the construction industry (primarily designers) and the fact that the system of qualification education of specialists in working professions is also not developed [20].

It should be noted that in papers [5, 21] we considered the issues of preserving the quality of labor resources in construction; in particular, it was proved that the lack of high-quality training creates a "platform" for the growth of injuries in the industry, since the personnel subsystem of this process is unsecured. These forecasts are fully confirmed. Nowadays professionals make up only 10 - 15% of the total number of construction workers [22].

Insufficient level of education does not allow a designer to issue high-quality projects at once. Truly safe projects are designed by him only years later when he gains experience analyzing his mistakes. Designers of specific facilities should put into practice computer-aided design technologies that are implemented in modern software products with high accuracy and model the behavior of formwork structures correctly (which is especially important when calculating structures with complex architectural forms, space-planning and structural solutions).

Today the training of professional personnel in working professions, the training of civil engineers, as well as advanced training and retraining of both are the primary tasks for national education. The training program for specialists in the field of technology and the organization of construction should be constantly updated, take into account modern requirements, have a practical orientation and be carried out in close interconnection with other general professional and special disciplines [12, 13].

Now we need to focus on the following tasks: improving international exchange of modern knowledge and technologies; development of innovative forms of higher education, science and construction based on the study of international experience; activation of multi-lateral mobility of students, postgraduates and teachers; integration with the global educational space.

We need to improve information support of construction. The essence of this problem is a thorough, highly professional investigation of the causes of accidents by independent engineering and consulting firms, publication of information about accidents in special publications accessible to the general engineering community, creation of special services, centers that conduct a statistical analysis of the causes of accidents and develop measures to prevent them. However, the reality is that: after an accident, commissions are created to find the "culprit" in a short time, and all information about their activities is not available even for professionals. This makes it impossible to collect all the information for conducting official statistics, analyzing the real causes of major accidents and making relevant decisions to prevent them. Gathering information about cases of accidents at construction sites, summarizing and systematizing this information, openness of data and a thorough analysis of the causes are very important scientific and practical work aimed at preventing similar situations from happening again in the future [23].

4. Conclusions

The various causes of collapse of formwork systems and injuries are closely related to each other and are so intertwined and mutually determined in most cases that it is impossible to imagine each of them in a separate form. Therefore, a single detailed classification of the causes of collapses and injuries cannot be made at all stages of in-situ concrete work, since the specificity of various processes requires taking into account well-defined initial events and technology features of the type of work under consideration.

The developer and the person executing construction works should fully control the safety of building structures, the quality and correct use of the formwork used and the work performed [24].

Accident prevention is a set of measures including professional selection of personnel, improving the quality of professional construction education, mastering advanced construction technologies and management methods of construction organizations [25], as well as organizing and implementing customer construction control (technical supervision) of the construction of facilities.

Specialists in the field of technology and organization of construction production should have a deeper education in order to thoroughly and qualitatively consider the issues of construction production from the point of view of innovative achievements, have an idea of the main provisions and directions for improving the technology, organization and planning of construction production, and advanced domestic and foreign methods of production of construction works. Only on the basis of a constant exchange of experience can they activate reserves for improving the efficiency and safety of construction production. Especially interesting in terms of gaining experience should be considered training and practical work on construction sites of landmark projects, both in Russia and abroad

References

[1] Report on the results of monitoring of labor conditions and safety in the Russian Federation in 2016. URL: http://www.trudcontrol.ru/files/editor/files/Мониторинг_2016.pdf
[2] Healthcare in Russia, Moscow: Rosstat. 2017: 170 p
[3] PRAVMIR news service. URL: https://www.pravmir.ru/bolee-1700-chelovek-pogibli-v-2017-godu-ot-neschastnyih-sluchaev-na-proizvodstve/
[4] Tomakov V I 2006 Industrial injuries in the construction industry. Life Safety (**3**):13-22.

[5] Tomakov V I 2006 Production and economic activity and labor safety in construction. Analysis and forecast, Kursk, 180 p
[6] Professional risks in construction [Electronic resource]. URL: http://www.kiout.ru/info/news/207
[7] Association "Monolith". URL: http://www.opalubka-system.ru/monolitnoe_stroitelstvo.html
[8] Eremin K I, Makhutov N A, Pavlova GA and Shishkina N A 2001 Register of accidents of buildings and structures 2001-2010, Moscow: RAASN, 318 p
[9] Shishkina N A 2014 Analysis of accidents of buildings and structures. Science and Safety; **4**(13): 34-50
[10] Dergunov K N, Telichenko V I 2013 The Safety of innovative formwork systems. Technology and organization of building production; (**1**): 19-21
[11] Volovik M V, Ershov M N, Ishin A A and others 2013 Modern formwork systems as a factor of quality assurance and reliability of construction objects: round table. Technology and organization of construction production; (**1**): 14-18
[12] Volovik M V, Ishin A A, Lapidus A A and others 2014 Problems of professional training in construction: round table. Technology and organization of construction production; (**1**)10-17
[13] Lapidus A A 2014 Actual problems of professional training of specialists in the field of technology and organization of construction production. Technology and organization of construction production; (**1**): 1
[14] Volovik M V 2014 Professional training of workers in the field of construction production. Technology and organization of construction production; **1**(6): 33-34
[15] Nazarov V V 2009 The problems of migrant workers in the construction industry: sociological aspect. Bulletin of the Chelyabinsk state University; **11**(149): 157-163
[16] Tomakov M V 2012 The legal basis in the system of labor protection management and industrial safety of the construction organizations. News of southwest state University. Series: Technology and engineering; (**2-3**): 248-252
[17] Ishin A V 2013 Mandatory certification of specialists of companies-a guarantee of safety and quality of work performed by them. Technology and organization of construction production; (**3**): 23-24
[18] Zhukov A N, Artyukhina O V 2014 The problem of qualified personnel in construction. Young scientist; (**20**): 129-131.
[19] The dominance of the ignorant: while the threat or already a reality? 2014 Construction; (**11**): 14-16
[20] Reports of the National Congress "Modernization of Russian industry: development Priorities". URL: http://www.nationalkongress.ru/otchety.html
[21] Tomakov V I 2006 Safety in construction and problems of preservation of the quality of labour resources. Proceedings of the Kursk state technical University; **1**(16): 98-106
[22] Expert: The Construction market suffers an acute shortage of skilled workers. URL: http://sroportal.ru/news/partners/ekspert-stroitelnyj-rynok-terpit-ostruyu-nexvatku-kvalificirovannyx-rabochix/
[23] Merkulov S I, Tomakov V I 2005 The Concept of safety of industrial buildings and structures. Safety of the construction Fund of Russia. Problems and solutions: proceedings of International academic readings. Kursk; 155-163 p
[24] Practical guide to the organization and implementation of construction control of the customer (technical supervision) for the construction of capital construction 2010 Moscow: JSC "Tsentrinvestproekt", 73 p
[25] Tomakova I A, Tomakov M V, Charkina Yu N and Brezhnev AV 2018 Main provisions of the modern policy of the organization in the field of occupational safety and health personnel. News of Southwest state University. Series: Engineering and technology; **2**(27): 85-98

IOP Conf. Series: Materials Science and Engineering 1138 (2021) 012042 doi:10.1088/1757-899X/1138/1/012042

Comparative analysis of two seismic energy dissipation solutions in the case of a steel structure

F Tepes-Onea[1*], D Vintila[1] and C Tudose[2]

[1] University Ovidius of Constanta, Faculty of Civil Engineering, Romania

* tflorin@univ-ovidius.ro

Abstract. The present paper aims at the comparative analysis of two seismic energy dissipation solutions for a steel structure in frames. We opted for non-braced frames and centrally braced frames. In the case of unbroken frames, the resistance structure consists of steel beams and columns. Potential plastic areas are formed mainly at the ends of the rulers and, to a lesser extent, in columns. Horizontal actions are mainly taken by bending the structural elements. To ensure the most efficient observance of the rigidity requirements for lateral actions, the solution of the rigid beam-column joint is required. The use of the concept of weak beam - strong column leads to the appearance of plastic joints in the beams, close to their joints with the columns. Centric braced frames can take over the horizontal actions, which are manifested in their plane, through axial forces. Energy dissipation occurs mainly by plasticizing the extended diagonals. The comparative analysis consists of highlighting the results obtained from the structural calculation with the help of a commercial automatic calculation program for the two solutions: non-braced frames and centrally braced frames. An office building with 2 levels was analyzed, located in Constanta, where the two types of the staff were provided in turn. The obtained results will be compared from the point of view of the amount of steel used, so from the economic point of view. Another comparison refers to the displacements to SLS and SLU and to the seismic energy dissipation capacity. Finally, another possible comparison refers to the over-resistance of the structural system.

1. Case study of steel frame structure

1.1. General data about the structure

The structure, which is to be analyzed, is made up of unbroken frames, with rigid nodes and is a steel frame with 2 stories, 3 openings and 4 beams (see figure 1). This structure is made of reinforced concrete floor, made of lost corrugated sheet steel formwork that supports a system of secondary beams.

The structure is located in Constanta and is to be designed according to the DCM ductility class. The gravitational loads on the floor are made up of the permanent load of 5 kN/m^2 and the payload of 3 kN/m^2.

The main features of the structure are summarized below: opening: L = 6 m; opening in the other direction: B = 5 m.

Floor height: h = 3.5 m; height: ground floor + 2 floors (3 levels); height: H = 10.5 m; plan size: 18 x 20 m; location: Constanta; permanent load on the floor, g_k = 5 kN/m^2.

The payload on the floor, qk = 3 kN/m^2; the structure will be designed according to the DCM ductility class; class of importance and exposure class III, according to table 4.2 of P 100-1/2013; the factor of importance: $\gamma_{I,e}$ = 1.0 [1].

IOP Conf. Series: Materials Science and Engineering 1138 (2021) 012042 doi:10.1088/1757-899X/1138/1/012042

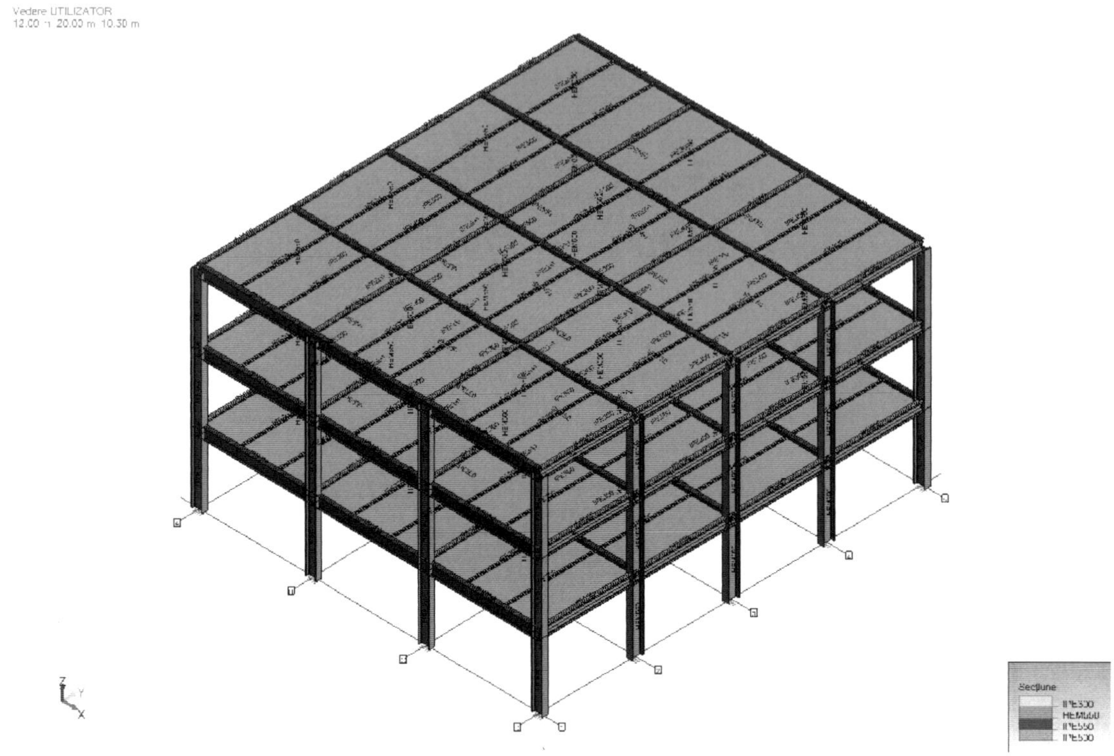

Figure 1. 3D spatial model of steel frame structure.

The structural calculation is performed with an automatic calculation program.

In this study, the second order calculation and imperfections, were considered. SR EN 1993-1-1 presents three methods for taking into account second-order effects and imperfections. The third method was adopted: for basic cases, by individual stability checks of equivalent bars, using buckling lengths corresponding to the overall mode of instability of the structure.

1.2. Calculation of strength and stability of the steel frame structure contains. Verification of the strength and general stability of the beams

SR EN 1993-1-1 [2] will be used for the verification and conformity of the beams to the general stability in the hypothesis that "only at one end a plastic joint has been formed" (1, 2, 3).

$$\frac{M_{Ed}}{M_{pl,Rd}} \leq 1.0 \tag{1}$$

$$\frac{N_{Ed}}{N_{pl,Rd}} \leq 0.15 \tag{2}$$

$$\frac{V_{Ed}}{V_{pl,Rd}} \leq 0.5 \tag{3}$$

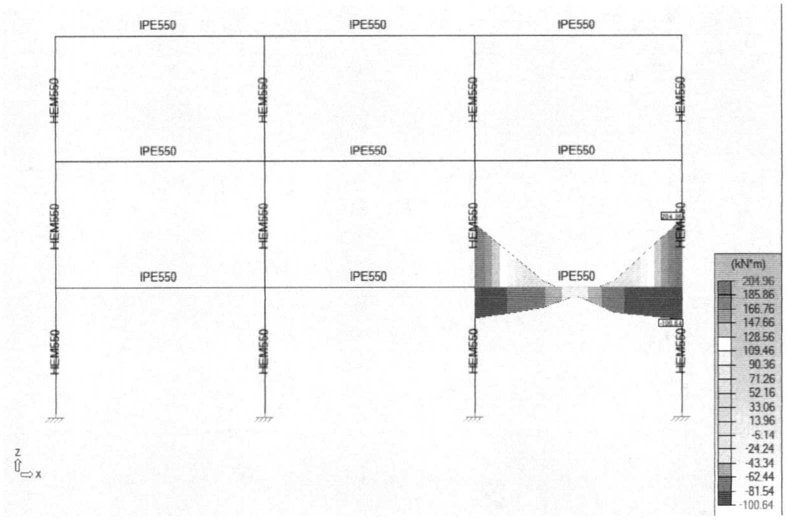

Figure 2. The bending moment diagram of the most requested beam.

1.3. Calculation of the super resistance of the structural system

The over-resistance (table 1, table 2) of the structural system, used in the calculation of the stresses in the non-dissipative components, is determined by the relation:

$$\Omega_T = 1.1 \cdot \gamma_{ov} \cdot \Omega^M \tag{4}$$

$$\Omega^M = \min(\Omega_i{}^M) \tag{5}$$

$$\Omega_i{}^M = \frac{M_{pl,Rd}}{M_{Ed}} \tag{6}$$

Ω_T it is the value of the super-resistance of the structural system.
Ω^M minimum value of $\Omega_i{}^M$calculated for all beams with the potential plastic area;
$M_{pl,Rd,i}$ is the plastic resistance of the projectors in the beam *i*;
$M_{Ed,i}$ this is the bending moment in the beam and in the load group (figure 2).

Table 1. Super-resistance calculus Ω_T, X direction.

Opening	Level	M_{Ed}	$M_{pl,Rd}$	$M_{Ed}/M_{pl,Rd}$	$\Omega_i{}^M = M_{pl,Rd}/M_{Ed}$	Ω^M	$\Omega_i{}^M$-Ω^M	$\Omega_i{}^M$-Ω^M (%)	Ω_T
	2	146.0	696.7	0.21	4.77		1.37	28.76	
A-B	1	194.9	696.7	0.28	3.57		0.18	4.91	
	P	204.7	696.7	0.29	3.40		0.00	0.12	
	2	129.2	696.7	0.18	5.39		1.99	36.94	
B-C	1	174.4	696.7	0.25	3.99	3.4	0.59	**14.89**	**4.86**
	P	183.3	696.7	0.26	3.80		0.40	10.55	
	2	145.3	696.7	0.20	4.79		1.39	29.06	
C-D	1	195.3	696.7	0.28	3.56		0.17	4.67	
	P	204.9	696.7	0.29	3.39		0.00	0.00	

Table 2. Super resistance calculus Ω_T, Y direction.

Opening	level	M_{Ed}	$M_{pl,Rd}$	$M_{Ed}/M_{pl,Rd}$	$\Omega_i^M = M_{pl,Rd}/M_{Ed}$	Ω^M	$\Omega_i^M - \Omega^M$	$\Omega_i^M - \Omega^M$ (%)	Ω_T
	2	165.71	548.5	0.302	3.310		0.52	15.84	
1-2	1	181.26	548.5	0.330	3.026		0.24	7.94	
	P	191.33	548.5	0.349	2.867		0.08	2.82	
	2	165.88	548.5	0.302	3.307		0.52	15.75	
2-3	1	187.96	548.5	0.343	2.918		0.13	4.54	
	P	196.09	548.5	0.358	2.797	2.79	0.01	0.41	**3.98**
	2	166.82	548.5	0.304	3.288		0.50	15.27	
3-4	1	189.08	548.5	0.345	2.901		0.12	3.97	
	P	196.89	548.5	0.359	2.786		0.00	0.00	
	2	164.31	548.5	0.300	3.338		0.55	**16.55**	
4-5	1	179.76	548.5	0.328	3.051		0.27	8.70	
	P	190.84	548.5	0.348	2.874		0.09	3.07	

In the X direction, the maximum and minimum values of Ω_i^M differ by 14.89% < 25%. Exceptions are made to the values obtained for the beams from the last level, which results in being oversized.

In the Y direction, the maximum and minimum values of Ω_i^M differ by 16.55% < 25%.

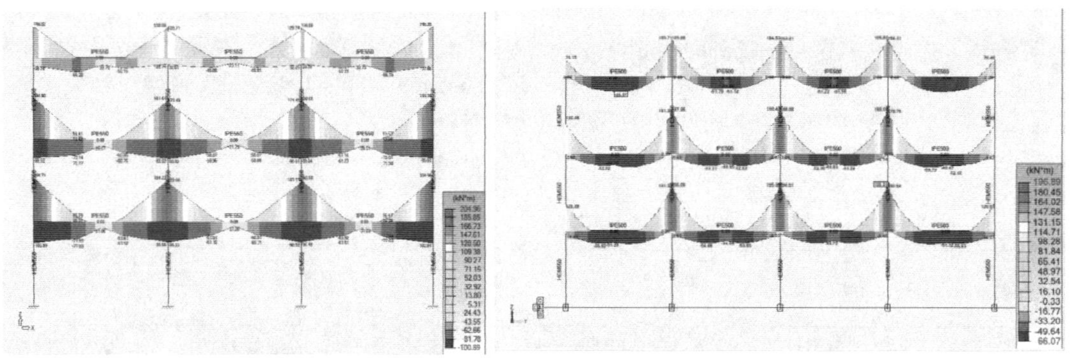

Figure 3. Bending moment diagram, M_{Ed}-Ax 5; Ax C.

1.4. Verification of the strength and general stability of the columns

The verification at ULS of the columns on the ground floor is simplified for the most requested element, located within axis 5.

Calculation forces for columns and determination according to the relations (7, 8, 9) from P 100-1/2013.

$$N_{Ed} = N_{Ed,G} + \Omega_T N_{Ed,E} \tag{7}$$

$$M_{Ed} = M_{Ed,G} + \Omega_T M_{Ed,E} \tag{8}$$

$$V_{Ed} = V_{Ed,G} + \Omega_T V_{Ed,E} \tag{9}$$

Ω_T is the value of the super-resistance of the structural system;

$\Omega_T = 3.0$ (P100-1/2013, Annex F- Table F.1)

We had the following results:

$M_{y,Ed,1} = 357.31\ kNm$; $M_{z,Ed,1} = 49.18\ kNm$; $M_{y,Ed,2} = 3.09 + 331.33 = 334.42\ kNm$; $M_{z,Ed,2} = 0 + 107.17 = 107.17\ kNm$; $N_{Ed} = 1.13 + 727.84 = 728.97\ kNm$; $V_{z,Ed} = 484.76 + 399.64 = 884.4\ kN$; $V_{y,Ed} = 9.26 + 66.82 = 76.08\ kN$ (figure 4).

The plastic resistances of the design of the section are:

$$M_{pl,Rd} = \frac{W_{pl,y} \cdot f_y}{\gamma_{M0}} = \frac{7933 \cdot 275}{1.1} = 1983.25\ kNm \quad (10)$$

$$N_{pl,Rd} = \frac{A \cdot f_y}{\gamma_{M0}} = \frac{35440 \cdot 275}{1.1} = 8860\ kN \quad (11)$$

$$V_{pl,z,Rd} = \frac{A_{v,z} \cdot f_y}{\sqrt{3} \cdot \gamma_{M0}} = \frac{13960 \cdot 275}{\sqrt{3} \cdot 1.1} = 2014.95\ kN \quad (12)$$

Thus, the relations of specific provisions P100 could be verified.

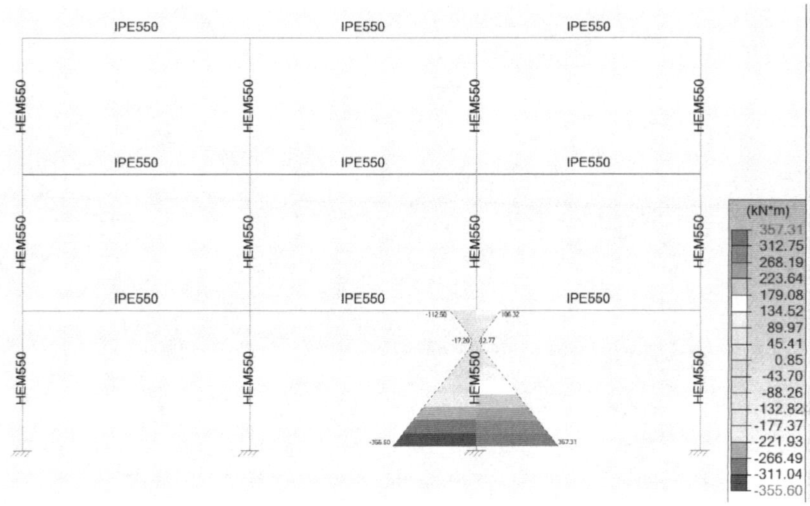

Figure 4. Bending moment diagram, $M_{y,Ed,1}$.

Other calculations made refer to: bending and compression checking; checking the buckling of the columns; checking the beam-column joints [3 - 6].

For M-N interaction: $k_{yy} = 0.79$; $k_{yz} = 0.53$; $k_{zy} = 0.42$; $k_{zz} = 0.78$ and applying the verification relationship results: 0.05 + 0.13 + 0.01 = 0.19 < 1.00 (19%) and 0.06 + 0.07 + 0.02 = 0.14 < 1.00 (14%).

1.5. Checking the beam-columns joints

The joints of the beams with the columns, at a projected structure that dissipates the seismic energy in the beams, must be designed to work in the elastic domain during the earthquake.

The total rotational capacity of the beam-column node, denoted by, must be ensured at cyclic loads, without producing degradations of strength and rigidity greater than 20%.

1.6. Vibration modes

10 own modes of vibration were taken into account. Own vibration periods T and effective modal masses M in relation to the total mass of the structure are presented in table 3.

It can be observed that the sum of the effective modal masses from the first 4 eigenmodes of vibrations exceeds 90% of the total mass of the structure, being fulfilled the requirement in section 4.5.3.3 of P 100-1.

The masses of the structure were calculated automatically from the gravitational loads applied to the structure.

Table 3. Mass on levels.

Level	φ	$\Psi_{2,i}$	$\Psi_{E,i}$	$G_{k,j}$ (kN/m^2)	$Q_{k,i}$ (kN/m^2)	Surface (m^2)	Mass (tons)
Ground floor	0.8	0.3	0.24	5	3	180	104.95
Floor 1	0.8	0.3	0.24	5	3	180	104.95
Floor 2	1	0.3	0.3	5	3	180	108.26
				Total			318.17

Table 4. Center of mass and rigidity on the floor.

Number of level	Level name	Center of mass X Y	Center of rigidity X Y
1	Ground floor	9.00 10.00	9.00 10.00
2	floor 1	9.24 10.29	9.00 10.00
3	floor 2	9.21 10.26	9.00 10.00

The deformed structure in the first proper way of vibration is presented in figure 5.

Modal mass on X direction is 999.74 t from 1202 t meaning 83% (vibration mode 1). Modal mass on Y direction is 994.9 t from 1202 t meaning 82.7% (vibration mode 2).

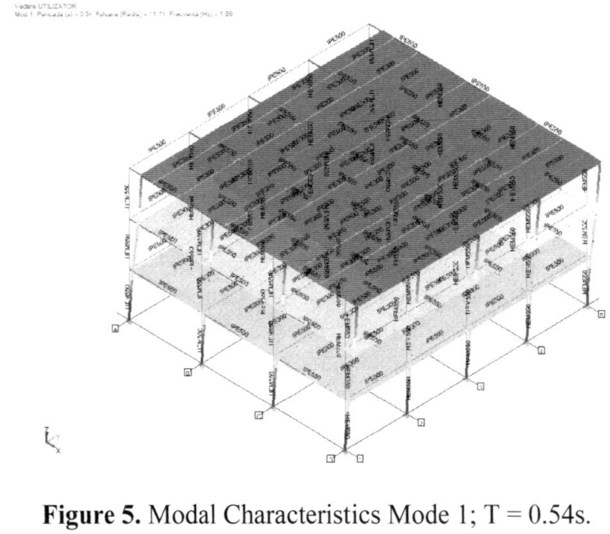

Figure 5. Modal Characteristics Mode 1; T = 0.54s.

Table 5. Check displacements at the service limit state (SLS).

level	d_{re} Direction X	d_{re} Direction Y	ν	q	d_r^{SLS} Direction X	d_r^{SLS} Direction Y	$d_{r,a}^{SLS}$	check X	check Y
floor 2	0.0125	0.0001	0.5	4	0.0250	0.0002	0.0263	ok	ok
floor 1	0.0131	0.0002	0.5	4	**0.0262**	0.0004	0.0263	ok	ok
Ground floor	0.0099	0.0004	0.5	4	0.0198	0.0008	0.0263	ok	ok

$$d_r^{SLS} = 0.0262 < d_{r,a}^{SLS} = 0.0263 \rightarrow is\ verified$$

Table 6. Check displacement at last limit state (SLU).

level	d_{re} DirectionX	d_{re} Direction Y	c	q	d_r^{ULS} Direction X	d_r^{ULS} Direction Y	$d_{r,a}^{ULS}$	Direction X	Direction Y
floor 2	0.0115	0.0001	1.074	4	0.0494	0.0004	0.0875	ok	ok
floor 1	0.0131	0.0002	1.074	4	0.0563	0.0009	0.0875	ok	ok
Ground floor	0.0099	0.0004	1.074	4	0.0425	0.0017	0.0875	ok	ok

$$d_r^{ULS} = 0.0563 < d_{r,a}^{ULS} = 0.0875 \rightarrow is\ verified$$

2. Calculation of strength and stability of the steel frame structure with diagonals

2.1. General data on structure

The structure, which is to be analyzed, is made up of a centrally braced X-shaped frame arranged in the transverse axis 1 and 5 (X direction) and in the longitudinal axis A and D (Y direction). The steel frame has 2 floors, 3 openings and 4 open beams (see figure 6). This structure is made of reinforced concrete floor, made of lost corrugated sheet steel formwork that supports a system of secondary beams. The structure is located in Constanta and is to be designed according to the DCM ductility class.

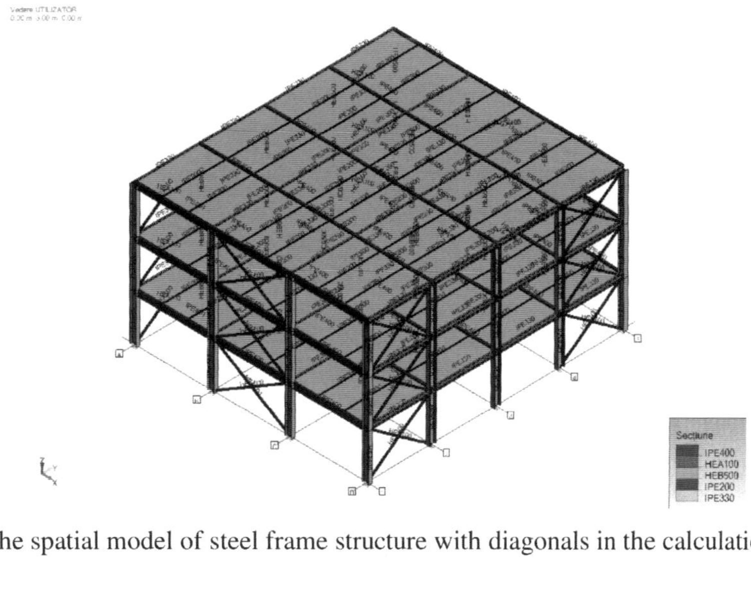

Figure 6. The spatial model of steel frame structure with diagonals in the calculation program.

The diagonals were dimensioned according to Eurocode [2].

2.2. Calculation of the super resistance of the structural system

Table 7. Calculation of the super resistance Ω_T , on the X direction.

Opening	level	N_{Ed}	$N_{pl,Rd}$	$N_{Ed}/N_{pl,Rd}$	$\Omega_i^N = N_{pl,Rd}/N_{Ed}$	Ω^N	$\Omega_i^N - \Omega^N$	$\Omega_i^N - \Omega^N$ (%)	Ω_T
B-C	2	117.22	531	0.221	4.53	3.00	1.53	33.87	**4.28**
	1	177.27	531	0.334	3.00		0.00	0.00	
	P	143.21	531	0.270	3.71		0.71	**19.21**	

Table 8. Calculation of the super resistance Ω_T , on the Y direction.

Opening	level	N_{Ed}	$N_{pl,Rd}$	$N_{Ed}/N_{pl,Rd}$	$\Omega_i^N = N_{pl,Rd}/N_{Ed}$	Ω^N	$\Omega_i^N - \Omega^N$	$\Omega_i^N - \Omega^N$ (%)	Ω_T
B-C	2	82.22	531	0.155	6.46	4.20	2.25	34.90	**6.01**
	1	126.29	531	0.238	4.20		0.00	0.00	
	P	109.61	531	0.206	4.84		0.64	**13.21**	

The difference between the maximum and minimum ratio Ω_i^N is less than 25% for each direction of the structure. In the X direction this difference is 19.21%, and in the Y direction it is 13.21%. Exceptions to the rule make diagonals to the 2nd floor, which results in being oversized.

Other calculations made refer to: the buckling of the column; compression and bending verification; calculation of joints of dissipative elements (bracing) [3, 4].

2.3. Joints of dissipative elements (braces)

A gusseted welded joint was adopted to allow buckling of the bracing in out of plan [5]. The buckling is realized by bending a portion of the gusseted with a width equal to 2 times its thickness, t_g. The joint must be designed to have a resistance of at least $1.1 \cdot \gamma_{ov}$ to the joined element (bracing), according to sections 6.7.3 and 6.5.5 of P 100-1.

The joint will be dimensioned according to SR EN 1993-1-8 to the axial force:

$$N_{Ed,imb} = 1.1 \cdot \gamma_{ov} \cdot N_{pl,Rd} = 1.1 \cdot 1.3 \cdot 531 = 759.33\ kN$$

3. Conclusions

One of the principles of anti-seismic design is the constraint of a favorable structural mechanism for energy dissipation (plasticization mechanism) under high-intensity seismic actions, thus preventing the collapse.

In the first case of a steel frame structure, the dissipation of the seismic energy takes place through the formation of the plastic joints at the ends of the beams, representing the main dissipative elements. The buckling of the element, the suppleness of the sections, and the presence of important forces of compression and/or shearing influence the appearance of the plastic capable bending and providing the capability of rotation. In order that the capable bending moments of the section and its rotational capacity are not diminished, the axial force was limited to 15% of the plastic axial force of the section, and the shear force to 50% of the capable plastic cutting force.

The large bending moments at the ends of the beams led to substantial deformations of the core panels of the columns at the right of the joint (by cutting force). These deformations led to important displacements, and to reduce lateral displacements and prevent structural damage, it was necessary to use elements (columns and beams) with larger dimensions.

The most unfavorable relative displacement, at SLS, is $d_r^{SLS} = 2.62\ cm$, and at ULS $d_r^{ULS} = 5.63\ cm$. In case two, steel frame structure with diagonals, the dissipative zones are located in diagonally, when they are stretched.

For steel structure with diagonals were used steel profiles HEM550 for columns, IPE 550 for main transverse beams, IPE500 for longitudinal main beams, and IPE 300 for secondary beams, resulting in a total weight of 124 tons. The period of the fundamental vibration mode is de T_1=0.54 s < T_C=0.7 s. The sum of the effective modal masses exceeds 90% of the total mass of the structure from the first 4 eigenmodes of vibrations.

The most unfavorable relative displacement, at SLS, is $d_r^{SLS} = 2.62\ cm$, and at ULS it is $d_r^{ULS} = 5.63\ cm$.

In the case of frames with diagonals, the dissipative zones are located in diagonal, when they are stretched. The bracing required for the stretching has been dimensioned in such a way that the capable plastic effort of the $N_{pl,Rd}$ cross-section is greater than the computational effort of the N_{Ed} seismic combination. The presence of bracing diagonals causes the system to be loaded mainly with axial forces, having smaller bending moments in the beams and implicitly smaller dimensions of these.

For the steel frame structure with diagonals, HEB 500 steel profiles were used for columns, IPE 400 for main beams, IPE330 for longitudinal beams, IPE 200 for total beam and for diagonal beams, and HEA100 for diagonals, resulting in a total weight of 81 tons.

The period of the fundamental vibration mode is T_1 = 0.48s < T_C = 0.7s. The sum of the effective modal masses exceeds 90% of the total mass of the structure in the first 5 eigenmodes of vibration.

The most unfavorable relative displacement, at SLS, $d_r^{SLS} = 1.96 cm$, and at ULS it is $d_r^{ULS} = 4.82\ cm$. The steel frame structure (first case) will have a better behaviour to the seismic action, it presents a high ductility, ensuring the dissipation of a large amount of energy, but it does not require larger sections of the elements. It is preferred by architects as it does not condition the positioning of gaps for windows on the facade.

The centrally braced structure has a limited ability to dissipate seismic energy. The existence of diagonals makes the beam sections smaller, and therefore an advantage in terms of economic value. It is not preferred by architects as it conditions the positioning of the gaps for windows on the facade.

Acknowledgments

Rapid Earthquake Damage Assessment Consortium-REDACt [BSB 966].
Contract Nr: MLPDA 88712/26.06.2020.

References

[1] P100-1/2013 *Design of structures for earthquake resistance, part 1*

[2] SR EN 1993-1-1/2006 *Eurocode 3 Design of steel structures. General rules and rules for buildings*

[3] Dubina D and Ungureanu V 2010 *Checking the stability of a steel element in accordance with SR EN 1993-1.1, Recommendations for calculating comments and application examples*

[4] Dubina D and Dinu F 2010 *The global structural design of the metallic structures. Recommendations, comments, and examples of an application according to SR EN 1993-1-1 and SR EN 1998-1*

[5] Dubina D and Grecea D 2009 *Calculation and projection of joints structural steel in conformity SR-EN 1993-1-8*

[6] Eric M. Lui Wai-Fah Chen 1988 *Behavior of braced and unbraced semi-rigid frames,* International Journal of Solids and Structures Volume 24 Issue 9 Pages 893-913

IOP Conf. Series: Materials Science and Engineering 1138 (2021) 012043 doi:10.1088/1757-899X/1138/1/012043

The impact of the decreasing number of users on the evolution of a centralized heating system

E Vitan[1], T Rus[1*], A Hotupan[1] and C Cilibiu[1]

[1] Technical University of Cluj-Napoca, Faculty of Building Services Engineering, Cluj-Napoca, Romania

* tania.rus@insta.utcluj.ro

Abstract: The paper presents an analysis of a centralized heating system located in a city in the west-central part in Romania. We examined the last 15 years of functioning of the system from a technical and economic point of view. One of the main factors that led to the inefficiency of this system, and in the end to collapse, was the reduction of the system's thermal efficiency due to the decrease of customers over time. In conclusion, we highlighted the necessity of a periodic diagnosis of system's efficiency and a risk analysis to avoid economic problems and collapse.

1. Introduction

The supply of thermal energy to consumers can be done in a centralized or decentralized system. In central heating systems energy is produced in one location (source) and is distributed to consumers through pipelines network, and in decentralized systems thermal energy is produced and consumed in the same location [1].

In Romania, in the period 1951-1989, the thermal energy system was formed and developed by introducing centralized thermal energy production systems for industry and for residential buildings in urban areas. Until the 90s, in Romania, the Centralized Thermal Energy Supply Systems (CTESS) were almost the only form of heating for apartment buildings, a solution imposed by the state for the controlled combustibles consumption [1, 2].

After 1991, the distribution of natural gas is taken over by the companies that aim to increase sales, and the demand for thermal energy in a centralized system began to decrease quite quickly by the disconnection of industrial consumers from CTESS, mainly for the following reasons:

- replacement of the centralized source with own one for cost control;
- reduction and reorientation of activities on industrial platforms.

This decrease in the demand for thermal energy has resulted in the shutdown of some production units within CTESS, with an important unfavorable influence on their economic indicators, the decrease of energy efficiency and the increase of the price of delivered energy.

After 1995, a second stage of disconnections began, that of residential consumers (population) who considered more attractive the solution of heat supply from wall mounted boiler, in their own apartment, powered with natural gas.

After 2003, heating plants began to be gradually subordinated to local municipalities. This created the possibility of unifying the production activity with the transport and distribution of thermal energy, thus creating the possibility of correlating modernization, investment and operation programs

IOP Conf. Series: Materials Science and Engineering 1138 (2021) 012043 doi:10.1088/1757-899X/1138/1/012043

throughout the system that could have led to increased economic efficiency of CTESS [1, 3]. However, in most cases these networks, over time, have become dysfunctional and local governments have abandoned them.

The number of localities connected to the centralized heat supply system in Romania decreased in the period 1989-2014 by approximately 78%, respectively from 315 to 70 localities [1, 4, 5] with the possibility, for some of them, to cease or close their activity in the years to come. Due to some unfavorable decisions and lack of support from the local and national authorities, in 2018 only 52 localities remained supplied with thermal energy and domestic hot water from centralized system with district heating plants (DH) [6].

This paper describes the involution of a centralized system with district heating plants, powered by natural gas, in order to understand that a periodic diagnosis of the efficiency of the system and a risk analysis that considers the development of competing alternative systems, can avoid huge economic problems for local public administrations which are unfortunately also associated with special social problems.

2. Materials and methods

2.1. District heating system location and description

The district heating system is in a city in the west-central part in Romania with a temperate climate. The climate features in Romania respect the temperate-continental type which is characteristic for Central Europe. Temperatures in winters vary from very cold to cool, summer range from very warm to hot, so seasons are very distinct. The climate is classified as Dfb by the Köppen-Geiger system [7].

As part of the infrastructural base of the city, the centralized heating system had 31 heating facilities, with a total installed thermal power of 129 Gcal/h. The distribution pipelines network totalized more than 25 km. The district heating system covered the demand for heat and domestic hot water for 16 635 apartments and 457 economic agents.

2.2. Methodology

The methodology used in this paper was through measurements and analysis of different indicators of the DH system. The main indicators followed in our investigation were: the system's thermal energy production and thermal efficiency, specific consumptions of utilities (electric energy and natural gas) and the number of consumers connected to the heating network.

2.3. Materials

The materials used in our research involved monitorizations and recordings of the produced and delivered thermal energy, utility bills, annual reports and official websites of the Romanian government [1, 8, 9]. The amount of analyzed materials and information were significant since the evaluation of the DH system focused on the last 15 years of functioning.

3. Results and discussions

3.1. The installed thermal power of the district heating system

The district heating system, with its 31 heating centrals, had, between 1997 and 2005, an installed thermal power of 129.25 Gcal/h. By 2006, with less consumers connected to the network, the installed thermal power had a significant decrease of 65%, reaching a value of 84.35 Gcal/h. In figures 1 and 2 are presented the evolutions of the installed thermal power and apartment disconnections over the last 15 years of functioning.

The reasons for consumers to disconnect were plenty. The main one that influenced the householders to separate from the centralized heating system was the liberalization of the natural gas market. There were companies who were selling natural gas to individual consumers at a competitive price compared to the one established by the DH system for the sold thermal energy. Another was the emerging market

IOP Conf. Series: Materials Science and Engineering 1138 (2021) 012043 doi:10.1088/1757-899X/1138/1/012043

and intense publicity of wall mounted boilers, whose sellers were promising better control of interior microclimate at smaller prices. The last factor was the consumers' wish and opportunity to control thermal comfort, namely the possibility to choose when to turn on or off the heat.

Thus, the evolution of consumers disconnections from the DH system, as it can be seen in figure 2, had a decreasing path, with an average diminishment rate between 1997 and 2003 of 12%/year. The maximum value of the decreasing rate was reached in 2004 with 71%. After this year, the collapse of the district heating system was predictable. From 2005 and the years to come, the disconnections rate continued to be around 50%.

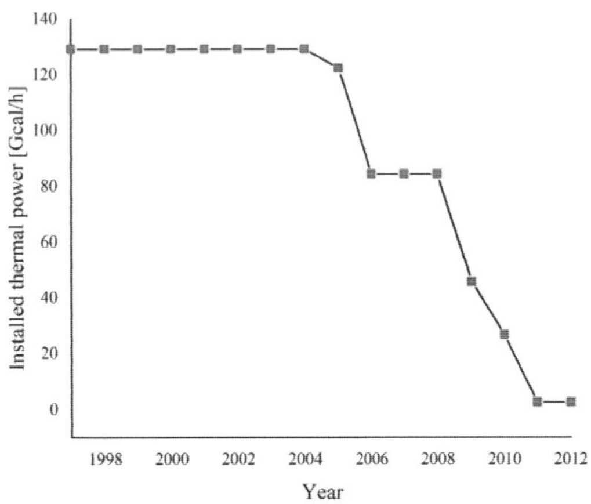

Figure 1. The installed thermal power of the district heating system.

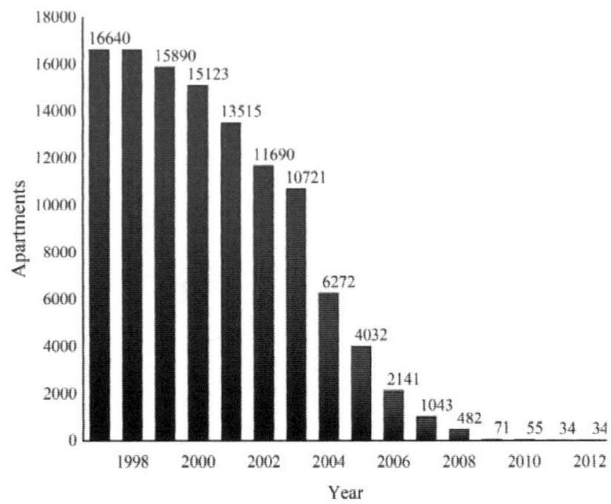

Figure 2. Evolution of the apartments connected to the DH system.

3.2. Thermal efficiency of the DH system

By the start of the '90, Romania faced a major economic crisis with many political and social changes. In short time, the industrial sectors demanded less and less thermal energy reaching the point to disconnect from DH systems, as the goods production decreased significantly. The massive disconnections put serious challenges on the district heating system as it was meant to operate an oversized energy supply, with different parameters from its design, but with less demand [5].

The DH system was seriously affected on one hand by the overall thermal efficiency and on the other hand by hydraulic misbalances of the distribution pipeline network. Almost immediately with the decline of the industrial sector, many householders disconnect from the district heating, forced by economic reasons. As it can be seen in figure 3 the thermal efficiency of the DH system decreases to the level of not being feasible.

Thermal energy losses exceeding 30% cannot be tolerated for this type of system (district heating system with combustible natural gas). The lower the thermal (consumer) density determines the lower efficiency of the system. Likewise, the decrease in demand (output reduction) can't be technically diminished with reduction in inputs leading to performance decline due to the partial use of heating facilities [10, 11].

After 2004, to maintain the DH system above the threshold of economic profitability, the owner of the system decided to shut down some heat facilities. Its evolution, or proper said its involution, with operative heating centrals is shown in figure 4.

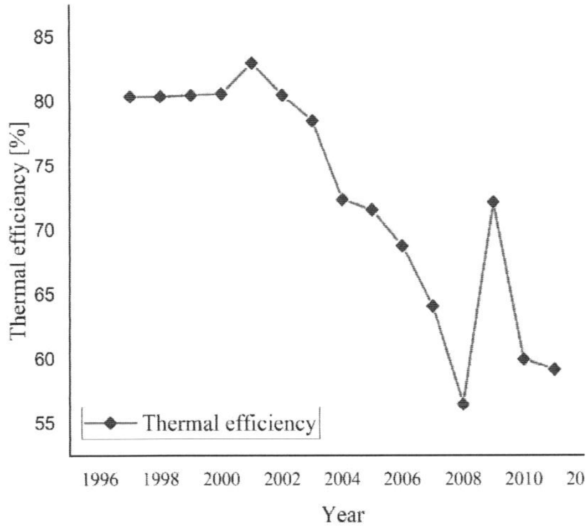

Figure 3. The evolution of the DH system's thermal efficiency.

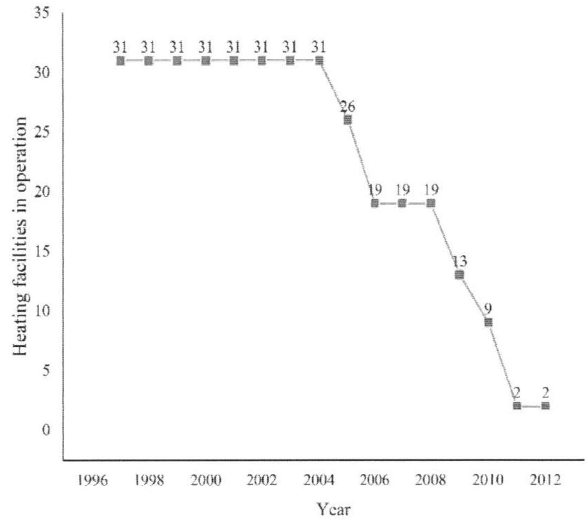

Figure 4. Heating facilities in operation.

At this stage of system's operation, we can state that a thermal efficiency and risk analysis of the system would have been required few years earlier. In addition, a market study would have been necessary to understand consumers' choices and needs.

In figure 4, the first 5 heating facilities were disconnected from the district heating system in 2005 after its thermal efficiency reached a level of 71.6% (see figure 3). Although other technical measures were taken, the performance of the system which was supposed to stagnate or increase, followed it's decreasing path. Along with this thermal efficiency drop and the fact that consumers continued to disconnect, the DH system is forced to close in the next year another 7 heating facilities. Likewise, the decreasing path of the performance DH system reaching a value of 56.5%, continued in the next three years until other 6 heating centrals were shut down. With only 13 operative heating facilities the DH system efficiencies finally started to increase until 72.2% (see figure 3) but still under the threshold of economic profitability. The performance of the DH system didn't last long and started to diminish again in 2010. From this point the collapse of the centralized heating system was imminent, a fact that happened in 2012.

3.3. Evolution of system's specific consumptions

One of the main reasons for the accentuated decrease in the overall DH system's thermal efficiency in 2008 (see figure 3), is related to the increase of consumption of specific electricity reported to sold thermal energy (see figure 5), reaching a level of 58.3 kWh/Gcal [9].

The continuous reduction of consumers with the preservation of less and less charged thermal networks leads to an alarming increase in the electricity consumption necessary for the production and distribution of thermal energy in the centralized system. Thus, if the production of the thermal energy unit (Gcal) in 1997 required 13.6 kWh, in 2004 it needed about three times more electricity (30.8 kWh/Gcal), making it clear that the system is facing serious problems.

To produce thermal energy, the DH system had an annual natural gas consumption of 30 441 024 m^3 in 1997 reaching only 163 441 m^3 in 2011 [8]. The consumption of specific natural gas reported to the sold thermal energy follows the same path of the consumption of specific electricity, i.e. an overall increase. If in '97 the amount of natural gas consumed by the DH system on the sold thermal energy was 154.4 m^3/Gcal, in 2004 increased to 171.5 m^3/Gcal, reaching a level of 219.7 m^3/Gcal.

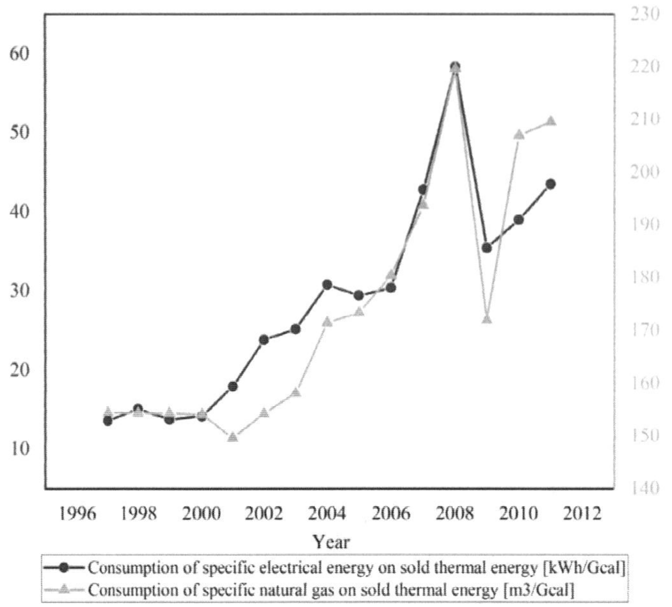

Figure 5. The evolution of the DH system's specific consumptions reported to annual sold thermal energy.

3.4. Analysis of the main annual indicators in the DH system

We analyzed the evolution of the main annual indicators through an overlap graph that is presented in figure 6. These indicators are the main factors that are considered for the establishment of the district heating system's economic efficiency, i.e. annual produced energy, annual consumption of natural gas and electricity and number of system's employees.

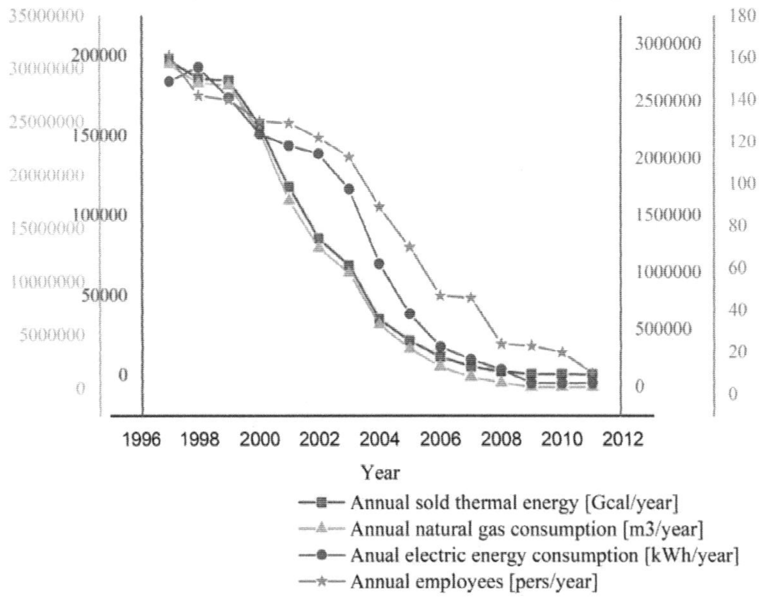

Figure 6. The evolution of the main analyzed indicators of the DH system.

As can be seen in figure 6, the measures taken by the management, being restricted by the safety rules of the installations, the system technology and the need to ensure the thermal energy of all customers remaining in the system, cannot make the slope of heat production decrease be followed by an identical slope of the main costs for the production of useful thermal energy.

In 1997 the DH system produced and distributed with each employee 1224.3 Gcal, in 2004 with the same employee it produced and distributed only 394.2 Gcal and in 2011 only 78 Gcal. Analyzing the graph, we can state that the efficiency of staff use was in an alarming decline throughout this period.

Given the performance indicators presented on the graph and the lack of clear regulations in the field, a steeper slope of the graph occurred after the year 2004, meaning after this year the financial balance of the business was broken and the collapse of the system was imminent.

As a result, the thermal energy tariff continues to rise steadily, reducing its competitiveness, and the subsidy system of local authorities can only delay an inevitable collapse.

The following graph (figure 7) represent an eloquent image of our research outcome, in which we determined a collapse threshold for the district heating system.

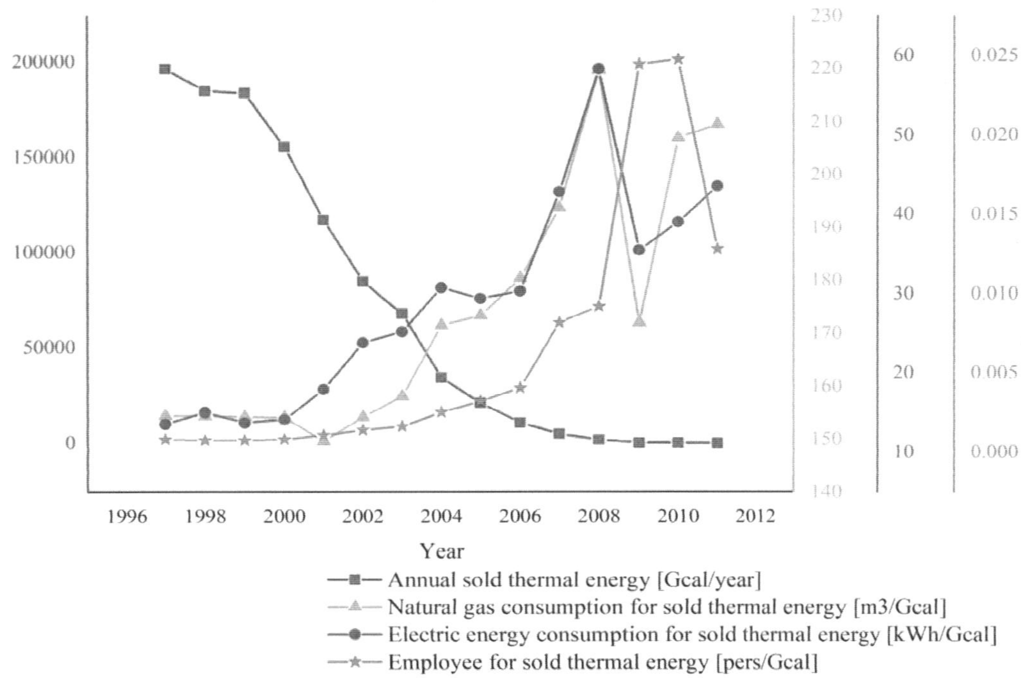

Figure 7. Relevant specific consumptions to the cost of sold thermal energy by DH system.

The collapse threshold for the studied district heating system is somewhere between the years 2002 and 2004. For that period, it can clearly be seen on the graph that while the annual sold energy slope decreases drastically, the slopes of consumption and operational costs with employees are rising almost exponentially. After that period, when the supply was exceeding the demand and the consumption of resources were higher and higher, the slope of sold thermal energy continued to decrease constantly while the costs of resources in order to produce the thermal energy were growing uncontrollably.

4. Conclusions

Given the analysis of the main indicators of the thermal system we can state that the collapse of the system was predictable, a long time before this happened.

The main scope of the study was to identify in the first stage a graphical threshold for the economic efficiency of the considered DH system, following that in a second stage we can frame the curves of the graphs in mathematical functions and thus they can be used for exact determinations of the critical points for the DH systems that are still in operation. To do that, we will expand this case study for other several DH systems who had the same sad fate.

Considering the above, the team lays the foundations for an assessment tool for determining the economic efficiency of a DH to apply it and forecast for existing DH systems.

To achieve that scope, the team identified the main technical-economic indicators, the variation of which can determine both the current state of a DH system and the forecasts for the coming years.

The main conclusions that can be highlighted from our analysis are:

1. the lack of performance and risk analysis for DH systems in operation can lead to their collapse with significant economic and social costs;
2. the performance and risk analysis must be carried out periodically through the care of the owner (usually the local public authority) by independent qualified consultants;
3. the risk analysis should focus on the alternative heat supply options available in the area.

References

[1] Romanian Ministry of Regional Development and Public Administration and Romanian Ministry of Energy 2015 *Raport privind evaluarea potențialului național de punere în aplicare a cogenerării de înaltă eficiență și a termoficării și răcirii centralizate eficiente* (*Report on the assessment of the national potential for the implementation of high-efficiency cogeneration and efficient district heating and cooling*) https://ec.europa.eu/energy/sites/ener/files/documents/Raport%20privind%20evaluarea%20potenialului.pdf accessed 20.08.2020

[2] Poputoaia D and Bouzarovski S 2010 Regulating district heating in Romania: legislative challenges and energy efficiency barriers. *Energy Policy* **38** (7), 3820–29

[3] Mandil C 2004 *Coming in from the cold: Improving district heating policy in transition economies* (McFarland) pp 1-262

[4] Chisalita D 2005 *Conversia energiei–Cogenerarea (Energy conversion – Cogeneration)* no 25 ed AGIR (Bucharest: Univers Ingineresc) pp 15-68

[5] Iacobescu F and Badescu V 2011 Metamorphoses of cogeneration-based district heating in Romania: A case study. *Energy Policy*, **39** (1), pp 269-280

[6] Romanian Energy Regulatory Authority 2019 *Situația serviciilor publice de alimentare cu energie termică în sistem centralizat - semestrul II 2018* (*The situation of public heat supply services in centralized system - second semester 2018*) Available: https://www.anre.ro/download.php?f=hqqBhA%3D%3D&t=vdeyut7dlcecrLbbvbY%3D, accessed 15.09.2020

[7] Kottek M, Grieser J, Beck C, Rudolf B and Rubel F 2006 World Map of the Köppen-Geiger climate classification updated *Meteorologische Zeitschrift* **15** (3) pp 259-263

[8] Romanian Energy Regulatory Authority 2011 *Annual monitoring report for the internal gas market – 2011* https://www.anre.ro/download.php?f=f6aEiA%3D%3D&t=vdeyut7dlcecrLbbvbY%3D accessed 14.09.2020

[9] Romanian Energy Regulatory Authority 2015 *Electricity market monitoring report* https://www.anre.ro/download.php?f=gqqEgw%3D%3D&t=vdeyut7dlcecrLbbvbY%3D, accessed 14.09.2020

[10] Team A and Baffert C 2015 Energy poverty and vulnerable consumers in the energy sector across the EU: analysis of policies and measures *Policy* **2**

[11] Delcea A, Bitir-Istrate I, Pătrașcu R and Minciuc E 2019 Statistical evaluation of heat carrier losses from district heating networks *Int Conf on energy and environment (CIEM)* IEEE pp. 333-337

Study on heat treatment and coating of copper screw female in friction pair of rock drill

Y Y Wang[1*], M Zhao[1] and X L Zhang[1]

[1] School of Intelligent Engineering, Shenyang City University, Shenyang, China

* Wyyg_2000@163.com

Abstract. Aiming at the copper screw female parts of the rock drill friction pair that are made of tin bronze and easy to wear, the material, hardness, the metallographic structure and other comparative studies have conducted in-depth research on the formation mechanism of the application of thermal spraying technology on the copper spiral female parts of the rock drill product. The copper spiral female in the friction pair of the rock drill after heat treatment and thermal spraying has been tested on site by the customer. With better results, it is hoped that it can be used as a reference for the study of other friction pairs in rock drill products.

1. Introduction

Pneumatic rock drill occupies a large share in the market of rock drilling machinery due to its low price. It is widely used in mining, railway and highway construction sites, and its working environment is generally in open air, pit or tunnel, etc [1]. The screw female is the most critical part of the rotary mechanism in pneumatic rock drill [2]. It is usually made of tin bronze, and its quality directly affects the service life of pneumatic rock drill. According to the field survey, the tin bronze screw female used by a certain type of pneumatic rock drill has a very low life, with a footage of only about 2000m, causing serious waste of scarce non-ferrous metals. Serious wear of the screw parent is bound to lead to serious wear of the matching screw rod, which will significantly reduce the drilling efficiency of pneumatic rock drill [3, 4]. Skalka P et al. [5] studied the effect of the waviness of the bond coat and the thickness of the thermally grown oxide (TGO)layer on the stability of plasma-sprayed thermal barrier coatings. Slámečka K et al. [6] based on the obtained results and experimental evidence from the literature, a role of the interfacial topography in the failure sequence is discussed for oxidized and thermally-cycled coatings. Yin Ao [7], Su X et al. [8] took tin bronze (QSn7-0.2) as the research object, and heat treated the high temperature extruded tin bronze at 200, 300, 400 and 500°C respectively. The results show that the grain size and microstructure distribution of tin bronze are more uniform after heat treatment and annealing at 350~450°C, and it has better hardness, strength and plasticity, which is more conducive to the subsequent processing process. Xiao H [9] studied the effects of heat treatment temperature on microstructure and mechanical properties of semi-solid extruded ZCuSn10P1 tin bronze. The optimal heat treatment process is 350°C for 120 min, at which the tensile strength, elongation and Brinel hardness of tin bronze are 402 MPa, 4.5% and 1360 MPa, respectively, which are increased by 3.88%, 60.71% and 6.25% compared with those before heat treatment.

Therefore, it is an urgent problem for pneumatic rock drill to improve the service life of screw female and make it consistent with the service life of screw rod. Female in this paper, the two manufacturers copper screw parts of contrast research, including material, hardness, microstructure of the thermal

IOP Conf. Series: Materials Science and Engineering 1138 (2021) 012044 doi:10.1088/1757-899X/1138/1/012044

spraying technique on the mother rock drill product copper screw parts then explores its formation mechanism research, the drill after heat treatment and thermal spraying of the friction pair copper screw female on customer site test, to reduce the spiral mother wear, improve its service life provides certain reference basis.

2. Copper screw base material and hardness analysis

The friction pair of copper nut (see figure 1) is the most important connecting transmission mechanism on pneumatic rock drill. The quality of copper screw base directly affects the service life of pneumatic rock drill.

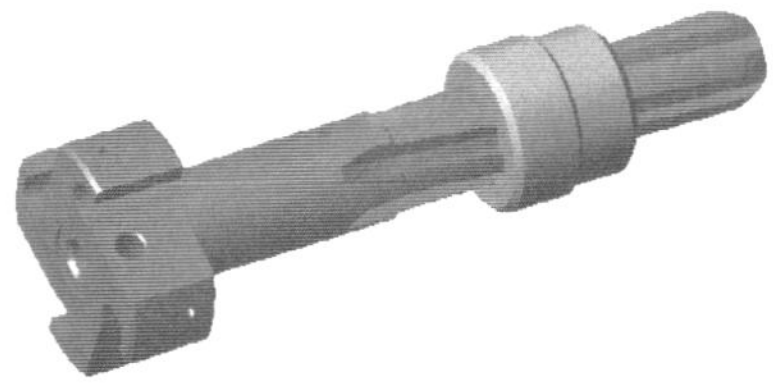

Figure 1. Copper nut friction pair.

As tin bronze QSN7-0.2 (chemical composition and properties are shown in table 1) has the advantages of high strength, good wear resistance, good corrosion resistance, good cutting ability, etc., the copper screw female is usually made of tin bronze.

Table 1. Qsn7-0.2 Chemical composition and mechanical properties of tin bronze.

Cu	Sn	Pb	P	Al	Fe	Si	Sb	Bi	Tensile strength (MPa)	Elongation δ10 (%)
6.0～8.0	⩽0.02	⩽0.25	⩽0.02	⩽0.05	⩽0.02	⩽0.002	⩽0.002	⩽0.15	⩾665	⩾2

Abrasion resistance refers to the ability to resist friction. The factors affecting this ability are not only dependent on the composition, structure and properties of copper, such as hardness carbide characteristics, quantity, shape and distribution, but also closely related to the conditions of use and tensile process, such as the adhesion of a large number of sand ash layer on the surface of wire rod.

Soft matrix + hard point is an ideal wear-resistant material (see figure 2). In tin bronze, the matrix is phase (Sn replacement solid solution in Cu, with a face-centered cubic structure), which has excellent plasticity and is suitable for cold and hot machining deformation. The hard point ($Cu_{31}Sn_8$, complex cubic lattice structure) is hard and brittle, and only + phase is usually observed in the cast and annealed tin bronze tissues.

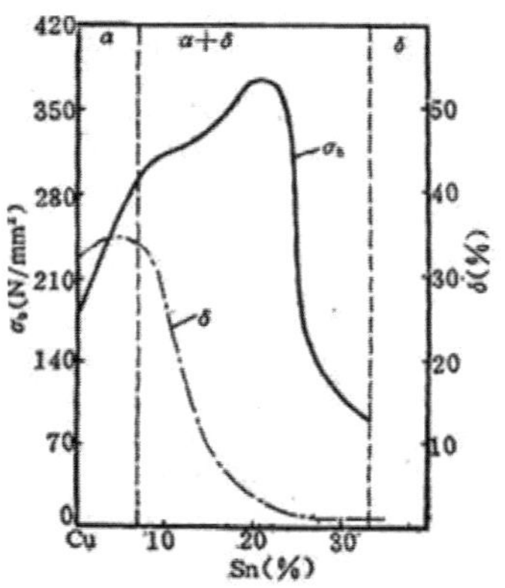

Figure 2. Relationship between tin content and mechanical properties of cast Copper-tin alloy [10].

Hardness is an important performance index to measure the degree of hardness of metal materials. It can be understood as the ability of material to resist elastic deformation, plastic deformation or failure, and can also be expressed as the ability of material to resist residual deformation and anti-failure. Under the same conditions (the same friction coefficient, composition, microstructure, environmental conditions, etc.), there is a nonlinear proportional relationship between hardness and wear resistance.

According to the field survey, the sites used are all iron ore. The footage of pneumatic drill products of A company and B company and the scrap standard of copper screw parent are shown in table 2. The material and hardness testing of copper screw parent of A Company, B company and C company are shown in table 3.

As can be seen from the above table, QSn7-0.2 tin bronze is selected as the spiral mother material for both company A and company b. the spiral mother hardness of company A is 195 HB, and that of company B is 124 HB.

Further research and improvement will be made on heat treatment and plasma thermal spraying of copper spiral mother of A company.

Table 2. Field footage and scrap standard for copper screw female.

Company	Working time	Working life (m)	Mode of action	Scrapping standard	Measuring position	Graphical representation
A	30 shifts	1380	2.5 m drill rod, 20 holes per shift	Spline profile Wear 2 mm	Measure the width w of spline keyway at 18.5 mm of large end face, and discard it when w ≥ 7.4 mm	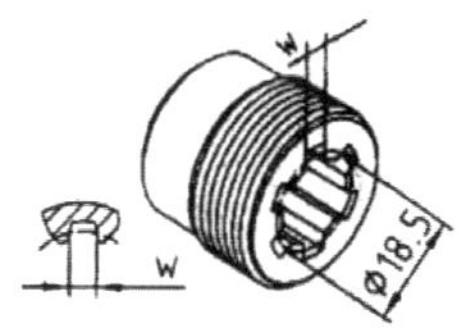
B	35 shifts	1610				

Table 3. Copper spiral parent material and hardness testing.

Manufacturer	Cu	Sn	Pb	P	Ni	Fe	Si	Ti	Hardness (HB)	Material
A	91.01	7.28		0.2		0.1	0.6	0.2	195	QSn7-0. 2
B	91.07	7.61			0.06	0.2	0.8	0.1	124	QSn7-0. 2
C	89.53	9.17	0.1	0.2	0.21	0.2	0.3	0.1	184	ZSn10

3. Research on plasma thermal spraying

Thermal spraying technology is through the flame, or plasma arc heat source, will be a linear or powdered material heated to melt or semi-molten state, and accelerate the formation of molten drop, high speed spray to form coating matrix, the properties of materials can be strengthened or regeneration, protection, and the components caused by corrosion or wear processing out-of-tolerance size reduced to repair. The surface of the material can also be given special properties.

Plasma spraying is a thermal spraying processing method in which rigid non-transfer plasma arc is used as heat source to heat the powder material to molten or semi-molten state and spray it to the surface of the pretreated workpiece through high-speed flame flow to form spray coating. Its principle is shown in figure 3.

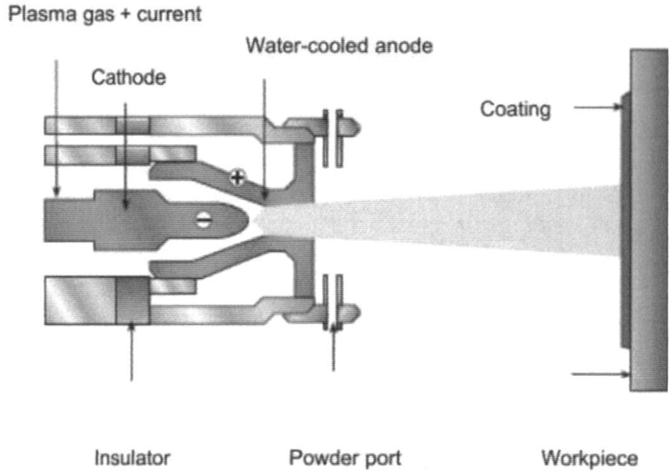

Figure 3. Diagram of plasma spraying [11].

In the process of coating formation, a single molten particle is the basic unit of coating formation, and its behavior reflects the characteristics of coating formation. The behavior of a single particle consists of three basic processes: first, the spraying material is sent to a heat source; then comes the interaction process between the spraying material and the heat source. Under the action of the heat source, the material is heated, melted and accelerated. At the same time, the interaction process between high temperature and high speed particles and the ambient atmosphere also occurs. Finally, the collision of high temperature and high speed molten particles with the matrix, horizontal flow flattening and rapid cooling solidification are discussed. The whole process is completed in a very short time of tens of microseconds.

The spray gun is actually a non-transfer arc generator, which is the most critical component. It concentrates the electricity, gas, powder and water of the whole system on the spray gun. The spray gun can be supplied with direct current by connecting to the power supply, which is usually a full-wave silicon rectifier. Spray gun can be replaced by nozzle and other parts to meet the requirements of spraying different materials. The spray gun is shown in figure 4. The top end of the spray gun is the powder conveying conduit, and the bottom end is the pipeline of cooling water and working gas. The powder feeder used is shown in figure 5.

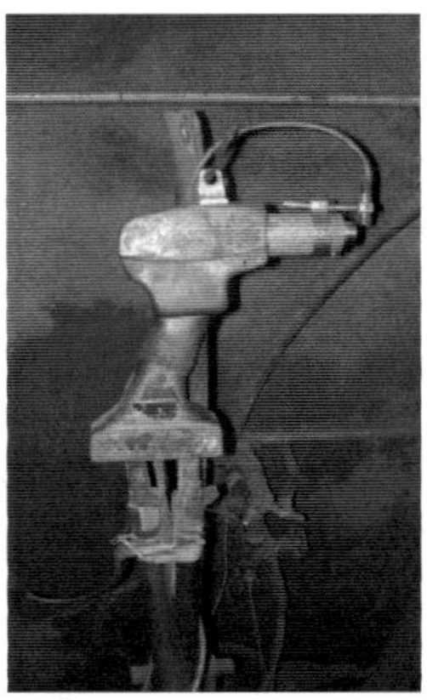

Figure 4. Plasma gun.

Figure 5. Powder feeder of plasma spray.

4. Copper spiral mother heat treatment and coating

Copper spiral mother is the raw material of tin bronze QSn7-0.2 in the manufacturing process, through the process of thermal deformation. After the alloy was heated and extruded, the structure property of billet was the best, especially the section shrinkage rate. It can be found from the study that the high temperature extrusion of tin bronze by induction heating has good plasticity and strength. However, in the processing process of high temperature extrusion, the grain structure of tin bronze is not evenly distributed, and some grains grow up unusually, resulting in the unstable plastic performance of tin bronze and the potential safety hazard. In order to solve this problem, it can be annealed after warm extrusion to make its structure homogenize, reduce the internal stress and improve its plastic property, which is beneficial to the subsequent spinning and other processing technology, and has a certain effect on reducing the stress concentration and improving the surface smoothness.

Tin bronze alloy is mainly composed of copper, tin, nickel, lead and other elements. The diffusion process of tin in solid state in copper is very slow, and it is easy to form in-grain segregation and a small amount of 8-phase in as-cast state, and it can be eliminated only after homogenizing annealing at high temperature. Both, nickel and copper are plane-centered cubic structures, and the difference in atomic radius is very small. Therefore, nickel can dissolve in unbounded solid, but nickel diffuses slowly in copper, and it is easy to form significant dendritic structure. Lead does not dissolve in copper and is generally distributed in a free phase between the crystal branches. By analyzing the main components of the alloy, combining with the phase diagram, changing the distribution of the eutectoid between the dendritic crystals and the dendrites by means of heat treatment can obviously change the properties of the alloy. Through the comparison of the above test results, it can be seen that the peak performance index of the alloy between 350 ~ 450°C appears, indicating that the conductivity, hardness and tensile strength of the alloy have been effectively improved at this time. The test data after heat treatment annealing are shown in table 4.

Table 4. Annealing heat treatment data sheet.

No.	Annealing temperature (°C)	annealing time (min)	external hardness (HB)	internal hardness (HB)	tooth root hardness (HB)
A	370	40	179	177	172
B	370	10	139	180	152
C	250	40	182	209	202
D	400	10	116	137	120

The metallographic structure after annealing is shown in figure. 6. The structure is α phase + δ phase, the black dendritic axis is copper rich solid solution (α phase), the white is tin rich solid solution (δ phase), the grain size and microstructure are uniformly distributed, and a large number of equiaxial twins can be seen. This is because static recrystallization occurs in tin bronze with the increase of annealing temperature.

Figure 6. Spiral mother QSN7-0.2 annealed metallographic structure.

Hard alloy is an excellent tool material and structural material. At present, its production raw material is mainly nano-tungsten-Cobalt (WC-CO) composite powder. Tungsten carbide (WC) is a kind of hard alloy material with high hardness, thermal stability and good wear resistance. Tungsten carbide insoluble in water, hydrochloric acid and sulfuric acid, easy to dissolve in nitric acid - hydrofluoric acid mixed acid. Pure tungsten carbide is brittle and can reduce brittleness if a small amount of metals such as titanium and cobalt are added. Tungsten carbide used as steel cutting tools is often added with titanium carbide, tantalum carbide or a mixture of them to improve the anti-explosion ability, and its chemical properties are stable. The powder is shown in figure 7.

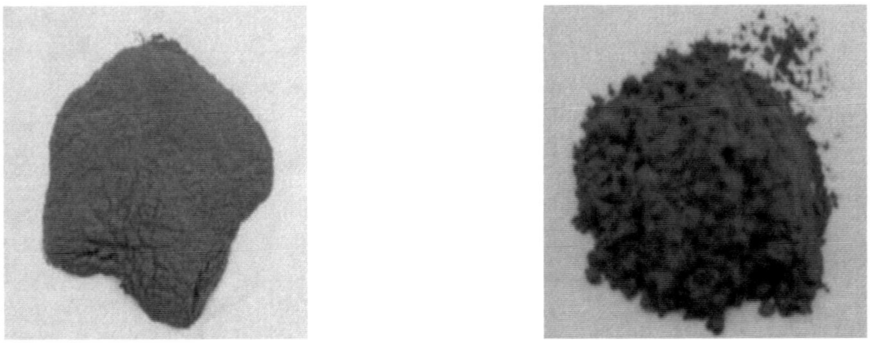

a. Unsintered powder b. Sintered powder

Figure 7. Nanometer Tungsten-Cobalt (WC-CO) composite powder.

Copper screw parent is shown in figure 8.

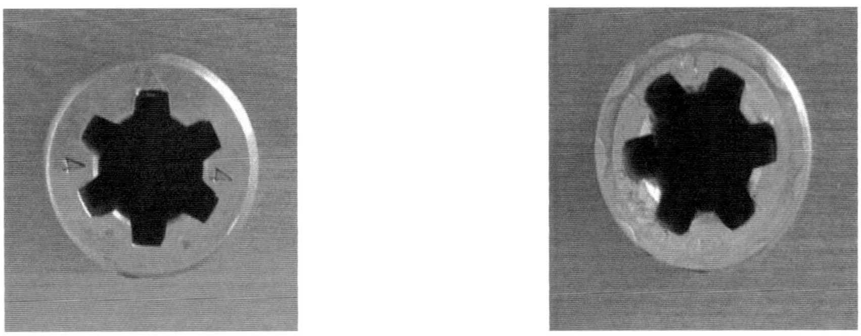

a. Copper screw before thermal spraying b. Copper screw after thermal spraying

Figure 8. Copper screw female.

After dealing with the annealing and thermal plasma spraying of copper screw female, with untreated copper screw female field test in some iron mine, test data as shown in table 5, can be seen from the table, after annealing and plasma thermal spraying, the copper screw female longevity by about 22.5%.

Table 5. Improved field test data table of copper screw parent.

Company	Mode of action	Working time	Working life (m)	Remarks
A	2.5 m drill rod, 20 holes per shift	31 shifts improvement	1426	before
A		38 shifts improvement	1748	after

5. Conclusion

The results show that the structure of the alloy tends to be homogenized with the reasonable annealing system, and the performance index peaks, but the elongation after fracture decreases. Therefore, by selecting a reasonable heat treatment process for the new type tin - bearing copper casting, the conductivity, hardness, tensile strength and other properties of the material can be improved completely, the study can be extended to other friction pair of pneumatic rock drill products.

References

[1] Liu E 2003 *J. Rock Drilling Mach. & Pneu. Tools* **153** 14
[2] Si J and Qu J 2002 *J. Lubr. Eng.* **27** 60
[3] Shen P 1976 *J. Rock Drilling Mach. & Pneu. Tools* **3** 27
[4] Chen B 1987 *J. Rock Drilling Mach. & Pneu. Tools* **47** 47
[5] Skalka P, Slámečka K, Pokluda J and Čelko L 2015 *J. Sur. & Coat. Tech.* **274** 26
[6] Slámečka K, Skalka P, Pokluda J and Čelko L 2016 *J. Sur. & Coat. Tech.* **304** 574
[7] Yin A, Feng Z, Hui R and Guo H 2018 *J. Foundry Tech. Tools* **39** 2854
[8] Su X, Feng Z, Gao C, and Hao W 2017 *J. Foundry Tech. Tools* **42** 156
[9] Xiao H, Duan Z, Li N, Xiong C, Zhou R, Lu D and Jiang Y 2019 *J. Rare Metal Mat. and Eng.* **48** 235
[10] Tian R and Wang Z 2002 *Copper alloy and its processing manual* (Chang Sha: Central South University Press)
[11] Xu B and Liu S 2002 *New technology in surface engineering* (Bei Jing: national defence industrial press)

IOP Conf. Series: Materials Science and Engineering 1138 (2021) 012045 doi:10.1088/1757-899X/1138/1/012045

Optimization of diamond circular saw blade by vibration noise analysis

T Wu[1*], D Wang[1] and M Zhao[1,2]

[1] Department of Intelligent, Shenyang City University, Shenyang City, China
[2] Shenyang Jianzhu University, Shenyang City, China

* 81836848@qq.com

Abstract. The vibration noise analysis and optimization of diamond circular saw blade is presented. The position of hole of the diamond circular saw blade matrix with diameter of 500 mm is established by analyzing the vibration and noise through experiment, and then making matrix topology optimization to the blade by ANSYS. The research content is divided into three parts. The noise analysis of diamond saw blade. Study on vibration characteristics of diamond saw blades. The optimization design of diamond saw blade. The study on the noise and vibration characteristics of diamond circular saw blade provides some reference for the design and technological parameter of blades. The successful application of the topology optimization has certain reference value to production and manufacturing of saw blades.

1. Introduction

Diamond circular saw blade [1] is the main tool for cutting stone, which is a thin plate tool with limited bending resistance. Moreover, because of its narrow sawing path, poor heat dissipation and uneven stress, thermal deformation and axial deformation often occur during operation, resulting in vibration [2]. If the vibration frequency is close to other external excitation frequencies, resonance will occur. If the vibration is too strong, noise pollution [3] will follow. The vibration and noise of diamond circular saw blades have serious adverse effects on operators, equipment and product quality. Therefore, analyzing the vibration and noise characteristics of diamond circular saw blades and trying to reduce the vibration and noise can improve the production quality and benefit people's health.

For now, most researches [4] on vibration and noise reduction of circular saw blades mainly focus on radial slots optimization and damping technology. However, the research on the internal structure of circular saw blade matrix, especially the method of determining the opening position [5] on the matrix, has rarely been reported in relevant literature.

Topology optimization technology has outstanding advantages in structural optimization, but few people had applied topology optimization technology in the structural optimization of saw blades. Therefore, topology optimization technology can also be used to study the optimization design of the substrate structure of circular saw blade.

In this study, a series of studies were carried out on the vibration and noise of diamond circular saw blades in use. The noise and vibration characteristics of diamond circular saw blade were analyzed through experimental [6] measurement results. The effects of saw blade structural parameters, cutting parameters and cutting materials [7] on saw blade noise were studied, and the natural frequencies and modal shapes of diamond circular saw blade were obtained. Under the condition of ensuring the stiffness of the saw blade, and without changing the whole manufacturing process of circular saw blade, the saw blade without holes in the matrix was optimized [8] through topology optimization technology, and the hole position on the saw blade matrix is determined and discussed in this paper.

2. Problem Description

In this study, the noise and vibration characteristics of diamond circular saw blade are analyzed, which is the theoretical basis for reducing noise and vibration of it. Topology optimization function of ANSYS Workbench [9, 10] was used, the topology optimization of diamond circular saw blade with no holes in the matrix was carried out, and the optimized diamond circular saw blade with a diameter of 500 mm was trial-manufactured. By measuring and comparing the noise value and natural frequency value of the diamond circular saw blade before and after optimization, which is used to verify whether the optimized diamond circular saw blade has certain vibration and noise reduction effect.

3. Experiment and analysis methods

3.1. The related basic theory vibration and noise

The vibration of diamond circular saw blade at work is usually [11] divided into transverse vibration, radial vibration and torsional vibration. As its thickness is much smaller than its diameter, so its bending stiffness is relatively low, and it is prone to bending deformation and transverse vibration. Among the three vibration forms of saw blade, the amplitude of transverse vibration [12] is far greater than that of radial vibration and torsional vibration. Therefore, transverse vibration is the main vibration form that leads to the failure of diamond circular saw blades.

The thickness of the diamond circular saw blade is much smaller than the diameter, so the diamond circular saw blade can be simplified as a thin plate. In the process of studying the transverse vibration of diamond, which based on Kirchhoff's simplified theory of bending thin plate with small deflection. The noise of diamond circular saw blade is steady-state and wide-band noise. Therefore, this study used A weighted sound pressure level to evaluate the noise.

3.2. The analysis about noise of diamond circular saw blade

In the experiment, TEC—1352H noise meter is used to measure noise. There are two ways of frequency weighting characteristics: A and C, the measurement accuracy is ± 1.5 dB, and the resolution is 0.1 dB. When measuring, the sound level meter was fixed on the tripod, and the microphone was 30 mm away from the surface of the cutting machine. In the process of measurement, the sampling interval was 1 second, and the slow gear was used for measurement, and the average reading was taken.

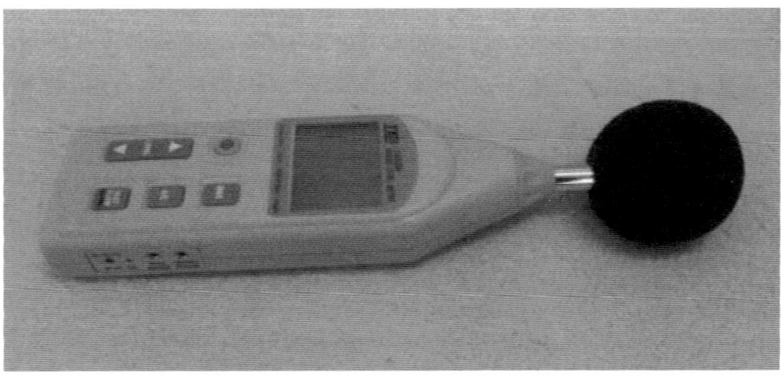

Figure 1. TEC—1352H sound level meter.

This noise experiment is carried out on QHC40 infrared bridge cutting machine. The saw blade consists of matrix and teeth, the material of the matrix is steel, and the teeth is made of diamond. The influence of different factors on the noise of diamond circular saw blade is studied in the following three aspects:

1. three kinds of stones, Lan-xing granite, Hei-jinsha granite and Mi-huang marble, were cut to study the influence of cutting materials on saw blade noise;

2. taking cutting marble as an example, the influence of cutting parameters on saw blade noise was studied by changing its rotation speed, feed speed and cutting depth and measuring noise;

3. by measuring the idling noise of diamond circular saw blades with diameters of 350 mm, 400 mm and 500 mm at different rotating speeds, the influence of the diameter of circular saw blades on noise was studied.

According to all the factors that may affect the noise of diamond circular saw blades, an orthogonal experiment with seven factors and two levels is designed to study the contribution of blade speed, feed speed, cutting depth, blade diameter, matrix thickness, segment number and friction coefficient with cutting material to noise. The experimental data shown in table 1 are finally obtained, MATLAB was used to fit eight groups of experimental data.

Table 1. Data of seven factors two levels orthogonal experiment.

Test No.	Rotation speed	Feed rate	Cutting depth	Blade diameter	Matrix thickness	Friction coefficient	Segment number	Noise
	V (r/min)	F (mm/r)	h (mm)	D (mm)	L (mm)	u	n	dB
1	300	30	5	400	3	0.7	20	30.76
2	300	30	5	500	4	1.2	24	34.91
3	300	45	10	400	3	1.2	24	38.62
4	300	45	10	500	4	0.7	20	39.44
5	800	30	10	400	4	0.7	24	42.53
6	800	30	10	500	3	1.2	20	53.86
7	800	45	5	400	4	1.2	20	45.2
8	800	45	5	500	3	0.7	24	48.08

The empirical formula of cutting noise is obtained as follows:

$$y = 5.381+0.02297x_1+0.154667x_2+0.775x_3+0.04795x_4-2.31x_5+5.98x_6+0.32x_7 \quad (1)$$

Where x_1 is rotation speed, x_2 is feed speed, x_3 is cutting depth, x_4 is blade's diameter, x_5 is matrix thickness, x_6 is friction coefficient, x_7 is segment number.

According to the empirical formula, it can be seen that increasing the thickness of the diamond circular saw blade can reduce the noise of the saw blade to a certain extent. The other six factors, i.e., saw blade speed, feed speed, cutting depth, saw blade diameter, segment number and friction coefficient with cutting material, will increase the noise in varying degrees.

Therefore, under the circumstance of satisfying cutting conditions, the noise can be reduced to a certain extent by selecting a lesser cutting depth, feed rate and rotating speed and a thicker saw blade. As for the cutting speed, relatively low cutting speed can be selected for low-speed cutting, and when cutting at medium and high speed, the highest noise speed point should be avoided (460 m/min in this experiment).

3.3. The analysis about vibration modal test analysis

Modal experiment was carried out on a diamond circular saw blade with a diameter of 500 mm, and the natural frequency and main mode shape of the saw blade were measured, thus the vibration characteristics of the saw blade were analyzed. In this experiment, the pulse excitation method was used, which applied exciting force to circular saw blade first, and then measured vibration. The force hammer with force sensor could send out exciting force, and the vibration measurement signal can be received by the sensor magnetically attracted on the circular saw blade. The measured signal was transformed by

FFT in the analyzer, and the modal parameters could be obtained according to the solved frequency response function. The specific model of the equipment is shown in table 2. The flange plate fixes the circular saw blade through the central hole, and the saw blade fixing mode and the saw blade grid division are shown as in figure 2.

Figure 2. Saw fixation and Meshing of diamond circular saw blades.

Table 2. Information of modal test equipment.

Instrument	Instrument model	Manufacturer	Effect
Hammer	FC43	Donghua dynamic testing company	Generating an excitation signal
Force sensor		Donghua dynamic testing company	Picking the force signals
Acceleration sensor	CQV9345	—	Picking vibration signals
Charge regulator	DH5857	—	Piezoelectric sensor signal transmission
Dynamic test analyzer	—	—	Collecting and analyzing signals
Analysis software	4633 dynamic signal test and analysis system	Donghua dynamic test technology co. ltd	Experimental modal analysis

The measured transverse vibration frequency of diamond circular saw blade with engineering value is mainly concentrated in low frequency band, so this study focuses on the low-order vibration frequency of diamond circular saw blade. According to the sampling theorem, the analysis frequency is between 0 and 1000 Hz, and the sampling frequency is between 0 and 2560 Hz.

After sampling, use modal analysis software to import the measured frequency response function into it. The superposition curve of frequency response function is shown in figure 3.

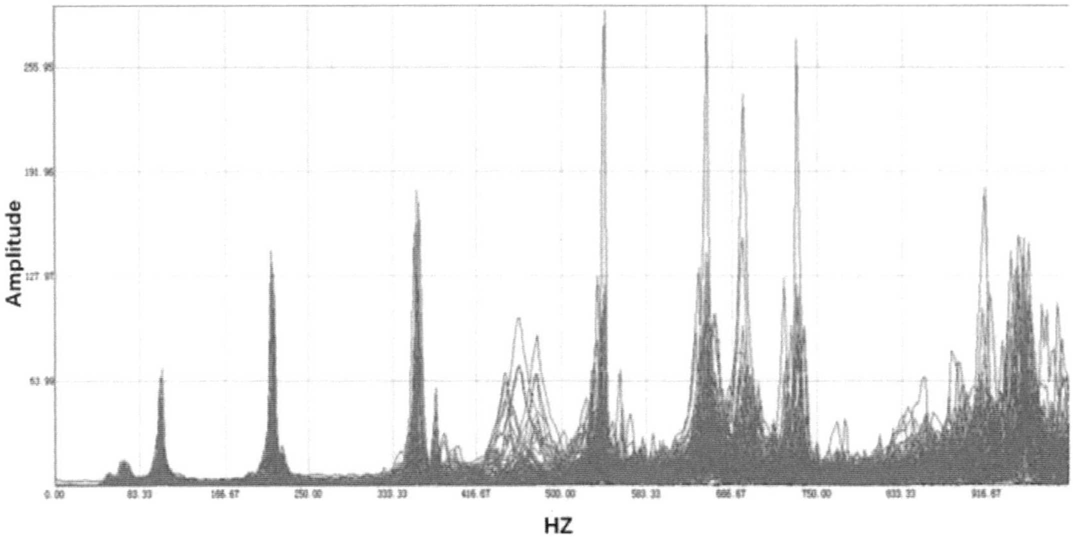

Figure 3. Superposition of frequency response function curve.

The frequency, damping ratio and modal shape of diamond circular saw blade can be obtained by searching the accumulated peak value of frequency response function with double cursor method. The natural frequencies and damping ratios of various modes of diamond circular saw blades are shown in table 3, and the modal shapes of various modes are shown in figure 4.

Table 3. Main modes of experimental modes of φ500 mm diamond saw blade.

Modes	1	2	3	4	5	6	7	8	9	10
Frequency/Hz	35	57.5.	67.5	152.5	285	312.5	357.5	445	565	614
Damping ratio	6.85	8.13	8.83	4.16	2.55	6.09	1.75	2.41	2.22	6.85

When the diamond circular saw blade works, its segments are impacted by the workpiece, and because the saw blade keeps rotating, the impact becomes periodic vibration excitation. When the excitation frequency is close to or equal to the natural frequency of the diamond circular saw blade, the saw blade will resonate. Diamond circular saw blade has a wide resonance frequency band, and has obvious resonance frequency in low frequency band, so it is easy to produce large vibration when working at this resonance frequency.

When resonance occurs, the amplitude of the saw blade increases, the vibration becomes more severe, and the damage to its matrix is greater. The red part in the figure represents the position on the saw blade with strong vibration and larger amplitude.

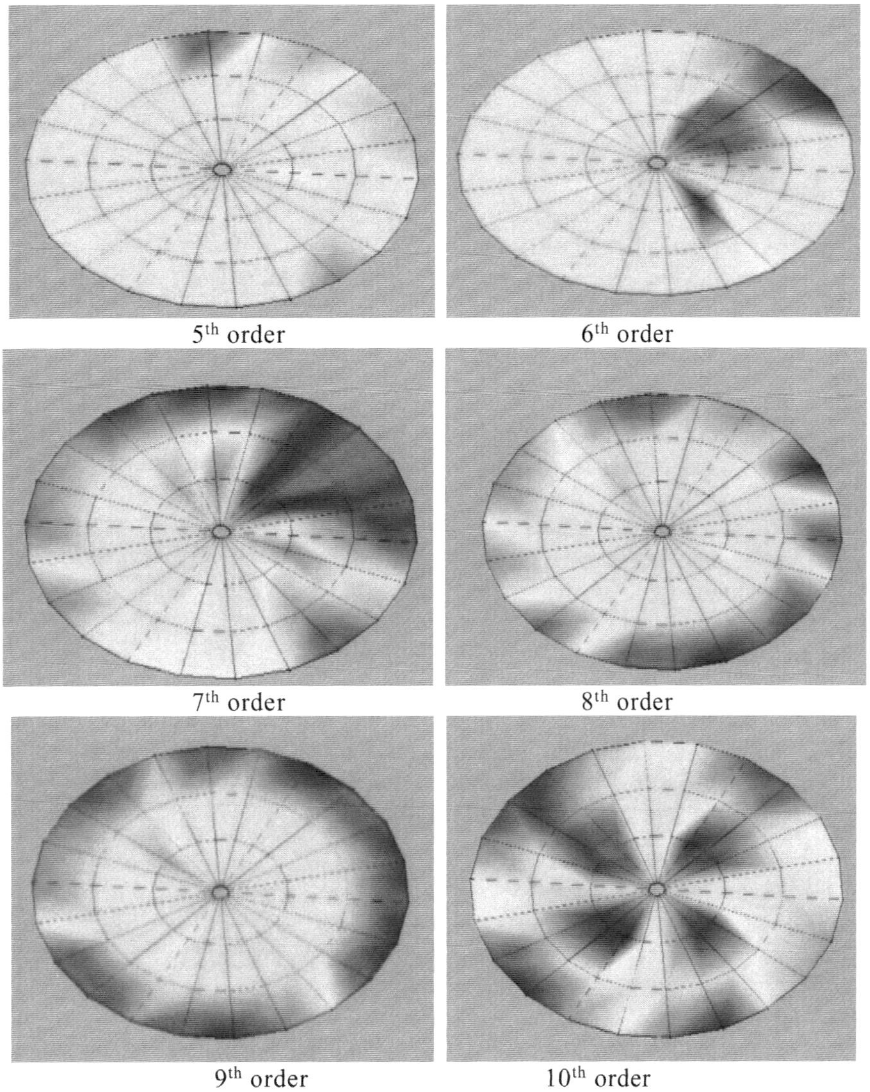

Figure 4. Vibration model under each order of experimental modal φ500 mm saw blades.

It can be seen from table 3 that the natural frequency of the diamond circular saw blade in the first three stages is not high, and the vibration frequency increases obviously with the increase of the order from the sixth stage. The frequencies of the first, second and third order modes are low and increase slowly, which indicates that the diamond saw blade only slightly swinged in the first three orders. It can also be seen that there are many bends from the 6th order, in which the natural frequencies of 6th and 7th order, 8th and 9th order are similar, which can be regarded as the multiple roots of the solution of vibration equation. The 8th and 9th orders belong to the edge vibration, while the 6th and 7th orders belong to the internal vibration of the matrix. In practice, the impact of external factors on the saw blade should be avoided or reduced as much as possible to ensure that the deformation of the saw blade is small. Then the relationship between the working speed of the diamond circular saw blade and its excitation frequency was compared.

If n is used as the rotation speed of saw blade, z is the number of segments, the excitation frequency $f = nz/60$ that can cause resonance. The higher the frequency, the higher the working speed of the diamond circular saw blade, and the greater the cutting force produced in this process, which is the

IOP Conf. Series: Materials Science and Engineering 1138 (2021) 012045 doi:10.1088/1757-899X/1138/1/012045

easiest to damage the saw blade. In this study, we didn't consider high-order frequencies. When the excitation frequency generated by the outside is close to the natural frequency, resonance is most likely to occur, and resonance will seriously affect the service life of the diamond circular saw blade. Therefore, in practical use, we should try our best to choose the working rotation speed with different external load excitation frequency and natural frequency to avoid the resonance of the saw blade.

According to the above analysis, by changing the diameter of diamond circular saw blade, the number of teeth and the thickness of matrix, the stiffness of circular saw blade can be improved, so as to achieve the purpose of vibration reduction and noise reduction. The experimental results can make preparations for the comparison of vibration characteristics of diamond circular saw blades before and after optimization.

3.4. Optimization design

With the medium strength topology optimization method of ANSYS Workbench, the deformation energy of the structure can be reduced under the condition of satisfying the structural constraints. Taking the maximum stiffness as the goal, the topology optimization of diamond circular saw blade was carried out. In the topology optimization design, it is not necessary to give the definition of parameters and optimization variables, as long as the parameters of the structure and the percentage of materials that can be omitted are given, the parameters and design variables are predefined by the software.

The analysis used three-dimensional SOLID95 solid elements, each of which has 20 nodes. The unit length of grid was 1 mm. The optimized area was the whole diamond circular saw blade matrix. Full restraint was imposed on the central hole of the blade at the flange, limiting the degrees of freedom in six directions. The force of 100 N is arranged on three adjacent saw teeth in the circumferential direction of the saw blade, which can simulate the force situation of the saw blade at a certain moment.

Through topology optimization of diamond circular saw blade with diameter of 500 mm, the saw blade has 24 segments, and every three adjacent segments can be divided into eight groups. After optimizing each group of segments, the optimized result picture shown in figure 5 was finally obtained.

The matrix material of diamond circular saw blade is 65Mn, and the material of segments are Cu+Co+Ni alloy.

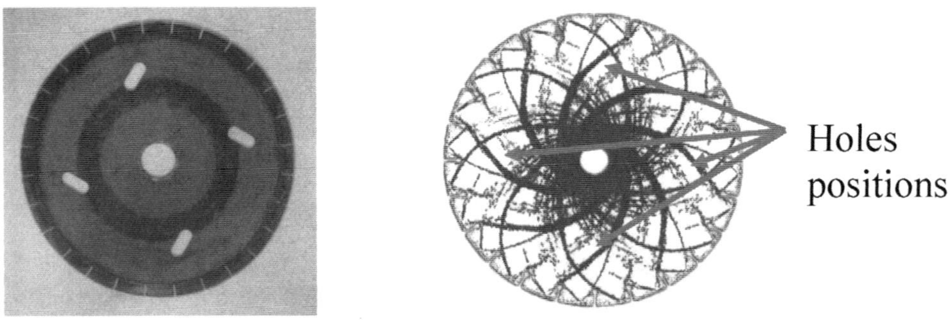

Figure 5. The positions of the holes.

It can be found from figure 5 that there are many materials reserved in places where loads were applied. On the whole, the transmission route of force is obvious, but when we look at the local structure carefully, a small quantity of materials is reserved in some places, but these materials are not connected with the surrounding materials, so we need to make some improvements in the manufacturing process of places similar to these details.

As topology optimization is a conceptual design and belongs to the theoretical stage, its results can provide us with a reference for structural optimization, but in order to design the final product, it needs to be determined according to the actual situation. According to the results of ANSYS topology optimization, combined with the actual shape, service performance, installation and convenience of the saw blade, considering the stability of the saw blade when rotating at high speed, and considering the overall stiffness and integrity of the actual saw blade, only part of the area is designated as the location for opening vibration and noise reduction holes, so the designed opening positions are at the four positions which are shown as figure 5. At the same time, in order to avoid stress concentration and fracture of collective materials, the holes were rounded. The saw blade matrix was processed according to the position in the figure by high-pressure water cutting machine, and the vibration and noise reduction saw blade is obtained.

4. Comparison of diamond circular saw blade before and after optimization

According to the topology optimization results, the position of the hole on the substrate of the diamond circular saw blade can be determined, and the noise and vibration characteristics of the diamond circular saw blade before and after optimization were compared by measuring the vibration and noise reduction of the diamond circular saw blade as below.

4.1. Comparison of noise before and after optimization

According to the noise value of the pre-optimized diamond circular saw blade at each rotating speed, under the same conditions, the noise value of the optimized diamond circular saw blade was obtained by the same method. The experimental data are shown in table 4.

Table 4. Comparison of idle noise before and after optimization of φ500 mm diamond circular saw blades.

Rotation speed/(r/mm)	150	300	450	600	800
Before optimization/(A)	68.95	73.88	78.56	77.11	79.62
After optimization/(A)	66.13	70.52	75.46	73.24	75.74

It can be seen from table 4 that when idling, the noise of the optimized diamond circular saw blade will gradually increase with the increase of cutting speed, but at the same speed, the noise value of the optimized saw blade is generally lower than that of the pre-optimized saw blade. In this experiment, the noise of the optimized saw blade can be reduced by 4.8% maximumly.

4.2. Comparison of vibration characteristics before and after optimization

The modal test of the optimized diamond circular saw blade is carried out. The result data of diamond circular saw blades before and after optimization are shown in table 5, from which we can see the contrast relationship between natural frequencies of saw blades before and after optimization.

Table 5. Comparison of natural frequency of φ500 mm diamond circular saw blade before and after optimization

Order number	1st	2nd	3rd	4th	5th	6th
Before optimization/Hz	35	57.5	67.5	152.5	285	312.5
After optimization/ Hz	32.7	58	70.6	161.8	291.3	326
Order number	7th	8th	9th	10th	11th	12th
Before optimization/ Hz	357.5	445	565	614	643	694
After optimization/ Hz	374.3	489.8	583.5	659.4	684.5	718.7

Table 5 shows the comparison of natural frequencies of saw blades before and after optimization. The natural frequencies of diamond circular saw blades before and after optimization are not much different in the first order, but they are obviously different from the second order. The natural frequency of circular saw blades with holes tends to increase. The higher the order, the more obvious the difference in natural frequency. From the 9th order, although the natural frequency of the optimized saw blade is still higher than that of the pre-optimized saw blade, the increasing trend has declined. Therefore, according to the topology optimization method described in this paper, determining the position and shape of the hole in the diamond circular saw blade matrix can indeed increase the natural frequency within a certain range of order, thus improving the stability of the saw blade and reducing the vibration and noise.

5. Conclusions

In this study, the noise and vibration characteristics of diamond circular saw blade were analyzed, and based on the analysis results, the corresponding measures to reduce noise and vibration were put forward. Using the topology optimization function of ANSYS Workbench, the topology optimization of diamond circular saw blades with no holes in the matrix was carried out, and the noise and natural frequency value of diamond circular saw blades before and after optimization were measured and compared again.

The test results showed that the noise of the optimized saw blade can be reduced by 4.8%, and the natural frequencies of the first 12 orders were increased in varying degrees, which proves that the vibration and noise of the saw blade were reduced after optimization. Therefore, it is proved that the optimization method described in this paper has guiding significance for the actual processing in factories.

References

[1] Marin J 1998 *Ind. Diamond Rev.* **58 (577)** 34
[2] Ersoy A and Atici U 2004 *Diam. Relat. Mater.* **3I** 223-7
[3] Wei X, Wang C and Zhou Z 2010 *J Mater. Process*. Tech. **139 (3)** 227-280
[4] Kvietkova M, Gaff M and Gasparik M 2015 *BioResources* **10 (1)** 1657-66
[5] Davim J, Reis P, Maranhao C, Jackson M, Cabral G and Gracio J 2010 *Int. J. Mater. Prod. Tec.* **37 (1)** 46-50
[6] Zhao M 2009 *Processing Tools and Technology of Stone* (Beijing: Publishing House of Electronics Industry)
[7] Xie J and Tamaki J 2007 *J Mater. Process Tech.* **5 (186)** 253-8
[8] Meng Y, Wei J, Wei J, Chen H and Gui Y 2019 *Couput. Electron. Agr.* **157** 38-48
[9] Li S, Wang C, Zheng L and Wang Y 2016 *J Mater. Process Tech.* **238** 108-123
[10] Yong C, Hang Z and Wang J 2019 *Eur. J. Mech-A-Solid* **75** 197-204
[11] Yurdakul M and Akdas H 2012 *Int. J. Rock Mech. Min.* **53** 38-44
[12] Shang X and Su J 2006 ANSYS/LS-DYNA*Method and Engineering Example of Dynamic Analysis* (Beijing: China Water&Power Press)

IOP Conf. Series: Materials Science and Engineering 1138 (2021) 012046 doi:10.1088/1757-899X/1138/1/012046

Reinforcement methods designed to increase the residual life expectation of rigid pavements

S Zghibarcea[1*], N Taranu[1], D Ungureanu[1], A Boboc[1] and V Boboc[1]

[1] "Gheorghe Asachi" Technical University, Iasi, Romania

* stefanzghibarcea@yahoo.com

Abstract. The lifespan of a road structure can be evaluated only based on a traffic forecast. After a road is open to traffic, degradations are starting to developed, even at light and moderate traffic levels. Thus, at a specific time moment during the structure's lifespan, the residual life expectation of a road can be determined as a function of the degradation state and the forecasted traffic. The degradation state can be evaluated either by using the approximated characteristics of the road construction materials or, based on the real material properties that are experimentally determined on core samples. If the residual life is estimated based on up-to-date traffic forecasts and real road materials properties, the common reinforcement methods used to prologue the lifespan of the structure can be optimized and adapted to the specific conditions of each type of pavement.

1. Introduction

The European road network consists of approximately 5.5 million km of which the majority are managed under local and regional road departments [1]. In this context, the annual maintenance spending in the European road infrastructures is about 30 billion euro. Also, according to the latest prognosis of the European Union Road Federation, in the upcoming future is expected that up to 90% of road works will involve the maintenance of the existing network and only 10% of investments will be used for construction of new roads [2].

If the residual life expectation of a road structure is estimated based on standard material characteristics and generic traffic forecasts (method referred to as *the simplified approach*), the road administrators can apply only common reinforcement methods to prologue the structure lifespan [3]. These methods are designed such that to be applied regardless the type and the intensity of the degradation state. Thus, in many cases, the common reinforcement methods are related to a deficit of efficiency. In addition, the lack of appropriate and timely maintenance induces an increase in the cost of both future reinforcement and vehicle operating costs. On the other hand, if the residual life expectation is estimated based on the real mechanical and elastic properties of the pavement materials and up to date traffic forecast (method referred to as *the accurate approach*), the reinforcement methods can be adapted to the overall characteristics of the road structure. Also, since only the necessary works are performed, the energy consumption and the greenhouse gas emission can be significantly reduced.

The residual life of a road, D_v (equation 1) represents the number of years for which the existing structure can sustain the forecasted traffic. In case of rigid pavements, the capacity to sustain the traffic is defined as the number of standard axles loads N_{adm} at which the road concrete layer can resist without developing excessive degradations, divided by the average annual traffic.

$$D_v = N_{adm} / N_{c,\ yearly} \quad (1)$$

where: $N_{c,\ yearly}$ – the average annual traffic, in millions of standard axles; is determined by dividing the traffic volume to the number of years corresponding to the period that is analysed.

This work consists in a case study related to a county road localised in the North – East part of Romania. For this pavement structure, the residual life expectation was computed using both approaches, the simplified and the accurate one and thus, two distinct reinforcement techniques were proposed, a common one and an optimised one.

2. The simplified approach

2.1. Tensile strength of concrete pavement

In the simplified approach, only the mechanical characteristics of the pavement materials are determined and the elastic properties are taken as provided in the guide norms. The mean tensile strength was experimentally determined by performing the split test on 5 cylindrical core samples which were extracted from different regions of the road (figure 1, table 1) [4].

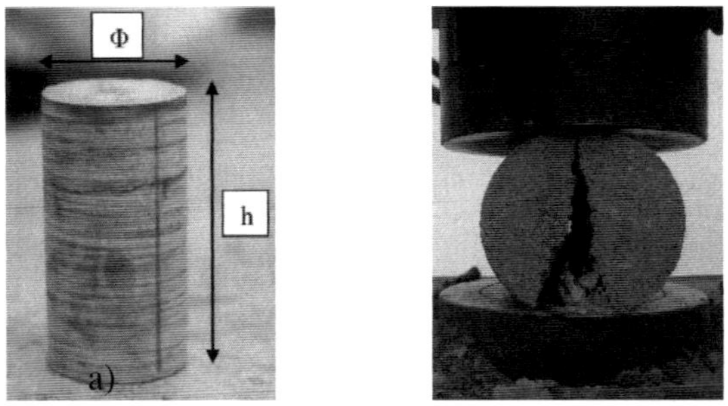

a) Cylindrical concrete core sample b) Failure mode

Figure 1. Tensile test of concrete pavement.

The tensile strength by splitting was computed using equation 2 [4].

$$R_B = 0.90 \cdot f_d \cdot R_{B,\ med} \quad (2)$$

where:

- 0.9 – partial factor for reducing the tensile strength of concrete (is chosen according to the severity of the degradation state);
- f_d - partial factor for reducing the tensile strength of concrete (is chosen according to the pavement degradation index);
- $R_{B,\ med}$ - average value of tensile strength by splitting, in MPa.

The characteristic tensile strength by splitting was computed using equation 3 [4].

$$R_{BC,\ car} = R_{med} - t_a \cdot s \quad (3)$$

where:

- t_a – coefficient that account for the number of samples for which the results may vary with more than 15% from to mean results;
- R_{med} – average value of the tensile strength;
- $R_{med} = \Sigma R_i / n$;
- n - number of samples, n = 5;
- R_i – tensile strength determined for sample *i*;

IOP Conf. Series: Materials Science and Engineering 1138 (2021) 012046 doi:10.1088/1757-899X/1138/1/012046

Table 1. Tensile strength of concrete pavement.

Sample number	h (cm)	h^2 (cm^2)	Φ (cm)	R_i (MPa)	R_i^2 (MPa)
1	17.50	306.25	9.30	2.82	7.95
2	17.80	316.84	9.40	3.14	9.86
3	17.30	299.29	9.30	3.76	14.14
4	15.00	225.00	9.30	4.16	17.31
5	18.00	324.00	9.40	3.27	10.69
		$R_B = 2.78\ MPa$			
		$R_{BC,\ car} = 2.75\ MPa$			

2.2. Modulus of elasticity of the concrete pavement
In the Romanian Norm NP 111-04 there are given rough estimations for the modulus of elasticity of concrete pavements, based on the average value of the tensile strength [5]. For the concrete pavement that was studied in this work, the approximative modulus of elasticity, E_{est} is 24000 MPa and the corresponding Poisson coefficient, μ is 0.15.

2.3. Numerical analysis of the road structure
The road structure lays on a foundation soil of type P5. The characteristics of P5 soils, according to the provisions given in the Romanian Norm NP 111-04 are listed in table 2 [5]. The hydrological regime (type 2b) was established based on the soil type and on the climate type of the Nord – East part of Romania, where the road is located [5].

Table 2. Characteristics of the foundation soil.

Soil category	Soil type	Classification	Plasticity index I_p	Clay (%)	Dust (%)	Sand (%)
Cohesive	P5	clay, dusty clay, sandy clay, clay, dusty sandy	Over 15	30…100	0…70	0…70

The numerical analysis of the pavement structure was performed using the Calderom 2000 software [6]. The latter is based on a rational computing method designed for the evaluation of the resilient stresses and strains in roadways pavements. Since the Calderom 2000 software is based on a rational computing method, it is possible to analyze a road pavement structure in two distinct approaches. The first one consists in a classical design of the structure, taken into consideration the given traffic and the corresponding service life.

The second approach is to analyze an existing road pavement structure by estimating the evolution of the degradation state over time in terms of cumulative probability of failure. For the pavement structure that was analyzed in this study it was applied the second approach. The road structure was modelled as a classical multi-layer elastic linear body. The properties of the road materials and the geometrical configuration of the layers are listed in table 3. The outcomes of the numerical analysis in terms of variation of stresses and strains over the structure's depth are listed in table 4.

Table 3. The properties of the road materials and the geometrical configuration of the layers.

No.	Layer	Thickness (cm)	Modulus of elasticity (MPa)	μ
1	Road concrete Bce4	16	24000	0.15
2	Ballast (foundation and capping layers)	30	169	0.27
3	Existing bedding soil	∞	65	0.42

Table 4. Stress and strain variation over the pavement depth.

Z (cm)	σ_r (MPa)	ε_r (μstrain)	ε_z (μstrain)
-16.00	.210E+01	.746E+02	-.281E+02
16.00	.113E-02	.746E+02	-.262E+03
-46.00	.159E-01	.975E+02	-.158E+03
46.00	-.217E-02	.975E+02	-.250E+03

As mentioned before, the capacity to sustain the traffic is determined based on the number of standard axles loads, N_{adm} at which the road concrete layer can resist without developing excessive degradations divided by the forecasted traffic. The concrete pavement which was analyzed in this work has the Bce 4 indicative. According to the provisions given in the Romanian Norm NP111-04, the corresponding exponent of Bce 4 is b = -1/15. The latter is a selective parameter based on which the equation for determining N_{adm} is chosen (equation 4) [5].

$$\log N_{adm}=15\,(\log R_{BC,\,car}-\log \sigma_{r,\,base})-6.30325 \tag{4}$$

where: $\sigma_{r,\,base}$ – the tensile stress at the base of the road concrete layer.

The residual life period was determined using equation 1. The obtained value (29×10^{-5} years) indicates that the analysed pavement structure can sustain the forecasted traffic for less than one year. Thus, in order to prolongate the lifespan of the structure and satisfy the safe and comfort requirements, the road pavement should be reinforced with an additional layer.

2.4. Numerical analysis of the reinforced road structure

The common method of reinforcement for the concrete pavements consists in an overlay structure compose of a top layer of 4 cm of asphalt concrete and a binding layer of 8 cm of asphalt concrete. The new road structure was modelled using the approach described in the previous section. The properties of the road materials and the geometrical configuration of the layers are listed in table 5. The outcomes of the numerical analysis in terms of variation of stresses and strains over the structure's depth are listed in table 6.

Table 5. The properties of the road materials and the geometrical configuration of the layers.

No.	Layer	Thickness (cm)	Modulus of elasticity (MPa)	μ
1	Top layer – asphalt concrete	4	4200	0.35
2	Binding layer – asphalt concrete	8	3600	0.35
3	Road concrete Bce4	16	24000	0.15
4	Ballast (foundation and capping layers)	30	169	0.27
5	Existing bedding soil	∞	65	0.42

Table 6. Stress and strain variation over the pavement depth.

Z (cm)	σ_r (MPa)	ε_r (μstrain)	ε_z (μstrain)
-12.00	-.423E+00	-.246E+02	-.606E+02
12.00	-.835E+00	-.246E+02	-.110E+02
-28.00	.129E+01	.460E+02	-.172E+02
28.00	.128E-02	.460E+02	-.154E+03
-58.00	.948E-02	.600E+02	-.101E+03
58.00	-.189E-02	.600E+02	-.158E+03

3. The accurate approach

3.1. Modulus of elasticity of the concrete pavement

In the accurate approach, both the mechanical and the elastic characteristics of the pavement materials are experimentally determined. For this purpose, 5 additional cylindrical core samples were extracted, profiled at the ends, instrumented with 2 micro-meter callipers with dial indicator and subjected to a series of compressive stress cycles up to approximately 40% of the ultimate compressive strength [7]. The modulus of elasticity of the concrete samples (E_c) was determined as the average slope of the stress-strain responses captured during the cyclic loading (table 7).

Table 7. Modulus of elasticity of concrete pavement.

Sample number	h (cm)	h^2 (cm^2)	Φ (cm)	R_i (MPa)	E_c (MPa)
1	16.70	278.89	9.40	2.82	33543.5
2	16.70	278.89	9.30	3.14	28394.8
3	15.93	253.65	9.30	3.76	43209.4
4	16.23	263.41	9.30	4.16	35493.4
5	16.39	268.63	9.40	3.27	35160
		$E_{C,\,avv}$ = 35160 MPa			

3.2. Numerical analysis of the road structure

The numerical analysis of the pavement structure was performed using the same approach as in section 2.3. The properties of the road materials and the geometrical configuration of the layers are listed in table 8. The outcomes of the numerical analysis in terms of variation of stresses and strains over the structure's depth are listed in table 9.

Table 8. The properties of the road materials and the geometrical configuration of the layers.

No.	Layer	Thickness (cm)	Modulus of elasticity (MPa)	μ
1	Road concrete Bce6	16	35160	0.15
2	Ballast (foundation and capping layers)	30	169	0.27
3	Existing bedding soil	∞	65	0.42

Table 9. Stress and strain variation over the pavement depth.

Z (cm)	σ_r (MPa)	ε_r (μstrain)	ε_z (μstrain)
0	-.242E+01	-.559E+02	.289E+01
-16.00	.229E+01	.554E+02	-.205E+02
16.00	.453E-03	.554E+02	-.200E+03
-46.00	.123E-01	.767E+02	-.127E+03
46.00	-.127E-02	.767E+02	-.201E+03

By taking into consideration the experimentally determined tensile strength and modulus of elasticity it results that the concrete pavement which was analyzed in this work has the Bce 6 indicative. According to the provisions given in the Romanian Norm NP111-04, the corresponding exponent of Bce 6 is b = -1/16 [5].

Therefore, by knowing the real concrete indicative and applying the accurate method the real residual life period of the road structure was determined (10^{-4} years).

3.3. Numerical analysis of the reinforced road structure

Since both the mechanical and elastic properties of the asphalt concrete are known, the reinforcing method can by optimized. Usually, the optimizing process consists in a gradual diminishing of the depths of the common reinforcing layers up to the limit where the tensile stresses at the bottom of the concrete layer reaches the admissible tensile strength.

The difference between the real modulus of elasticity of the concrete pavement (E_c = 35160 MPa) and the estimated one (E_{est} = 24000 MPa) is close to 50%. Thus, in the first part, it was cheeked if the top layer of the common reinforcement can be reduced from the beginning at 50%. In this case, the reinforcing layers consists in 4 cm asphalt concrete top layer and 4 cm asphalt concrete binding layer. The new road structure was modelled using the same approach that was described in the previous sections. The properties of the road materials and the geometrical configuration of the layers are listed in table 10. The outcomes of the numerical analysis in terms of variation of stresses and strains over the structure's depth are listed in table 11.

The maximum tensile stress at the base of the road concrete layer (total depth = -24 cm) is 1.75 MPa, less than 1.77 MPa (admissible tensile strength, computed according to the provisions given in [3]). Thus, the optimization of the common reinforcing method (which consists in 50% diminishing of the top layer depth) is suitable for the characteristics of the analyzed road structure.

Table 10. The properties of the road materials and the geometrical configuration of the layers.

No.	Layer	Thicknes s (cm)	Modulus of elasticity (MPa)	μ
1	Top layer – asphalt concrete	4	4200	0.35
2	Binding layer – asphalt concrete	4	3600	0.35
3	Road concrete Bce4	16	35160	0.15
4	Ballast (foundation and capping layers)	30	169	0.27
5	Existing bedding soil	∞	65	0.42

Table 11. Stress and strain variation over the pavement depth.

Z (cm)	σ_r (MPa)	ε_r (μstrain)	ε_z (μstrain)
-8.00	-.486E+00	-.325E+02	-.634E+02
8.00	-.144E+01	-.325E+02	-.385E+01
-24.00	.175E+01	.425E+02	-.157E+02
24.00	.542E-03	.425E+02	-.150E+03
-54.00	.939E-02	.599E+02	-.102E+02
54.00	-.204E-02	.599E+02	-.160E+03

All the numerical analyses that were presented in this paper were performed using Calderom 2000 software. The latter is a well-known tool that enable the rational design of a pavement structure, which is part of the Romanian county or national road network [8]. For road structures that are different compared to the common pavement systems, which are used in Romania, this software may provide inaccurate results. This fact is due to several limitations, from which the most significant one is related to the impossibility of modelling the interface condition for adjacent layers. Thus, it should be noted that the results that were presented in this paper refers strictly to a specific road pavement structure and, if changes in the structural compositions are to be done, different assumptions should be defined and, in some cases, a specific software should be used.

This work is part of an extensive research project, related to the methods deigned to prolongue the road structures lifespans and to the modern rehabilitation and recycling methods of pavement layers. More information regarded to the experimental programs that were performed in the frame of this research project can be consulted in a recently published paper [9].

4. Conclusions

The evaluation of the road resilient life expectation is one of the core parts that characterize the transport network reliability. In Romania, the rigid road pavements are constantly exposed to large traffic volumes, high temperature changes and aggressive influence of deicing agents during winter period, which decrease the durability of concrete.

At a particular level of degradation, the residual life expectation can be determined by applying either the simplified method or the accurate evaluation method. In the second case, both the mechanical and elastic properties of the concrete pavement are experimentally determined. Also, up-to-date traffic forecasted are used, which are based on stochastic behavioral models and risk models designed for estimating the travel choice behavior over time. In this work, both methods of evaluating the residual life expectation were applied for a county road with rigid pavement structure.

The reinforcing layers that were designed based on the accurate method implies less materials and work compared to the ones designed based on the common reinforcing method. In particular, the top layer depth is half than the same layer that is indicated in the common reinforcing method. Beside the diminishing of the intervention costs, by applying the accurate method, only the necessary works are performed and thus, the energy consumption and greenhouse gas emission can be significantly reduced. Also, it can be concluded that the road reinforcing solutions that are designed based on the accurate method are results-oriented and can improve the efficiency of the transportation industry.

References

[1] Di Graziano A, Marchetta V and Cafiso S 2020 Structural health monitoring of asphalt pavements using smart sensor networks: A comprehensive review *Journal of traffic and transportation engineering* **7(5)** 639-51

[2] European Union Road Federation (ERF) 2014 *Road Asset Management-an ERF Position Paper*

for Maintaining and Improving a Sustainable and Efficient Road Network European Union Road Federation, Bruxelles

[3] Florida Department of Transportation, Office of Design 2019 Rigid pavement design manual *Pavement Management Section*

[4] SR EN 12390-6/2010 Testing hardened concrete. Part 6: Stress resistance by splitting test pieces

[5] NP 111-04 Normative for the calculation of the base layers made of cement concrete for road structures

[6] http://www.cnadnr.ro/ro/comunicare/publicatii/proiectare (accessed at 06.01.2021)

[7] SR EN 12390-13/2014 Testing hardened concrete. Part 13: Determination of the secant modulus of elasticity in compression

[8] PD 177-2001 Normativ pentru dimensionarea sistemelor rutiere supple şi semirigide (Metoda analitică), (in Romanian)

[9] Ungureanu D, Ţăranu N, Hoha D, Zghibarcea Ş, Isopescu D N, Boboc V, Oprişan G, Scutaru M C, Boboc A and Hudişteanu I 2020 *Accelerated testing of a recycled road structure made with reclaimed asphalt pavement material* Construction and Building Materials **262** 120658

IOP Conf. Series: Materials Science and Engineering 1138 (2021) 012047 doi:10.1088/1757-899X/1138/1/012047

Study of two-phase transportation mode of liquefied natural gas through a pipeline by the gravitational method

V Zhmakin[1*] and A Samoylov[1]

[1] Southwest State University, Kursk, Russia

* va-zhmakin@yandex.ru

Abstract. The given paper considers the choice of method and technology of liquefied natural gas transportation from an isothermal LNG storage facility via a pipeline to a transport cryogenic tank. Theoretical studies have been performed, including mathematical modeling of thermal and hydro-dynamic processes during the transportation of liquefied natural gas with partial evaporation through a pipeline. A new method for calculating the transportation of liquefied natural gas with partial evaporation through a pipeline using the gravitational method has been proposed in the paper.

1. Introduction

At present, a significant share of hydrocarbon energy resources market is occupied by liquefied natural gas (LNG). Countries with natural gas deposits have developed technologies that allow natural gas to be liquefied, stored in special isothermal tanks, and transported in cryogenic tanks on water methane carriers, as well as by air, rail, and road transport.

The largest LNG producers and exporters are Algeria, Libya, Brunei, Abu Dhabi, Indonesia, Malaysia, Australia, Qatar and the United States. The largest LNG importer is Japan, which accounts for about 60% of global LNG consumption. Other major importers are South Korea, France, Spain, and Belgium. One of the most promising LNG markets in the world is the United States' LNG market which is both an importer and an exporter [1].

Russia possesses the largest natural gas deposits, which are found in remote areas (the Barents Sea, the Kara Sea shelf, Sakhalin Island, etc.) and which are unfavorable for the construction of main gas pipelines [2]. Therefore, two large natural gas liquefaction plants have already been built on Sakhalin Island and the Yamal Peninsula and they continue to increase their capacity [1].

Countries that import LNG must also have receiving terminals and storage tanks for LNG, as well as equipment and communications for transporting it to consumer tanks.

This paper considers the choice of method and technology for transporting liquefied natural gas from a stationary isothermal LNG storage via a pipeline to a transport cryogenic tank.

LNG transportation from one container to another container via a pipeline is possible by several methods: using submersible self-priming pumps, displacement by an inert gas with a lower boiling point, displacement by its own LNG vapor obtained in the evaporator [3, 4].

According to some sources [5 - 9], LNG transportation is most appropriate via a pipeline using the gravitational method, i.e. due to the action of gravity, provided that there is a large difference in the height of the tank being emptied and filled. In addition to that, there are no energy costs, expensive cryogenic pumps are not used, and there is no need to raise high pressure in the emptied storage, which can lead to the rupture of its walls. The gravitational method is widely used in the discharge and filling of petroleum products from railway tanks [3, 4].

The main feature of LNG transportation through a pipeline is that LNG in this process can boil to form a vapor-liquid flow. LNG is a cryogenic liquid with a boiling point of minus 161.5°C at atmospheric pressure, which at a positive ambient temperature causes intense heat flows to LNG and, subsequently, the evaporation of part of LNG. The appearance of LNG vapor-liquid flow in the pipeline leads to a significant decrease in its through capacity and an increase in its hydraulic resistance, to large losses of gas into the atmosphere, and a longer duration of the pumping process [10 - 12, 19, 20].

At the initial stage of LNG transportation, the "warm" pipeline is cooled and filled. LNG enters the pipeline which has ambient temperature, where all the liquid quickly evaporates and the pipeline is filled with steam. The steam pressure in the pipeline increases quickly enough and so much that it can be equalized with and even exceed the pressure at the inlet to the pipeline, thereby leading to the release of liquid back into the tank being emptied. This phenomenon is known as the "geyser" effect [5, 10 - 13].

As the initial section of the pipeline cools, the liquid front moves towards the outlet and gradually the entire pipeline acquires an operating temperature close to the temperature of the transported liquid. After cooling the pipeline up to cryogenic temperatures, LNG is transported in a single-phase state or with partial evaporation [5, 6, 10, 11].

To prepare a cryogenic pipeline for LNG transportation in steady-state, the "warm" pipeline can be pre-purged with another inert cryogenic liquid, such as liquid nitrogen, which has a boiling point even lower than that of LNG – minus 196°C [10 - 13].

2. Materials and methods

When operating cryogenic pipelines, LNG transportation parameters are selected in such a way as to prevent the formation of a vapor phase in the pipeline. There are a number of ways to achieve that:1) increasing pressure in the system above the saturated vapor pressure; 2) preliminary supercooling of the liquid relative to the saturation temperature; 3) reducing heat flows through the pipeline walls [10, 11].

- During LNG transportation, a certain pressure must be created above the liquid surface in the tank being emptied. It must exceed the pressure of saturated vapors. Raising pressure in the tank increases the subcooling of LNG up to the saturation temperature, providing a single-phase flow in the pipeline [10, 11].
- LNG must be sufficiently supercooled relative to the saturation temperature at the pipeline pressure. To perform this, LNG is sent by a pressure pump to a heat exchanger being cooled by the pumped LNG at atmospheric pressure [10, 11].

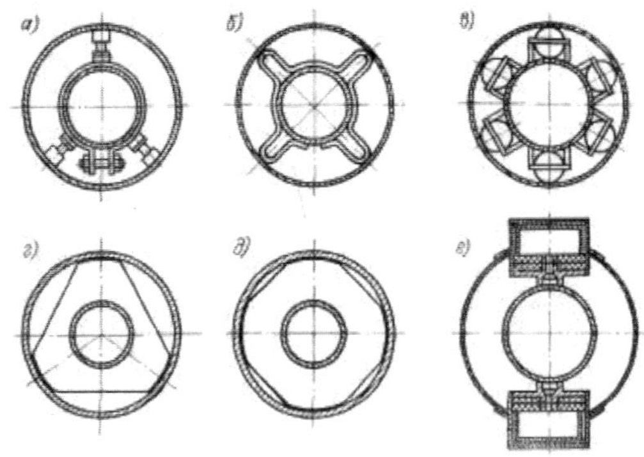

Figure 1. Schematic designs of inner tube supports in the casing: a) finger-type; b) wire; C) ball; d), d) disk; e) trunnion.

- The use of vacuum types of thermal insulation – vacuum-powder, layered-vacuum or pure vacuum for cryogenic pipelines (figure 1) allows reducing heat flows up to the minimum values [10 - 13].

Pipelines with bulk-type thermal insulation are also widely used: loose, porous and fibrous thermal insulation (figures 2, 3). Domestic and foreign experts recommend foamed materials as thermal insulation: polyurethane foam, expanded polystyrene, cork, foamed epoxy resin, foam glass [13, 14].

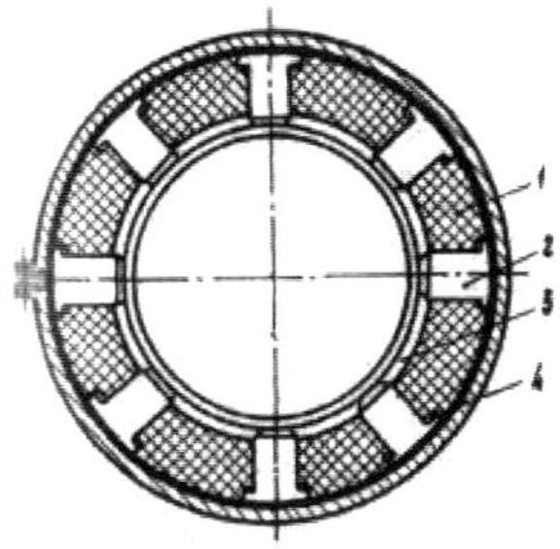

Figure 2. Sliding construction of heat-insulated pipeline coating: 1 – heat-insulation material; 2 – sliding elements; 3 – pipeline; 4 – casing.

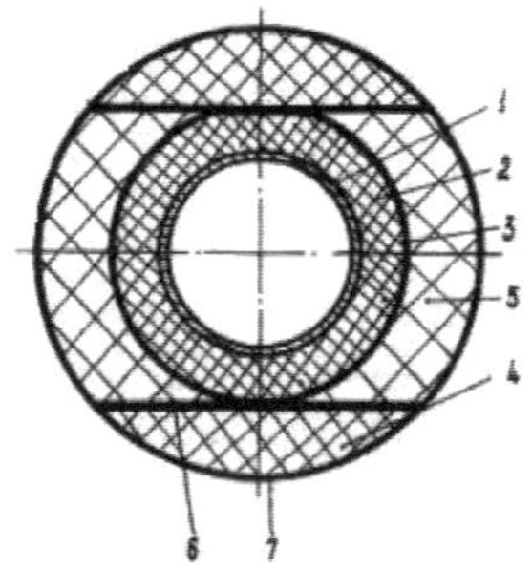

Figure 3. Heat-insulation construction for pipe bend angle: 1 – pipe; 2 – inner heat-insulation layer; 3 – steam tight layer; 4 – rigid segments; 5 – soft segments; 6 – antifriction layer; 7 – outer casing.

However, these methods of keeping LNG in a single-phase state directly depend on the chosen method of LNG transportation (by pump, by pressurization of inert liquid vapors, by own LNG vapors, or by the gravitational method).

In [15, 16], a mathematical model of LNG transportation by the gravitational method in a single-phase state is proposed, and the calculation technique proposed on its basis allows finding the length of a single-phase LNG flow section and the parameters of its transportation: pressure p, temperature T, mass flow G (figure 4).

The parameters of LNG transportation through a low-pressure pipeline are found near the saturation line, so even minor changes in pressure and temperature can lead to the evaporation of part of LNG and the appearance of a two-phase vapor-liquid flow, which significantly reduces the efficiency of LNG transportation by the gravitational method. Therefore, when calculating the parameters of LNG transportation through a pipeline, changes in the thermal properties of LNG should be taken into account [5- 7, 17].

LNG is transported through the pipeline under the influence of available pressure difference in the storage and in the receiver tank and is characterized by the following parameters: pressure Pvx, temperature Tvx, mass flow Gvx = G1.

The parameters of LNG transportation through a low-pressure pipeline are found near the saturation line, so even minor changes in pressure and temperature can lead to the evaporation of part of LNG and the appearance of a two-phase vapor-liquid flow, which significantly reduces the efficiency of LNG

IOP Conf. Series: Materials Science and Engineering 1138 (2021) 012047 doi:10.1088/1757-899X/1138/1/012047

transportation by the gravitational method. Therefore, when calculating the parameters of LNG transportation through a pipeline, changes in the thermal properties of LNG should be taken into account [5 - 7, 17].

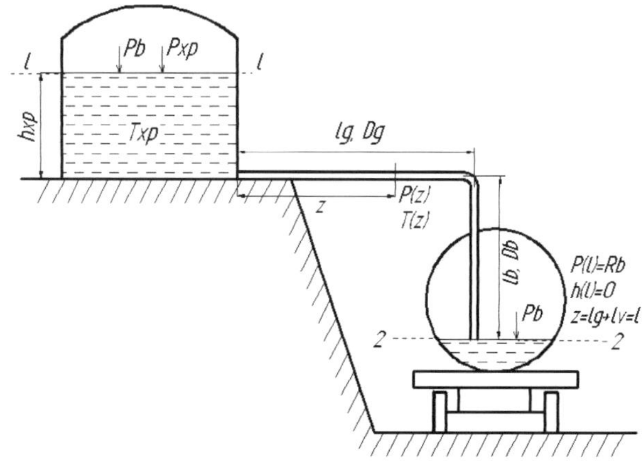

Figure 4. Design model of LNG transportation from one tank to another tank by the gravitational method.

LNG is transported through the pipeline under the influence of available pressure difference in the storage and in the receiver tank and is characterized by the following parameters: pressure Pvx, temperature Tvx, mass flow Gvx = G1.

The pipeline consists of horizontal and vertical sections and has a thermal insulation thickness δ with a thermal conductivity coefficient λ. For example, these may be patented technical solutions [8, 9]. The ambient temperature is assumed to be a constant value Tn = const.

3. Results

The calculation of heat input to LNG is performed according to the well-known laws of the theory of thermal conductivity and convective heat transfer: Fourier's law and Newton-Richman equation [18].

The well-known equations of hydraulics: Bernoulli and Darcy-Weisbach [18] are used to calculate pressure loss in the section of a single-phase LNG flow.

As a result of heat flows q from the external environment and pressure drop in the pipeline Δpv due to the hydraulic resistance of the LNG pipeline at a certain distance from the inlet z = l/ the state of saturation is reached with parameters T(l/) = T/, p(l/) = p/, G1 [15, 16].

Based on the above said, the length of a single-phase LNG flow section through the pipeline is found by the expression [15, 16]:

$$l^{/} = \frac{103 + 0{,}000083 \cdot p_{vh} - T_{vh}}{0{,}000083 \cdot \kappa_g + a_1 - a_2 T_{vh}} \tag{1}$$

where a_1, a_2 – complex values that take into account the influence of heat flows from the environment and the parameters of LNG transportation on its temperature:

$$a_1 = \frac{\pi D \lambda T_n}{G_1 c_P \delta} + \frac{\lambda_{mp} G_1^2}{2DA^2 \rho^2 c_P} - \frac{g Cos\alpha}{c_P}, \tag{2}$$

$$a_2 = \frac{\pi D \lambda}{G_1 c_p \delta} \tag{3}$$

κ_g — specific pressure loss in the horizontal section of the pipeline, Pa/m;

$$\kappa_g = \frac{\lambda_{mp} \rho W_\Gamma^2}{2D_g} \quad (4)$$

Below this section, a two-phase LNG flow occurs with parameters T_g,p_{cm} , G_{cm} , x. The calculation of pressure loss in the section of a two-phase LNG flow using the formulas for a single-phase flow leads to significantly understated results [10, 11].

In engineering practice, simplified methods are usually used for calculating two-phase flows that are not related to the consideration of specific flow modes: a model with phase slip (Martinelli - Lockhart - Nelson correlation) and a homogeneous model [10,11,18].

When calculating a two-phase flow according to the model with slip phase, the pressure loss in the turbulent flow mode is found by the equation [11]:

$$\left(\frac{\Delta p}{\Delta l}\right)_{cm} = \left(\frac{\Delta p}{\Delta l}\right)_{j0} \cdot (1-x)^{1.75} \cdot \Phi_j^2, \quad (5)$$

where $(\Delta p/\Delta l)_{j0}$ – pressure drop per length unit of the pipeline when the liquid phase moves with full flow $G = G_g + G_j$, It is found by the well–known Darcy – Weisbach equation, Pa/m; x – mass vapor content of the flow; Φ_j – function expressing the ratio of pressure loss of a two-phase mixture to pressure loss of a single-phase liquid.

The influence of heat flows on the flow hydrodynamics is taken into account by changing mass vapor content along the length of the pipeline.

Φ_j function in expression (5) is found graphically (figure 5) depending on parameter χ [11, 18]:

$$\chi = \left(\frac{\rho_g}{\rho_j}\right)^{0.5} \cdot \left(\frac{\mu_j}{\mu_g}\right)^{0.125} \cdot \left(\frac{1}{x} - 1\right)^{0.875} \quad (6)$$

where ρ_g, ρ_j – density of vapor and liquid phases, respectively, kg/m^3;
μ_g , μ_j – dynamic viscosity of vapor and liquid phases, respectively, Pas.

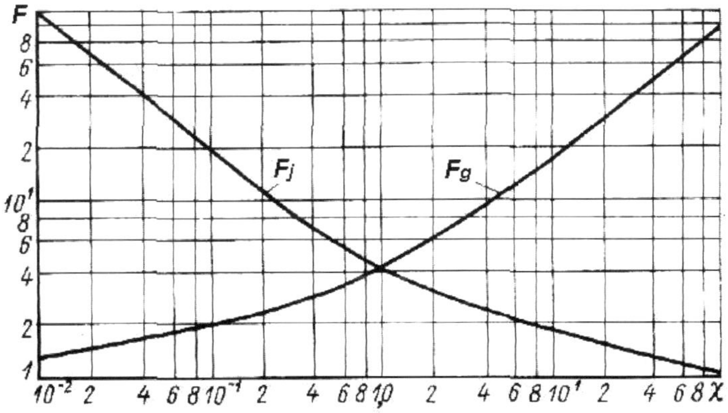

Figure 5 – Ratio between $\Phi_g(\chi)$ and $\Phi_j(\chi)$ in turbulent phase flow.

With low vapor content of the flow and high mass velocities, the homogeneous model has the greatest agreement with experimental data. The essence of the model is that a two-phase flow is considered as a single phase. Its specific volume in each section of the flow is related to mass vapor content and specific volume of each phase [10]:

$$v_{cm} = v_j(1-x) + v_g \cdot x \quad (7)$$

The equation for calculating friction loss is written in a form similar to Darcy – Weisbach equation [10]:

$$\left(\frac{\Delta p}{\Delta l}\right)_{cm} = \frac{\lambda_{cm}(\rho W)^2 \upsilon_{cm}}{2D} \tag{8}$$

where λ_{cm} – resistance coefficient of a two-phase flow which can be represented by the resistance coefficient of a single-phase flow λ_{tr} and some empirical function ψ of two-phase flow parameters, i.e.

$$\lambda_{cm} = \psi \lambda_{tr} \tag{9}$$

Due to the complex nature of dependence ψ on the operating parameters of a two-phase flow, engineering calculations take $\psi = 1$.

The basic assumptions of the homogeneous model consist in the assumption of linear velocities equality of steam and liquid, thermodynamic equilibrium of phases, applicability of dependencies for calculating the friction coefficient of a single-phase flow to two-phase flow [10].

After analyzing the known mathematical models of two-phase fluid flows, the authors suggest the following technique for calculating a two-phase LNG flow.

The initial data for calculating the parameters of a two-phase LNG flow are the values of parameters T', p', l', obtained from the calculation of a single-phase flow [15, 16].

The section of a pipeline $l_{cm} = l - l'$, where liquid vaporization occurs is divided into equal segments Δl. Within each segment, mass vapor content x and the parameters of each phase are assumed to be constant. At the end of each segment, vapor content and pressure loss are calculated.

Mass vapor content in section l_{i+1} is determined by the vapor content of flow in the previous section li and the increase in vapor content in segment $\Delta l = l_{i+1} - l$ due to heat inflow and pressure drop over the saturated liquid:

$$x_{i+1} = x_i + \frac{4q \cdot \Delta l}{D \cdot \rho \cdot W \cdot r} + \frac{i_S(p_i) - i_S(p_{i-1})}{r} \tag{10}$$

where $i_S(p_i)$ – enthalpy on LNG saturation line at a pressure of p_i, J/kg; r – the specific heat of vapor formation, J/kg; q – the amount of heat input to LNG, which can be found by the formula:

$$q_l = \frac{2\pi \lambda_{from}(T_н - T_{LNG})}{ln\left(^{D_{from}}/_{D}\right)} \tag{11}$$

The pressure at the end of each segment in section $li+1$ is connected with the pressure in section li by the following ratio:

$$p_{i+1} = p_i - \Delta p_{cm.tr} - \frac{\Delta l \cdot \cos\alpha}{\upsilon_{cm}}, \tag{12}$$

where α – the angle of inclination of segment Δl to the vertical, υ_{cm} – the specific volume of vapor-liquid mixture, m^3/kg.

The necessary data on the physical and thermodynamic properties of LNG (specific volume, enthalpy on the saturation line, heat of vaporization, heat capacity and others) are found according to table dependencies in [17].

When moving to the next segment, changes in vapor content and density of vapor phase due to heat inflow and pressure drop in the previous segment are taken into account.

To calculate pressure loss due to friction $\Delta p_{cm.tr}$ in a two-phase flow at segments Δl, the ratios of both the slip flow model and the homogeneous model can be assumed.

The calculation is correct if the following condition is met:

$$\frac{p_{p.vykh} - p_{vykh}}{p_{vykh}} < \varepsilon, \tag{13}$$

where p_{pvykh} – the estimated pressure at the end of the pipeline, Pa; p_{vykh} – pre-determined pressure magnitude, Pa; ε – the specified relative error of calculation.

If the estimated pressure happens to be higher than the pre-determined value, in that case it is necessary to increase flow rate and repeat the calculation; if the estimated pressure is lower, then one should reduce flow rate and repeat the calculation. The values ρW and x_{vykh} corresponding to ratio (13) will be the ones being sought for.

4. Conclusions

The main conclusions of the paper are as follows:

- when LNG is delivered from a storage facility by the gravitational method, LNG subcooling at the pipeline inlet is provided by the hydrostatic pressure of liquid column and the excess pressure of LNG vapors in the storage;
- in the process of transportation, the temperature of LNG along the length of the pipeline changes slightly due to heat inflows, taking into account the use of effective thermal insulation, and the state of LNG saturation is achieved mainly due to pressure loss caused by the hydraulic resistance of the pipeline;
- if a pipeline is long, LNG boils at a certain section of it and then is piped by a two-phase flow, creating vapor lock in the vertical pipeline section with the possibility of circulation reversal, the so-called "geyser" effect;
- a method for calculating pressure loss in a two-phase LNG flow section has been proposed. According to the method, a two-phase LNG flow section is divided into segments, within which vapor content x and the parameters of each phase are assumed to be constant. To calculate pressure loss due to friction $\Delta p_{cm.tr}$ at the segments Δl, the ratios of both the slip flow model (5-6) and the homogeneous model (7 - 12) can be adopted.

References

[1] Barmin I, Yu C and Kunis I 1996 Liquefied natural gas - an alternative energy carrier and affordable fuel *Refrigeration* No. **3**

[2] Inkov A, Skorodumov B and Darbinyan R 2003. Decentralized energy supply using liquefied natural gas, *Ventilation, heating, air conditioning, heat supply and building thermal physics (AVOK)* **No. 2**

[3] Shapira I and Moscow V 2005 Oil and gas construction: a textbook for students specializing in "Management in the oil and gas industry" Ed. Mazura

[4] Preobrazhensky N 1975 Liquefied petroleum gases. - Leningrad, "Nedra", 279p

[5] Polozov A, Kobelev N and Zhmakin V 2005 Dynamics of the two-phase flow of liquefied natural gas during gravity transmission through pipelines *Vibration machines and technologies: Sat. scientific*. tr. - Kursk: KSTU, S. 140-142

[6] Polozov A, Shaptala V and Zhmakin V 2007 Investigation of the effective transfer of LNG through technological low-pressure pipelines of storage and delivery systems of LNG // Scientific research, nanosystems and resource-saving technologies in the construction industry: Coll. reports of XVIII international scientific-practical. conf. h. 5. Belgorod 18.09.2007. - Belgorod, S. 155-162

[7] Polozov A and Zhmakin V 2007 Effective scheme of low-pressure LNG transmission through technological pipelines //Gas industry, M: Gazoil press, pp. 68-70

[8] Zhmakin V 2006 Patent 54135 RF, F17D 1/13, F16L 9/18. Cryogenic pipeline. / (Russia). - No. 2005136946/22; Stated 11/28/2005; Publ. 06/10/2006, Bul. No. 16

[9] Polozov A and Zhmakin V 2006. Patent 58658 RF, F17D 1/13, F16L 9/18. Pipeline for the transfer of cryogenic liquid. / (Russia). - No. 2006116911/22; Stated May 16, 2006; Publ. 27.11.2006, Bul. No. 33

[10] Filin N and Bulanov A 1985 Liquid cryogenic systems Mechanical engineering *Leningrad. department*, 245 p

[11] Fastovsky V, Petrovsky Y and Rovensky A 1967 Cryogenic technology *Moscow: Energiya*

[12] Ilyinsky A 1976 Transport and storage of industrial liquefied gases *Moscow Chemistry*

[13] Trunk pipelines of refrigerated and liquefied natural gas 1988 Ed. A. Polozova. - Syktyvkar
[14] Thermal insulation 1976 Ed. G.F. Kuznetsova. Ed. 3rd, rev., Moscow: Stroyizdat, -- 439 p.
[15] Zhmakin V 2007 Development of methods and technical solutions for the transportation of liquefied natural gas through low-pressure pipelines: dissertation for the degree of candidate of technical sciences *Scientific Research Institute of Natural Gases and Gas Technologies* Moscow, 117 p
[16] Zhmakin V 2019 Development of energy-saving and environmentally friendly technology for transportation of liquefied natural gas through technological pipelines of the station for receiving, storing and delivering liquefied natural gas to consumers: monograph Kursk: Publishing house of CJSC "University Book"
[17] Zagoruchenko V and Zhuravlev A 1969 Thermophysical properties of gaseous and liquid methane. Moscow: Publishing house of the Committee of standards at the Council of Ministers of the USSR
[18] Kutepov A, Sterman L and Styushin N 1986 Hydrodynamics and heat transfer during vaporization: Textbook for universities-3rd ed., Revised - Moscow: Higher. shk
[19] Shchedrina G and Shchedrina D 2019 Modeling of Hydrodynamic and Heat and Mass Transfer Processes in Absorption of the Drying Air Emissions. *Journal of Applied Engineering Science*, 17, 623, pages 408-412
[20] Ezhov V 2018. Heat Generator Plant as a Heat and Chemical Installation. *Chemical and Petroleum Engineering*, **54**

AUTHOR INDEX